Chirality and Wobbling in Atomic Nuclei

The book provides an introduction to both theoretical and experimental results on chirality and wobbling in atomic nuclei.

It details the achievements in the study of chirality over the past 25 years since the first prediction of this mode of collective motion in nuclei, as well as those on the wobbling motion. It offers a detailed review of the most relevant theoretical developments on both types of collective motion and the experimental results supporting or not the theoretical predictions.

Different views on wobbling are included and confronted with the contradicting experimental results on low-spin wobbling. It is intended to foster further the research on these types of exotic collective motion in nuclei. Which and how these exotic collective motions occur in nuclei, which are their predicted fingerprints and how they are supported by the experimental facts will be presented. Polemics, debates and ambiguities of the interpretation of the experimental results will be exposed.

The reader will have the opportunity to have together different views on the two phenomena which animated the scientific activity in low-energy nuclear physics in many laboratories around the world. The book will be a valuable reference for PhD students, post-docs and researchers in addition to universities and research institutions.

Key Features:
- The first book on chirality and wobbling in nuclei.
- Contains a comprehensive review of topics related to chirality and wobbling, including both theoretical and experimental aspects.
- Contains chapters from leading researchers in the field.

Costel Marian Petrache is a Professor of Physics at University Paris-Saclay, CNRS/IN2P3 IJClab, France. After completing his PhD in Physics at the University of Bucharest, Petrache has held several research positions in Romania, France and Italy. His research interests include nuclear spectroscopy of nuclei far from the valley of stability, nuclear structure in extreme conditions (high spins and large deformations), symmetries in nuclei, and Physics applied to art and archaeology. Petrache is a scientific referee for several scientific journals, including Physical Review Letters, Physics Letters B, and Physical Review C. He organized the series of conferences Shapes and Symmetries in Nuclei: from Experiment to Theory (SSNET), and most recently organized the conference "Chirality and Wobbling in Atomic Nuclei" (Huizhou, China, 2023).

Chirality and Wobbling in Atomic Nuclei

Edited by
Costel Marian Petrache

CRC Press
Taylor & Francis Group
Boca Raton London New York

CRC Press is an imprint of the
Taylor & Francis Group, an **informa** business

Designed cover image: Dr. Isabelle Deloncle (IJC Lab, CNRS). Credit: Earth's Eastern Hemisphere NASA GSFC.

First edition published 2025
by CRC Press
2385 NW Executive Center Drive, Suite 320, Boca Raton FL 33431

and by CRC Press
4 Park Square, Milton Park, Abingdon, Oxon, OX14 4RN

CRC Press is an imprint of Taylor & Francis Group, LLC

ISBN: 978-1-032-69128-2 (hbk)
ISBN: 978-1-032-69160-2 (pbk)
ISBN: 978-1-032-69163-3 (ebk)

DOI: 10.1201/ 9781032691633

Publisher's note: This book has been prepared from camera-ready copy provided by the authors.

Dedicated
to my lovely wife DANIELA,
to
my parents LUCRETIA and RADUCANU,
and to the younger members of the family,
JOCELYN and MIRCEA

Contents

Foreword

The rotational motion of triaxial systems is an old problem that has been discussed for classical rotors. If the rotational axis is different from any of the principal axis of the intrinsic reference systems, the system is distinguishable from its mirror image and many interesting phenomena show up. The situation is much more complicated in quantum-mechanical many-body systems, such as atomic nuclei, because the concept of deformation and the concept of an intrinsic system are applicable only as approximations for heavy nuclei. Bohr and Mottelson have already introduced in the seventies the nuclear wobbling motion in analogy to vibrational excitations of a triaxial top. In this model, a series of rotational bands that can be distinguished by the wobbling quantum number exists.

However, it is not clear from the beginning that chirality and wobbling are realized in nuclei and their rotational spectra. Nonetheless, Frauendorf and Meng proposed 1997 in a pioneering work that even in quantum mechanical nuclei, chirality could be observed in experiments by two degenerate rotational bands with $\Delta I = 1$ and the same parity. Starting from that moment, a new field in nuclear spectroscopy has been established and a large number of experimental and theoretical investigations have been devoted to the existence of chirality and wobbling in nuclei.

This book describes the basic ideas of this fascinating field and discusses details recently proposed by experimental and theoretical investigations. It is divided into 17 chapters written by the leading scientists in this field. In the first chapter, Costel Petrache discusses the history of chiral motion and wobbling, their experimental features, and the possible developments of this concept in the future. Additional chapters are devoted to the concepts of transverse and longitudinal wobbling, as well as to the various experimental methods and their applications. On the theoretical side, nearly all the methods used successfully in heavy nuclei have also been applied and are discussed extensively. Examples are various phenomenological models, such as triaxial particle rotor models, boson models, and boson-fermion models. They allow algebraic descriptions of such spectra. Besides phenomenological models, several microscopic theoretical methods are discussed, such as the random phase approximation, the triaxial projected shell model, and semiclassical approximations. Of course, covariant density functional theory, one of the leading methods for microscopic descriptions of heavy nuclei, has also been applied to chiral and wobbling motion. Some of the chapters are devoted to specific regions in the periodic table, where such motion is observed.

In summary, this book provides an excellent introduction and a rather complete survey of the field of chirality and wobbling in nuclear motion.

Peter Ring

Preface

The present book collects a number of contributions on chirality and wobbling in atomic nuclei, with the aim to offer the reader a synthesis of what has been achieved in these research fields after 25 and 20 years, respectively, and to emphasize the questions which remain open and need further investigations.

It is known that degenerate quantum states arise from fundamental symmetries, which are often broken in finite many-body systems like atomic nuclei. Chirality, a word first used by Lord Kelvin in 1894, is a property of a system that is distinguishable from its mirror image. In nuclei, the chirality only appears when the nucleus is asymmetric (triaxial) and rotates around an axis which does not lie in any of the principal planes of the intrinsic reference system. To realize such a three-dimensional rotation, there must be at least two active nucleons with angular momenta oriented in directions different from that of the core. Experimentally the chirality should manifest by two nearly degenerate $\Delta I = 1$ sequences of states with the same parity. Frauendorf and Meng suggested the existence of chiral bands in nuclei in 1997. Since then, tens of articles have been published which confirmed the existence of chirality in several regions of the Segré chart of nuclei. The present book gives an overview of the experimental results and theoretical developments over the 25 years of researches devoted to chirality in atomic nuclei.

The nuclear wobbling motion can be considered by the analogy with the spinning motion of an asymmetric top, where perturbations are superimposed on the main rotation around the principal axis with the largest moment of inertia. It was introduced in 1970s by Bohr and Mottelson and considered since then as the hallmark for quantal rotation of triaxial nuclei. The wobbling degree of freedom induces sequences of bands with an increasing number of wobbling quanta. The description of the experimental observations depends sensitively on the ratios between the three moments of inertia of the triaxial rotor. The wobbling bands have been first observed in the years 2000 in well deformed odd-even nuclei and their properties described using the particle rotor model, in which a parallel coupling of the angular moments of the odd nucleon and of the nuclear core was adopted. More recently, in 2014, Frauendorf and Dönau introduced a new type of coupling between the angular momenta of the odd nucleon and the core, called "transverse", which is different from that originally proposed, called "longitudinal", and corresponds to the perpendicular coupling of the single-particle and core angular momenta. This triggered an intense experimental search for the two newly introduced types of wobbling bands, exhibiting different properties not only among them, but also from those of the known wobbling bands in the well deformed rare earth nuclei: the new wobbling bands are predicted at low spin, where the original high-spin approximation of Bohr and Mottelson is not valid. A series of results confirming the existence of such bands have been published, but also results contradicting their wobbling nature have been reported. Theoretical models also lead to contradicting conclusions on the nature of the claimed low-spin wobbling bands. The present book gives a up-to-date overview of the wobbling bands and their interpretation.

I warmly acknowledge all contributors to this book, in which the tremendous progress on chirality and wobbling is presented, from both experimental and theoretical points of view. It can help young researchers to get a global view on the two topics and trigger their interest on the investigation of the microscopic world, in particular that of the atomic nucleus.

Orsay, France   Costel Marian Petrache

Contributors

Gow-Har Bhat
Government Degree College Shopian
Kashmir, India

Radu Budaca
National Institute for Physics and Nuclear
    Engineering
Magurele, Romania

Duo Chen
Jilin University
Changchun, China

Qi-Bo Chen
East China Normal University
Shanghai, China

Hua-Ming Dai
East China Normal University
Shanghai, China

Stefan Frauendorf
Notre Dame University
Notre Dame, Indiana

Ernest Grodner
National Centre for Nuclear Research
Warsaw, Poland

Song Guo
Institute of Modern Physics
Lanzhou, China

Sheikh Jehangir
Islamic University of Science and Technology
Kashmir, India

Hong-Di Jiang
Liaoning Normal University, Dalian
Dalian, China

Elena Atanasova Lawrie
iThemba LABS
Cape Town, South Africa

Jian Li
Jilin University
Changchun, China

Ying-Zhi Ji
East China Normal University
Shanghai, China

Bing Feng Lv
Institute of Modern Physics
Lanzhou, China

Jie Meng
Peking University
Peking, China

Kosuke Nomura
Hokkaido University
Hokkaido, Japan

Costel Marian Petrache
Paris-Saclay University
Orsay, France

Robert Poenaru
UNESCO International Centre for Advanced
    Training and Research in physics
Magurele, Romania

Bin Qi
Shandong University
Weihai, China

Apolodor Aristotel Raduta
National Institute for Physics and Nuclear
    Engineering
Magurele, Romania

Cristian Mircea Raduta
National Institute for Physics and Nuclear
    Engineering
Magurele, Romania

Javid Ahmad Sheikh
Kashmir University
Srinagar, India

Kazuko Sugawara-Tanabe
RIKEN Nishina Center
Wako, Saitama, Japan

Kosai Tanabe
RIKEN Nishina Center
Wako, Saitama, Japan

Ya-Kun Wang
Peking University
Peking, China

Yi-Ping Wang
Peking University
Peking, China

Yu Zhang
Liaoning Normal University, Dalian
Dalian, China

Peng-Wei Zhao
Peking University
Peking, China

1 History and future perspectives on chirality and wobbling in atomic nuclei

Costel Marian Petrache

Université Paris-Saclay, CNRS/IN2P3, IJCLab, Orsay, France

1.1 INTRODUCTION

Since the prediction of chiral bands in nuclei by Frauendorf and Meng in 1997, impressive progress was done from both experimental and theoretical points of view, in particular in the identification and characterization of multiple chiral bands in a single nucleus. The isotopes with the richest sets of chiral bands are in the $A \approx 130$ mass region, with a record of five chiral doublets candidates in ^{136}Nd. New regions of chiral nuclei have been proposed, while those already established have been extended. Chiral doublet bands in the presence of additional broken symmetries like space-reflexion or pseudo-spin have been identified. The wobbling motion in nuclei introduced in 1975 by Bohr and Mottelson, intensively studied around year 2000 in nuclei of the $A \approx 160$ mass region, returned recently in the focus of the experimental and theoretical researches after the introduction by Frauendorf and D$^{\prime}$onau in 2015 of the transverse wobbling mode. Important theoretical developments were accomplished in the description of one- and two-quasiparticle wobbling bands. The experimental results are instead contradictory since the original reported predominance of electric over magnetic character of the transitions connecting the wobbling partners is not confirmed by recent measurements.

The properties of the chiral doublets and of the wobbling bands have been extensively investigated, but the experimental information is far from being complete. Measurements with state-of-the-art setups and new methods of investigation are highly required. The aim of the present review is to summarize the progress and achievements in the study of the chirality and wobbling in atomic nuclei from an experimental perspective.

The chiral motion in atomic nuclei is a collective excitation present if the nuclear shape is not axially symmetric, like an ellipsoid. For a nucleus with an axially symmetric shape, the rotation can only take place around an axis perpendicular to the symmetry axis, while for a nucleus with shape without axial symmetry, the rotation can take place around each of the three principal axes of the intrinsic reference system. For a nucleus composed of a core plus two or more quasiparticles, the rotation can even take place around an axis out of the three principal planes of the intrinsic system. This type of nuclear rotation can give rise to chiral bands if the core has a pronounced triaxiality and rotates with high enough frequency, as first discussed by Frauendorf and Meng in 1997 [1]. Since then, the chiral phenomenon in nuclei has been one of the topics intensively studied both theoretically and experimentally. In this contribution I will try to review what has been done in 25 years of research on chirality from an experimental perspective. The following contributions will be dedicated to the theoretical developments and achievements.

Another type of collective excitation of a triaxial nucleus is the wobbling motion, which was first predicted in the high-spin regime of even-even nuclei by Bohr and Mottelson [2], and, together with

DOI: 10.1201/9781032691633-1

1

the chiral motion, is considered as a fingerprint of a triaxial nuclear shape. It is a collective excitation of a deformed nucleus with three unequal moments of inertia, which rotates about the axis with the largest moment of inertia, whose orientation executes a low-amplitude oscillation. This type of motion was observed experimentally 20 years ago in the high-spin regime of several odd-A nuclei of the $A \approx 160$ mass region. All observed wobbling bands were based on triaxial strongly deformed (TSD) shapes with a substantial deviation from axial symmetry ($\varepsilon_2 \approx 0.4, \gamma \approx 20°$) and have been interpreted using particle-rotor model calculations by assuming parallel angular momenta of the odd nucleon and the triaxial core [3, 4]. From angular correlations and polarization measurements, it was deduced that the $\Delta I = 1$ transitions between the one- and zero-phonon wobbling bands are predominantly electric [5, 6], being in agreement with the wobbling motion interpretation which involves the entire nuclear charge.

Recently, the wobbling motion in odd-even nuclei was revisited by Frauendorf and Dönau [7], and two types of coupling of the angular momenta of the odd nucleon and of the triaxial core have been proposed, with parallel and orthogonal geometry, which leads to transverse and longitudinal wobbling, respectively. The excitation energy of the one-phonon wobbling band relative to the zero-phonon wobbling band decreases (increases) with increasing spin in the transverse (longitudinal) wobbling mode. Many theoretical works have been devoted to these new types of wobbling motion, and experiments to search for the predicted longitudinal and transverse wobbling bands have been performed. There is, however, an intense debate on the mere existence of the wobbling mode in the low-spin regime of triaxial normal-deformed (TND) nuclei, since contradicting experimental results have been published on the electromagnetic character of the transitions connecting the one- and zero-phonon wobbling bands. In the present contribution, I will present the achievements on wobbling at high and at low spin, in strongly-deformed and normal-deformed triaxial nuclei from an experimental perspective.

1.2 CHIRALITY: A REVIEW OF THE MAIN EXPERIMENTAL ACHIEVEMENTS

The application of the tilted axis cranking model to a system composed of two particles coupled to a triaxial core published by Frauendorf and Meng in 1997 [1] is at the origin of chirality in atomic nuclei. They discussed the conditions necessary to have a rotation around an axis inside or outside the principal planes of the triaxial density distribution and concluded that the aplanar solutions manifest as pairs of degenerate $\Delta I = 1$ bands with the same parity, which differ by the chirality of the principal axes with respect to the angular momentum vector. The simplest nuclear system presenting chirality is an odd-odd triaxial nucleus, which has three non-coplanar angular momentum vectors – of the core, of the proton, and of the neutron. As one of the angular momentum vectors is that of the rotating core, the nearly degenerate states are observed only above a critical frequency when sufficient collective rotation develops. As the exact degeneracy corresponds to vanishing matrix elements connecting the left- and right-handed systems, one expects that the connecting transitions between the chiral partners become weaker with increasing spin and the tunneling between the two systems disappears.

A compilation of the data on chiral doublet bands reported before 2019, published by Xiong and Wang in Ref. [9], contains the most complete information on the known chiral doublet candidates. The present review is not a detailed and complete discussion of all the aspects and achievements of chirality in nuclei but only a personal synthesis of the results that I consider to be most relevant.

The nuclei of the $A = 130$ and $A = 190$ mass regions, for which the Fermi surfaces are close to high-j orbitals for both protons and neutrons, were indicated as favored chiral candidates. The odd-odd ^{134}Pr nucleus, studied by myself one year before the introduction of chirality in nuclei [8], having the two lowest lying bands nearly degenerate above spin 14 and a large triaxiality of $\gamma = 20°$ was one of the best candidates (see Fig. 1.1). The two bands have assigned configurations with one $h_{11/2}$ proton particle with the angular momentum aligned along the short axis and one $h_{11/2}$ neutron hole with the angular momentum aligned along the long axis, being therefore orthogonal among

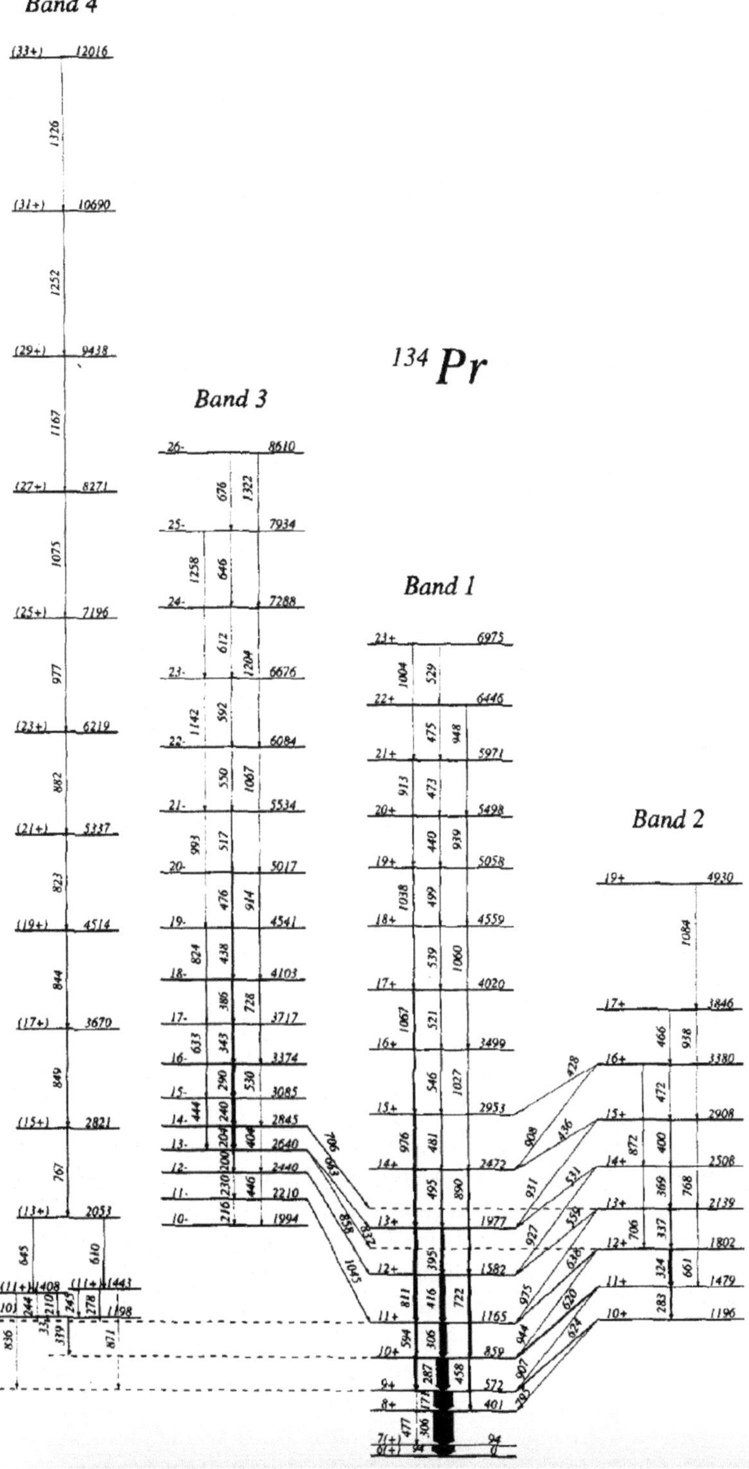

Figure 1.1 Decay scheme of ^{134}Pr, in which the first chiral doublet composed of bands 1 and 2 has been proposed. (Reprinted from Nuclear Physics A **597**, C. M. Petrache *et al.*, "Rotational bands in the doubly odd nucleus ^{134}Pr", p. 106–126, Copyright (1996).)

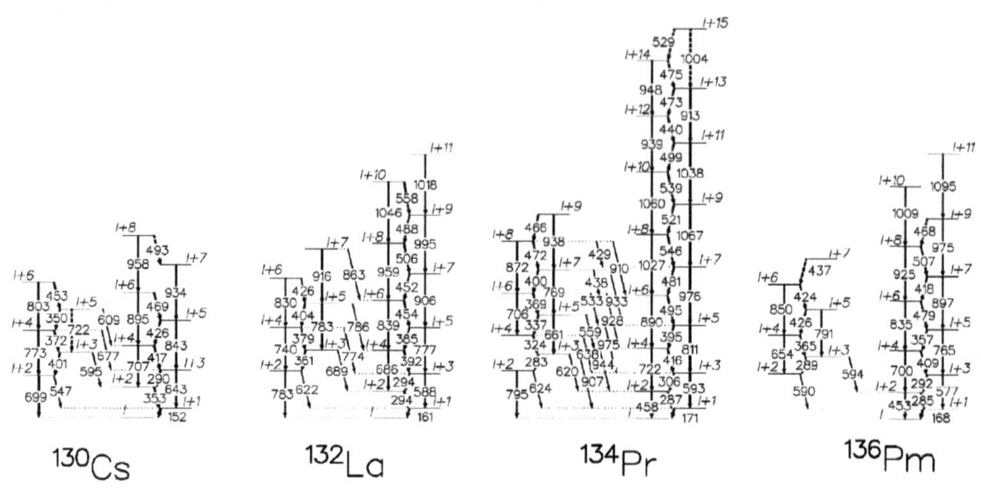

Figure 1.2 Partial level schemes of ^{130}Cs, ^{132}La, ^{134}Pr, ^{136}Pm with the chiral doublet bands. (Reprinted from K. Starosta *et al.*, 2000, "Chiral Doublet Structures in Odd-Odd N = 75 Isotones: Chiral Vibrations", Physical Review Letters **86**, 971-974.)

them and orthogonal to the core angular momentum, as required for a chiral geometry. However, the predicted γ-softness of the core and the presence of a band crossing at spin 14, which made the chiral interpretation questionable.

The experimental studies of chirality had a sudden intensification in 2001 when five papers reported candidate chiral bands in odd-odd nuclei of the $A = 130$ mass region [10–14]. Important results have been obtained by the Stony Brook and Yale groups, which performed a systematic study of the $N = 75$ and $N = 73$ odd-odd isotones around ^{134}Pr [11, 13] (see Figs. 1.2 and 1.3). Side bands with the same $\pi h_{11/2} \otimes \nu h_{11/2}$ configuration as the yrast band were identified in several nuclei. The similar properties of the $\Delta I = 1$ doublet bands observed systematically in a series of odd-odd nuclei suggested a common nature, related to the restoration in the laboratory frame of chiral symmetry broken in the body-fixed frame. The small energy differences between the doublet bands were interpreted as a weakening of the chiral symmetry breaking, which induces a collective chiral vibration. The systematic observation of chiral doublets in the $A = 130$ mass region opened the way to more extensive experimental investigations, aiming to establish the Z, N boundaries of this region of chirality and to find chiral bands in other mass regions.

In the early stage of the search for chiral doublet bands, the investigations were focused on odd-odd nuclei. Chiral candidates have been identified in the $A = 130$ mass region from Cs ($Z = 55$) to Eu ($Z = 63$), in the $A = 100$ mass region in Rh ($Z = 45$) and in the $A = 190$ mass region in Ir ($Z = 77$). Interesting to note, a tentative chiral interpretation of two nearly degenerate four-quasiparticle bands in the even-even ^{136}Nd nucleus was also proposed [15], but later invalidated by the measured transition rates which were different in the two bands [16]. The first observation of a three-quasiparticle chiral doublet bands was reported in ^{135}Nd [17] with a configuration involving two $h_{11/2}$ proton particles and one $h_{11/2}$ neutron hole, which differ from that of the odd-odd neighbor ^{134}Pr by one additional $h_{11/2}$ proton (see Fig. 1.4). It was followed by the measurement of the electromagnetic transition probabilities and tilted axis cranking plus random phase approximation calculations, which clearly revealed the chiral character of the bands and its evolution with increasing spin from chiral vibration to chiral rotation [18].

An important progress in the study of chirality in odd-odd nuclei was realized when the selection rules for electromagnetic transition probabilities in chiral bands were derived in 2004 [19]. The

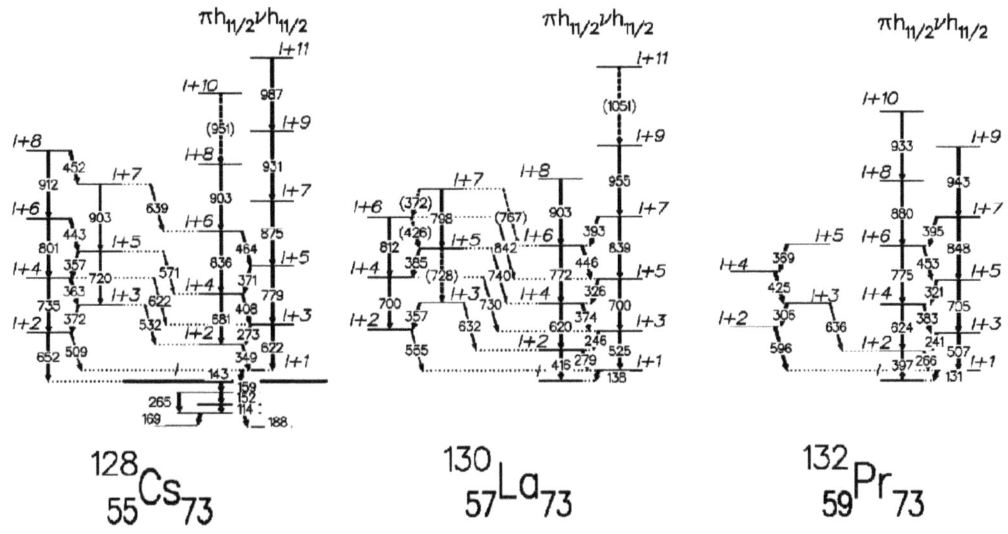

Figure 1.3 Partial level schemes of ^{128}Pr, ^{130}Pr, ^{132}Pr, in which the first chiral doublet composed of bands 1 and 2 has been proposed. (Reprinted from T. Koike *et al.*, 2001, "Observation of chiral doublet bands in odd-odd N=73 isotones", Physical Review C **63**, 061304(R).)

$B(E2)$ and $B(M1)$ values were predicted to be similar in the chiral partners. A staggering of the in-band $B(M1)$ values in both chiral partners as well as in the $B(M1)$ values of the inter-band transitions has been predicted. A first test of the predicted transition probabilities of the chiral doublet bands in odd-odd nuclei was realized in 2006 for ^{134}Pr [20]. The $B(M1)$ values in both partner bands resulted to be similar, while the intra-band $B(E2)$ strengths were different, those in the yrast band being 2 to 3 times larger than in the side band (see Fig. 1.5). These results are incompatible with the pure chiral picture (static chirality) where the intra-band $B(E2)$ transition strengths must be almost equal, and suggest a very soft vibrational regime. In order to investigate the effects of the shape fluctuations, the measured transition probabilities have been compared with the predictions of the two quasiparticle plus triaxial rotor (TQPTR) and interacting boson-fermion-fermion (IBFFM) models. The comparison with the experimental data showed a clear disagreement with the TQPTR calculation and a definitely better agreement with the IBFFM calculation, underlying the importance of the shape fluctuations for the description of the two bands.

At the same time, another important result on the chiral doublet bands in ^{134}Pr was obtained in Ref. [21] from the analysis of the observed experimental information.

When the excitation energies of bands 1 and 2 were plotted as a function of spin, it was observed that: (i) band 1 is crossed by band 2 between I = 15 and 16; (ii) band 1 is crossed by a 4qp-band 1 at I=19; (iii) band 2 is crossed by a 4qp-band 1 at I = 21; and (iv) band 2 is crossed by a 4qp-band 2 at I = 23 or higher. It was then easy to understand the reason why the two bands of ^{134}Pr remain nearly degenerate in a certain spin range: because band 1 after the crossing with band 2 is crossed by a 4qp band which, having a larger alignment, approaches band 2 and crosses it again at a higher spin. The overall result of these crossings is that the observed energy difference between the yrast and yrare levels for I = 14-22 is less than 200 keV. It was, therefore, obvious that the observed near degeneracy of the bands in ^{134}Pr for 13 < I<19 cannot be regarded as a fingerprint of the chiral regime. The energy signature staggering drawn as a function of spin in Fig. 1.6c is larger in band 1 than in band 2 at low spins, while the opposite is observed above spin 18. If two bands are chiral partners, the signature staggering should be equal. From the analysis of the branching ratios of transition probabilities $B(E2; I \rightarrow I - 2)_{in}$ and $B(E2; I \rightarrow I - 2)_{out}$ around the crossing between the

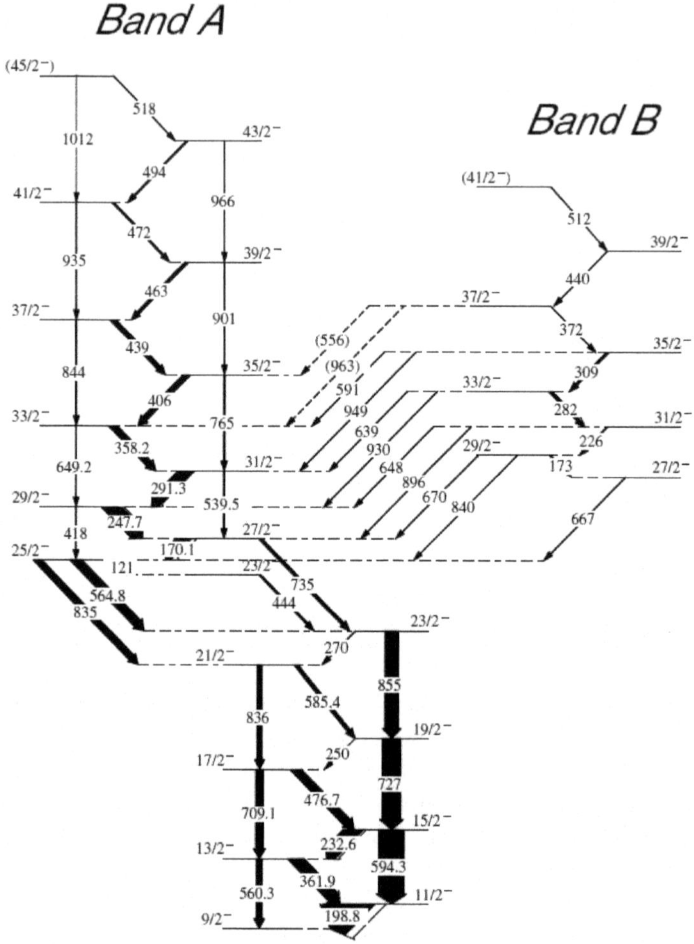

Figure 1.4 Partial level scheme of ^{135}Nd showing the chiral doublet bands. (Reprinted from S. Zhu *et al.*, 2003, "A Composite Chiral Pair of Rotational Bands in the Odd-A Nucleus ^{135}Nd", Physical Review Letters **91**, 132501.)

proposed chiral candidates, a value of $Q_{0,1}/Q_{0,2} = 2.0(4)$ was deduced for the ratio of the transition quadrupole moments of the two bands, which implies a considerable difference in the nuclear shape associated with the two bands, which is in sharp disagreement with the chiral interpretation. This result was, in fact, in agreement with the conclusion drawn from the lifetime measurements [20], which showed that the intra-band $B(E2)$ strengths in the yrast band are 2–3 times larger than in the side band. However, the fact that ^{134}Pr is one of the best chiral nuclei is not questioned. In fact, in a detailed spectroscopic study of ^{134}Pr reported several years later by Timár *et al.* [22], the good chiral doublet has been identified, involving the previously proposed side band (band 2) as yrast chiral partner and a new band (band 3) as the chiral side band. This new chiral doublet in ^{134}Pr was also theoretically supported two years later by Ikuko Hamamoto, who performed particle-rotor model calculations that predict the existence of multiple-chiral-pair bands in odd-odd nuclei with the same intrinsic configuration [23].

The second test of the predicted transition probabilities in the chiral nuclei was reported at the end of 2006 in ^{128}Cs and ^{132}La [24] (see Fig. 1.7), soon after the negative one in ^{134}Pr [21]. The

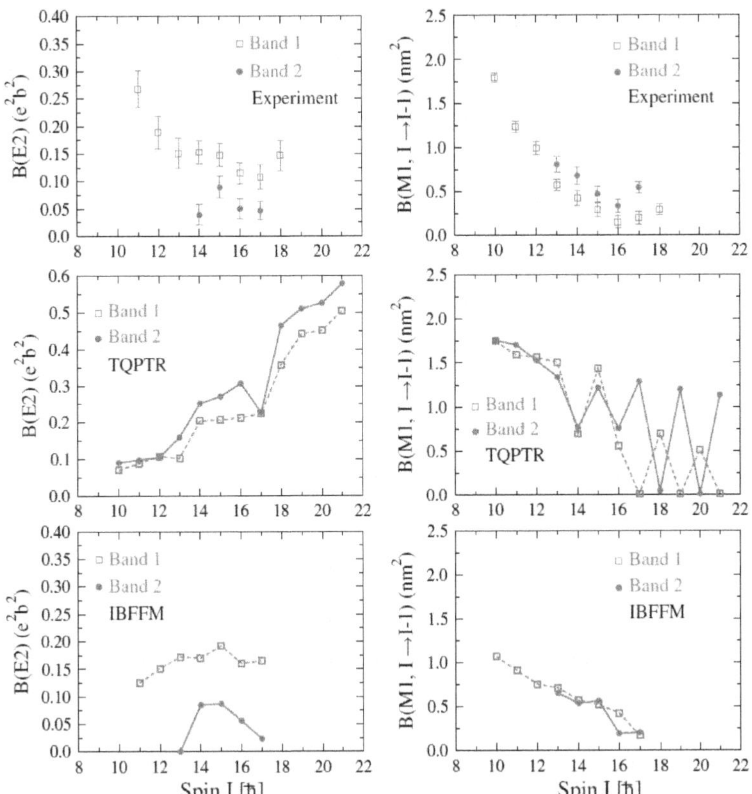

Figure 1.5 Experimentally determined and theoretically calculated $B(E2)$ and $B(M1)$ transition strengths of the candidate chiral bands of ^{134}Pr. (Reprinted from D. Tonev *et al.*, 2006, "Transition Probabilities in ^{134}Pr: A Test for Chirality in Nuclear Systems", Physical Review Letters **96**, 052501.)

lifetime measurements showed a striking difference between the two nuclei: ^{128}Cs exhibit the main features predicted for chiral doublet bands, including the staggering of the $B(M1)$ values in both chiral partners, whereas ^{132}La showed different $B(E2)$ and $B(M1)$ values in the supposed chiral partner bands, being thus in disagreement with the chiral interpretation. The authors of the paper on lifetimes in ^{128}Cs and ^{132}La triumphally stated that ^{128}Cs nucleus is the best-known example revealing the chiral symmetry-breaking phenomenon. Five years later they also published the results of lifetime measurements in ^{126}Cs [25], which showed again a qualitative agreement with all selection rules predicted for the strong chiral symmetry-breaking limit. In particular, the $B(M1)$ staggering along the partner bands, which remain observed only in ^{126}Cs and ^{128}Cs, is inferred to be due to an additional symmetry of the nuclear Hamiltonian, named S symmetry [26].

After the first investigations of chirality in the $A = 130$ mass region, the search for chiral bands in other mass regions started. In 2004, chiral doublet bands have been identified in ^{104,105,106}Rh and in ^{188}Ir, establishing two new regions of chirality around $A = 110$ and $A = 190$, respectively. The configurations in the $A = 110$ region involve the $\pi g_{9/2}^{-1}$ and $\nu h_{11/2}$ orbitals, while in the $A = 190$ mass region the $\pi h_{9/2}$ and $\nu i_{13/2}^{-1}$ orbitals, accompanied by another orbital close to the Fermi surface in the case of odd-even nuclei. In 2011, the first candidate for chiral nuclei in the $A = 80$ mass region was reported, ^{80}Br, with a configuration involving the $\pi g_{9/2}$ and $\nu g_{9/2}^{-1}$ orbitals [27].

The next achievement in the searches for chiral bands has been obtained in 2013, when evidence for multiple chiral doublet bands was obtained in ^{133}Ce, in which two distinct chiral doublet bands

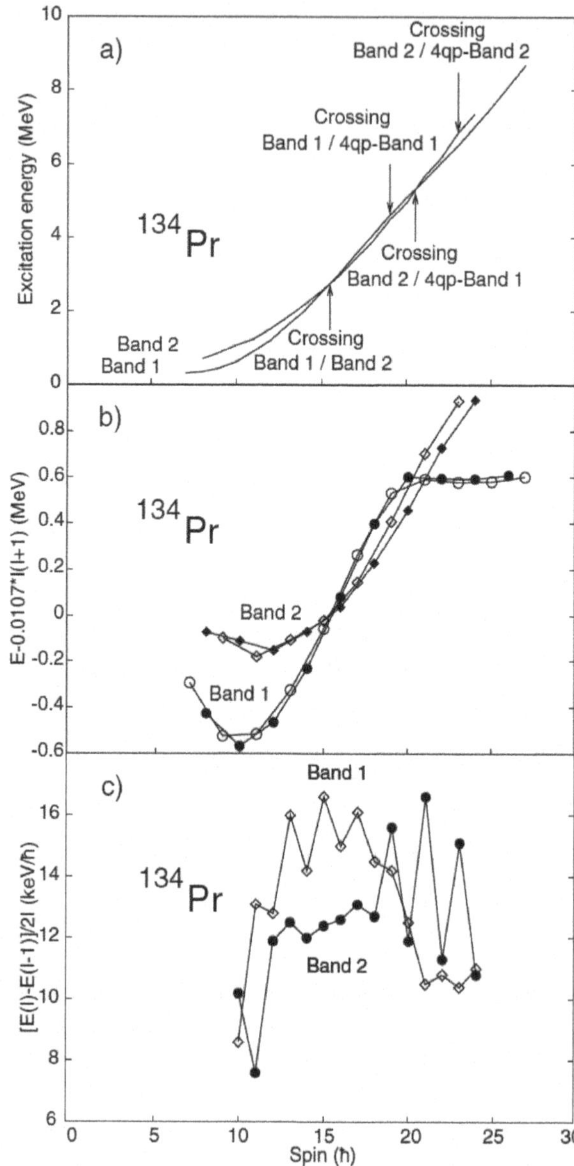

Figure 1.6 Data on the two bands in ^{134}Pr: (a) excitation energies vs spin; (b) excitation energies relative to a rigid-rotor reference; (c) energy signature staggering. (Reprinted from C. M. Petrache *et al.*, 2006, "Risk of Misinterpretation of Nearly Degenerate Pair Bands as Chiral Partners in Nuclei", Physical Review Letters **96**, 112502.)

based on three-quasiparticle configurations have been observed [28] (see Fig. 1.8). One year later two chiral doublet bands have been identified in ^{103}Rh [29]. Predicted by relativistic mean field (RMF) theory, the multiple chiral doublet bands are named with the acronym MχD. Multiple chiral bands have been also identified in ^{135}Nd and ^{137}Nd [30, 31] seven years later. But the record of multiple candidate chiral doublet bands is realized in ^{136}Nd, in which 5 chiral doublet bands have been reported [32] (see Fig. 1.9). In addition, ^{136}Nd is the only even-even nucleus in which

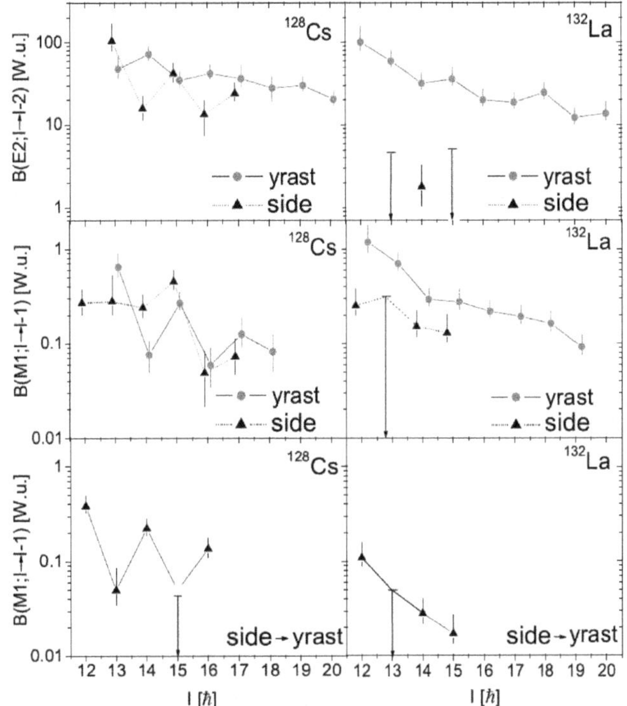

Figure 1.7 $B(E2)$ and $B(M1)$ reduced transition probabilities of ^{128}Cs (left part) and ^{132}La (right part). Top and middle: $B(E2)$ and $B(M1)$ values for in-band transitions of ^{128}Cs and ^{132}La. Bottom: $B(M1)$ values for inter-band side $\rightarrow$ yrast transitions. (Reprinted from E. Grodner *et al.*, 2006, "^{128}Cs as the Best Example Revealing Chiral Symmetry Breaking", Physical Review Letters **97**, 172501.)

multi-quasiparticle chiral doublet bands have been identified. State-of-the-art calculations supported the chiral interpretation of the observed bands, which correspond to shapes with nearly maximum triaxiality induced by different multi-quasiparticle configurations (see Fig. 1.10).

In 2016, another astonishing discovery was reported: two pairs of positive- and negative-parity doublet bands and strong electric dipole transitions linking their yrast positive- and negative-parity bands have been identified in ^{78}Br [33] (see Fig. 1.11). Four years later three nearly degenerate pairs of doublet bands have been identified in ^{131}Ba [34] (see Fig. 1.12)). Two of them, with positive-parity, are interpreted as pseudo-spin chiral quartet bands. This was the first time that a complete set of chiral doublet bands built on the pseudo-spin partners $\pi(d_{5/2}, g_{7/2})$ was observed. The chiral bands with opposite parity built on 3-quasiparticle configurations are directly connected by many $E1$ transitions. It was the first time that multiple chiral bands in the presence of enhanced octupole correlations and pseudospin symmetry were observed.

Concluding this short history of the main achievements in chirality of atomic nuclei, we can say that the research on this field is very intense even after 25 years, both experimentally and theoretically. The new state-of-the-art experimental setups for γ-ray detection, like AGATA and GRETA, will offer the opportunity to measure with high precision the intensities and electromagnetic properties for the in-band and out-of-band transitions of the weakly populated side bands, allowing to discern which are the best chiral candidates, and to what extent the selection rules of the transition probabilities are satisfied. Guided by the theoretical predictions, new bands will be identified in the already established regions of chirality, and possibly new regions in which the chiral symmetry is broken will be discovered.

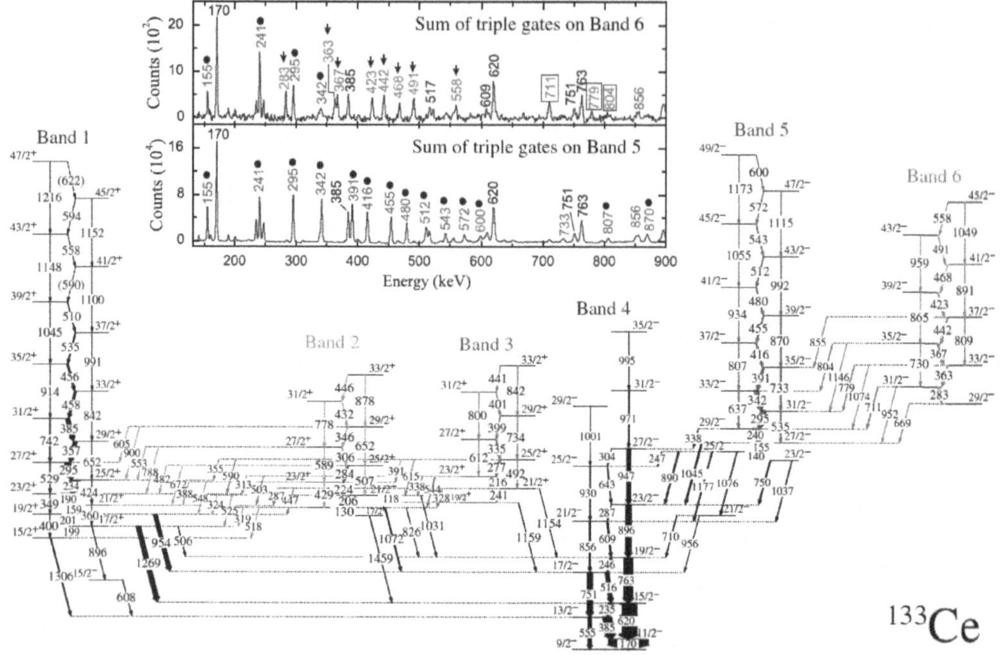

Figure 1.8 Partial level scheme of ^{133}Ce showing the chiral doublet bands 2-3 and 5-6. The inset shows coincidence spectra from different gate combinations between bands 5 and 6. (Reprinted from A. D. Ayangeakaa *et al.*, 2013, "Evidence for Multiple Chiral Doublet Bands in ^{133}Ce", Physical Review Letters **110**, 172504.)

1.3 WOBBLING IN NUCLEI

1.3.1 WOBBLING IN TSD NUCLEI: EXPERIMENTAL RESULTS, OLD AND NEW INTERPRETATIONS

The wobbling motion was first presented by Bohr and Mottelson at high spin in even-even nuclei [2]. They considered the motion of a triaxially deformed classical rigid rotor with small orientation fluctuations of the one-dimensional rotation about the principal axis of the largest moment of inertia. They quantize the rotor Hamiltonian under the condition $|I_1| \approx |I| >> 1$ (I is the total angular momentum and I_1 is the x component of the angular momentum vector), by using the technique similar to the second quantization of the harmonic oscillator. Under this conditions the excited levels corresponding to the nuclear wobbling motion have a vibrational character on the rotational band, $E(I;n_w) = \frac{I(I+1)}{2J_1} + \hbar\omega_w(n_w + 1/2)$, where $\hbar\omega_w = \hbar\omega_{rot}\sqrt{(J_1 - J_2)(J_1 - J_3)/(J_2 J_3)}$, J_1, J_2, J_3 are the moments of inertia around the x, y, z-axes, respectively, and n_w is the number of wobbling quanta.

The wobbling motion introduced by Bohr and Mottelson at high spin in even-even nuclei [2], called later by Franuendorf and D'onau "simple wobbling" to distinguish it from the wobbling modes with quasiparticles invloved [7], was never observed experimentally. Instead, it was observed in odd-A nuclei of the $A \approx 160$ mass region, ^{161}Lu [35], ^{163}Lu [5, 36], ^{165}Lu [37], ^{167}Lu [38], and ^{167}Ta [39]. All observed bands were interpreted by Hamamoto and Hagemann [3, 4] as arising from configurations with a high-j aligned particle, which is equivalent to the longitudinal wobbling.

The first experimental evidence of the wobbling mode in nuclei was published by Ødegård *et al.* in 2001 [5] (see Fig. 1.13). A new band was identified in ^{163}Lu, called TSD2, which was connected to band TSD1 by nine $\Delta I = 1$ transitions with mixed $M1/E2$ multipolarity deduced from direction correlation of γ rays from oriented states (DCO ratios) and from angular distribution ratios, from

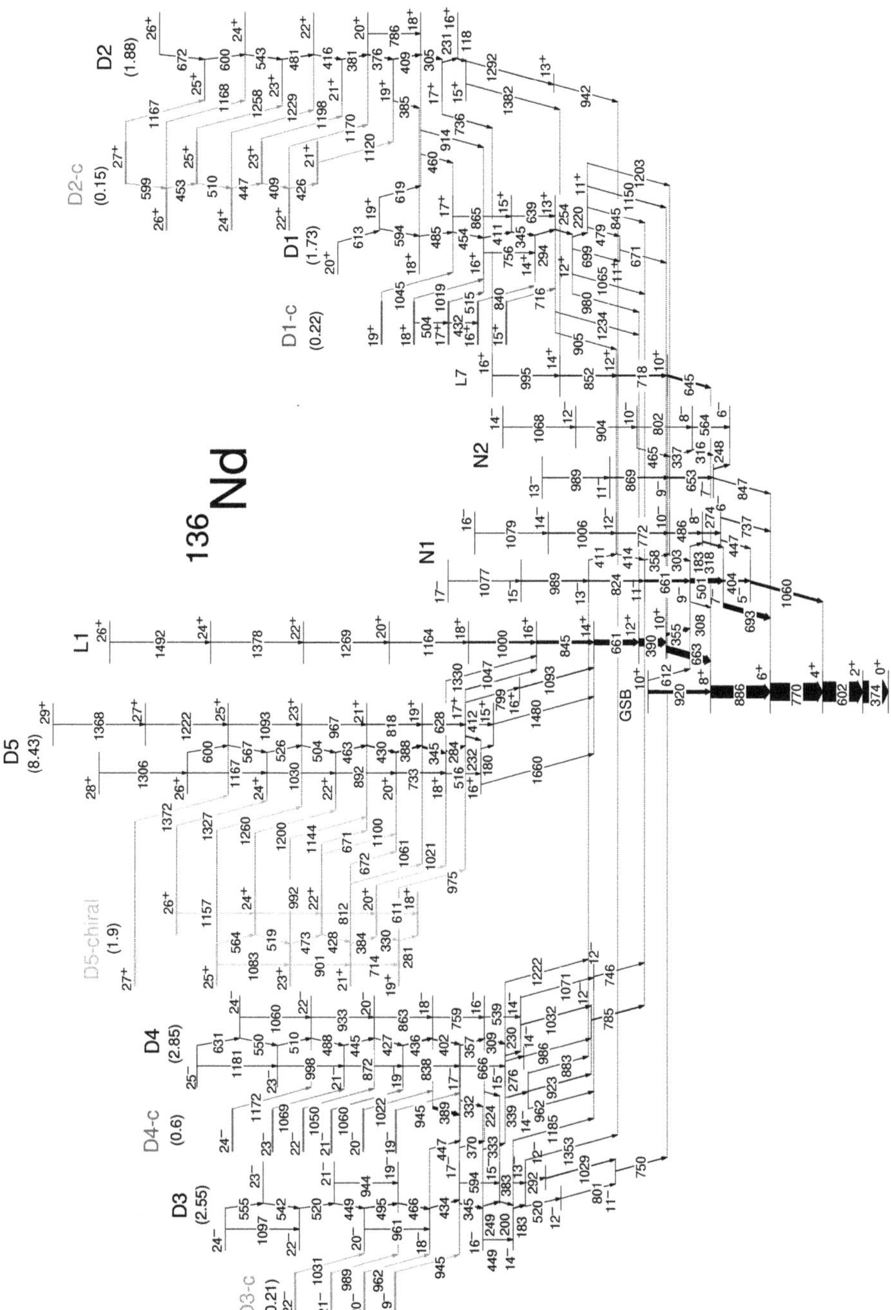

Figure 1.9 Partial level scheme of ^{136}Nd showing the newly identified doublet bands and their decay toward low-lying levels. (Reprinted from C. M. Petrache *et al.*, 2018, "Evidence of chiral bands in even-even nuclei", Physical Review C **97**, 041304(R).)

which mixing ratios of $\delta = -3.10^{+0.36}_{-0.44}$ or $-0.22^{+0.05}_{-0.03}$ were obtained. The low mixing ratio was rejected based on the polarization results which definitely yielded electric character for the connecting transitions, with $(90.6 \pm 1.3)\%$ $E2$ and $(9.4 \pm 1.3)\%$ $M1$ components. Band TSD2 is excited

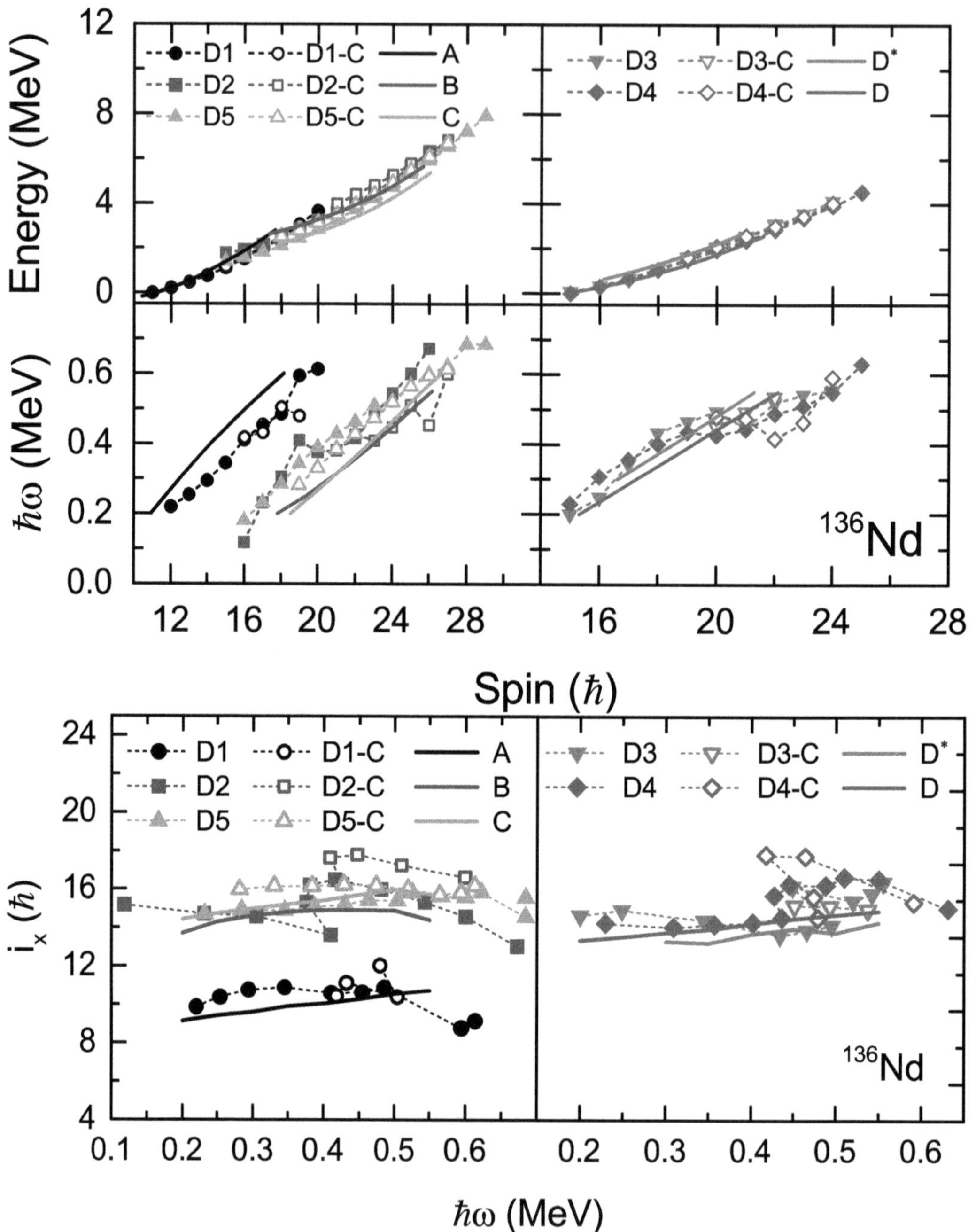

Figure 1.10 Excitation energies and $\hbar\omega$ vs I and quasiparticle alignments calculated by tilted axis cranking covariant density functional theory (TAC-CDFT) for the chiral rotational bands of ^{136}Nd. Solid and open circles with the same color represent experimental data of one pair of nearly degenerate bands, and different lines denote the theoretical results based on different configurations. (Reprinted from C. M. Petrache *et al.*, 2018, "Evidence of chiral bands in even-even nuclei", Physical Review C **97**, 041304(R).)

by only 250-300 keV relative to TSD1 and its excitation decreases as the spin increases. Therefore, the ratio of the wobbling and rotational frequencies $\hbar\omega_w/\hbar\omega_{rot}$ decreases from 1.5 to 0.5 with

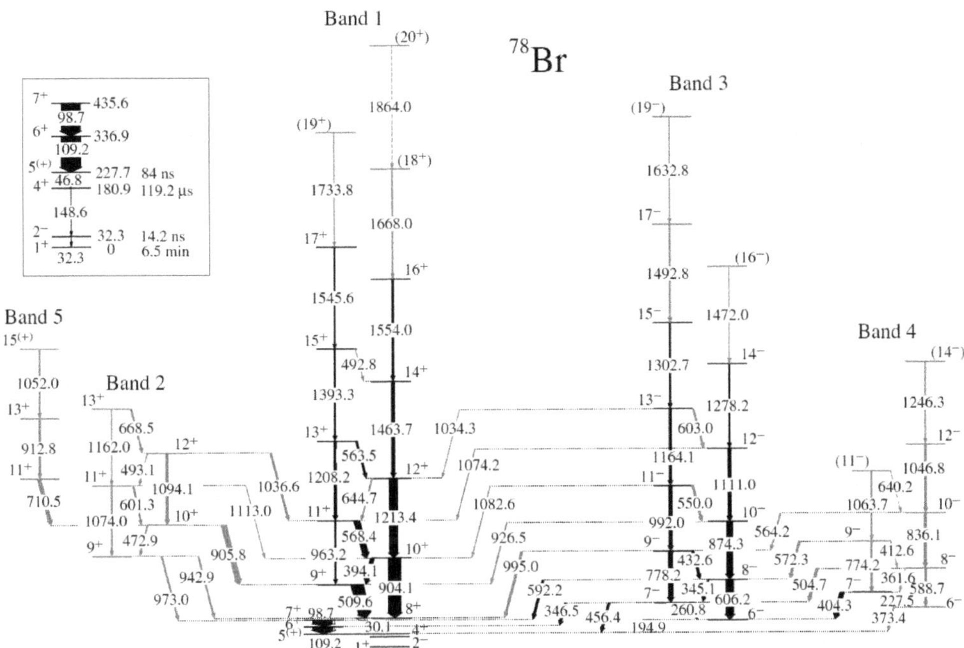

Figure 1.11 Level scheme of ^{78}Br showing the two pairs chiral doublet bands 1-2 and 3-4. The inset shows an expanded view of the lower part of the level scheme. (Reprinted from C. Liu *et al.*, 2016, "Evidence for Octupole Correlations in Multiple Chiral Doublet Bands", Physical Review Letters **116**, 112501.)

increasing spin, indicating a gradual change of the three moments of inertia. The two bands are built on the $\pi i_{13/2}$ orbital and a triaxial strongly-deformed shape with ($\varepsilon_2 \approx 0.40$ and $\gamma \approx \pm 20°$). From the comparison of the experimental results with those of cranking and particle-rotor model calculations (see Fig. 1.14), it was concluded that the properties of band TSD2 are in agreement with a wobbling excitation in the presence of an aligned particle built on band TSD1, and that the alternative interpretations as signature partner or three-quasiparticle excitation can be rejected. The possible phonon excitations in ^{163}Lu have been studied theoretically one year later by Ikuko Hamamoto [3] using the particle-rotor model in which one quasiparticle in the high-j ($i_{13/2}$) shell is coupled to the core of triaxial shape (see Fig. 1.15). It appeared that the wobbling bands built on the yrast band may be experimentally identified by very similar moments of inertia and spin alignments, by strong $B(E2;I \rightarrow I-1)$ values which dominate over the $M1$ components. The most easily to observe wobbling phonon excitations is in nuclei with most favorable triaxiality of $\gamma \approx +20°$, for which the Fermi level lies below the high-j shell, and in which aligned particles appear in the yrast line at spins as low as possible.

1.3.2 WOBBLING IN TND NUCLEI: CONTRADICTING EXPERIMENTAL RESULTS, DEBATED INTERPRETATIONS AND TERMINOLOGY

The two types of wobbling motion, with transverse (TW) and longitudinal (LW) coupling of the active particles, have been recently reported in odd-even nuclei: TW bands in ^{135}Pr [7, 40, 41] and in ^{105}Pd [42], LW in ^{133}La [43], ^{187}Au [44], ^{183}Au [45], and ^{127}Xe [46]. The TW nature of the low-spin non-yrast bands was supported by angular distribution measurements, and in some cases also by polarization measurements, like in ^{135}Pr [40], in which the reported absolute values of the mixing ratios of the transitions connecting the $n_w = 1$ wobbling band to the yrast band were larger than

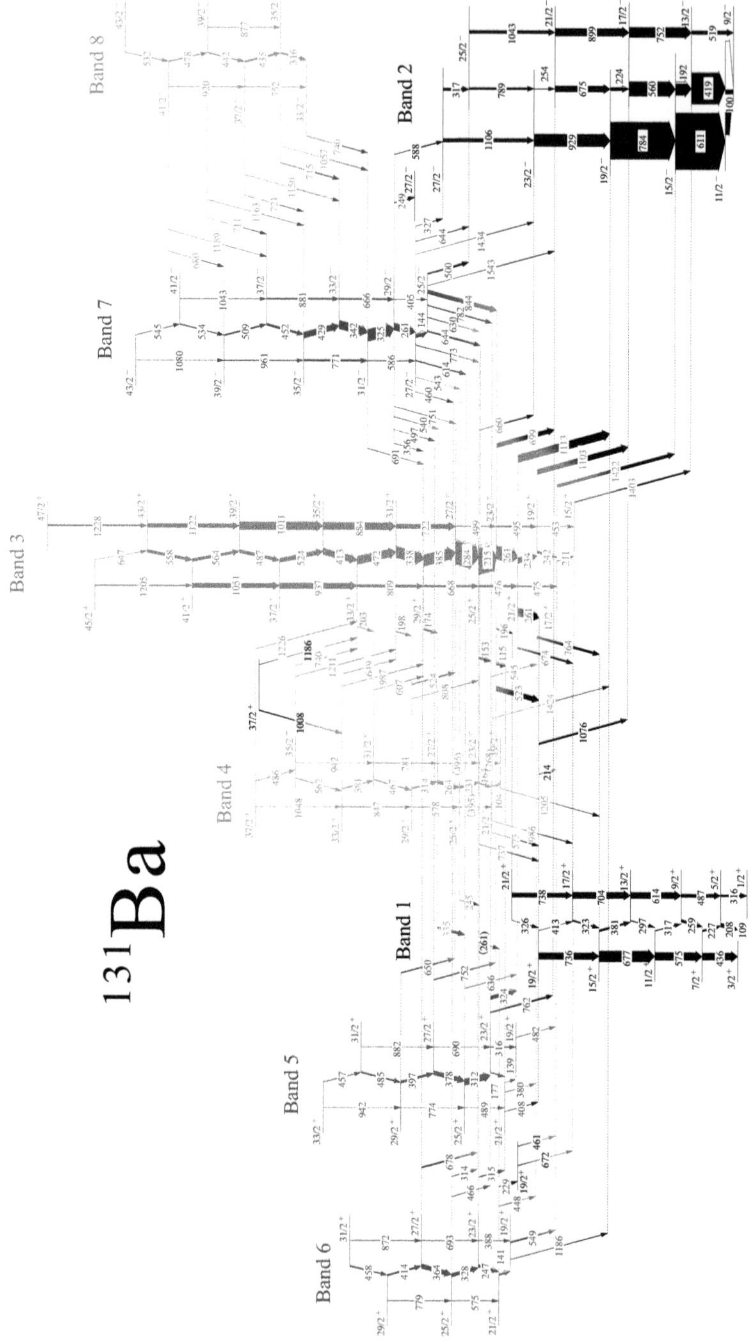

Figure 1.12 Partial level scheme of ^{131}Ba showing the three pairs chiral doublet bands 3-4, 5-6 and 7-8. (Reprinted from S. Guo *et al.*, 2020, "Evidence for pseudospin-chiral quartet bands in the presence of octupole correlations", Physics Letters B **807**, 135572.)

one, implying predominant $E2$ character in agreement with the wobbling interpretation (see Fig. 1.16). In ^{187}Au the reported LW band was supported by only angular distribution measurements. However, the wobbling interpretation of the non-yrast bands in both ^{135}Pr and ^{187}Au is contested

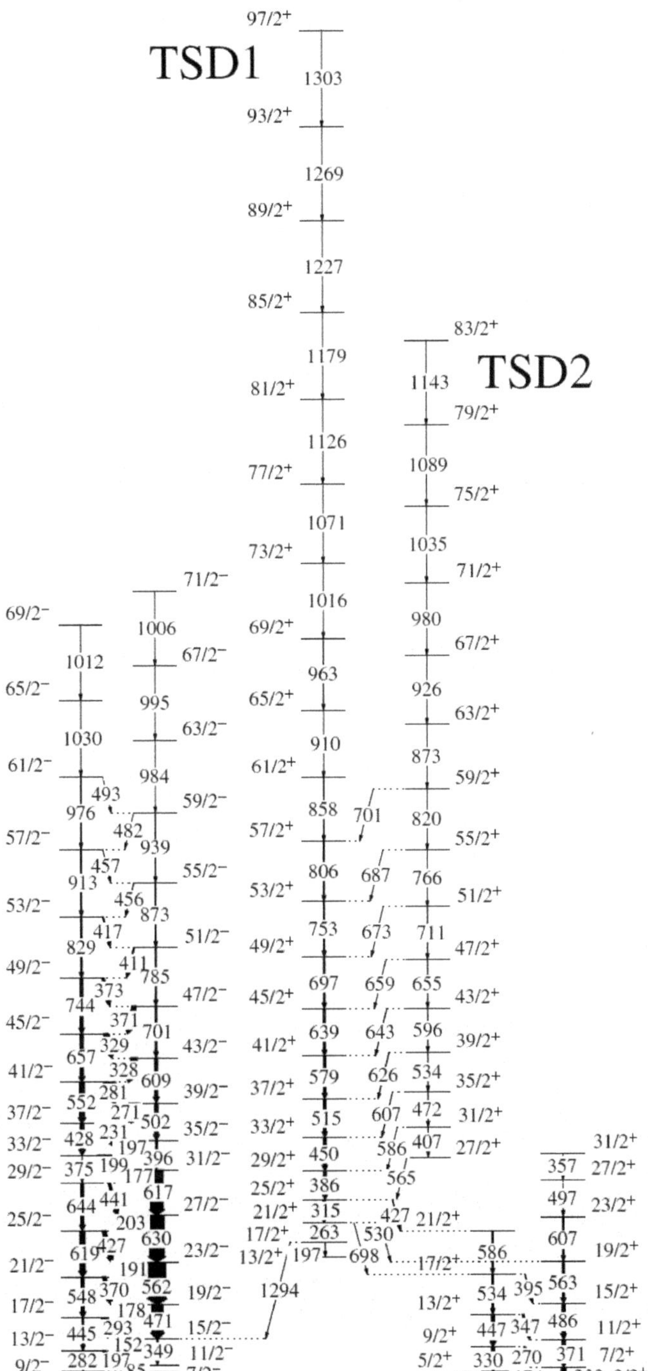

Figure 1.13 Partial level scheme of ^{163}Lu showing the two TSD bands together with the connecting transitions and the ND structures to which the TSD states decay. (Reprinted from S. W. Ødegård *et al.*, 2001, "Evidence for the Wobbling Mode in Nuclei", Physical Review Letters **86**, 5866.)

by more recent and complete experimental results. In ^{135}Pr the partial level scheme including the proposed wobbling bands was re-drawn by respecting all observed transitions and the rotational

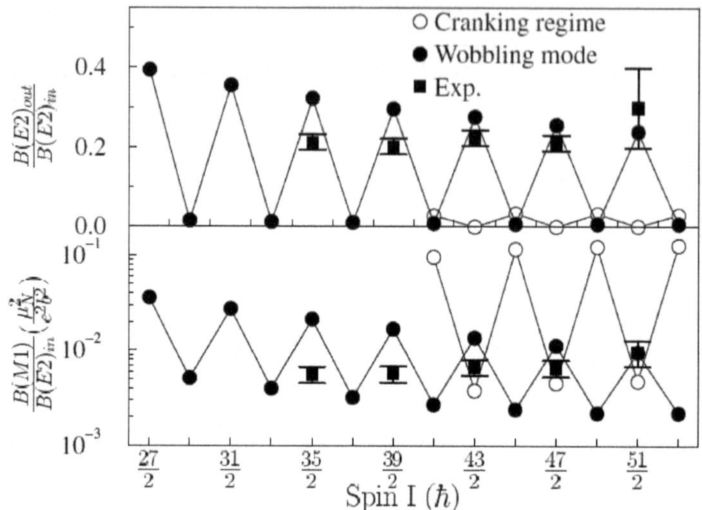

Figure 1.14 Experimental and calculated electromagnetic properties of the connecting transitions between bands TDS2 and TSD1. (Reprinted from S. W. Ødegård *et al.*, 2001, "Evidence for the Wobbling Mode in Nuclei", Physical Review Letters **86**, 5866.)

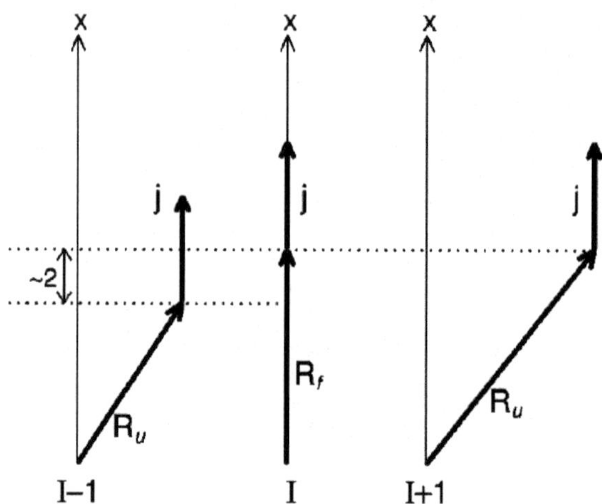

Figure 1.15 Schematic illustration of the coupling scheme of the angular momenta of the high-j particle $\vec{j}$ and the core $\vec{R}$ in the yrast favored (I) state and the yrast unfavored ($I \pm 1$) states of the wobbling excitation ($n_w = 1$). The x axis is the axis of the largest moment of inertia of the core, about which collective rotation is energetically cheapest. The total angular momentum is $\vec{I} = \vec{R} + \vec{j}$. (Reprinted from I. Hamamoto *et al.*, 2002, "Wobbling excitations in odd-A nuclei with high-j aligned particles", Physical Review C **65**, 044305.)

character of the bands, and the absolute values of mixing ratios of the connecting transitions from the $n_w = 1$ and $n_w = 2$ bands to the $n_w = 0$ band extracted from combined angular correlations and polarization measurements, were smaller than one, in contradiction with the wobbling interpretation (see Fig. 1.17). In ^{187}Au the reported signature partner of the yrast band resulted to be, in fact, a contamination from the ground state band of ^{188}Pt, and the absolute values of the mixing ratios of the connecting transitions from the $n_w = 1$ band to the $n_w = 0$ band extracted from combined

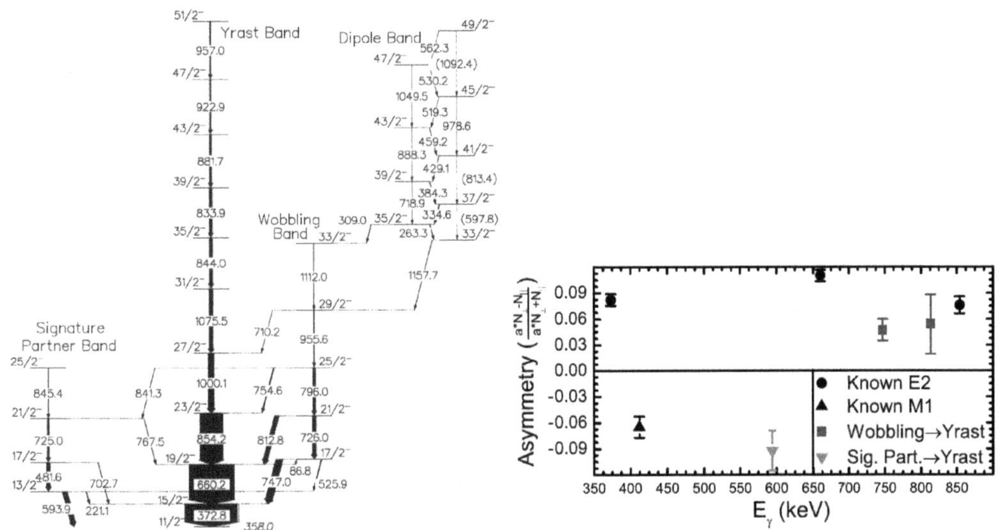

Figure 1.16   Partial level scheme of ^{135}Pr showing the previously known yrast band ($n_w = 0$), the signature partner band, the wobbling band ($n_w = 1$), and the dipole band. The asymmetries (the filled squares) for the first two transitions (747.0 and 812.8 keV) from the $n_w = 1$ band to the $n_w = 0$ band, as extracted from polarization data. The filled circles represent known $E2$ transitions from the yrast band and the black triangle a known 412.3 keV $M1$ transition from a level sequence in ^{135}Pr not displayed in the level scheme. Also shown is the measured polarization asymmetry for the 593.9 keV transition linking the signature partner and yrast bands. (Reprinted from J. T. Matta *et al.*, 2015, "Transverse Wobbling in ^{135}Pr", Physical Review Letters **114**, 082501.)

angular correlations and polarization measurements, were again smaller than one, in contradiction with the wobbling interpretation (see Fig. 1.18). From the theoretical point of view, as all these

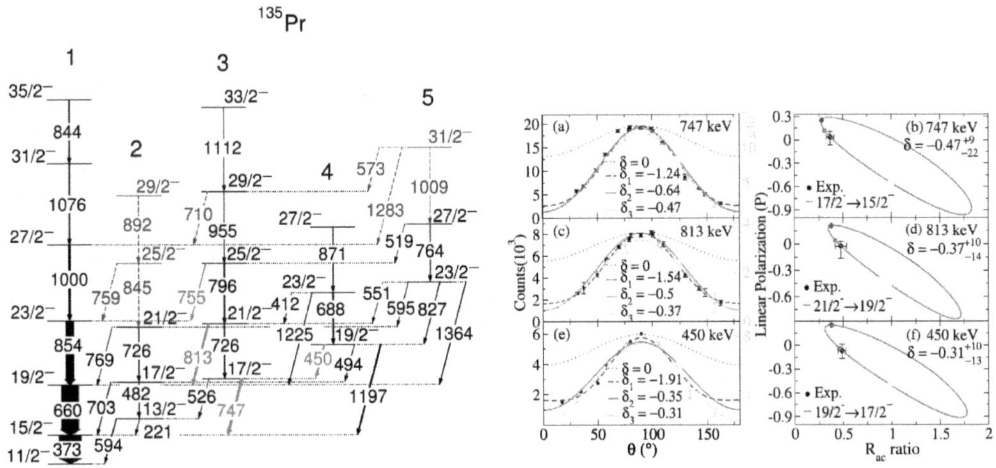

Figure 1.17   Partial level scheme of ^{135}Pr showing the low-spin negative-parity bands. The transitions reported in previous works [40, 41] , but not observed in the present data, are drawn with dashed lines. (Reprinted from B. F. Lv *et al.*, 2022, "Probing the nature of the conjectured low-spin wobbling bands in atomic nuclei", Physics Letters B **828**, 137010.)

recently reported wobbling bands develop at low spin, the high-spin approximation employed by Bohr and Mottelson when deriving the properties of the wobbling bands is not fulfilled [2], contrary to the wobbling bands in the A $\approx$ 160 mass region. In fact, the existence of the low-spin transverse wobbling in nuclei is under current debate, see e.g. Refs. [47–49]: it was pointed out that the frozen approximation proposed in Ref. [7] appears unrealistic, since it leaves out the effect of the Coriolis force on the quasi-particle which has the angular momentum coupled transversely to that of the core. A new concept of tilted Precession (TiP) motion was proposed in the framework of the quasiparticle-triaxial-rotor (QTR) model for triaxial nuclei by Lawrie *et al.* [47]: the three-dimensional rotation can be represented as a precession of the total angular momentum around a certain axis and at a given tilt. It was proven that the zero- and one-quasiparticle bands can be approximated with wobbling motion only at high spin. Currently, based on new experimental data and new theoretical calculations, the recently reported low-spin wobbling bands are seriously questioned [47–57].

Compared with the wobbling motion in odd-A nuclei, the evidence of the wobbling mode in even-even systems is extremely scarce, even though it has been proposed about half a century ago. Experimentally, the only simple wobbler candidates are low-spin bands in the even-even nuclei ^{112}Ru [58] and ^{104}Pd [59] reported around a decade ago. However, the wobbling interpretation of these bands is not fully supported experimentally, the crucial observables for the assignment of the wobbling nature – mixing ratios and transition probabilities of the connecting transitions – being not measured. In addition, these nuclei are expected to have a soft shape. In such cases the γ vibrations are not easy to disentangle from wobbling motion. At medium spin, after the breaking

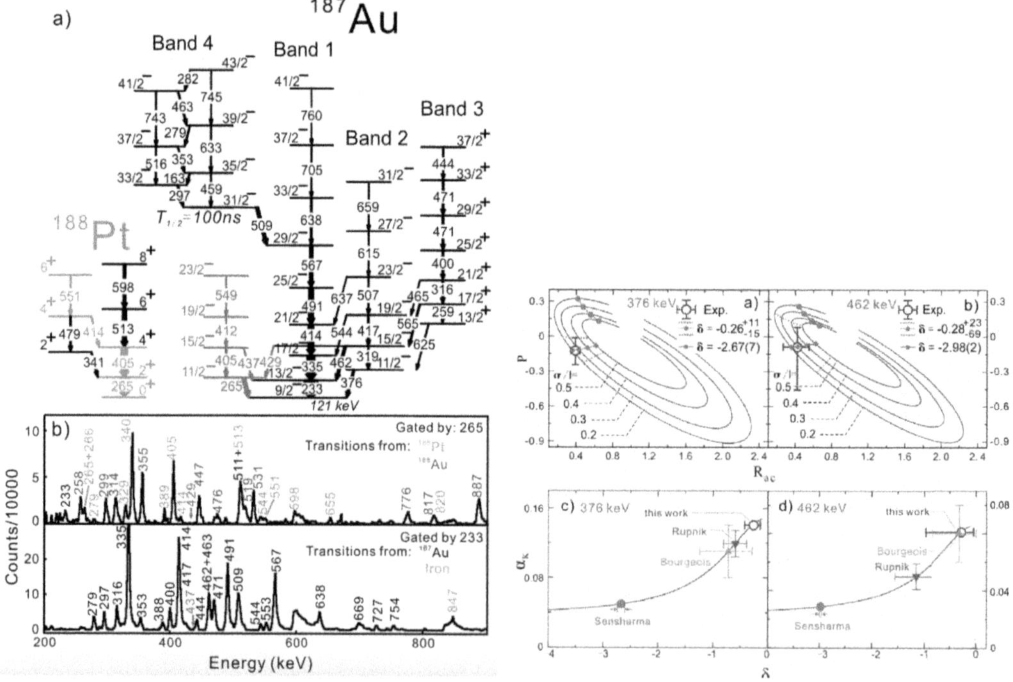

Figure 1.18   Right: a) Partial level scheme of ^{187}Au and ^{188}Pt. The cascade 405-412-549 and its connecting transitions reported in Ref. [44] were not observed in Ref. [53]. b) Spectra gated by the 265- and 233-keV transitions. Left: a) and b): Plots of the polarization P versus R_{ac} for different values of δ and σ/I, together with experimental results and estimated results for certain δ values. c) and d): Internal conversion coefficients α_K as a function of δ, together with results from several experiments. (Reprinted from S. Guo *et al.*, 2022, "Evidence against the wobbling nature of low-spin bands in ^{135}Pr", Physics Letters B **824**, 136840.)

of a nucleon pair, the nucleus acquires a more stable triaxial shape under the effect of rotation and of the polarization induced by the unpaired particles. The existence of one-phonon and possible two-phonon transverse wobbling bands at medium spin (above $I = 10$) was proposed for the first time in two even-even nuclei, ^{134}Ce and ^{136}Nd [60]. However, this conjecture was also not well supported experimentally, due to the lack of measured mixing ratios and transition probabilities of the connecting $\Delta I = 1$ transitions between the proposed wobbling bands. Very recently, the first case of transverse wobbling bands built on the two-quasiparticle $\pi h^2_{11/2}$ configuration in the even-even nucleus ^{130}Ba was reported [62]. The experimental wobbling excitation energy E_{wob}, and the electromagnetic transition probability ratios of band S1 and S1' previously observed in ^{130}Ba [61] (see Fig. 1.19) were in excellent agreement with the constrained triaxial covariant density functional theory combined with quantum particle rotor model calculations. It is important to note that, at

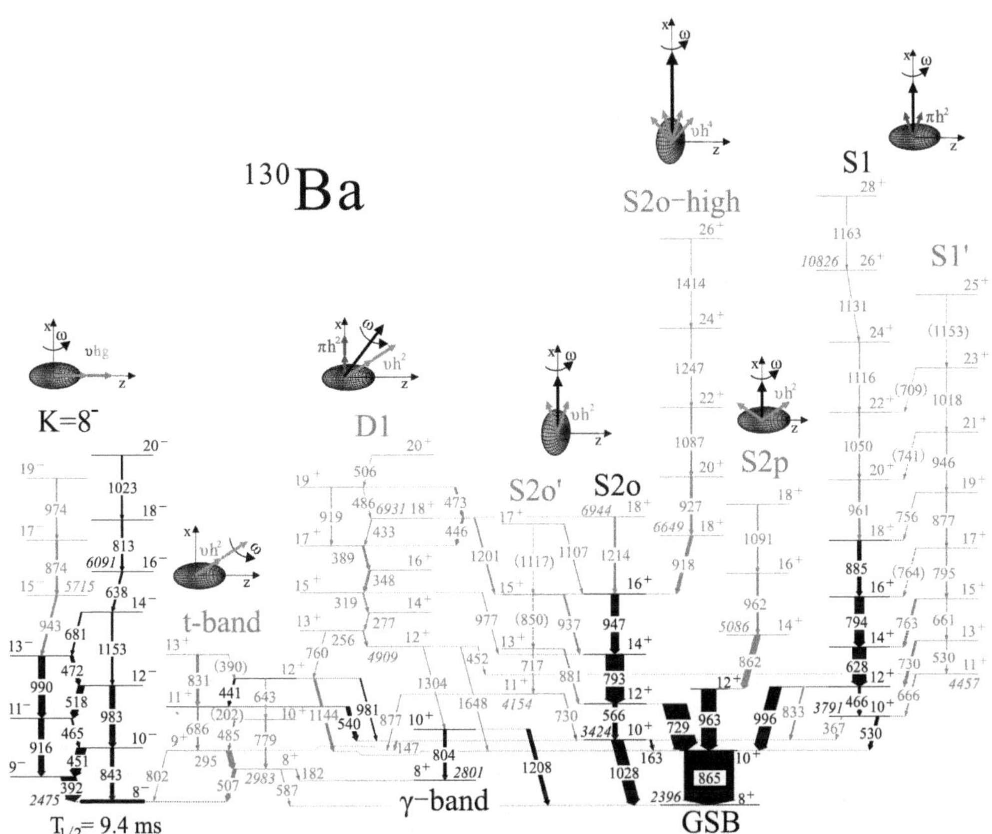

Figure 1.19 Partial level scheme of ^{130}Ba showing the newly observed bands feeding through the 8^+ GSB member at 2396 keV, as well as $n_w = 0$ and $n_w = 1$ 2-qp wobbling bands S1 and S1', respectively. The assigned configuration and corresponding shape are schematically sketched above each band. The rotation axes and the orientation of the angular momenta of the nucleons involved in each configuration are also indicated. (Reprinted from C. M. Petrache *et al.*, 2019, "Diversity of shapes and rotations in the γ-soft ^{130}Ba nucleus: First observation of a t-band in the $A = 130$ mass region", Physics Letters B **795**, 241.)

difference from the one-quasiparticle wobbling bands for which the connecting $\Delta I = 1$ transitions between the bands have predominant electric character, in the case of two-quasiparticle wobbling bands the connecting transitions can be predominantly magnetic, because the total gyromagnetic factor of two nucleons can lead to significantly larger magnetic component. The detailed analysis of the bands properties, that is the probability density distributions of the orientation of the angular momenta with respect to the body-fixed frame, and the angular momentum geometry, demonstrated the transverse geometry of the angular momenta of the unpaired nucleon and the triaxial core. Thus, it supported the presence of transverse wobbling motion in two-quasiparticle bands developing at medium spin in even-even nuclei [62]. At the same time, the medium-spin bands of ^{136}Nd have

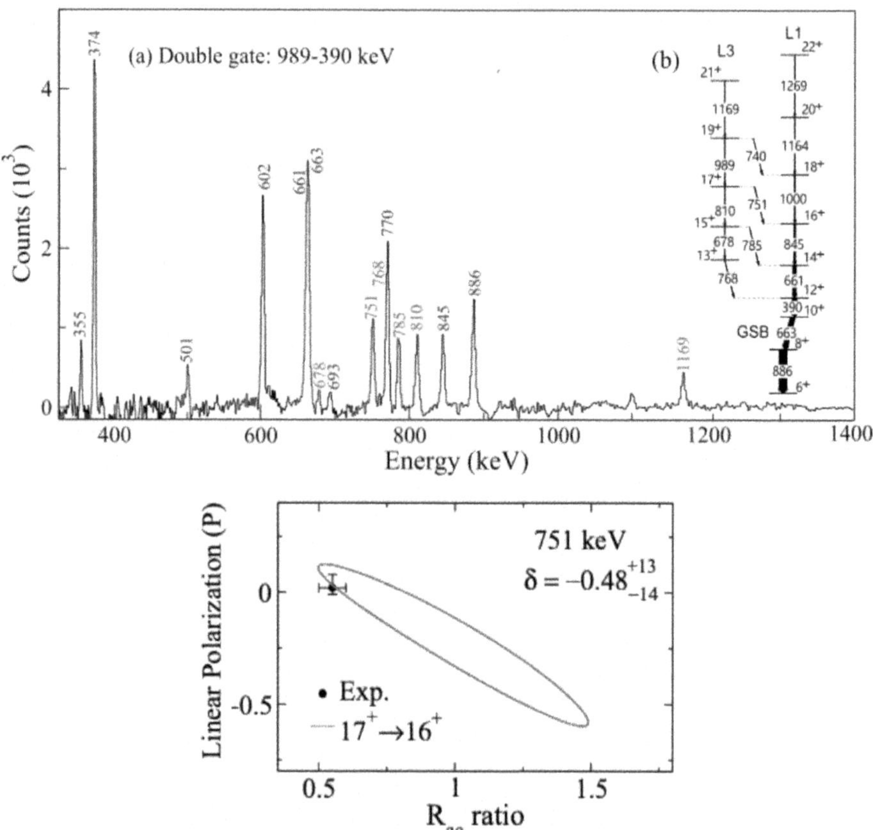

Figure 1.20   Upper panel: a) Double-gated spectrum on the 989- and 390-keV transitions showing the transitions in bands L1 and L3. The transitions of the ground state band and of band L1 are indicated. b) Partial level scheme of ^{136}Nd. Lower panel: Linear polarization P versus R_{ac} ratio for the 751-keV transition. (Reprinted from B. F. Lv *et al.*, 2022, "Experimental evidence for transverse wobbling bands in ^{136}Nd", Physical Review C **105**, 034302.)

been studied in detail [63], and a band with odd-spins and positive parity has been identified (band L3), which decays to the band with even spins (band L1) built on the $\pi h^2_{11/2}$ configuration (see Fig. 1.20)). These two-quasiparticle bands were also theoretically investigated in detail with the triaxial projected shell model, proposing their interpretation in terms of transverse wobbling motion [64]. The ^{136}Nd nucleus became then the second candidate for two-quasiparticle transverse wobbling bands in even-even nuclei. The proposed two-quasiparticle wobbling bands in ^{136}Nd are similar to the corresponding bands in ^{130}Ba, but their evolution with increasing spin is different, the wobbling character being eroded faster in ^{136}Nd. To confirm the wobbling character of the odd-spin band L3 we deduced the mixing ratio for one of the $\Delta I = 1$ transitions linking the two bands L1 and L3 from a combined angular correlation and polarization measurement and compared it with the predictions of a newly developed particle-rotor approach in which the angular momenta of the two quasiparticles and of the triaxial core are in an orthogonal geometry, and found a good agreement. Similar good agreement was obtained for two-quasiparticles bands in ^{130}Ba, ^{134}Ce and ^{138}Nd, suggesting their wobbling nature [65] (see Fig. 1.21).

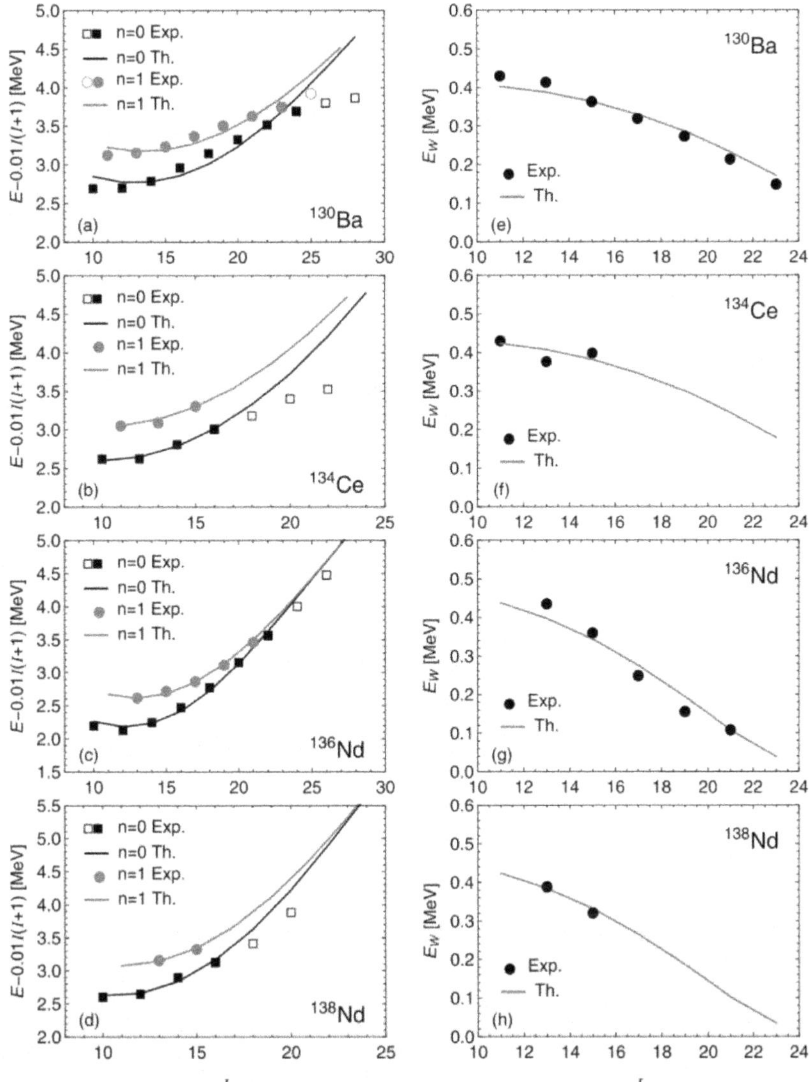

Figure 1.21 Calculated energies (minus a rigid rotor reference) of the yrast and excited bands for ^{130}Ba, ^{134}Ce and ^{136,138}Nd are compared to experimental data in the left column. Open symbols designate data points which were not considered in the fitting procedure. The corresponding theoretical and experimental wobbling energies are compared in the right column. (Reprinted from R. Budaca and C. M. Petrache, 2022, "Beyond the harmonic approximation description of wobbling excitations in even-even nuclei with frozen alignments", Physical Review C **106**, 014313.)

In conclusion of this section one can say that the research aimed to the investigation of the wobbling motion in nuclei continues to fascinate theoreticians and experimentalists. On the experimental side precise and complete information on the transitions between the wobbling bands are needed to firmly their electromagnetic character, as well as lifetime measurements of the in-band transitions of the multiple-phonon wobbling bands. Such challenging measurements can be performed with next generation high-efficiency γ-ray arrays like AGATA and GRETA which will allow the collection of sufficient statistics to extract the polarization necessary for determining the mixing ratios and the line shapes from which the lifetimes can be determined.

Bibliography

1. S. Frauendorf and J. Meng, Nucl. Phys. A **617**, 131 (1997).
2. A. Bohr and B. R. Mottelson, *Nuclear Structure* (Benjamin, New York, 1975), Vol. I.
3. I. Hamamoto, Phys. Rev. C **65**, 044305 (2002).
4. I. Hamamoto and G. B. Hagemann, Phys. Rev. C **67**, 014319 (2003).
5. S. W. Ødegård, *et al.*, Phys. Rev. Lett. **86**, 5866 (2001).
6. D. R. Jensen, *et al.*, Phys. Rev. Lett. **89**, 142503 (2002).
7. S. Frauendorf and F. Dönau, Phys. Rev. C **89**, 014322 (2014).
8. C. M. Petrache, *et al.*, Nucl. Phys. A **597**, 106 (1996).
9. B. W. Xiong and Y. Y. Wang, Atomic Data and Nuclear Data Tables, **125**, 193 (2019).
10. R. A. Bark, *et al.*, Nucl. Phys. A **691**, 577 (2001).
11. K. Starosta, *et al.*, Phys. Rev. Lett. **86**, 971 (2001).
12. A. A. Hecht, *et al.*, Phys. Rev. C **63**, 051302(R) (2001).
13. T. Koike, *et al.*, Phys. Rev. C **63**, 061304(R) (2001).
14. D. J. Hartley, *et al.*, Phys. Rev. C **64**, 031304(R) (2001).
15. E. Mergel, *et al.*, Eur. Phys. J. A **15**, 417 (2002).
16. S. Mukhopadhyay, *et al.*, Phys. Rev. C **78**, 034311 (2008).
17. S. Zhu, *et al.*, Phys. Rev. Lett. **91**, 132501 (2003).
18. S. Mukhopadhyay *et al.*, Phys. Rev. Lett. **99**, 172501 (2007).
19. T. Koike, K. Starosta, and I. Hamamoto, Phys. Rev. Lett. **93**, 172502 (2004).
20. D. Tonev *et al.*, Phys. Rev. Lett. **96**, 052501 (2006).
21. C. M. Petrache, G. B. Hagemann, I. Hamamoto, and K. Starosta, Phys. Rev. Lett. **96**, 112502 (2006).
22. J. Timár *et al.*, Phys. Rev. C **84**, 044302 (2011).
23. I. Hamamoto, Phys. Rev. C **88**, 024327 (2013).
24. E. Grodner *et al.*, Phys. Rev. Lett. **97**, 172501 (2006).
25. E. Grodner *et al.*, Phys. Lett. B **703**, 46 (2011).
26. S. G. Rohoziński, L. Próchniak, K. Starosta, and Ch. Droste, Eur. Phys. J. A **47**, 90 (2011).
27. S. Y. Wang *et al.*, Phys. Lett. B **703**, 40 (2011).
28. A. D. Ayangeakaa, *et al.*, Phys. Rev. Lett. **110**, 172504 (2013).
29. I. Kuti, *et al.*, Phys. Rev. Lett. **113**, 032501 (2014).
30. B. F. Lv, *et al.*, Phys. Rev. C **100**, 024314 (2019).
31. C. M. Petrache, *et al.*, Eur. Phys. J. A **56**, 208 (2020).
32. C. M. Petrache, *et al.*, Phys. Rev. C **97**, 041304(R) (2018).
33. C. Liu, *et al.*, Phys. Rev. Lett. **116**, 112501 (2016).
34. S. Guo *et al.*, Phys. Lett. B **807**, 135572 (2020).
35. P. Bringel *et al.*, Eur. Phys. J. A **24**, 167 (2005).
36. D. R. Jensen, *et al.*, Phys. Rev. Lett. **89**, 142503 (2002).
37. G. Schönwaßer *et al.*, Phys. Lett. B **553**, 197 (2003).
38. H. Amro *et al.*, Phys. Lett. B **552**, 9 (2003).
39. D. R. Hartley, *et al.*, Phys. Rev. C **80**, 041304 (2009).
40. J. T. Matta *et al.*, Phys. Rev. Lett. **114**, 082501 (2015).
41. N. Sensharma *et al.*, Phys. Lett. B **792**, 170 (2019).
42. J. Timár *et al.*, Phys. Rev. Lett. **122**, 02501 (2019).
43. S. Biswas *et al.*, Eur. Phys. J. A **55**, 159 (2019).
44. N. Sensharma *et al.*, Phys. Rev. Lett. **124**, 052501 (2020).

45. S. M. Nandi *et al.*, Phys. Rev. Lett. **125**, 132501 (2020).

46. S. Chakaraborty *et al.*, Phys. Lett. B **811**, 135853 (2020).

47. E. A. Lawrie, O. Shirinda, and C. M. Petrache, Phys. Rev. C **101**, 034306 (2020).

48. B. F. Lv *et al.*,, Phys. Rev. C **103**, 044308 (2021).

49. A. A. Raduta, R. Poenaru, and C. M. Raduta, Phys. Rev. C **101**, 014302 (2020).

50. K. Tanabe and K. Sugawara-Tanabe, Phys. Rev. C **95**, 064315 (2017).

51. K. Tanabe and K. Sugawara-Tanabe, Phys. Rev. C **97**, 069802 (2018).

52. B. F. Lv *et al.*, Phys. Lett. B **824**, 136840 (2022).

53. S. Guo *et al.*, Phys. Lett. B **828**, 137010 (2022).

54. S. Guo, http://arxiv.org/abs/2011.14364.

55. S. Guo and C. M. Petrache, http://arxiv.org/abs/2007.10031v2.

56. R. Garg *et al.*, Phys. Rev. C **100**, 069901 (2019).

57. K. Nomura and C. M. Petrache, Phys. Rev. C **105**, 024320 (2022).

58. J. H. Hamilton *et al.*, Nucl. Phys. A. **834**, 28c (2010).

59. Y. X. Luo *et al.*, in Exotic Nuclei: Exon-2012: Proceedings of the International Symposium (World Scientific, Singapore, 2013).

60. C. M. Petrache and S. Guo, http://arxiv.org/abs/1603.08247.

61. C. M. Petrache *et al.*, Phys. Lett. B 795, 241 (2019).

62. Q. B. Chen, S. Frauendorf, and C. M. Petrache, Phys. Rev. C **100**, 061301 (2019).

63. B. F. Lv *et al.*, Phys. Rev. C **98**, 044304 (2018).

64. Q. B. Chen and C. M. Petrache, Phys. Rev. C **103**, 064319 (2021).

65. R. Budaca and C. M. Petrache, Phys. Rev. C **106**, 014313 (2022).

2 Wobbling motion in triaxial nuclei

Stefan Frauendorf

Notre Dame University, Notre Dame, Indiana, USA

2.1 INTRODUCTION

In this chapter I review the work on the wobbling mode from my perspective. Section 2.2 discusses the triaxial rotor (TR) model. Based on its correspondence with the rotation of the classical TR, the TR eigenstates are classified as "wobbling" and "flip" states. The accuracy of this classification is justified. Section 2.3 is devoted to the particle+triaxial rotor (PTR) model. The classification of the PTR eigenstates as "transverse wobbling" (TW) and "longitudinal wobbling" (LW) based on the topology of the precession cones of the corresponding classical the angular momentum with respect to the principal axes is introduced. Subsection 2.3.1 reviews the evidence for TW in strongly deformed nuclei, and discusses the details for a selected case. In the same way Subsection 2.3.2 reviews the evidence for TW in normal deformed nuclei. Three selected nuclei are discussed in detail, where the controversy about the TW interpretation is addressed. The instability of the TW mode and the transition to LW is discussed in Subsection 2.3.3. The evidence for LW in normal deformed nuclei is reviewed in Subsection 2.3.4. One selected case is presented in more detail. Subsection 2.3.5 presents my perspective on the various approximate solutions of the PTR model. Section 2.4 reviews the studies of the wobbling mode by means of the microscopic random phase approximation (RPA). Section 2.5 gives a short introduction to the microscopic description of wobbling in the frame work of the triaxial projected shell model (TPSM). It presents its application to two nuclei only because the TPSM is comprehensively reviewed in Chapter 11. Section 2.6 addresses wobbling in nuclei with soft triaxiality.

2.2 THE TRIAXIAL ROTOR

Nuclei are composed of identical nucleons. This implies that the orientation of the nuclear density with respect to a symmetry axis cannot be specified, and collective rotation about this axis does not exist. The rational bands of deformed axial nuclei reflect the collective rotation about the axis perpendicular to the symmetry axis. The axial rotor Hamiltonian is simply $(\hat{R}^2 - \hat{R}_3^2)/2\mathcal{J}$ with 3 being the symmetry axis and $\mathcal{J}$ the moment of inertia (MoI). For reflection-symmetric nuclei the rotational band built on the ground state is the sequence $I(I+1)/2\mathcal{J}$, $I = 0, 2, 4, \ldots$. The odd I do not appear because a rotation $\mathcal{R}_{\perp}(\pi)$ about an axis perpendicular to the symmetry axis leaves the intrinsic state of the rotor invariant. In their textbook [1], Bohr and Mottelson discussed in detail the rotational features of the axial rotor when the intrinsic state contains trins excited quasiparticles.

For triaxial shape, specified by the deformation parameters β and γ, (see [1]) collective rotation about the third axis becomes possible. The new degree of freedom leads to a new class of collective excitations – the "wobbling" mode, which is described by the triaxial rotor Hamiltonian

$$H_{\mathrm{TR}} = \sum_{i=1,2,3} A_i \hat{J}_i^2, \quad A_i = \frac{1}{2\mathcal{J}_i}, \tag{2.1}$$

where i refers to the three principal axes of the triaxial density distribution.

DOI: 10.1201/9781032691633-2

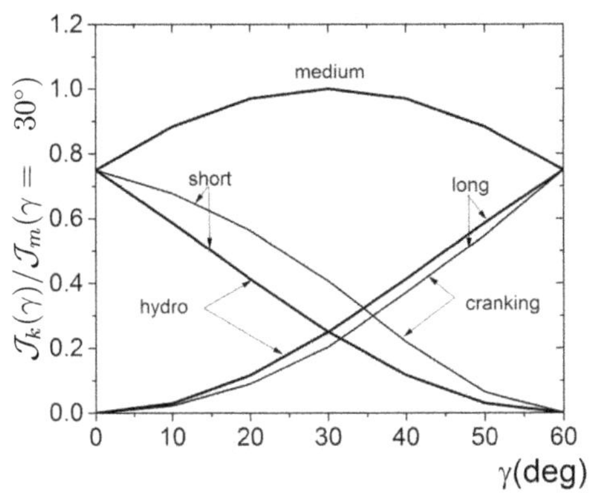

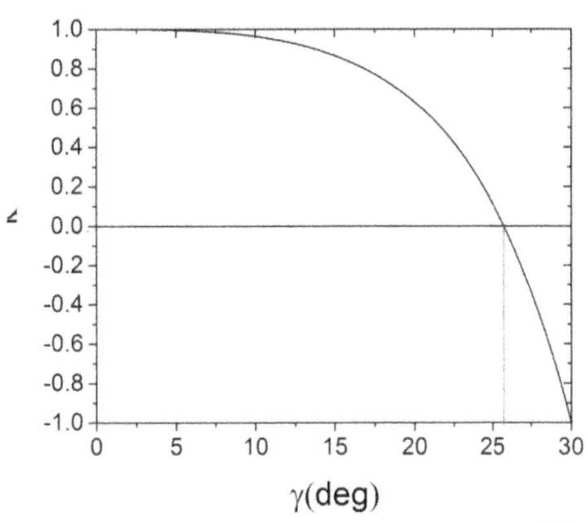

Figure 2.2 The asymmetry parameter $\kappa(\gamma)$ Eq. (2.2) for irrotational flow MoI's.

Figure 2.1 Thick curves: Irotational flow MoI's as functions of the triaxiality parameter γ (Copenhagen convention). Thin curves: Microscopic cranking calculations for $Z = 68$, $N = 96$, $\Delta_p = 1.1$ MeV, $\Delta_n = 1.0$ MeV, $\varepsilon = 0.25$. Reproduced with permission from Ref. [3] (where the Lund convention with the opposite sign of γ was used).

The model parameters are the three MoI's $\mathcal{J}_i$. Their choice is restricted by the fundamental condition that collective rotation about a symmetry axis is not allowed. More specifically, the more the density of a principal axis deviates from rotational symmetry the larger is its MoI, which requires the order $\mathcal{J}_m > \mathcal{J}_s > \mathcal{J}_l$. In the following, we associate the axes $i = 1, 2, 3$, with the short, medium, and long principal axes of the density distribution, respectively, using the short-hand notation $s-$, $m-$ and $l-$ axes.

The eigenstates are determined by the asymmetry parameter of the MoI's

$$\kappa = \frac{2A_1 - A_2 - A_3}{A_2 - A_3}, \quad A_2 \leq A_1 \leq A_3. \tag{2.2}$$

The inertia ellipsoid changes from a prolate spheroid for $\kappa = -1$, $A_1 = A_3$ to an oblate spheroid for $\kappa = 1$, $A_1 = A_2$ with maximal triaxiality at $\kappa = 0$.

The condition is obeyed by the assumption that the ratios between MoI's agree with the ones of irrotational flow of an ideal liquid [1] Eq. (6B-17),

$$\mathcal{J}_i(\beta, \gamma) = \mathcal{J}_0(\beta) \sin^2\left(\gamma - \frac{2\pi i}{3}\right), \tag{2.3}$$

where $\mathcal{J}(\beta)$ sets the scale. Fig. 2.1 shows the irrotational-flow MoI's and Fig. 2.2 the pertaining asymmetry parameter as functions of γ. Below $15°$, where $\mathcal{J}_3$ is small, the inertia ellipsoid does not change much. For $\gamma = 25.76°$ it is maximal triaxial. Above, it rapidly develops to a prolate spheroid with $\mathcal{J}_2 >> \mathcal{J}_1 \approx \mathcal{J}_3$.

The analysis of of the rotational energies and E2 matrix element from COULEX experiments [2] demonstrated that the dependence of MoI's on the triaxiality parameter γ is well accounted for by Eq. (2.3). Microscopic calculations by means of the cranking model [3] demonstrated that the MoI's approach the ratios (2.3) for strong pairing correlations. Fig. 2.1 includes the MoI's for realistic pairing. For weak pairing substantial deviations appear. However, the deviations do not indicate an approach to the ratios of rigid body flow, because these contradict the no-rotation-about-a-symmetry-axis restriction of the nuclear many body system. The moment of the medium axis is always the largest because the density distribution maximally deviates from axial symmetry.

Not all solutions of the triaxial rotor (TR) Hamiltonian are accepted for nuclei. Restricting the discussion to the case that the intrinsic triaxial ground state is invariant with respect to the three rotations $\mathcal{R}_i(\pi)$ only the TR solutions that are symmetric with respect to the three rotations are rotational excitations of the intrinsic ground state (completely symmetric representations of the D_2 point group) are allowed.

The TR Hamiltonian is diagonalized in the basis of the axial rotor states $|IMK\rangle$, where M is the angular momentum projection on the laboratory z-axis and K the projection the 3-axis of the rotor. The TR eigenstates

$$|IM\nu\rangle = \sum_{K=-I}^{I} C_{IK}^{(\nu)} |IMK\rangle \tag{2.4}$$

are given by the amplitudes $C_{IK}^{(\nu)}$, which depend only on K, the angular momentum projection on one of the body-fixed principal axes. The symmetry restricts K to be even and requires $C_{I-K}^{(\nu)} = (-1)^I C_{IK}^{(\nu)}$.

The right panel of Fig. 2.3 shows the TR energies with the MoI's $\mathcal{J}_m = 30\ \hbar^2/\text{MeV}$, $\mathcal{J}_s = 10\ \hbar^2/\text{MeV}$, $\mathcal{J}_l = 5\ \hbar^2/\text{MeV}$. Their ratios correspond to $\kappa = 0.24$, which is close to 0 of maximal asymmetry oft the inertia ellipsoid. It is instructive to classify the states using quasi-classic correspondence. The left hand panel shows the classical orbits of the angular momentum vector **R** with respect to the principal axes of the triaxial shape, where

$$R_1 = R\sin\theta\cos\phi, \ R_2 = R\sin\theta\sin\phi, \ R_3 = R\cos\theta. \tag{2.5}$$

The orbits are determined by the conservation of the angular momentum $R^2 = I(I+1)$ and the energy $E_{TR}(I)$, Eq. (2.1), which is set equal to the quantal energies in right panel. The panel shows the orbits for the states $I_\nu = 8_\nu$ which are connected by the vertical line at $I = 8$ in the right panel. There are two types of orbits, which are separated by the separatrix shown as the dashed orbit in the left panel. For the states inside the separatrix the angular momentum revolves around the m-axis with the maximal MoI, and for the ones outside it revolves around the l-axis with the minimal MoI. The orbits are numbered by their energy in ascending order. For the lowest states the motion is harmonic as seen by the elliptic shape of the orbit.

Using semiclassical correspondence the states are classified by their topology. On page 190 of their textbook Nuclear Structure II [1] Bohr and Mottelson introduced the name "wobbling" as "...the precessional motion of the axes with respect to the direction of I; for small amplitudes this motion has the character of a harmonic vibration... ". This clearly describes the states that correspond to the orbits inside the separatrix. Hence it is appropriate to call these states wobbling excitations. Accordingly, the states labeled by $n = 0, 1, 2, ...$are called zero, single, double, ... wobbling excitation, respectively. In Ref. [5] (see Chapter 6), the authors suggested restricting the name "wobbling" to the elliptic (harmonic) precession cones and call the other tilted precession (TiP) cones (see the detailed discussion below). In my view this is an unnecessary complication. It does not comply with Bohr and Mottelson, who associate wobbling with a general precession and then add harmonic vibration as a special case. I recommend using the simple topological classification with respect to the separatrix.

It is instructive indicating the axis about which the angular momentum vector precesses. In the TR model one has the m-axis wobbling states which correspond to the orbits inside the separatrix and the l-axis wobbling states which correspond to the orbits outside it.

Classically, the separatrix represent the unstable rotation about the s-axis with the intermediate MoI. The orbits in its vicinity represent a number of revolutions about the unstable axis and then a rapid change to rotation about the opposite orientation of the unstable axis for another number of revolutions and back to the original orientation, and so on. This behavior of the triaxial top is also known as the Dzhanibekov effect after the Russian cosmonaut who observed it when he unscrewed a wing nut, which flipped off his hand and floated around. Instructive videos can be

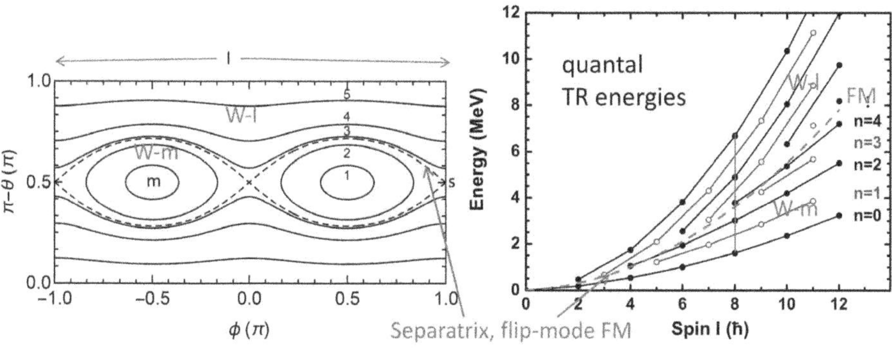

Figure 2.3 Left panel: Classical orbits of the triaxial rotor states that are connected by the vertical line at $I = 8$ the right panel. The energies are equal to their quantal energies in the right panel, labeled in ascending order. Right panel: Energies of the triaxial rotor as functions of the angular momentum R. The states are labeled by the wobbling quantum number n. The dashed line represents the energy of the separatrix $I(I+1)/(2\mathcal{J}_s)$. Reproduced with permission from Ref. [4] and [27] and augmented.

found in Wikipedia. In Ref. [4], Chen and Frauendorf suggested the name "flip mode" (FM) for the states corresponding to orbits close to the separatrix. The right panel of Fig. 2.3, demonstrates how new FM states are born at the energy of the separatrix while the lowest wobbling states become more and more harmonic when the angular moment increases. The anharmonicity of the mode increases with the wobbling number n. There is no sharp borderline between the wobbling and flip regimes.

The Spin Coherent State (SCS) representation distills the semiclassical features from the TR states. The probability of the angular momentum orientation is given by

$$P(\theta\phi)_{Iv} = \frac{2I+1}{4\pi}\sin\theta$$

$$\times \sum_{K,K'=-I}^{I} D_{IK}^{I*}(0,\theta,\phi)\rho_{KK'}^{(v)}D_{K'I}^{I}(0,\theta,\phi). \tag{2.6}$$

In the case of the TR the density matrix is

$$\rho_{KK'}^{(v)} = C_{IK}^{(v)}C_{IK'}^{(v)*} \tag{2.7}$$

with $v = n$ being the wobbling number. Plots of $P(\theta\phi)_{Iv}$ (SCS maps) have first been produced for the two-particle triaxial rotor model [6] (see Fig 2.4) in order to visualize the appearance of the chiral geometry of the quantal results. In these cases $\rho_{KK'}^{(v)}$ is the reduced density matrix. Subsequently Chen *et al.* [7] published the visualization technique under the name "azimuthal plots" for the projected shell model. Later on it has been used quite a bit to illustrate the chiral and wobbling geometry.

The SCS map in Fig. 2.5 shows that the distributions are fuzzy rims tracing the classical orbits, which are included as black curves. As discussed in detail in Ref. [4], the quantal values of $P(\theta\phi)_{Iv}d\sin\theta d\phi$ correspond to the ratio dt/T of the time dt the classical orbit spends within the area $d\sin\theta d\phi$ and T the time for one revolution.

Inspecting the figure, it is instructive to remember the physical pendulum, for which the elongation angle φ and the momentum $p = \mathcal{J}d\varphi/dt$ are canonical variables. In analogy, the angle ϕ and $R_3 = R\cos\theta$ are the conjugate position and momentum of the classical TR. For small energy the motion is harmonic (mathematical pendulum), and the phase space curve is an ellipse. With

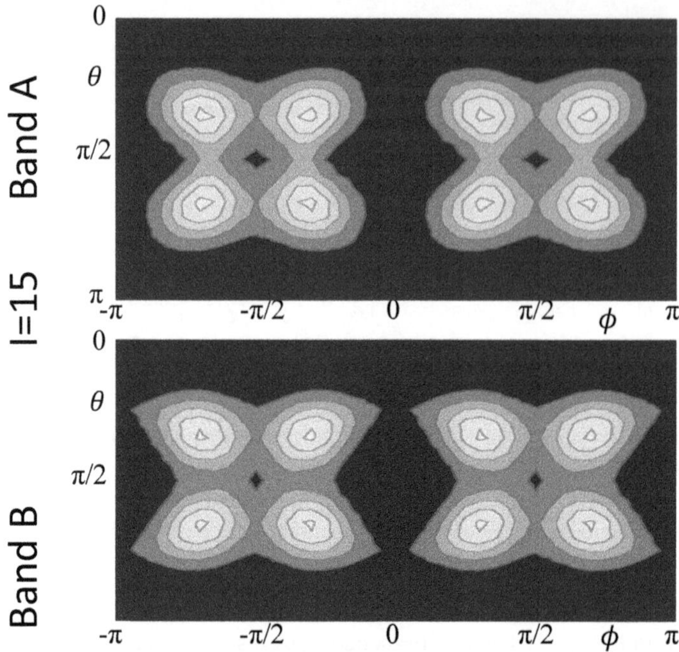

Figure 2.4 Probability distribution of the angular momentum orientation $P(\theta\phi)_{Iv}$ generated from the reduced density matrix of the particle-hole plus TR calculation for ^{134}Pr in Ref. [?]. The SCS map shows that the distributions of the chiral partner bands are almost the same at $I = 15$ where the two bands are nearly degenerate. Reproduced with permission from Ref. [6].

increasing energy the curve deforms. Along with this, the time near the two turning points increases and the time near $\phi = \pi/2$ or $\varphi = 0$ deceases. At the critical energy of the separatrix the pendulum takes the unstable position $\varphi = \pm\pi$ and the TR the position $\phi = 0$ corresponding to the unstable rotation about the s-axis. For energies slightly below the pendulum stays a long time at the turning points to swing rapidly through $\varphi = 0$. For energies slightly above the pendulum very slowly passes over $\varphi = \pi$ and keeps going, i.e. it rotates about its axis. The relative passage time through $\varphi = \pi$ decreases with a further increase of the energy. In analogy, for energies above the critical value the TR angular momentum precesses around the l-axis instead about the m-axis as for the energies below. For energies near the critical value the TR angular momentum vector stays very long near the s-axis and moves rapidly through $\phi = \pi/2$, like the rapid passage of $\varphi = 0$ of the pendulum.

The state 8_3 has an energy somewhat above the critical value. Its angular momentum stays long near the s-axis $\phi = 0$, rapidly moves (flips) to the negative s-axis $\phi = \pm\pi$, stays long, flips back to the positive s-axis $\phi = 0$, and so on. Accordingly it is called the "flip mode" and labeled by FM in Fig. 2.3, where the lower states are labeled as W-m and the higher as W-l according to their precession axes.

"Experimental" wobbling energies of the TR spectrum are defined as the excitation energies above the yrast states,

$$E_w(I) = E(I) - \left(E(I-1)_{yrast} + E(I+1)_{yrast} \right)/2 \tag{2.8}$$

Fig. 2.6 shows $E_w(I)$ of the TR with irrotational-flow MoI's for different γ. For the lowest states and sufficiently high angular momentum, located far enough below the separatrix, the harmonic

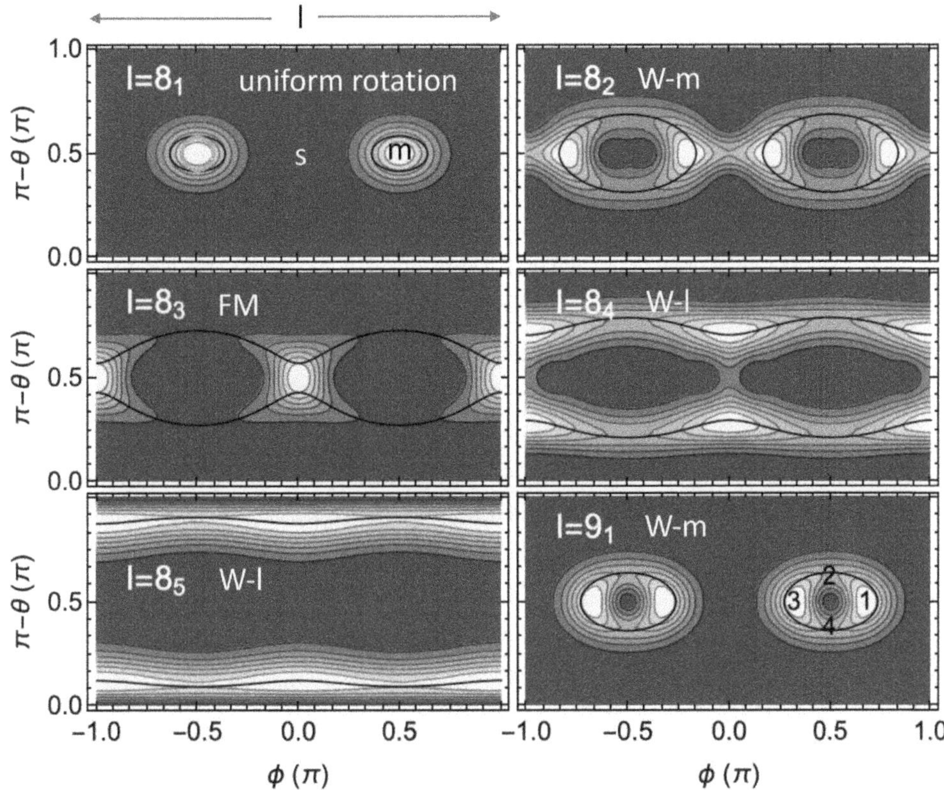

Figure 2.5 SCS probability densities of the angular momentum orientation $P(\theta\phi)_{Iv}$ of the triaxial rotor states $I_v = 8_1, \dots 8_5$ and 9_1 shown in Fig. 2.3. The states are labeled by I_v in direction of ascending energy. The densities are normalized. The classical orbits are shown as black full curves. The energies are equal to their quantal energies in Fig. 2.3. The numbers indicate the turning points. The green circle indicates the finite resolution of the plot caused by the non-orthogonaliy of the SCS basis. Reproduced with permission from Ref. [4].

approximation [1], p. 190 ff, [27] becomes valid. The TR energies are given by

$$E_n(I) = A_3 I(I+1) + \left(n + \frac{1}{2}\right)\hbar\omega_w, \quad A_i = 1/2\mathcal{J}_i \tag{2.9}$$

where n is the number of wobbling quanta and the wobbling frequency $\hbar\omega_w$ is equal to

$$\hbar\omega_w = 2I\left[(A_1 - A_2)(A_3 - A_2)\right]^{1/2}. \tag{2.10}$$

Reduced transition probabilities are given in Refs. [1] and [27] (where the different axes assignment $A_1 \geq A_2 \geq A_3$ was used). An example is the case $\gamma = 22°$ with $\kappa{=}0.47$ in Fig. 2.6, where for $I > 15$ the distance between lowest wobbling excitations $n{=}1, 2, 3$ is about the same.

For $\gamma = 30°$ the irrotational-flow MoI's have the ratios $\mathcal{J}_m : \mathcal{J}_s : \mathcal{J}_l = 4 : 1 : 1$. The TR is a symmetric top (Meyer-ter-Ven limit [8]) with the energy expression

$$E_n(I) = I(I+1)A_s + K^2(A_m - A_s), \tag{2.11}$$

where $K = I - n$ is the angular moment projection on the m-axis, which takes even values between I and $-I$. The expressions for the transition probabilities can be obtained from Ref. [1] by relabeling the axes.

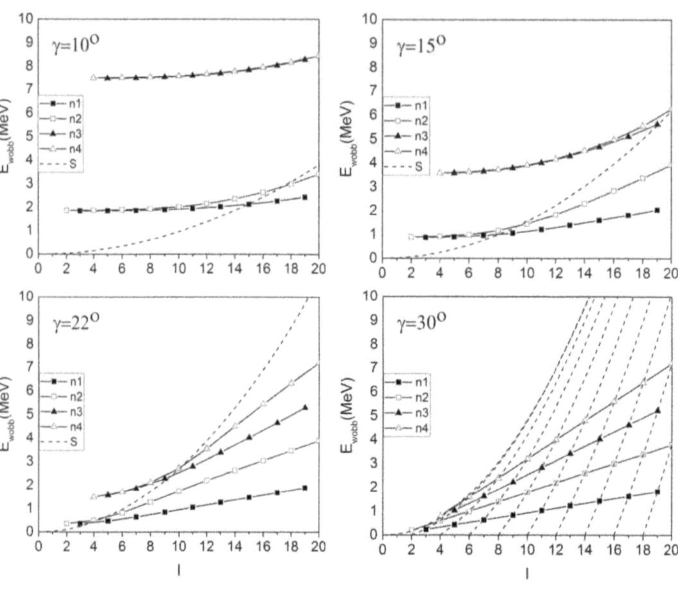

Figure 2.6 Experimental wobbling energies Eq. (2.8) for $\gamma = 10°$, $15°$, $22°$, $30°$ assuming irrotational-flow MoI's Eq.(2.3). The respective inertia asymmetry parameters are $\kappa = 0.96$, 0.87, 0.47, -1. The energies are normalized to $E(2_1^+)$. The dashed curves show the separatrix, except for $\gamma = 30°$, where they connect the sequences with the same projection K on the m-axis.

The states group into sequences of fixed K, which are displayed by the dashed curves in Fig. 2.6. The full lines display the lowest bands to be seen experimentally. As usual, these states are connected by strong E2 transitions. Their corresponding classical motion is a precession of the angular momentum around the m-axis, where the orbit in the s-l-plane is a circle with the radius $n = I - K$, which is a special case of wobbling. When γ increases toward $30°$ the elliptical orbits develop into circles and the separatrix approaches the $K = 0$ sequence.

Lawrie *et al.* [5] (see Chapter 6) introduced a different terminology for the classical orbits. They restrict the name wobbling to the elliptical orbits which can be approximated as harmonic oscillations. All other orbits are called Tilted Precessions (TiP). The qualitative properties of the TiP orbits are described as the ones of the $\gamma = 30°$ symmetric top, and the authors use the angular momentum projection on the m-axis K to label the TiP bands. The TR Hamiltonian (2.1) is more general, it encompasses different MoI's for all three axes for which the TiP approximation does not hold. On the other hand, the harmonic approximation for wobbling by Eqs. (2.9, 2.10) and the pertaining expressions for the transition probabilities hold up to small terms of the order of $1/I$ for the lowest bands of the $\gamma = 30°$ TR (see Fig.1 of Ref. [5]), which makes the TiP terminology inconsistent. Moreover "tilted precession", as meant by the authors, is a tautology, because in a precession cone the angular momentum vector is tilted with respect to the axis it revolves by definition.

In my view, it is better keeping the terminology simple and call all the states that correspond to the orbits which revolve around the m-axis wobbling states, irrespective if the precession is harmonic or not. The name is quite appropriate, because the change of the rotational axis of the earth, which is quite irregular, and the tumbling of the rotational axis of a baseball are both called wobbling motion. It is also consistent with the way Bohr and Mottelson introduced the name into nuclear physics (see above). The first phrase indicates a general precessional mode, the second phrase indicates the special case of small amplitude motion, which is worked out in the book.

In Fig. 2.6 $\gamma = 10°$ illustrates the other case of small asymmetry of the inertia tensor $\kappa = 0.96$, with the MoI's ratios $\mathcal{J}_m : \mathcal{J}_s : \mathcal{J}_l = 29 : 19 : 1$. The separatrix is close to the yrast line, such that the $n = 1$ wobbling state appears only for $I \geq 17$. For smaller I the lowest excited states form a doublet, which correspond to a narrow precession cone around the l-axis. The SCS maps show maxima at $\theta = \pi/2$, $\phi = 0$, $\pm\pi$ for even I and $\theta = \pi/2$, $\phi = \pm\pi/2$, which represent the near-axial density distribution $\propto |D_{I,2}^I(0,\theta,\phi) + (-1)^I D_{I,-2}^I(0,\theta,\phi)|^2$. The doublet sequence has the characteristics of the γ band of axial nuclei. As already pointed in Ref. [1], the properties of the γ band are well accounted for by a slightly asymmetric TR. Examples are the experimental information on the E2 matrix elements in ^{168}Er (TR $\gamma = 7°$) [9] and the work in Ref. [2]. The γ vibration represents a wave that travels around the l-axis of the axial nucleus, which is equivalent with the $K = 2$ excitation of a TR with weak triaxiality.

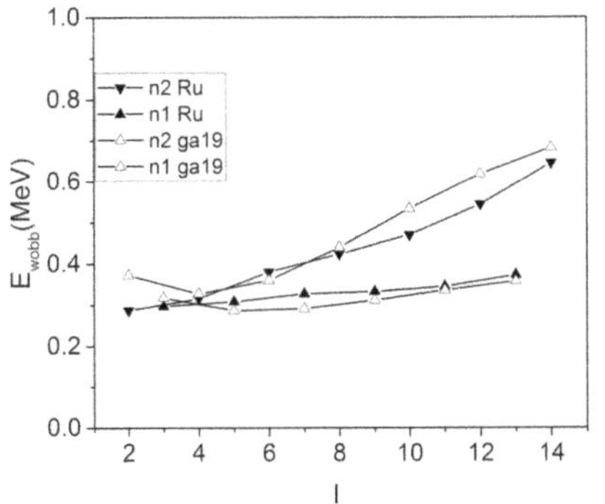

Figure 2.7 Experimental wobbling energies Eq. (2.8) for ^{112}Ru (full symbols) compared with the TR calculation (open symbols) using irrotational-flow MoI's for $\gamma = 19°$ and $\mathcal{J}_0 = 30(1+0.2I)\hbar^2/$MeV.

Fig. 2.7 compares the $n = 1$, 2 wobbling bands in ^{112}Ru [10] with the a TR calculation assuming irrotational flow, Eq. (2.3) with $\gamma = 19°$, and the variational moment of inertia (VMI) $\mathcal{J}_0 = 30(1+0.2\,I)\hbar^2/$MeV. With increasing I the structure changes from the a flip mode to anharmonic wobbling. The use of the VMI is decisive for the good agreement with the experimental wobbling energies. However, it renders too low energies for the ground band. It is not possible to adjust γ and $\mathcal{J}_0$ such that good agreement for both the wobbling and yrast energies is achieved.

The case of ^{112}Ru together with ^{76}Ge [11] and ^{192}Os[10] are the scarce examples of wobbling excitations on the ground state band. There is a larger number of even-even nuclei for which the even-I sequence of the quasi γ band is lower than the odd-I one, while for wobbling the order is opposite. This "staggering of the γ band" indicates whether, respectively, the deviation from axial shape is dynamic or static or, as commonly said, the nuclei are γ soft or rigid (see e. g. Refs. [12, 13]). In Refs. [13–15], the authors demonstrated that the microscopic triaxial projected shell model (TPSM) [16] very well describes the experimental energies and E2 matrix elements of both γ rigid and soft nuclei, provided the coupling to the quasiparticle excitations is taken into account (see Section 2.6). This microscopic approach removes the problem with the yrast energies of the phenomenological TR description of ^{112}Ru (see Figs. 5 and 7 and Tab. 3 of Ref. [13]). It would be interesting to investigate the angular momentum geometry of γ soft nuclei.

2.3 PARTICLES COUPLED TO THE TRIAXIAL ROTOR

The experimental evidence for wobbling is clearer when one or more high-j particles are coupled to the TR, because their interaction stabilizes the triaxial shape. The high-j particles act like gyroscopes, which change the wobbling motion. The Particle plus Triaxial Rotor (PTR) model [8] describes the coupled system and has become a standard tool for interpretation. The corresponding Hamiltonian is [1, 8]

$$H_{\text{PTR}} = \sum_{i=1,2,3} \frac{(\hat{J}_i - \hat{j}_i)^2}{2\mathcal{J}_i} + h_p(\gamma), \qquad (2.12)$$

where $\hat{J}_i = \hat{R}_i + \hat{j}_i$ is the total angular momentum, $\hat{j}_i$ the angular momentum of the particle, $\hat{R}_i$ the one of the triaxial rotor, and $h_p(\gamma)$ describes the coupling of the particle with the triaxial potential.

If necessary, the pair correlations are taken into account by applying the BCS approximation to h_p. In this case the model is called the quasiparticle triaxial rotor (QTR) model. In the following, I will refer to both PTR and QTR even when the pair correlations are unimportant. The coupling of the high-j particles $f_{7/2}$, $h_{11/2}$ $i_{13/2}$, $j_{15/2}$ is often described by the simplified Hamiltonian [17] which assumes that their angular momentum is conserved

$$h_p(\gamma) = \frac{\kappa}{j(j+1)}\left[(3j_3^2 - j(j+1))\cos\gamma + \sqrt{3}\left(j_1^2 - j_2^2\right)\sin\gamma\right].\tag{2.13}$$

The coupling strength κ is determined by adjusting the single particle energies generated by $h_p(\gamma)$ to the ones of a realistic triaxial nuclear potential of the same triaxiality parameter γ. As the shape of the neutron and proton densities are very similar the same γ value determines the intrinsic charge quadrupole moments and the $E2$ transition probabilities as well. The inclusion of pairing and the generalization to quasiparticles is straightforward [17]. The MoI's are input parameters which are confined by the conditions of the TR discussed in Section 2.2. In some studies the irrotational ratios (2.3) for the same value of γ as for the particle potential are used, and only $\mathcal{J}_0$ is adjusted to the rotational energies. In most studies some deviation of the inertia asymmetry κ (not to be confused with κ in Eq. (2.13)) from the irrotational value is accepted as long as the general conditions $\mathcal{J}_m > \mathcal{J}_s > \mathcal{J}_l$ are met. (see Chapter 10 for examples.)

The PTR eigenstates are represented in the basis $|IIK\rangle|jk\rangle$, where $|IIK\rangle$ are the rotor states for half-integer I and $|jk\rangle$ the high-j particle states,

$$|IIv\rangle = \sum_{K,k} C^{(v)}_{IKk}|IIK\rangle|jk\rangle.\tag{2.14}$$

The coefficients $C^{(v)}_{IKk}$ are restricted by requirement that rotor-core states must be symmetric representations of the D_2 point group, which implies

$$C^{(v)}_{I-K-k} = (-1)^{I-j}C^{(v)}_{IKk}, \quad K-k \text{ even.}\tag{2.15}$$

Expression for the transition probabilities are given for example in Ref. [42]. The reduced density matrices

$$\rho^{(v)}_{kk'} = \sum_K C^{(v)}_{IKk}C^{(v)*}_{IKk'}\tag{2.16}$$

and

$$\rho^{(v)}_{KK'} = \sum_k C^{(v)}_{IKk}C^{(v)*}_{IK'k}\tag{2.17}$$

are inserted into Eq. (2.6) to generate SCS maps, which illustrate the corresponding motion of the particle $\mathbf{j}$ and total angular momentum $\mathbf{J}$, respectively.

2.3.1 TRANSVERSE WOBBLING IN TRIAXIAL STRONGLY DEFORMED NUCLEI

Clear evidence for wobbling in triaxial strongly deformed (TSD) nuclei was established by the Copenhagen group. They found in $^{161-167}$Lu [18–22] and ^{167}Ta [23] very regular bands in the spin range $I \approx 15 - 40$, which are based on the $i_{13/2}$ proton orbital and a deformation of $\beta \approx 0.4$. The bands TSD2 $\rightarrow$ TSD1 and TSD3 $\rightarrow$ TSD2 are interconnected by strong $I \rightarrow I-1$ collective $E2$ transitions, which are the hallmark of the wobbling motion of the charge density of the whole nucleus. The bands TSD1, TSD2, TSD3 were assigned to carry $n =$0, 1,2 wobbling quanta, respectively. Hamamoto and Hagemann [19, 24, 25] carried out PTR calculations. They assumed $\gamma = 20°$, irrotational flow MoI's and adjusted $\kappa\mathcal{J}_0$ to account for the average wobbling energy. However, the authors exchanged the MoI such that $\mathcal{J}_s$ is largest and $\mathcal{J}_m$ is second in contrast to the order implied by the principles of spontaneous symmetry breaking.

The PTR system has the particle degrees of freedom in addition to the orientation of the total angular momentum. The authors illustrated the pertaining types of excitations in Fig. 2.8. The $i_{13/2}$ proton has the lowest energy when its angular momentum $\mathbf{j}$ aligns with the s-axis, because this orientation generates the best overlap with the triaxial potential (see the discussion of particle alignment in Ref. [9]). In the yrast band (TSD1) the TR angular momentum $\mathbf{R}$ aligns with the s-axis as well. In panel (a) the proton $\mathbf{j}$ is tilted from the s-axis about which it precesses while $\mathbf{R}$ stays aligned. This is the excitation type that the cranking model describes as a signature partner band (The signature is defined as $\alpha = I + 2n$.) In panel (b) the proton $\mathbf{j}$ stays aligned while the $\mathbf{R}$ is tilted from the s-axis about which it precesses. This the first wobbling excitation. It generates strong E2 radiation, which is not the case for the signature partner band. The difference can be used to identify the bands as wobbling or cranking excitations.

The PTR $\frac{B(E2,I\to I-1)_{out}}{B(E2,I\to I-2)_{in}}$ ratios in Fig. 2.9 agree well with the experimental ones. The ratio of 0.2 indicates a strong collective enhancement of the inter band E2 transitions, which the hallmark of the wobbling mode. However, the PTR energy difference between the $n_w = 0$ yrast band the $n_w = 1$ wobbling bands increases with I, which is in contrast to the decrease of the distance between the bands TSD2 and TSD1. The authors noticed that the natural order $\mathcal{J}_s < \mathcal{J}_m$ leads to a decrease of the wobbling energy, but dismissed it, because it was too rapid.

The studies of ^{161,165,167}Lu [18, 21, 22] rendered similar wobbling behavior as ^{163}Lu. The accordance of $\frac{B(E2,I\,n=2\to I-1,\,n=1)_{out}}{B(E2,I\to I-2)_{in}}$ ratios with the PTR calculations is the same as for ^{163}Lu.

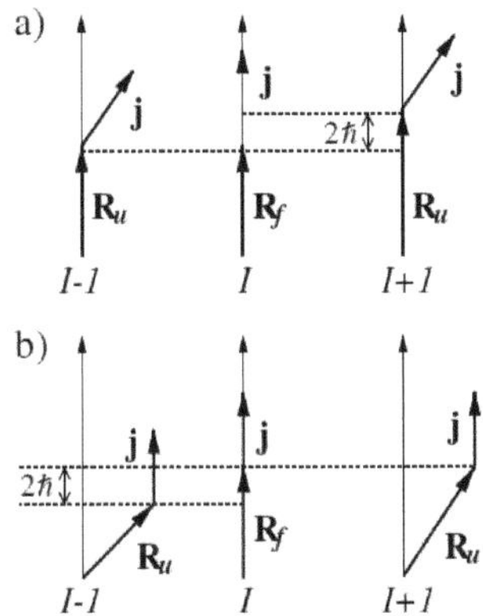

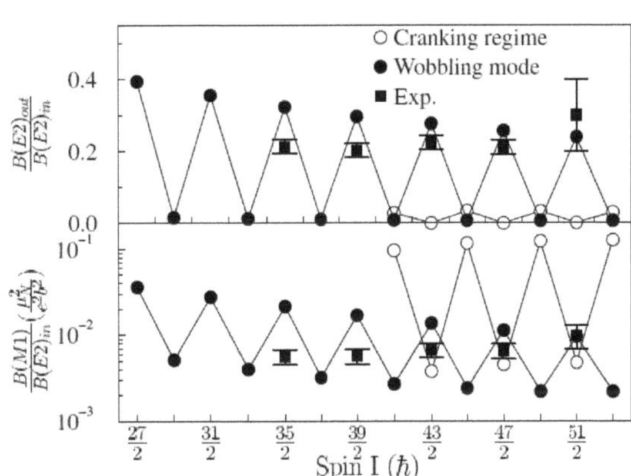

Figure 2.8 Schematic coupling scheme of the particle and core angular momenta in the favored (I) and unfavored ($I \pm 1$) states for (a) the cranking regime and (b) the wobbling mode ($n_w = 1$). Reproduced with permission from Ref. [19].

Figure 2.9 Ratios $\frac{B(E2,I\to I-1)_{out}}{B(E2,I\to I-2)_{in}}$ and $\frac{B(M1,I\to I-1)_{out}}{B(E2,I\to I-2)_{in}}$ between the bands TSD2 and TSD1 compared with the PTR calculation in Refs. [19, 24, 25]. Reproduced with permission from Ref. [19].

The exchange of the two MoI's appears to be a serious problem because it violates a basic principle of spontaneous symmetry breaking. Analyzing their PTR calculations, Frauendorf and Meng discovered that the rotational motion of odd-odd nuclei may attain a chiral character [26] (c.f. Figs. 2b and 5 of Ref. [26] and Chapter 3 of this book). The chiral geometry emerges as the combination of the angular momenta of a proton particle, a neutron hole and the TR core aligned with the s-axis, the l-axis and the m-axis, respectively. That is, chirality appears because the m-axis has the larges MoI. Fig. 2a of Ref. [26] demonstrates that the rotational spectrum is analogous when both the proton and the neutron are holes with their angular momentum aligned with the l-axis. The distance between the lowest even and odd I bands decreases with I. The same kind of decrease is expected in the spectrum of the TR combined with one proton with its angular momentum aligned with the

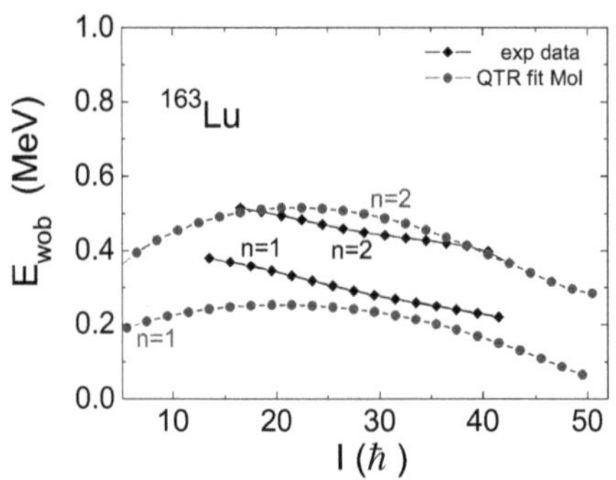

Figure 2.10 Experimental wobbling energies Eq. (2.8) of the bands TSD2 and TSD3 in 163Lu compared with the PTR (labeled by QTR) calculations in Ref. [27] using the parameters $\varepsilon = 0.4$, $\gamma = 20°$, $\mathcal{J}_{m,s,l} = (64, 56, 13)\hbar^2/\text{MeV}$, respectively. Reproduced with permission from Ref. [27].

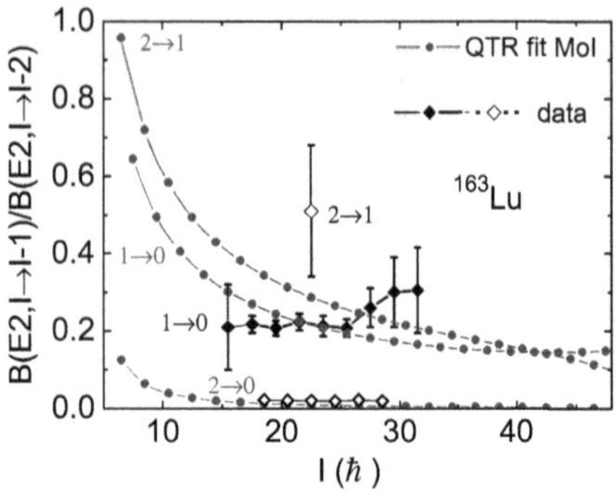

Figure 2.11 Ratios $\frac{B(E2,I\to I-1)_{out}}{B(E2,I\to I-2)_{in}}$ and $\frac{B(M1,I\to I-1)_{out}}{B(E2,I\to I-2)_{in}}$ between the bands TSD2 and TSD1 compared with the QTR calculations in Ref. [27]. Reproduced with permission from Ref. [27].

s-axis, which is observed in the Lu isotopes. To arrive at an interpretation which is consistent with the extended work on chirality by then, Frauendorf and Dönau [27] reanalyzed the TSD bands of the Lu isotopes by means of the QTR, requiring that the m-axis has the largest MoI.

The authors carried out QTR calculations[1] with the same deformation parameters $\varepsilon = 0.4$ and $\gamma = 20°$ as in Refs. [19, 24, 25]. Three sets of MoI's were investigated which obey the requirement that $\mathcal{J}_m$ is the largest. Figs. 2.10 and 2.11 compare the energies and $\frac{B(E2,I\to I-1)_{out}}{B(E2,I\to I-2)_{in}}$ ratios with the ones from TSD2 and TSD3. The ratios between the TR MoI's were adjusted to obtain the best agreement of the wobbling energies with experimental ones. The corresponding inertia asymmetry of $\kappa = 0.93$ is close to $\kappa = 0.91$ for the microscopic MoI's obtained by the cranking model, which gave results very similar to the ones in Figs. 2.10 and 2.11. The irrotational flow MoI's for $\gamma = 20°$ correspond to $\kappa = 0.64$. The enhanced asymmetry of the inertia ellipsoid causes a too steep decrease of the wobbling energy with I. Thus, the features of the TSD wobbling bands are well accounted for by the QTR applying it in consistency with the studies of chirality. The only caveat is an overestimate of the $\frac{B(M1,I\to I-1)_{out}}{B(E2,I\to I-2)_{in}}$ ratios.

In order elucidate the physics behind the decreasing wobbling energy Dönau and Frauendorf [27] drastically simplified the PTR by assuming that the proton $\mathbf{j}$ is rigidly aligned with the s-axis, which they called the frozen alignment (FA) approximation.

$$H_{FA} = A_1(\hat{J}_3 - j)^2 + A_2\hat{J}_1^2 + A_3\hat{J}_2^2, \tag{2.18}$$

where j is a number.

Applying the small-amplitude approximation

$$\hat{J}_3 = (\hat{J}^2 - \hat{J}_1^2 - \hat{J}_2^2)^{1/2} \approx J - \frac{\hat{J}_1^2}{2J} - \frac{\hat{J}_2^2}{2J}, \quad J = \sqrt{I(I+1)}$$

to the FA Hamiltonian the harmonic FA Hamiltonian (HFA) is obtained,

$$H_{HFA} = A_1(J - j)^2 + (A_2 - \bar{A}_1)\hat{J}_2^2 + (A_3 - \bar{A}_1)\hat{J}_3^2, \quad \bar{A}_1 = A_1(1 - j/I). \tag{2.19}$$

---

[1]The actual calculations were carried out in the framework of the core-quasiparticle-coupling model (QCM), which for the TR core is just the PTR or QTR reformulated in the laboratory frame. See Section 2.6 and Ref. [27].

Replacing A_1 by $\bar{A}_1$, the HFA Hamiltonian agrees with the one in Ref. [1] from which Bohr and Mottelson derived the harmonic wobbling energies and transition probabilities. The wobbling frequency (2.10) becomes

$$\hbar\omega_w = 2I\left[(A_2 - \bar{A}_1(I))(A_3 - \bar{A}_1(I))\right]^{1/2}. \tag{2.20}$$

The expressions for the transition probabilities are given in Refs. [1, 27].

The linear increase of $\hbar\omega_w$ with I is counter acted by the decrease of the product under the square root caused by the increase of $A_3(I)$, which results in the HFA curve shown in Fig. 2.10. At the critical spin

$$I_c = j\mathcal{J}_2/(\mathcal{J}_2 - \mathcal{J}_1) \tag{2.21}$$

the HFA becomes unstable because $A_2 - \bar{A}_1(I)) = 0$, the wobbling frequency is zero, and the HFA breaks down.

The presence of the particle $\mathbf{j}$ aligned with the s-axis leads to the replacement of the rotational parameter A_1 by the effective one $\bar{A}_1 = A_1(1 - j/I)$, which is the smallest as long as $0 < \bar{A}_1 < A_2, A_3$. The wobbling motion represents the precession of the total angular momentum $\mathbf{J}$ about the s-axis, which is perpendicular to the m-axis with the largest TR MoI. Frauendorf and Dönau suggested the name "transverse wobbling" (TW) in order to indicate that the particle $\mathbf{j}$ and the axis of the precession cone are aligned with the s-axis, which is perpendicular to the m-axis with the maximal MoI. In the FA approximation the orientation of the precession cone agrees with the one of $\mathbf{j}$. The axis of the $\mathbf{J}$ precession cone provides general classification criterion that encompasses the case when $\mathbf{j}$ is no longer narrowly aligned with the s-axis (see below).

The presence of a hole $\mathbf{j}$ aligned with the l-axis leads to the replacement of the rotational parameter A_3 by the effective $\bar{A}_3 = A_3(1 - j/I)$, which is another case of transverse wobbling with the hallmark of decreasing wobbling energies. A midshell quasiparticle has the tendency to align its $\mathbf{j}$ with the m-axis. The HFA limit corresponds to the replacement of A_2 by $\bar{A}_2 = A_2(1 - j/I)$, which remains to be the smallest. Frauendorf and Dönau classified this case as "longitudinal wobbling" because $\mathbf{j}$ and the axis of the precession cone have the direction of the m-axis with the maximal MoI. The wobbling frequency (2.10) replaced with $\bar{A}_2$ increases more rapidly with I than for constant A_2, which is the hallmark of LW.

It should be stressed that Frauendorf and Dönau introduced the notations LW and TW in order to *classify the exact PTR* results. Contrary to the claim by Lawrie *et al.* [5] they did not intend to restrict LW and TW to the I range where the HFA is a good approximation. Below I will be expound how to use the classification scheme beyond the I range where the HFA is valid.

For ^{163}Lu the critical spin is $I_c \approx 50$, which is just reached by the QTR calculations. The QTR $n = 1$ wobbling energy is small there as seen in Fig. 2.10. Although the calculation shows the hallmark of TW, the decrease of the wobbling energies at large I, it does not reproduce the almost linear decline seen in experiment. Fig. 2.11 shows that the QTR ratios $\frac{B(E2,I \to I-1)_{out}}{B(E2,I \to I-2)_{in}}$ monotonically decrease with I while the experimental ones first stay constant to increase at high spin. The calculation underestimates the experimental $\frac{B(E2,I\ n=2 \to I-1,\ n=1)_{out}}{B(E2,I \to I-2)_{in}}$ ratio of about two times the value for the $n = 1 \to 0$ transitions, which seems to point to a more harmonic nature of the wobbling mode. The QTR ratios $\frac{B(M1,I \to I-1)_{out}}{B(E2,I \to I-2)_{in}}$ are about two orders of magnitude larger than in experiment. The discrepancies of the PTR with the experiment go away for the TPSM calculations (see Chapter 11).

2.3.2 TRANSVERSE WOBBLING IN TRIAXIAL NORMAL DEFORMED NUCLEI

The nucleus ^{134}Pr was presented as the first example for chirality [26]. Having in mind the similarity with TW, Frauendorf and Dönau [27] carried out QTR calculations for ^{135}Pr, which has only one odd $h_{11/2}$ proton. They used the mean field equilibrium deformation parameters $\varepsilon = 0.16$, $\gamma = 26°$ and fitted the MoI to the observed band energies. Fig. 2.12 shows that observed wobbling energies decrease until $I = 29/2$ and then increase. The QTR $n = 1$ curve has the same shape with a less

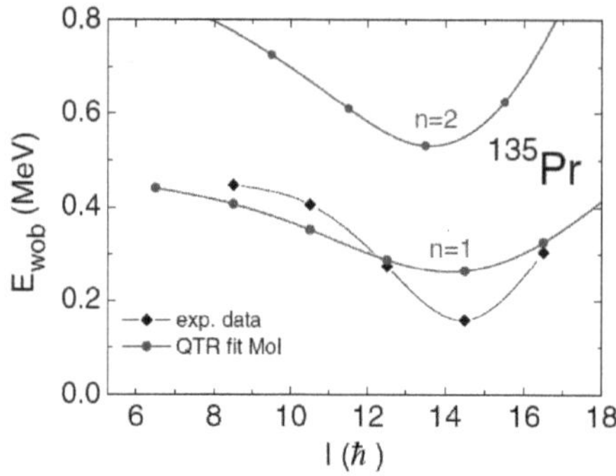

Figure 2.12 Experimental wobbling energies Eq. (2.8) of ^{135}Pr compared with the QTR calculations in Ref. [27] using the parameters $\varepsilon = 0.16$, $\gamma = 26°$, $\mathcal{J}_{m,s,l} = (21,13,4)\hbar^2/$MeV, respectively. Reproduced with permission from Ref. [27].

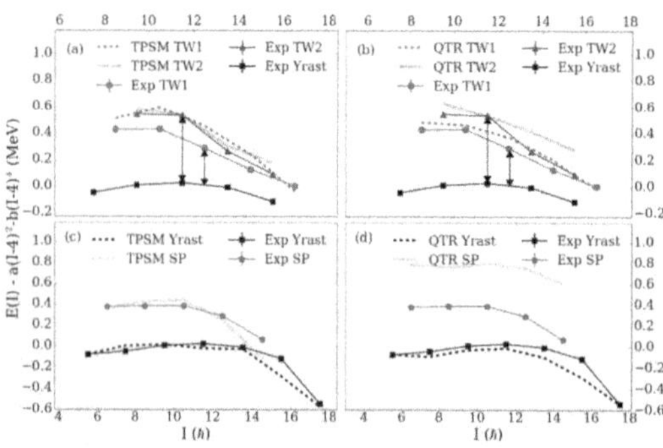

Figure 2.13 Experimental level energies minus a rotor contribution *vs.* spin, for the TW1, TW2, Yrast and SP bands in ^{135}Pr compared with the corresponding TPSM and QTR (labeled as QTR) calculations. The double pointed arrows between the experimental TW2, TW1 and Yrast points indicate the wobbling energy for the TW2 and the TW1 bands, respectively. Reproduced with permission from Ref. [30].

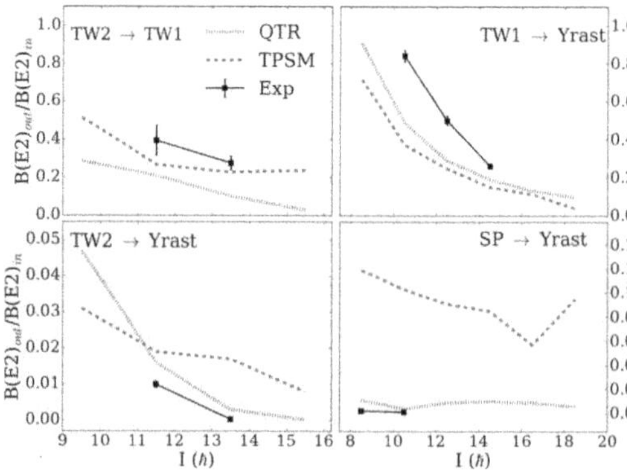

Figure 2.14 B(E2)$_{out}$/B(E2)$_{in}$ *vs.* spin for the TW2 $\rightarrow$ TW1, TW1 $\rightarrow$ Yrast, TW2 $\rightarrow$ Yrast and the SP $\rightarrow$ Yrast transitions. (TPSM (red dashed line), QTR (dotted line) and experiment (squares connected by solid line). Reproduced with permission from Ref. [30].

pronounce minimum around the critical spin $I_c = 14.4$, where the HFA becomes unstable. The authors interpreted the results as follows. On the down side the nucleus is in the TW regime and on the up side it is in the LW regime.

The fitted MoI's correspond to an inertia asymmetry of $\kappa = 0.71$. For $\gamma = 27°$ irrotational flow gives the smaller value of $\kappa = -0.17$, which causes a too early instability of the TW regime. Deviations from the irrotational flow ratios may have various reasons. There are shell effects which modify the MoI's (see Fig. 2.1). Fluctuations of γ may result in a smaller effective value. In Ref. [28], it was noted that the γ value of the modified oscillator potential corresponds to a smaller γ of the density distribution, which determines the MoI's. The irrotational flow value $\kappa = 0.69$ corresponds to $\gamma = 19°$, which is about the value that corresponds to $\gamma = 26°$ used for the potential in the QTR calculation. (see Ref. [28].)

The calculations motivated U. Garg's group to measure the mixing ratios of the $I \rightarrow I - 1$ transitions between the suggested wobbling bands [29–31]. Figs. 2.13 and 2.14 compare the results of the measurements with a slightly modified QTR calculation. As in Fig. 2.12, the excitation energy of the first (TW1) wobbling state decreases below and increases above I_c. The distance between TW1 and the second (TW2) wobbling state is much smaller than the excitation energy of TW1, which indicates strong anharmonicities. The ratios $\frac{B(E2,I \rightarrow I-1)_{out}}{B(E2,I \rightarrow I-2)_{in}}$ show the enhancement that

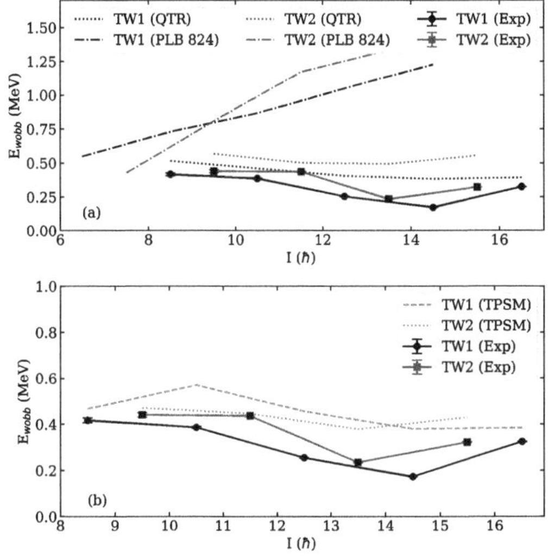

Figure 2.15 Wobbling energies $E_W(I)$ for ^{135}Pr compared with calculations. The labels TW1 and TW2 refer to the first and second band above the yrast band. QTR labels the calculations of Ref. [30], which uses fitted MoI's. Ref. [35]. PLB 824 labels the QTR calculations of Ref. [35], which used irrotational flow MoI. The pertaining calculated small ratios $\frac{B(E2,I\rightarrow I-1)_{out}}{B(E2,I\rightarrow I-2)_{in}}$ ratios indicate that TW1 has the character of a signature partner band. TPSM labels the triaxial projected shell model calculations from Ref. [30]. Reproduced with permission from Ref. [36].

signifies the collective nature of the wobbling mode. The ratios $\frac{B(E2,TW2\rightarrow TW1)}{B(E2,TW1\rightarrow yrast)} \sim 1$ indicate strong anharmonicities as well. The ratios $\frac{B(E2,TW2\rightarrow yrast)}{B(E2,TW1\rightarrow yrast)} < 0.05$ are small as expected. The QTR calculations also describe the excitation of the odd particle, the signature partner (SP) band. As it should be, its $\frac{B(E2,I\rightarrow I-1)_{out}}{B(E2,I\rightarrow I-2)_{in}}$ ratios are very small like in the experiment. However, the QTR places the SP band at too high energy.

Lv *et al.* [35] reinvestigated ^{135}Pr. In contrast to Refs. [29–31] they determined small mixing ratios $|\delta| < 1$, which indicate small $\frac{B(E2,I\rightarrow I-1)_{out}}{B(E2,I\rightarrow I-2)_{in}}$ ratios and large $\frac{B(M1,I\rightarrow I-1)_{out}}{B(E2,I\rightarrow I-2)_{in}}$ ratios. They presented the results as "Evidence against the wobbling nature of low-spin bands in ^{135}Pr", because large collective $E2$ transitions between the bands are the decisive signature for TW. In measuring the mixing ratios the challenge is to decide between the two minima of χ^2 from the fit to the angular distributions. The authors of Ref. [35] chose the minimum with small $|\delta|$ the authors of Refs. [29–31] chose the one with large $|\delta|$. In Ref. [36], Sensharma *et al.* present additional details of their fit procedure which support their choice of the large $|\delta|$.

The QTR calculations in Ref. [35] account for the small $\frac{B(E2,I\rightarrow I-1)_{out}}{B(E2,I\rightarrow I-2)_{in}}$ ratios, which reassign the TW1 band as a signature partner band. The new assignment is supported by composition of the QTR states discussed in supplementary material of Ref. [35]. The difference between the two QTR calculations consists in the inertia asymmetry, which is $\kappa = 0.71$ for the fitted MoI's of Refs. [29–31] and $\kappa = -0.17$ for the irrotational-flow MoI's of Ref. [35]. The small κ causes a very early instability of the TW regime which moves the SP band below the wobbling band. The SP band is characterized by the steady increase of $E_W(I)$ seen in the upper panel of Fig. 2.15. [2] In my view, the stark contrast with the observed decrease $E_W(I)$ makes the SP assignment unbelievable.

The TPSM calculations (see Section 2.5) in the lower panel of Fig. 2.13 evaluate the rotational response microscopically. There is no freedom in choosing the MoI's. The similarity between the QTR of Refs. [29–31] and the TPSM results lends credibility to the adjusted inertia asymmetry ratio of $\kappa = 0.71$.

When the TW1 band were the SP band then it would be difficult to explain the presence of second nearby band with SP properties, which was observed in Refs. [29–31]. Finally there is the far-reaching analogy between ^{135}Pr and ^{105}Pd (see next paragraph), for which the carefully measured

---

[2] Private communication of the QTR energies by E. Lawrie is acknowledged.

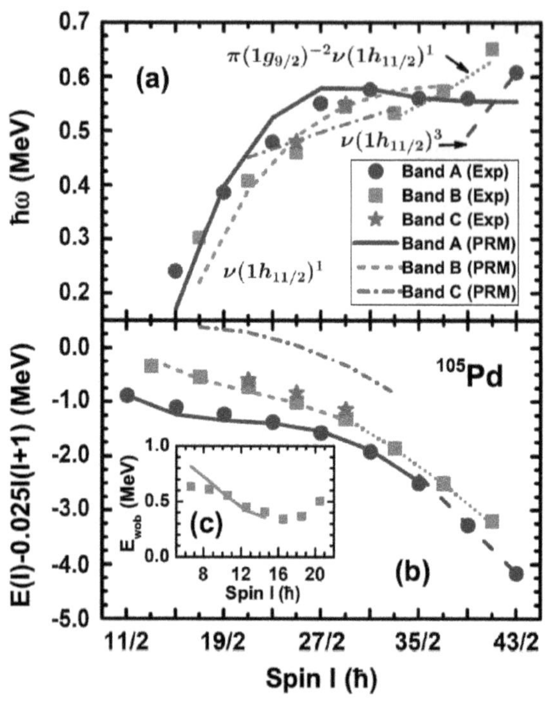

Figure 2.16 Experimental values derived from the energies of lowest bands in ^{105}Pd compared with the QTR (labeled as PRM) calculations: (a) rotational frequency, (b) energies minus a rotor reference with the wobbling energies $E_W(I)$ as the inset (c). The bands A, B and C are assigned to, respectively, $n = 0$ TW, $n = 1$ TW and SP. Reproduced with permission from Ref. [32].

mixing ratios indicate large $\frac{B(E2,I \to I-1)_{out}}{B(E2,I \to I-2)_{in}}$ ratios, which are consistent with the ones of Refs. [29–31].

In my opinion all the mentioned facts favor the TW interpretation of the bands. A new experiment would be needed to safely establish the experimental δ values.

Another example for TW is ^{105}Pd, which was studied in Ref. [32] . The authors carried out QTR (called PRM) calculations using a triaxial deformation of $\gamma = 25°$ derived from mean field calculations and fitted MoI's, which correspond to an inertia asymmetry of $\kappa = 0.69$. Fig. 2.16 compares the experimental results with QTR calculations (denoted as PRM), where the rotational frequency is defined as

$$\hbar\omega(I-1/2) = (E(I)-E(I-2))/2. \tag{2.22}$$

The experimental (PRM) $\frac{B(E2,I \to I-1)_{out}}{B(E2,I \to I-2)_{in}}$ ratios for the $n = 1 \to 0$ trrasitions are $I = 17/2$: 0.66 ± 0.18 (0.736), $I = 21/2$: 0.60 ± 0.09 (0.465), $I = 25/2$: 0.34 ± 0.07 (0.329).

From the PTR perspective the nucleus is much like ^{135}Pr, just that the odd $h_{11/2}$ proton is replaced by the odd $h_{11/2}$ neutron. The slightly lower inertia asymmetry as compared to ^{135}Pr is reflected by the somewhat higher value of $I_c = 33/2$ of the minimum of the wobbling energy. The QTR predicts energy of the SP bands too high as in the case of ^{135}Pr.

A fourth band was reported in Ref. [33]. The experiment could not substantiate the nature of the band. Karmakar et $al.$ [34] measured mixing ratios for the transitions connecting it with the TW1 band, which did not indicate a strong $E2$ component. The finding is in contrast to ^{135}Pr where the connecting transitions of the analog band with TW1 show a strong $E2$ component, which allowed the authors of Ref. [30] to identify it as the second wobbling excitation TW2. In ^{105}Pd band 4 represents a TW excitation on top of the SP band. The reason for the difference is the lower energy of the SP band in ^{105}Pd than in ^{135}Pr. As a consequence, in ^{105}Pd the TW excitation on the SP band has a smaller energy than the second wobbling excitation, while in ^{135}Pr the second wobbling excitation has a smaller energy than the TW excitation on the SP band. The microscopic TPSM calculations (see Section 2.5) account for the different nature of band 4, while the QPR predicts

TW2 for band 4 in both nuclei. The latter is not surprising because the QTR overestimates the energy of the SP band in both nuclei.

Mukherjee *et al.* [37] identified in ^{151}Eu a TW band built on the proton $h_{11/2}$ yrast band together with the pertaining SP band. See Section 2.6 for an analysis in the framework of the TPSM.

Nandi *et al.* [38] identified in ^{183}Au a TW band built on the proton $h_{9/2}$ band and another TW band built on the proton $i_{13/2}$ band. The wobbling energies from the $h_{9/2}$ pair decrease rapidly corresponding to a $I_c \approx 14$. The wobbling energies of the $i_{13/2}$ pair slowly increase like the PTR E_W in Fig. 2.10 far below $I_c \approx 30$. The authors used different MoI's to reproduce the different I_c and the resulting I-dependences of $E_W(I)$, which also provided the correct $\frac{B(E2,I\rightarrow I-1)_{out}}{B(E2,I\rightarrow I-2)_{in}}$ ratios. Remarkably, the microscopic TPSM reproduces these quantities of both TW bands with nearly the same set of input deformations. See Fig. 12.11 and 12.14 of Chapter 12 and Fig. 14 of Ref. [37].

Rojeeta Devi *et al.* [39] confirmed the prediction [27] that TW appears for high-hole states as well, where the **J** precesses about the *l*-axis. They identified in ^{133}Ba the first and second TW bands built on the neutron $h_{11/2}$ hole band and carried out a qualitative analysis in HFA approximation.

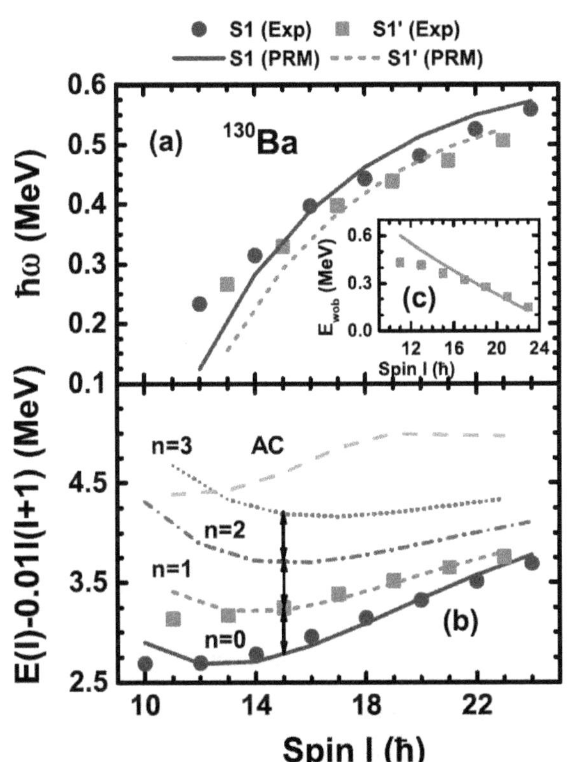

Figure 2.17 Experimental values derived from the energies of the bands S1, S1' and AC in ^{130}Ba compared with the QTR (labeled as PRM) calculations: (a) rotational frequency, (b) energies minus a rotor reference with the wobbling energies $E_W(I)$ as the inset (c). The QTR calculations used the parameters $\varepsilon = 0.24$, $\gamma = 21°$, $\mathcal{J}_s/\mathcal{J}_m/\mathcal{J}_l = (0.73, 1, 0.43)\hbar^2$/MeV, respectively. Reproduced with permission from Ref. [40].

The TW regime is more stable when two particles align their angular momentum with the s-axis, as the value of j in I_c doubles. The first example of TW on a $\pi(h_{11/2})^2$ configuration was reported by Chen *et al.* [40] for ^{130}Ba. The even-I band S1, which is AB in CSM notation, was interpreted as the $n = 0$ TW band. The odd-I band S1' was assigned to the $n = 1$ TW band. The second odd-I band, called AC in CSM notation, represents the SP band, because the quasiproton B is replaced by C with the opposite signature. They compared the experimental values with QTR (labeled as PRM) calculations using the deformation parameters from a mean field calculation and adjusting the MoI's to fit the energies of the $n = 0$ and 1 bands.

Fig. 2.17 compares the rotational frequencies and the energies of the bands S, S' and AC with the PRM results, which agree rather well. The experimental mixing ratios of the $I \rightarrow I-1$ transitions between the $n = 1$ and $n = 0$ bands allowed the authors determining $\frac{B(E2,I \to I-1)_{out}}{B(E2,I \to I-2)_{in}}$ ratios between 0.3 and 0.4, which signify the wobbling character of the $n = 1$ band. They are consistent with the PRM ratios $\frac{B(E2,I \to I-1)_{out}}{B(E2,I \to I-2)_{in}}$=0.51, 0.42, 0.35, 0.29, 0.25 for I=13, 17, 19, 21, respectively. The $E2$ component of the transition from the AC band to the $n = 0$ (AB) band was found to be small as expected for the SP band.

The wobbling energy in Fig. 2.17 shows that that the TW regime is stable within the displayed spin range (I_c=37). Fig. 2.18 displays the SCS probability distributions for $I = 14$ and 15. The $n = 0$ and AC states correspond to uniform rotation about the s-axis. The $n = 1$ and $n = 2$ states show the precession orbits around the s-axis, which characterize the TW regime.

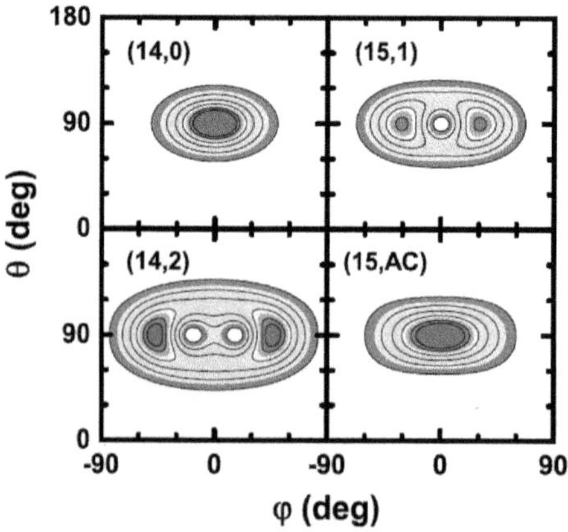

Figure 2.18 SCS probability distributions $P(\theta, \phi)$ of the angular momentum orientation **J** with respect to the body-fixed frame for the four lowest bands in Fig. 2.18. In accordance with Fig. 2.17, the panels are labeled by (I, n) with n being TW wobbling number or AC. Reproduced with permission from Ref. [40].

A second example for TW on the $\pi(h_{11/2})^2$ configuration is ^{136}Nd, which was studied by Chen and Petrache [41].

2.3.3 TRANSITION FROM TRANSVERSE TO LONGITUDINAL WOBBLING IN TRIAXIAL NORMAL DEFORMED NUCLEI

The FA approximation provides only a first qualitative TW-LW classification. The **j** of the odd particle reacts to wobbling motion of the nucleus, which must be taken into account to understand the region where the TW becomes unstable and beyond. For this purpose Chen and Frauendorf [4] invoked the classical adiabatic energy. The classical energy is defined by replacing in the PTR Hamiltonian (2.12,2.13) the operators by their corresponding classical numbers,

$$\hat{J}_3 \to J\cos\theta, \quad \hat{J}_1 \to J\sin\theta\cos\phi, \quad \hat{J}_2 \to J\sin\theta\sin\phi, \tag{2.23}$$

$$\hat{j}_3 \to j\cos\vartheta, \quad \hat{j}_1 \to j\sin\vartheta\cos\varphi, \quad \hat{j}_2 \to j\sin\vartheta\sin\varphi. \tag{2.24}$$

The adiabatic classical energy $E_{ad}(\theta, \phi)$ is obtained by minimizing the classical energy $E_{class}(\theta, \phi, \vartheta, \varphi)$ with respect to ϑ and φ for given θ, ϕ, that is, let the particle react to the Coriolis force. It is displayed Fig. 2.19 for the PTR Hamiltonian of ^{135}Pr. The energy contours represent the classical orbits when **j** adiabatically follows **J**, which is the case when the energy scale of the particle is large compared to the scale of the wobbling mode. As demonstrated in Ref. [42], adiabticity is a reasonable approximation for the lowest three bands. See Figs. 2.24 and 2.29 below.

The I dependence reflects the competition between the Coriolis force, which tries to keep the angle between **J** and **j** small, the triaxial potential, which tries to align **j** with the s-axis and the energetic preference of the m-axis with the maximal MoI. Fig. 2.20 shows the angles ϕ and φ of the minimum of the adiabatic energy.

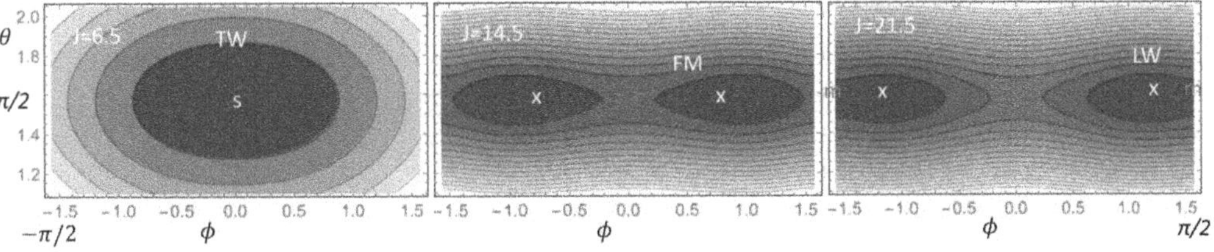

Figure 2.19 Adiabatic classical energy $E_{ad}(\theta,\phi)$ in ^{135}Pr for the TW, FM and LW regimes. The contours of constant energy represent the classical orbits with the corresponding energy and the indicated angular momentum. Minima and saddles are marked by crosses. The letters s and m indicate the location of the s-and m-axes.

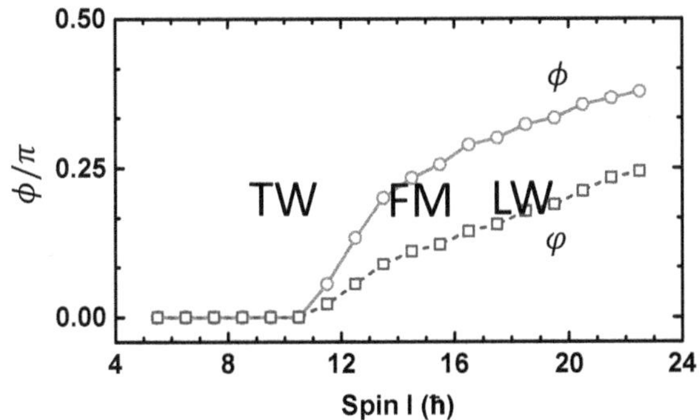

Figure 2.20 Location of the minima of the adiabatic classical energy $E_{ad}(\theta,\phi)$ in ^{135}Pr on the line $\phi = \pi/2$ in Fig. 2.19 (marked by the crosses). The pertaining angles of $\mathbf{j}$ are $\vartheta = \pi/2$ and φ. Reproduced with permission and adapted from Ref. [4].

At low I the classical orbits revolve the s-axis, which is the TW regime. At the critical angular momentum of $J_c = 10.5$ the $\phi = 0$ minimum changes into a maximum and two minima at $|\phi| > 0$ emerge. The rotation about the s-axis becomes unstable. The TW regime changes into the flip mode (FM), which becomes well established at $J = 14.5$. The FM represents flipping between the narrow precession cones around the two tilted axes. Quantum mechanically, the $n = 0$ state is an even and the $n = 1$ state an odd superposition of two states localized near the tilted axes. As the two minima at ϕ_0 approach $\pm\pi/2$ the orbits merge with the ones at $\pi \mp \phi_0$ such that the precession cones revolve around the m-axis, which means the region of LW has been reached.

The flip regime appears around $I = 14.5$ where E_W has its minimum in Fig. 2.13. At larger I values $E_W(I)$ increases with I. As seen in Fig. 2.20, the angle φ of the particle $\mathbf{j}$ remains far below $\pi/2$ in this region. Therefore, the authors of Ref. [4] suggested a generalization of the TW-LW classification of Ref. [27], which is based on the topology of the classical orbits that correspond to the quantal states:

Classification of the lowest bands	$\mathbf{J}$ revolves around axis	$\mathbf{j}$ revolves around axis	$E_W(I)$	$\frac{B(E2,I\to I-1)_{out}}{B(E2,I\to I-2)_{in}}$
transverse wobbling (TW)	short or long	short or long	decreases*	order one
longitudinal wobbling (LW)	medium	short, long, tilted	increases	order one
flip mode (FM)	tilted	tilted	constant	order one
signature partner (SP)	short, medium, long	short, medium, long	increases	small

* Depending on I_C there may be a slight increase at low I.

The scheme is quite simple from the experimental point of view and has been used in Refs. [27, 29, 30] in classifying the experimental results. Fig. 2.21 illustrates the angular momentum geometry in a schematic way.

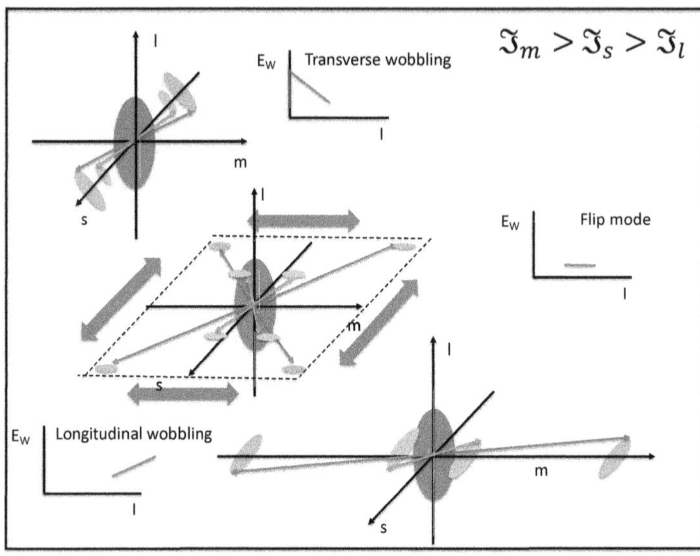

Figure 2.21 Schematic illustration of the collective wobbling modes. Arrows show the precession cones of the particle **j** and of the total angular momentum **J**.

The energy difference between the $n = 0$ and $n = 1$ states of the flip mode reflects the mixing of the states localized near the tilted axis. It disappears when the coupling goes to zero. This limit represents uniform rotation about a tilted axis, which breaks the signature symmetry and leads to a $\Delta I = 1$ band (see discussion of the rotating mean field symmetries in Ref. [43]).

Fig. 2.22 shows the probability distribution of the **J** with respect to the principal axes for the lowest three bands $n = 0, 1, 2$. The classical orbits (as shown in in Fig. 2.19) for the quantal energies of the states are included.

The SCS maps in the upper row show rims that are centered at the s-axis, which are the hallmark of TW.

The middle row shows the four spots near the tilted axes between the FM flips. The $27/2_1$ state represents the even superposition (large $P(0)$ and the $29/2_1$ state the odd superposition ($P(0) = 0$) of the flip states. The $27/2_2$ state flips between the s- and m-axes, which are both unstable. The structure is analog to the flip mode of the unstable s-axis of the TR discussed in Sec. 2.2.

The lower row displays the LW regime. The very elongated orbits enclose the m-axis at $\pm\pi/2$, which is the topological characteristics of LW. They differ substantially from the HFA-LW limit, because the minima of E_{ad} are still away from the m-axis (see Fig. 2.20).

For the classical TR the angular momentum component J_3 and the angle of **J** with the s-axis in the s-m-axes plane represent the momentum and coordinate of a pair of canonical variables. This provides a complementary perspective on the structure of the PTR states. It employs the correspondence between the classical motion in a potential and the probability density of the corresponding quantal Hamiltonian, which is familiar from the textbooks. The classical adiabatic energy at $E_{ad}(\pi/2, \phi)$ represents the potential $V(\phi)$. For not too high energy one can expand the classical adiabatic energy up to second order in $\theta - \pi/2$, which gives the classical adiabatic Hamiltonian

$$H_{ad} = T + V(\phi), \quad T = \frac{J_3^2}{2B(\phi)}, \quad \frac{1}{B(\phi)} = \frac{1}{I^2}\frac{\partial^2 E_{ad}(\theta,\phi)}{\partial \theta^2}\bigg|_{\pi/2}. \tag{2.25}$$

The potential is shown in Fig. 2.23 together with bars at the PTR energies relative to the minimum of $V(\phi)$. The difference represents the kinetic energy T. If $T > 0$ the wave function oscillates with a wavelength that decreases with T. If $T < 0$ the wave function decreases exponentially with $|T|$. The value of T defines the TW-FM-LW classification. In the TW regime ($I=13/2, 23/2$) the s-axis is in the range $T > 0$, and in the LW regime ($I=43/2$) the m-axis is in the range $T > 0$. In the FM the n

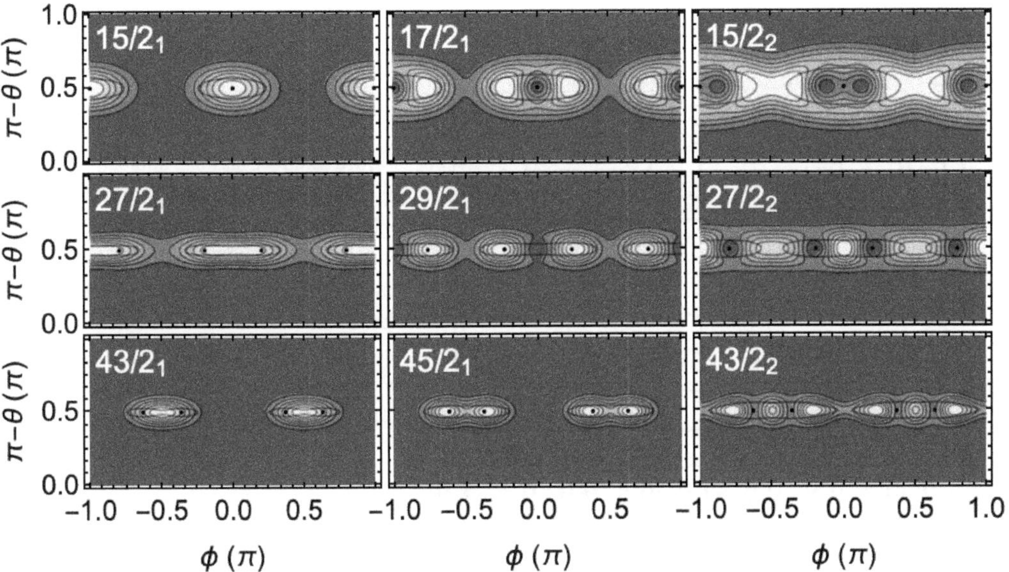

Figure 2.22 Probability distribution of the angular momentum orientation $P(\theta\phi)_{Iv}$ generated from the reduced density matrix of the PTR calculation for ^{135}Pr in Ref. [27]. The three rows show the SCS maps for the TW, FM and LW regimes. The black curves show the classical orbits obtained by setting the adiabatic classical energy equal to the quantal energy of the states. The author thanks Q. B. Chen for preparing the figure.

=0 and n=1 states are localized around $\phi \approx \pm\pi/4$. The mode flips between rotation about the two orientations of the tilted axis.

Chen and Frauendorf introduced the Spin Squeezed State (SSS) representation of the reduced density matrix [42].

$$P(\phi)_v = \frac{1}{2\pi} \sum_{K,K'=-I}^{I} e^{-i(K-K')\phi} \rho_{KK'}^{(v)}, \tag{2.26}$$

where ϕ is the angular momentum angle with the s-axis in the s-m-plane. The over-complete, non-orthogonal basis comes as close to a continuous wave function as possible for the finite Hilbert space of dimension $2I+1$. It is generated by rotating the localized state

$$|II0\rangle = (2I+1)^{-1/2} \sum_{K=-I}^{I} |IIK\rangle \tag{2.27}$$

around the 3-axis. It has a smaller width of $\sim 3/2I$ as compared to the width $\sim 1/\sqrt{2I}$ of the generating state of the SCS basis. Fig. 2.24 shows the SSS plots of some states for ^{135}Pr, some of which are displayed in the form of SCS maps in Fig. 2.22.

The TW states $I < 25/2$ have the profile of oscillations within the potential centered at the s-axis. The $n = 0$ states have a maximum at $\phi = 0$. The $n = 1$ states have a minimum at $\phi = 0$, and the $n = 2$ states have two minima at $\phi = \pm\pi/6$. If the states were pure, the minima would be the zeros of the collective wave function that characterize the oscillations. There is a certain degree of incoherence caused by the deviations from the adiabatic approximation, which lead to a finite probability density at the minima (see discussion in Sec. 2.3.5).

The LW states $I > 37/2$ have the profile of oscillations within the potential centered at the m-axis: The $n = 0$ states have maxima at $\phi = \pm\pi/2$. The $n = 1$ states have minima at $\phi = \pm\pi/2$, and the $n = 2$ states have two minima near $\phi = \pm4/6\pi$.

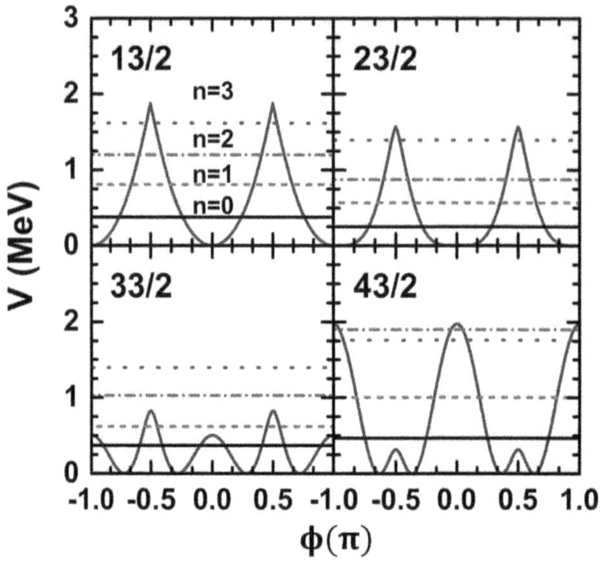

Figure 2.23 Classical adiabatic energy of ^{135}Pr $V(\phi) = E_{\mathrm{ad}}(\pi/2, \phi)$ for selected spins. The horizontal lines show the PTR energies $E_n - E_{ad}(\phi_0)$ relative to the minimum of E_{ad} at ϕ_0. For the two values of n that do not appear for I the line is located at $(E_n(I-1) + E_n(I+1))/2$. Reproduced with permission and adapted from Ref. [4].

The FM states around $I = 31/2$, $33/2$ have flip character. The $n = 0$ and 1 are the respective even and odd linear combinations of the two states localized at $\phi \approx \pm\pi/4$. The $n = 2$ states are standing waves with probability maxima at the the tops of the barriers where the classical motion is slowest.

The SSS curves are similar to the profile of the SCS maps along the line $\theta = \pi/2$. This reflects the fact that the SCS states along this path are generated by rotating the localized state $|IIK_s = I\rangle$ around the l-axis instead of the SSS basis state, (2.27), which is also centered at the s-axis just being more narrow. The over-completeness of the SCS and SSS basis sets results in a certain ambiguity in visualizing of the structure.

2.3.4　LONGITUDINAL WOBBLING IN TRIAXIAL NORMAL DEFORMED NUCLEI

Biswas *et al.* [44] reported the first case of LW for ^{133}La based on their observation of large $\frac{B(E2,I \to I-1)_{out}}{B(E2,I \to I-2)_{in}}$ ratios and increasing wobbling energy. Following this, evidence for LW was reported for ^{187}Au by Sensharma *et al.* [45] and ^{127}Xe[46] by Chakraborty *et al.*

The wobbling energies of ^{133}La in Fig. 2.25 increase with I, which is the hallmark of LW. Biswas *et al.* [44] measured the mixing ratios and found that $\frac{B(E2,I \to I-1)_{out}}{B(E2,I \to I-2)_{in}}$ ratios were collectively enhanced. The right panel shows the isotone ^{135}Pr with the discussed TW behavior. As in both nuclei the proton occupies the bottom of the $h_{11/2}$ shell, TW ought to appear in ^{133}La. The authors of Ref. [44] explained the appearance LW as follows. TW in ^{135}Pr is the result of $\mathcal{J}_s < \mathcal{J}_m$, which is seen as a rapid increase of the rotational frequency with I along the yrast band. In ^{133}La a much slower increase is observed, which is caused by the gradual alignment of the angular momentum of a pair of positive parity protons with the s-axis. The difference $\mathcal{J}_m - \mathcal{J}_s(I)$ becomes quickly small, such that rotation about the l-axis remains energetically favorable, and the wobbling energy increases with I. The authors identified the role of the proton pair by comparing with the rotational response of ^{134}Pr where the alignment of the proton pair is blocked. The QTR calculations in Fig. 2.25 were carried out in the framework of the quasiparticle-core-coupling model (see Sec. 2.6) using a TR core with the MoI's $\mathcal{J}_s = (9.12 + 0.66\,R)\,\hbar^2/MeV$ for ^{133}La and $\mathcal{J}_s = (9.34 + 0.10\,R)\,\hbar^2/MeV$ for ^{135}Pr and for both nuclei $\mathcal{J}_m = 21\,\hbar^2/MeV$, $\mathcal{J}_l = 4\,\hbar^2/MeV$ and a coupling strength corresponding to $\varepsilon = 0.16$, $\gamma = 26°$.

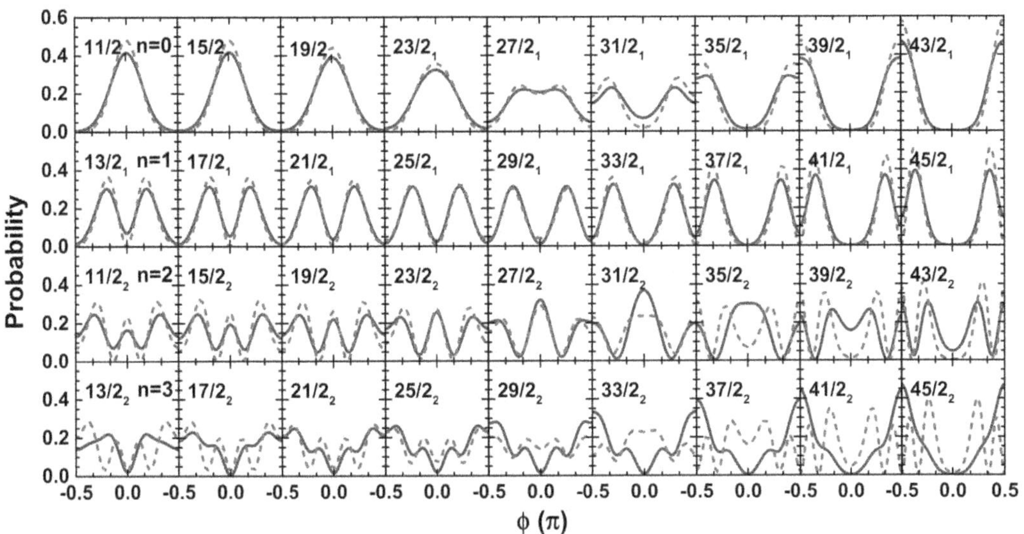

Figure 2.24 Full curves show the probability density of the SSS for the $n = 0, 1, 2, 3$ states calculated by PTR for ^{135}Pr, while the dashed curves display the ones from the collective Hamiltonian (CH). Reproduced with permission Ref. [42].

The TPSM calculations (see Fig. 2.40 in Sec. 2.5) reproduce the change from TW to LW for, respectively, $Z = 61$, 59 in the $N = 74$ isotones using nearly the same triaxial mean field, which confirms the microscopic origin of the change. The energies of the $h_{11/2}$ bands in the $Z = 57$, $N = 74$ isotone indicate TW again, which is reproduced by the TPSM. It would be interesting to substantiate this by measuring the mixing ratios.

Based on their highly collective $\frac{B(E2,I \rightarrow I-1)_{out}}{B(E2,I \rightarrow I-2)_{in}}$ ratios, Sensharma $et\ al.$ [45] assigned LW to the $h_{9/2}$ pair of bands. They carried out QTR calculations using irrotational-flow MoI's with $\gamma = 23°$, which correspond to an asymmetry of $\kappa = 0.37$ of the inertia ellipsoid. Good agreement with experimental energies and the $\frac{B(E2,I \rightarrow I-1)_{out}}{B(E2,I \rightarrow I-2)_{in}}$ ratios was obtained. The TPSM calculations in Chapter 12 (see Figs. 12.11 and 12.14) also gave LW with $\frac{B(E2,I \rightarrow I-1)_{out}}{B(E2,I \rightarrow I-2)_{in}}$ ratios very close to the experimental ones reported by Sensharma $et\ al.$ [45].

Guo $et\ al.$ [47] challenged the LW assignment for ^{187}Au. They disagree with the authors of Ref. [45] on which of the two minima of the χ^2 fit should be accepted in determining the mixing ratios δ. Guo $et\ al.$ [47] chose the one with small $|\delta|$ at variance with Sensharma $et\ al.$ [45], who chose the one with large $|\delta|$. This led to small $\frac{B(E2,I \rightarrow I-1)_{out}}{B(E2,I \rightarrow I-2)_{in}}$ and large $\frac{B(M1,I \rightarrow I-1)_{out}}{B(E2,I \rightarrow I-2)_{in}}$ ratios, which indicate SP instead of LW for the excited band. Guo $et\ al.$ [47] carried out QTR calculations using irrotational-flow MoI's with $\gamma = 12°$. The weak asymmetry of $\kappa = 0.97$ of the inertia ellipsoid is the reason that their QTR calculations assigned a signature partner pair to the two $h_{9/2}$ bands in good agreement with the experimental results based their choice of $|\delta|$.

The authors of Ref. [45] are preparing an extended publication of their experimental results, which will address the controversy. The relevant experimental information is already published in Ref. [48]. In my view, the TPSM calculations, for which the MoI's cannot be adjusted to the data but are fixed by the microscopic structure, represent clear evidence that supports the LW assignment.

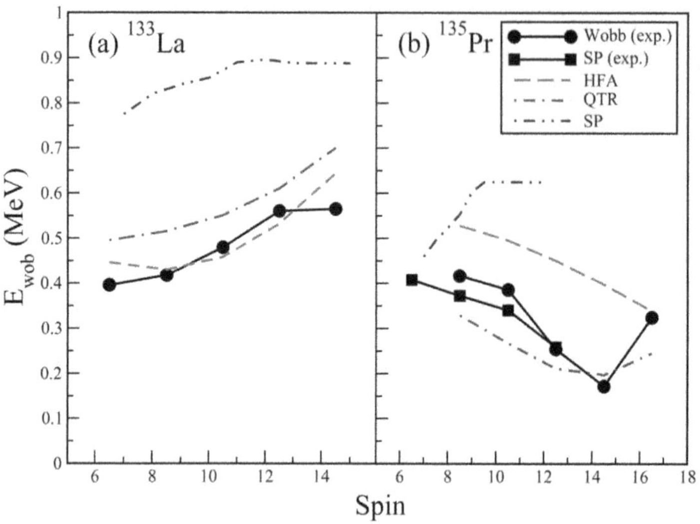

Figure 2.25 Wobbling and SP band energies in ^{133}La and ^{135}Pr compared with QTR calculations. Reproduced with permission from Ref. [44] .

2.3.5   APPROXIMATE SOLUTIONS OF THE PTR MODEL

2.3.5.1   Frozen alignment approximation

Approximate solutions of the PTR problem have been constructed in order to elucidate the physics of the coupled system. The HFA approximation was introduced in Ref. [27] to define the TW-LW classification. Budaca [49] (see Chapter 13) extended the HFA beyond the critical angular momentum I_c, given by Eq. (2.21). Above I_c the minimum of the FA energy lies at the finite angle

$$\cos\phi_0 = \frac{2A_1 j}{(2I-1)(A_1 - A_2)} \tag{2.28}$$

in the case that **j** is aligned with the l-axis.[3] The yrast states correspond to uniform rotation about an axis tilted by ϕ_0 into the 1-2-plane. The author applied the harmonic approximation with respect to the tilted axis and presented simple algebraic expressions for the energies and reduced $E2$ and $M1$ transition probabilities for $I > I_c$, which complement the ones of Ref. [27]. Fig. 2.26 shows the HFA wobbling energies for $j = 11/2$ and several γ values. At the critical angular momentum I_c the wobbling energy turns zero and then increases, which is the hallmark of LW. The model fails near near the instability I_c. Above I_c the ellipses around the axes with the tilt angles ϕ_0 and $\pi - \phi_0$ soon overlap (see Fig. 2.19). It remains to be seen how well the harmonic wobbling excitations about a tilted axis describe the FW regime.

The model is a handy tool for a quick exploration of experimental results. The problems around I_c may not become too serious when the distance between the discrete I points is large enough, like in the case of ^{135}Pr displayed in Fig. 2.27. The pertaining values of $\frac{B(E2,I\rightarrow I-1)_{out}}{B(E2,I\rightarrow I-2)_{in}}$ ratios of 0.2–0.3 deviate substantially from the experimental and QTR values but confirm the collective enhancement.

The model has three parameters: γ, which controls the ratios of the irrotational flow MoI's, $\mathcal{J}_0$, which sets the scale of the rotational energies and j the particle angular momentum aligned with one of the principal axes. The author assumes that j remains frozen to the axis, which is not

---

[3]The alignment with other axes is discussed as well.

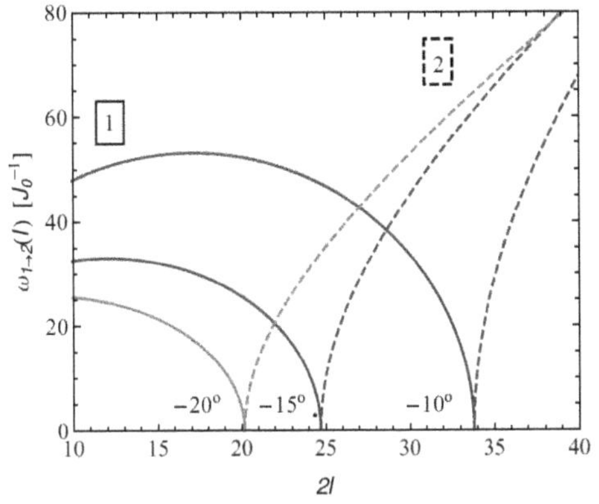

Figure 2.26 Wobbling energies in HFA approximation for different values of the triaxiality parameter γ. Reproduced with permission from Ref. [49].

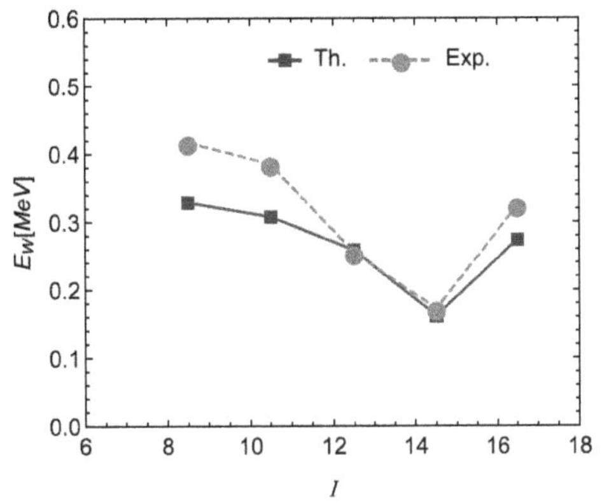

Figure 2.27 Wobbling energy adjusted to the experimental energies of ^{135}Pr. Reproduced with permission from Ref. [49].

realistic as demonstrated by Fig. 2.20. This suggests considering a fixed angle of $\mathbf{j}$ with the l-axis as another parameter. In Ref. [50], the authors studied this generalization. Unfortunately, no algebraic expressions for a handy application are given. Their derivation appears straightforward. The only complication is the solution of the transcendental equation for the angle of the minimum of the classical energy, which could be tabulated or displayed.

2.3.5.2 Small amplitude approximation

Tanabe and Sugawara-Tanabe [51] (see Chapter 4) developed an algebraic approximation to the PTR. They applied the Holstein-Primakoff boson expansion to both operators $\hat{\mathbf{J}}$ and $\hat{\mathbf{j}}$ and kept the terms up to forth order in the boson number. In second order the PTR Hamiltonian represents two coupled harmonic oscillators. The eigen modes are found by solving a biquadratic equation, which has a simple analytic solution. Analytic expressions for the energies and transition probabilities in terms of the three MoI's and the strength κ of the triaxial potential are given. The fourth-oder terms are taken into account by first order perturbation theory in the eigenmode basis. The expression for the additional terms are given as well but are rather complex. The bands can be classified by the oscillator quantum numbers of the two eigenmodes.

Without coupling, there is a pure $\mathbf{J}$ oscillation, which is the wobbling mode and a pure $\mathbf{j}$ oscillation, which is the cranking or SP mode. The coupling mixes the modes. In the TW regime the mixed modes retain their character as being predominantly wobbling or cranking, and the n_{wobb} and n_{crank} oscillator quantum numbers can be assigned to the bands. The HFA represent the limit of a very stiff (large κ) cranking mode. The harmonic approximation breaks down when the minimum of the classical adiabatic energy at the angles $\theta = \pi/2$ and $\phi = 0$ changes into a maximum. This is the angular momentum where the angle ϕ deviates from zero in Fig. 2.20.

The authors applied the model to the TSD wobbling bands in the Lu and Tm isotopes assuming rigid body MoI's corresponding to the equilibrium deformation. As the respective ratios are similar to the ones in Ref. [25] comparable good agreement with the experimental $\frac{B(E2,I\to I-1)_{out}}{B(E2,I\to I-2)_{in}}$ and $\frac{M1,I\to I-1)_{out}}{B(E2,I\to I-2)_{in}}$ ratios was obtained. For rigid body MoI's the short axis has the largest MoI. Because this implies LW, their wobbling energies increase with I. In order to remove the discrepancy with the observed decrease they made the scale of the MoI's increasing with I as

$$\mathcal{J}_0(I) = \mathcal{J}_0 \frac{I - 0.60}{I + 23.5}. \tag{2.29}$$

They attributed this increase to a gradual decrease of the pair correlations along the bands. The TDS bands cover the range of $I = 13/2 - 97/2$. The scaling gives $\mathcal{J}_0(97/2)/\mathcal{J}_0(13//2) = 3.14$. However, the experimental ratio of $\mathcal{J}^{(1)}$ is 73/65=1.12 and of $\mathcal{J}^{(2)}$ is 60/70=1.16 [53]. Obviously, the pairing hypothesis is not convincing. The experimental change of the MoI's cannot reverse the I dependence of the wobbling energy. As discussed at the beginning of Section 2.2, the rigid body MoI's ratios contradict the general principle of spontaneous symmetry breaking. The conclusions have been discussed at the beginning of Sec. 2.3.1.

In Ref. [54], the authors applied the model to ^{135}Pr assuming irrotational flow MoI's and a tri-axialty parameter $\gamma = 26°$ as used in Ref. [27]. They found that the harmonic vibrations become unstable for $I > 13/2$ and claimed that the TW mode does not appear. Their claim has been refuted by Frauendorf [3] (see also their reply in Ref. [55]) for the following reasons. As discussed in the beginning of Sec. 2.3.2, the PTR calculation with irrotational flow MoI's for $\gamma = 26°$ result in an early transition of the TW into the FM around $I = 17/2$ (see Fig. 16 of Ref. [27]). In the small-amplitude approximation the TW becomes unstable when the minimum of the classical energy at $\phi = 0$ disappears. This happens at a smaller angular momentum than the transition to the FM in the exact PTR calculation. (Compare $J = 10.5$ in Fig. 2.20 with $I = 25/2$ in Fig. 2.15.) Hence the very early instability of the TW mode found in Ref. [54] results from the combination of using irrotational flow MoI's with the near maximal inertia asymmetry of $\kappa = -0.17$ and the harmonic approximation. It cannot be used to question the existence of the TW mode in normally deformed triaxial nuclei.

Raduta *et al.* [56, 57] (see Chapter 8) re-expressed the PTR Hamiltonian in the basis of the SCS, which becomes a continuous function of two variables that represent a canonical pair of momentum and coordinate. They approximated the Hamiltonian by expanding it up to second order in momentum and coordinate and solved the pertaining eigenvalue problem on the classical and quantal level. In essence, the approach is the two-oscillator model of Ref. [51] (no fourth order terms). The authors studied ^{163,165,167}Lu. They assumed the rigid-body ratios of the MoI's for the equilibrium deformation and adjusted the scale of the MoI's and the strength of the deformed potential to the observed energies of the TSD bands. As one can expect, the results were similar to the ones of Ref. [51]. The model well describes the $\frac{B(E2,I\rightarrow I-1)_{out}}{B(E2,I\rightarrow I-2)_{in}}$ and $\frac{M1,I\rightarrow I-1)_{out}}{B(E2,I\rightarrow I-2)_{in}}$ ratios but gives wobbling energies which increase with I. Hence, my remarks to the work [51] apply accordingly.

In Ref. [58], Raduta *et al.* suggested that the TSD2 band can be interpreted as uniform rotation about the s-axis like the TSD1 band, where in the former case the core angular momentum $R = 3$, 5, 7, ... and in the latter $R = 2$, 4, 6, ... In Eq. (25) of Ref. [58], they assigned to both bands the energies $E_{classical}(I) + \hbar\omega_W(I)/2$ with $I = 13/2, 17/2, 21/2, ...$ for TSD1 and $I = 15/2, 19/2, 23/2, ...$ for TSD2. The two terms $E_{classical}(I)$ and $\hbar\omega_W(I)$, derived in Refs. [56, 57], are smooth functions of I. Therefore, the model merges the two bands into one $\Delta I = 1$ sequence. In other words, the wobbling energies are zero in stark contrast to the experiment. (A plot of the published energies confirms this[4].) In Ref. [59], Raduta *et al.* applied the same approach to ^{135}Pr with the modification assuming a fixed orientation of the particle's **j**, which gave zero wobbling energies as well. In my view, the characterization of the wobbling mode discussed in Refs. [58, 59] on the basis of this approach cannot be trusted.

2.3.5.3 Collective Hamiltonian

Frauendorf and Chen [42] quantized the classical adiabatic Hamiltonian (2.25) using the Pauli pre-scription for replacing the classical kinetic energy T by the corresponding operator $\hat{T}$. The resulting collective Hamiltonian (CH) was diagonalized in the basis of the axial rotor states $|IMK\rangle$. As the CH is invariant with respect to the D_2 group, all representations of the point group appear as eigenstates. Only the ones that are similar to the full PTR states are accepted.

---

[4] Private communication of the accurate energies by A. A. Raduta is acknowledged.

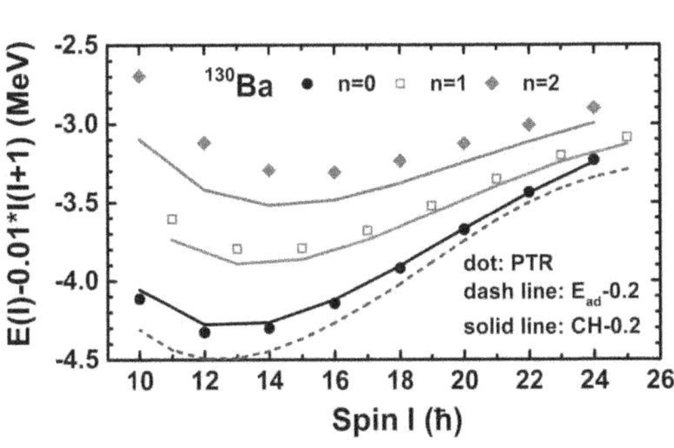
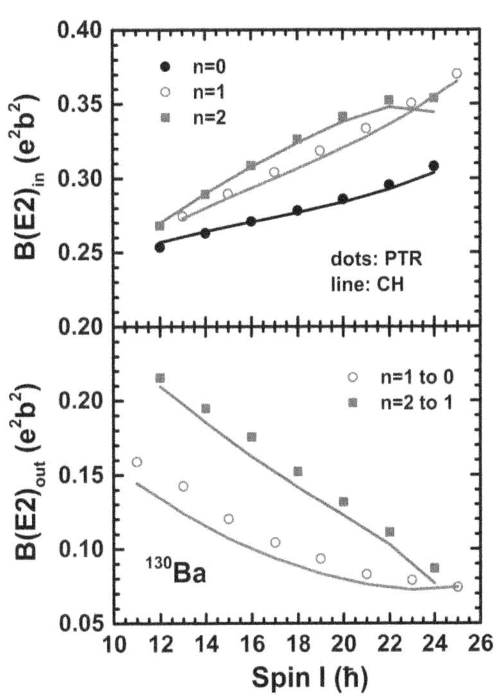

Figure 2.28 Comparison of the energies (left panel) and $B(E2)$ values (right panel) of ^{130}Ba calculated by means of the full PTR model (symbols) and the CH (solid curves). The adiabatic classic energy is shown by the dashed curve. All the energies are down shifted by 0.2 MeV. Reproduced with permission from Ref. [42].

Fig. 2.28 compares the energies and $B(E2)$ values calculated by means of the PTR and of the CH for ^{130}Ba. The CH reproduces the exact PTR values rather well. Fig. 2.29 compares the SSS plots generated from the reduced density matrices of the PTR solutions with the SSS plots generated from the corresponding eigenstates of the CH (c. f. Eq. (2.7). The CH densities for the $n = 0$, and 1 states agree very well with the PTR densities. At the PTR minima, they become zero, which characterizes an oscillating pure wave function. The PTR density has only minima because the coupling to the two particles generates some degree of incoherence. The degree of incoherence increases with n, which is seen as a progressive washing out of the wave-function pattern. The pattern changes from the oscillations of the lowest Hermite polynomials in the harmonic TW region near $I=10$ to the onset of the flip regime near $I=24$.

For ^{135}Pr, the second example studied, the agreement of the observables between CH and PTR is comparable. As seen in Fig. 2.24, the incoherence increases more rapidly with n. This is expected, because one $h_{11/2}$ proton is weaker bound to the triaxial core than a pair of them. The $n=0$ and $n=1$ SSS-CH plots correlate very well with the SSS-PTR plots through the TW-FM-LW transition, which allows one to associate the structures with partially coherent quantum oscillations. For the states $n = 2$, $I \geq 39/2$ and $n = 3$ the CH plots seriously deviate from the PTR plots, which indicates the break down of the adiabatic approximation.

There is no problem in generating the numerical solutions for the PTR Hamiltonian. There are several codes available, both with the reduced high-j and the full version the triaxial potential in the particle Hamiltonian h_p. One motivation to approximate the PTR model by a CH is the intuitive interpretation in terms of wave functions derived from a one-dimensional Hamiltonian of the form $T + V$. A second is that the study provides the proof of principle that one can construct a realistic CH Hamiltonian from the energy surface generated by a Tilted Axis Cranking calculation. The TAC approach is widely used for the interpretation of high-spin experiments. The TAC model accounts only for uniform rotation about the tilted axis, which defines the minimum of the energy. A CH constructed from microscopic TAC calculations will allow one to study excitations with wobbling nature in addition to the TAC solutions with $n=0$ nature. In the preceding work [60, 61] a CH was

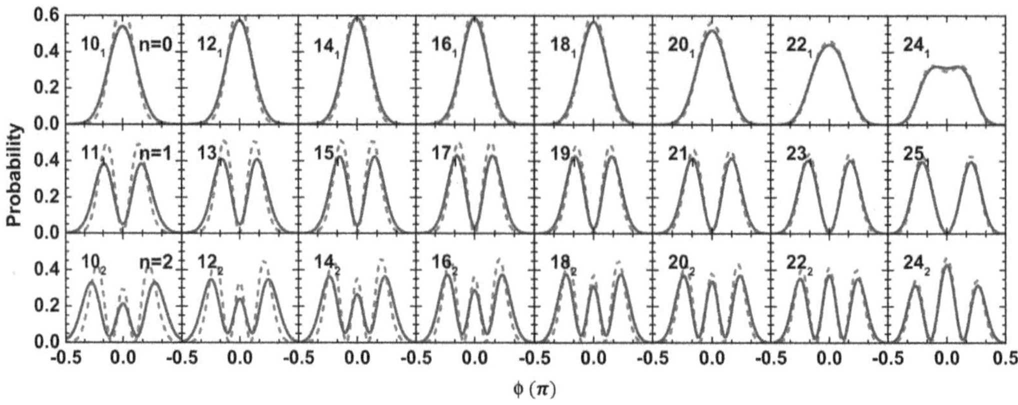

Figure 2.29   Full curves show the probability density of the SSS for the $n = 0, 1, 2, 3$ states calculated by PTR for ^{130}Ba, while the dashed curves display the ones from the collective Hamiltonian (CH). Reproduced with permission from Ref. [42].

constructed with V being the Routhian at a fixed rotational frequency and T estimated by the HFA applied at the minimum of the Routhian, which also provided promising results. The feasibility of extending the TAC along this avenue should be investigated.

2.3.6   VIRTUES AND LIMITS OF THE PTR

The PTR model is simple from the conceptual point of view and has lead to the intuitive pictures of the underlying physics, which have been discussed so far. The computational effort is negligible. However, there is the question of how to choose the input parameters. The triaxiality of the shape determines the potential that binds the particles to the rotor and the ratios between the MoI's, the interplay of which produces the wobbling behavior. Microscopically the two elements are tied together. On the PTR level they are to some extend independent, which has the advantage to achieve a good description of the experiment by adjusting them. On the other hand, the freedom of the parameter choice may lead to qualitative different interpretations as discussed in the preceding sections. The parameter choice is restricted by the fundamental principle of spontaneous symmetry breaking, which requires the irrotatotional flow-like oder $\mathcal{J}_l < \mathcal{J}_s < \mathcal{J}_m$. The expressions (2.3) with the triaxility parameter γ taken from some mean field calculations should be used with caution, because the equilibrium values usually represent the minimum of a shallow energy surface and often differ between various versions of the mean field approaches. The asymmetry parameter κ of the MoI's depends sensitively on γ between $20°$ and $30°$ (see Fig. 2.2), which introduces an element of uncertainty. Adjusting the MoI's to reproduce the energies of observed wobbling bands turned out to be the best strategy for interpreting specific experiments. (See also Chapter 10.)

In the microscopic approaches to be presented in the next two sections the deformed mean field determines both elements in a unique way, which reduces the number of parameters. In addition they avoid a basic problem of the PTR, which assumes that the collective orientation angles of the TR are completely independent of the particle degrees of freedom. Microscopic calculations by means of the cranking model show that this is only correct to some extend (see discussion in Section 2.6).

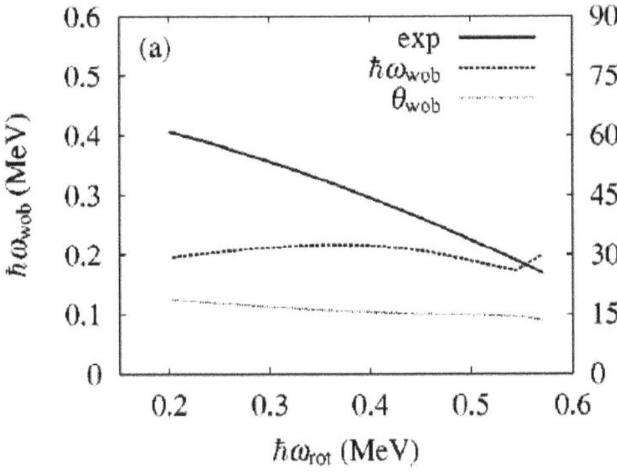

Figure 2.30 Wobbling energies in RPA approximation for ^{163}Lu. The opening angle ot the precession cone around the s-axis is θ_{wob} with the scale (in degree) on the right axis. Reproduced with permission from Ref. [65].

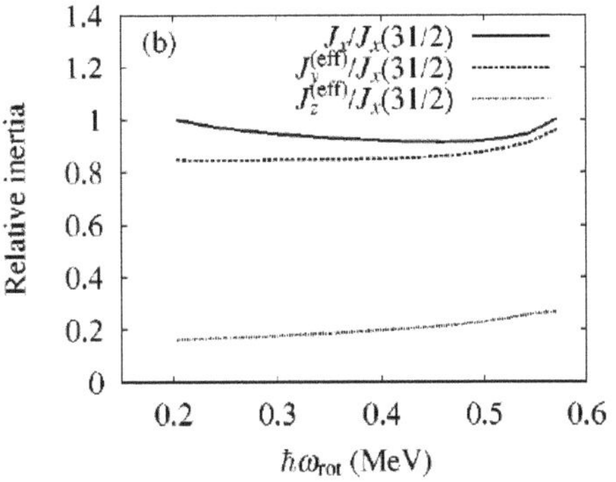

Figure 2.31 The moments of inertia (2.36) in RPA approximation for ^{163}Lu. The labels x, y, z correspond to 1, 2, 3. Reproduced with permission from Ref. [65].

2.4 RANDOM PHASE APPROXIMATION

The microscopic descriptions of wobbling is based on the pairing-plus-quadrupole-quadrupole many-body Hamiltonian

$$\hat{H}_{PQQ} = \sum_{\tau=\pi,\nu} [\,\hat{h}_\tau^\circ - G_\tau(\hat{P}_\tau^\dagger \hat{P}_\tau) - \lambda_\tau \hat{N}_\tau\,]$$
$$- \frac{\kappa_0}{2} \sum_{m=-2,2} (-1)^m \hat{Q}_m \hat{Q}_{-m}. \tag{2.30}$$

The operator $\hat{h}_\tau^\circ$ is the spherical part of the Nilsson Hamiltonian, where the isospin index $\tau = \pi, \nu$ distinguishes the neutron and proton contributions, respectively, and $\hat{P}_\tau^\dagger$ and $\hat{P}_\tau$ are the familiar monopole pair operators. As pairing is treated in BCS approximation the terms $\lambda_\tau \hat{N}_\tau$, containing the particle number operators $\hat{N}_\tau$, are introduced to attain the average particle numbers. The following term in Eq. (2.30) is the quadrupole-quadrupole interaction, where $\hat{Q}_m = \sum_{\tau=\pi,\nu} \hat{Q}_m(\tau)$. The BCS approximation is applied to H_{PQQ}, which leads to the quasiparticle Hamiltonian

$$\hat{h}_\tau = \hat{h}_\tau^\circ - \Delta_\tau(\hat{P}_\tau^\dagger + \hat{P}_\tau) - \lambda_\tau \hat{N}_\tau$$
$$- \hbar\omega_0 \frac{2}{3}\varepsilon \left(\cos\gamma \hat{Q}_0 - \frac{\sin\gamma}{\sqrt{2}}(\hat{Q}_2 + \hat{Q}_{-2}) \right). \tag{2.31}$$

and the selfconsistency relations

$$\kappa_0 \langle \hat{Q}_0 \rangle = 2/3 \hbar\omega_0 \varepsilon \cos\gamma,$$
$$G_\tau \langle P_\tau^\dagger \rangle = \Delta_\tau, \quad \langle \hat{N}_\tau \rangle = N_\tau, \tag{2.32}$$

where $\hbar\omega_0 = 41A^{-1/3}$ MeV. The average $\langle ... \rangle$ is taken with the quasiparticle configuration of interest. In the applications the deformation parameters ε, γ and the pairing gaps Δ_τ are taken as input, which fix by the relations (2.32) the coupling constants κ_0 and G_τ. The last equation is fulfilled by the appropriate choice of the Fermi energy λ_τ.

The random phase approximation (RPA) to the wobbling mode has been developed by Michailov and Janssen [62, 63] and Marshalek [64]. For the TW regime the RPA starts with finding the mean

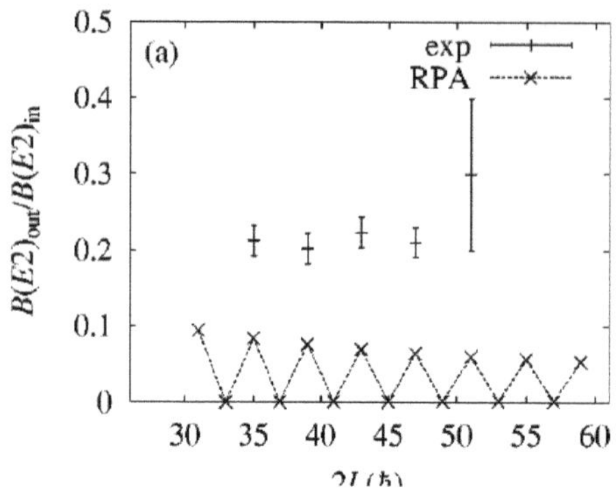

Figure 2.32 The $\frac{B(E2,I \to I-1)_{out}}{B(E2,I \to I-2)_{in}}$ ratios in RPA approximation for ^{163}Lu. Reproduced with permission from Ref. [65].

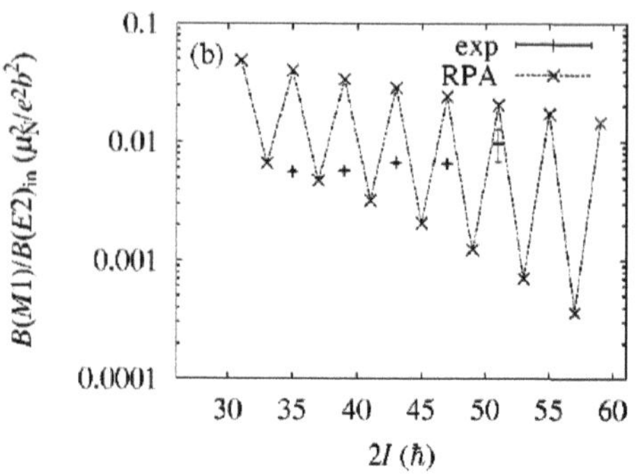

Figure 2.33 The $\frac{B(M1,I \to I-1)_{out}}{BE2,I \to I-2)_{in}}$ ratios in RPA approximation for ^{163}Lu. Reproduced with permission from Ref. [65].

field solutions to the TAC Routhian

$$\hat{H}'_{PQQ} = \hat{H}_{PQQ} - \omega_1 \hat{J}_1 - \omega_2 \hat{J}_2 - \omega_3 \hat{J}_3, \tag{2.33}$$

with $\omega_1 = \omega_{rot}$, $\omega_2 = \omega_3 = 0$. The solution represent uniform rotation about the s-axis, which is the yrast band, where the rotational frequency is chosen such $\langle J_1 \rangle = I + 1/2$. The solutions are given as configurations of quasiparticles generated by the quasiparticle Routhian

$$\hat{h}'_\tau = \hat{h}_\tau - \omega_{rot}\hat{J}_1, \quad [\hat{h}'_\tau, \alpha^\dagger_m] = E'_m \alpha^\dagger_m, \tag{2.34}$$

where E'_k are the quasiparticle energies in the rotating frame (quasiparticle routhians) and $\alpha^\dagger_m$ the pertaining quasiparticle creation operators. (For details of the TAC approach see e .g. Ref. [43].)

Then it is assumed that $\omega_2(t)$ and $\omega_3(t)$ execute small harmonic oscillations, the wobbling mode. The solutions are found in first order of time-dependent perturbation theory. The wobbling frequency is given by the expression

$$\omega_W = \omega_{rot} \left[\frac{(\mathcal{J}_1 - \mathcal{J}_2(\omega_w))(\mathcal{J}_1 - \mathcal{J}_3(\omega_w))}{\mathcal{J}_2(\omega_w)\mathcal{J}_3(\omega_w)} \right]^{1/2}, \tag{2.35}$$

which has the form of the original expression (2.10) from Ref. [1]. The expressions for the transition propabilities are the same as well. The microscopic MoI's are similar to the ones from the cranking model

$$\mathcal{J}_1 = \frac{\langle J_1 \rangle}{\omega_{rot}}, \quad \mathcal{J}_{i=2,3} = 4 \sum_{k,m} \frac{|\langle |J_1 \alpha^\dagger_k \alpha^\dagger_m| \rangle|^2 (E'_k + E'_m)}{(E'_k + E'_m)^2 - \omega_W^2}, \tag{2.36}$$

where the indices k, m run over all quasiparticle excitations of the yrast configuration $|\rangle$. As both sides depend on ω_W, Eq. (2.36) is the RPA dispersion relation, which must be solved numerically.

Matsuzaki *et al.* [65, 66] applied the method to the wobbling mode in ^{163,167}Lu. They used the deformation $\varepsilon = 0.43$, $\gamma = 20°$ and weak pairing of $\Delta_n = \Delta_p = 0.3$ MeV. The one-quasineutron configuration $\alpha^\dagger_{i_{13/2}}|0\rangle$ was the yrast band, which implies the replacement $E_m \to -E_m$, $\alpha^\dagger_m \to \alpha_m$ for $m = i_{13/2}$ in the sum (2.36). Their results are compared with the experimental ones in Figs. 2.30–2.33.

As the wobbling energies are constant with a slight downward trend and the $\frac{B(E2,I \to I-1)_{out}}{B(E2,I \to I-2)_{in}}$ ratios are collectively enhanced, the results represent the first example for a microscopic description of

TW. The authors explained the angular momentum dependence of $E_{(I)}$ in the same way as Frauendorf and Dönau when they introduced the TW-LW classification based on the HFA approximation [27]. The linear increase $\propto \omega_{rot}$ is overcompensated by the decrease of the factor $\mathcal{J}_1 - \mathcal{J}_2$ in Eq. (2.35), which is caused by the decrease of $\mathcal{J}_1$. The origin of the latter is the presence of the particle's $\mathbf{j}$ being aligned with the s-axis. In such a case the angular momentum $J_1 = j + \omega_{rot}\mathcal{J}_1^{(2)}$ and $\bar{\mathcal{J}}_1 = J_1/\omega_{rot} = \mathcal{J}_1^{(2)}/(1 - j/J_1)$, which agrees with the $\bar{A}_1$ in Eq. (2.19). Assuming that $j = 13/2$, one can estimate from the value of $\mathcal{J}_1$ in Fig. 2.31 that the TR MoI is $\mathcal{J}_1^{(2)} = 0.67$ in the units of the figure at $\hbar\omega_{rot}=0.2$ MeV. The RPA calculations confirm the qualitative requirement $\mathcal{J}_m > \mathcal{J}_s > \mathcal{J}_l$ in a quantitative way.

Compared with the experiment, the wobbling energy is too small at low angular momentum and stays almost constant. The $\frac{B(E2,I\to I-1)_{out}}{B(E2,I\to I-2)_{in}}$ are by a factor of two too small and the $\frac{B(M1,I\to I-1)_{out}}{B(E2,I\to I-2)_{in}}$ ratios by a factor 5–10 too large.

The wobbling motion represent oscillations of $\mathbf{J}$ with respect to the principal axes of the triaxial shape. In the laboratory frame $\mathbf{J}$ stands still, and the triaxial shape oscillates. For small amplitudes wobbling is a harmonic oscillation of the mass quadrupole moment $Q_{1-}(t) = (Q_1(t) - Q_{-1}(t))/\sqrt{2}$. Application of the time-dependent perturbation theory leads to the well known RPA for separable interactions, which is equivalent with the just discussed RPA in the principal axes system [64].

In Ref. [67], Frauendorf and Dönau re-investigated ^{163}Lu by means of the standard RPA, because the available computer code allowed them to include other separable interactions. They kept κ_0 constant and adjusted it to obtain $\varepsilon \approx 0.4$ over the range of ω_{rot}, which gave $\gamma \approx 10°$ from the self-consistency condition for Q_2. Weak pairing gaps of $\Delta_n = 0.35$, $\Delta_p = 0.45$ MeV and standard effective charges of $e_p = e(1+Z/A)$ and $e_n = eZ/A$ charges were used. The results are labeled as QQ$_{sc}$ in Figs. 2.34 and 2.35. The results are quite similar to the ones of Matzusaki *et. al.* with the exception that the wobbling mode becomes unstable around $\hbar\omega = 0.45$ MeV.

Using the deformations $\varepsilon \approx 0.4$ and $\gamma \approx 20°$ from Matzusaki *et al.* marginally changed the results. Including the isovector QQ and spin-spin interactions changed the results insignificately. However, a current-current interaction of the type $\mathbf{L}\cdot\mathbf{L}$ did. As displayed by the curves QQ$_{sc}$ + LL in Figs. 2.34 and 2.35, the wobbling energies agree very well with the experimental values, and the $\frac{B(M1,I\to I-1)_{out}}{BE2,I\to I-2)_{in}}$ ratios are reduced to the observed ones, while the $\frac{B(E2,I\to I-1)_{out}}{BE2,I\to I-2)_{in}}$ ratios did not change much.

The LL interaction couples the wobbling mode to the scissors mode, which is the isovector version of the wobbling mode: the proton system and the neutron system precess with anti-phase around the total angular momentum $\mathbf{J}$. The RPA calculation for the neighbor ^{164}Yb with the same coupling strength give the position and $B(M1)$ strength observed for the scissors mode in the nuclei of the region. A more detailed investigation of the relation between the scissors and wobbling modes seems interesting because the PTR calculations systematically overestimate the $\frac{B(M1,I\to I-1)_{out}}{BE2,I\to I-2)_{in}}$ ratios.

2.5 TRIAXIAL PROJECTED SHELL MODEL

The triaxial projected shell model (TPSM) [68] has been the other microscopic approach for studying the wobbling mode. More details are presented in chapters 12 and 14. The TPSM diagonalizes the pairing-plus-quadrupole-quadrupole many-body Hamiltonian (2.30) (augmented by a quadrupole pairing interaction) in the basis of quasiparticle configurations which are projected onto

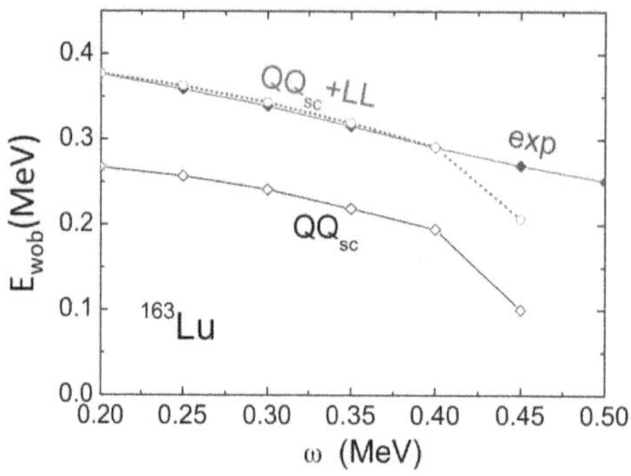

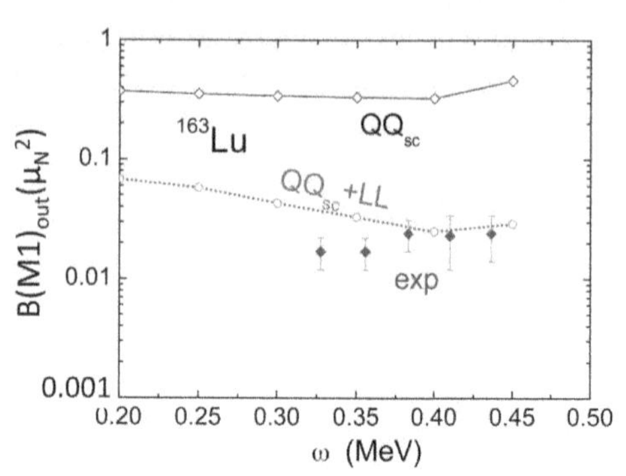

Figure 2.34 Wobbling energies in RPA approximation for ^{163}Lu. Reproduced with permission from Ref. [67].

Figure 2.35 The $\frac{B(M1,I\to I-1)_{out}}{BE2,I\to I-2)_{in}}$ ratios in RPA approximation for ^{163}Lu. Reproduced with permission from Ref. [67] .

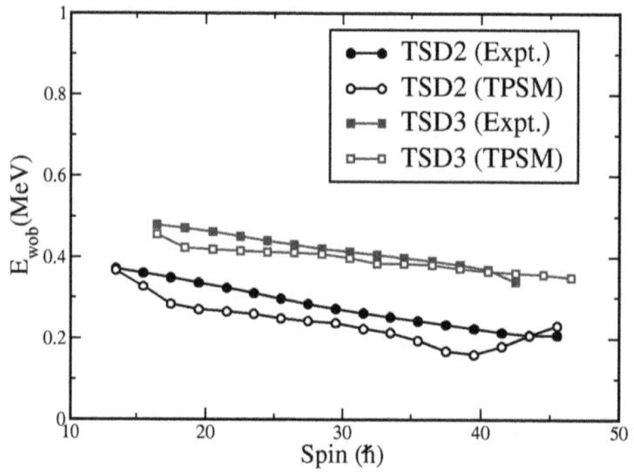

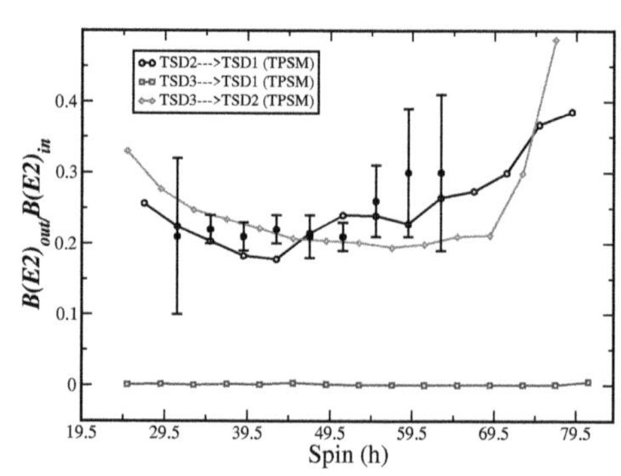

Figure 2.36 Wobbling energies in TPSM approximation for ^{163}Lu. From Chapter 12, private communication by G. Bhat is acknowledged.

Figure 2.37 The $\frac{B(E2,I\to I-1)_{out}}{BE2,I\to I-2)_{in}}$ ratios in TPSM approximation for ^{163}Lu. From Chapter 12, private communication by G. Bhat is acknowledged.

good angular momentum,

$$
\begin{aligned}
&\hat{P}^I_{MK}|0\rangle, && \hat{P}^I_{MK}\,a^\dagger_{p_1}|0\rangle, \\
&\hat{P}^I_{MK}\,a^\dagger_{p_1}a^\dagger_{p_2}|0\rangle, && \hat{P}^I_{MK}\,a^\dagger_{n_1}|0\rangle, \\
&\hat{P}^I_{MK}\,a^\dagger_{n_1}a^\dagger_{n_2}|0\rangle, && \hat{P}^I_{MK}\,a^\dagger_{p_1}a^\dagger_{n_1}a^\dagger_{n_2}|0\rangle, \\
&\hat{P}^I_{MK}\,a^\dagger_{p_1}a^\dagger_{p_2}a^\dagger_{n_1}a^\dagger_{n_2}|0\rangle, && \hat{P}^I_{MK}\,a^\dagger_{n_1}a^\dagger_{p_1}a^\dagger_{p_2}|0\rangle, \\
&\quad\ldots\ldots && \hat{P}^I_{MK}\,a^\dagger_{p_1}a^\dagger_{p_2}a^\dagger_{p_3}a^\dagger_{n_1}a^\dagger_{n_2}|0\rangle, \\
& && \hat{P}^I_{MK}\,a^\dagger_{n_1}a^\dagger_{n_2}a^\dagger_{n_3}a^\dagger_{p_1}a^\dagger_{p_2}|0\rangle, \\
& && \quad\ldots\ldots
\end{aligned}
\tag{2.37}
$$

where $|0\rangle$ represents the triaxial quasiparticle vacuum state, and $\hat{P}^I_{MK}$ projects the quasiparticle configuration on to good angular momentum states $|IMK\rangle$. In the majority of the nuclei, the near-yrast spectroscopy up to I=20 is well described using the above basis space. The extension is straight forward.

The input parameters are the pairing gaps and the deformation parameters ε, $\varepsilon' = \varepsilon \tan \gamma$. The PQQ coupling constants are fixed by means of the selfconsistency relations (2.4). Usually ε is taken from experiment or a mean field calculation and ε' is adjusted to the experimental band energies.

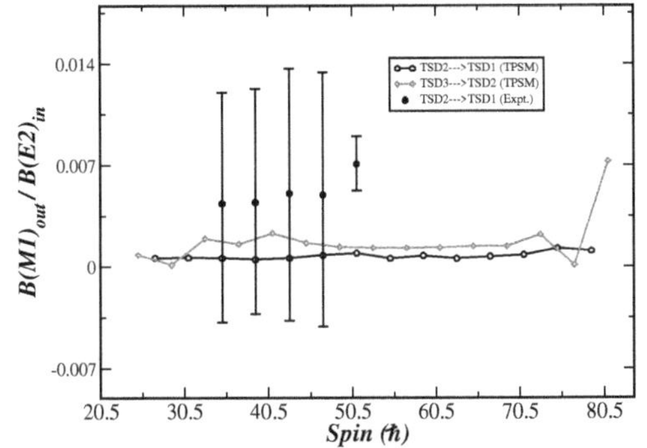

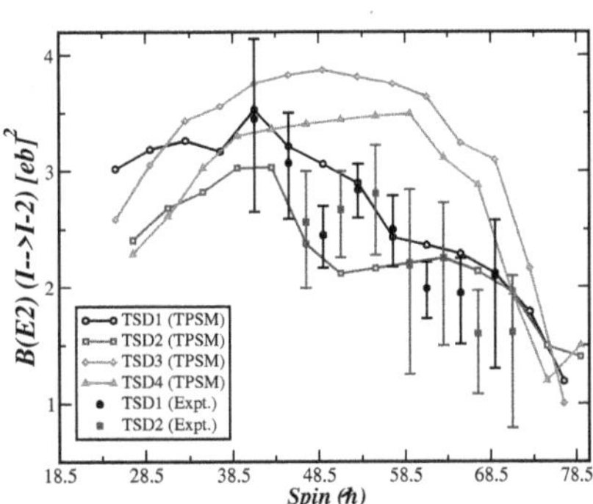

Figure 2.38   The $\frac{B(M1,I \to I-1)_{out}}{BE2,I \to I-2)_{in}}$ ratios in TPSM approximation for ^{163}Lu. From Chapter 12, private communication by G. Bhat is acknowledged.

Figure 2.39   Intra band $BE2, I \to I-2)_{in}$ values in TPSM approximation for ^{163}Lu. From Chapter 12, private communication by G. Bhat is acknowledged.

Sheikh presents of the systematics of theTPSM results for wobbling in Chapter 12. I discuss only few examples which demonstrate the new aspects of the microscopic TPSM as compared to the semi-phenomenological PTR approach.

Sheikh and coworkers studied wobbling in the ^{163}Lu and the other Lu isotopes. (see Chapter 12.) They used $\Delta_n = 0.7527$ MeV, $\Delta_p = 0.9637$ MeV and $\varepsilon = 0.40$ and $\gamma = 15°$, which was adjusted to the experimental value of $E_W(21/2)$. The results displayed in Figs. 2.36–2.39 agree very well with the experimental data. The bands TSD1, TSD2, TSD3 are assigned to the bands with $n=0, 1, 2$ wobbling quanta. The band TSD4 contains the $h_{9/2}$ proton. The close-to-harmonic TW picture is substantiated. The wobbling energy decreases with the correct average slope. There are wiggles not seen in experiment. The $\frac{B(E2,I \to I-1)_{out}}{BE2,I \to I-2)_{in}}$ ratios for the $n \to n-1$ transitions show the collective enhancement of collective wobbling. The transition $n = 2 \to 1$ is suppressed.

Remarkably, the $\frac{B(M1,I \to I-1)_{out}}{BE2,I \to I-2)_{in}}$ ratios in Fig. 2.38 agree with the experimental values. These ratios are systematically overestimated in the PTR calculations. As discussed in Sec. 2.4, this also the case for the RPA calculations. Only the inclusion of the coupling to the scissors mode coupling reduced the ratios to their experimental values. It would be interesting to see if there is a connection between coupling to the scissors mode and the mutli-quasiparticle states of the TPSM basis (4.7), which include scissors configurations.

Fig. 2.40 compares the experimental wobbling energies of the three $N = 74$ isotones $Z = 57, 59, 41$ with the TPSM calculations, which used, respectively, $\varepsilon = 0.14, 0.15, 0.16$ and $\varepsilon' = 0.10$ (corresponding to $\gamma = 35°, 33°, 32°$) as input. The fact that the TPSM with nearly constant ε, ε' accounts for the observed TW-LW-TW sequence along the isotone chain indicates that the change originates from the change of the quasiparticle orbitals at the Fermi level. The details are discussed in Sec. 2.3.2. Figs. 2.41 and 2.14 demonstrate the collective enhancement of the $n \to n-1$ transitions between the wobbling band in ^{133}La and ^{135}Pr, respectively.

Figs. 2.25 displays two examples for a general problem: The QTR predicts the SP band too high when locating the wobbling bands at the experimental energies. As seen in Fig. 2.13, TPSM locates both the SP and wobbling bands close to their experimental positions. The treatment of the TR core and the of the odd proton on equal footing in the microscopic TPSM seems to be decisive.

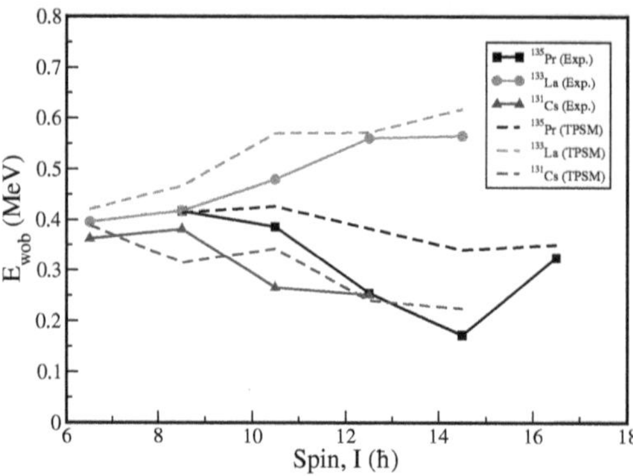

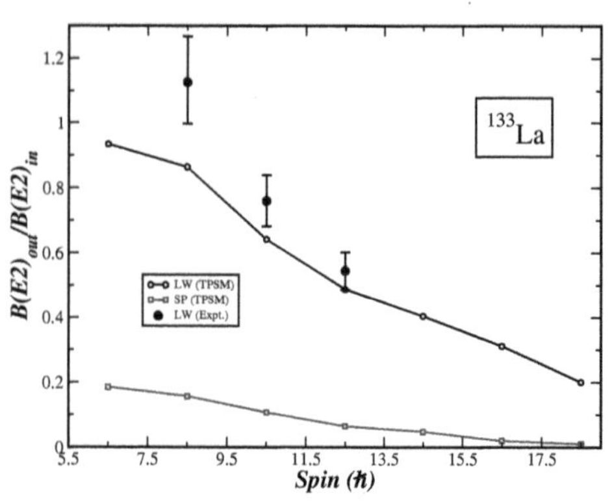

Figure 2.40 Wobbling energies in TPSM approximation for ^{131}Cs, ^{133}La, ^{135}Pr. From Chapter 12, private communication by G. Bhat is acknowledged.

Figure 2.41 The $\frac{B(E2,I\rightarrow I-1)_{out}}{BE2,I\rightarrow I-2)_{in}}$ ratios in TPSM approximation for ^{133}La. From Chapter 12, private communication by G. Bhat is acknowledged.

Chen and Petrache [69] (see Chapter 14) carried out TPSM calculation for the two-quasiproton configuration $(h_{11/2})^2$ in ^{136}Nd. Fig. 2.42 compare the TPSM energies with the ones of bands L1 and L3, which are interpreted as the n=0 and n=1 TW bands. Fig. 2.43 displays a selection of the SCS probability distributions generated from their TPSM results. For $I = 10$ one sees the n=0 characteristics: uniform rotation around the s-axis and the ground state fluctuations around it. For $I = 11$ the precession cone enclosing the s-axis is seen as the elliptical rim around it, which characterizes the n=1 TW state. The I=16 state is topologically classified as $n = 0$ with large fluctuations in direction of the m-axis. The I=17 state is topologically classified as a strongly perturbed $n = 1$ TW excitation. It encloses the s-axis, where the probability around $\phi = 0$ is strongly reduced. The approach of the FM is apparent in form of the two maxima at $\phi = \pm 50°$. The I=20 state has the characteristics of the $n =0$ FM. It flips between $-120°$, $-60°$, $60°$, $120°$ with no phase change, which indicated by the finite probability densities between the blops. The relatively large probabilities at $\phi = \pm 90°$ herald the transition to the LW regime. The $I = 21$ state has the characteristics of the $n = 1$ FM with phase change between the flips. The approach of the LW mode is not apparent, because the symmetry requires $P(\theta = 90°, \phi = \pm 90°) = 0$.

2.6 SOFT CORE

In even-even nuclei triaxiality is indicated by the low energy ratio $E(2_2^+)/E(4_1^+)$ ratio and the splitting between the even- and odd- spin members of the quasi γ band. A staggering parameter

$$S(I) = \frac{[E(I) - E(I-1)] - [E(I-1) - E(I-2)]}{E(2_1^+)} . \tag{2.38}$$

is defined, which measures the relative distance of the even- and odd-spin branches. When $S(I)$ is negative for even I the even-I sequence is lower than the odd-I one. Then the nucleus is classified as "γ-soft", i. e. the triaxiality appears as a large-amplitude oscillation around around a near-axial shape. When $S(I)$ is negative for odd I the odd-I sequence is lower than the even-I one. Then the nucleus is classified as "γ-rigid, i. e. the shape does not fluctuate much around some finite value of γ. The aspects of γ-softness are reviewed Ref. [12] and discussed in detail in Ref. [70].

Most even-even nuclei with a low ratio of $E(2_2^+)/E(4_1^+)$ are γ-soft. The few γ-rigid cases include the examples for wobbling discussed in Sec. 2.2. In these cases $S(I)$ is equal to two times the difference between the single and double wobbling bands divided by the energy of the 2_1^+ state.

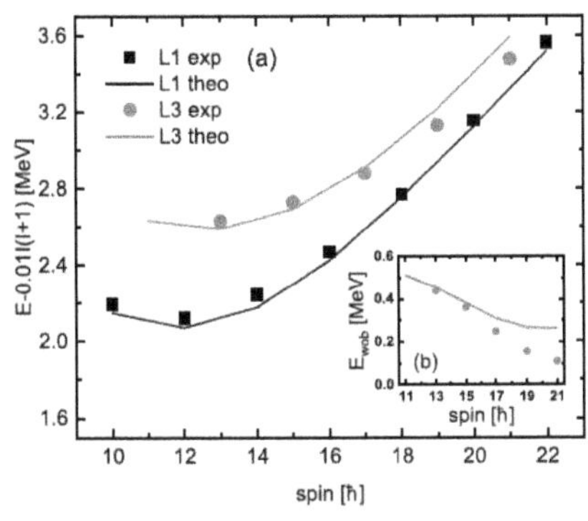

Figure 2.42 Energies in TPSM approximation for the n=0 (L1) and n=1 (L3) TW candidate bands in ^{136}Nd. The inset shows the wobbling energies. Reproduced with permission from Ref [69].

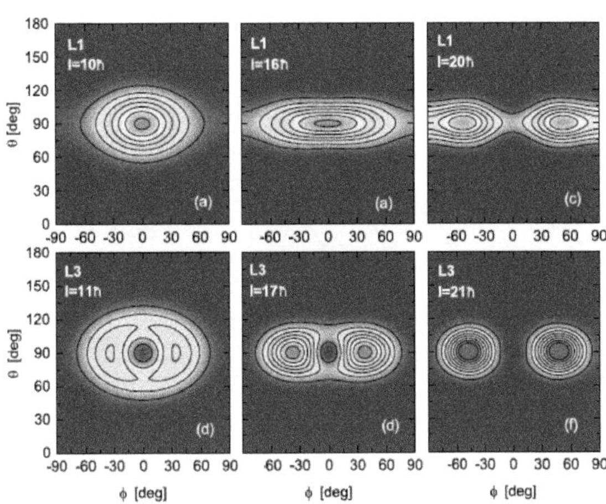

Figure 2.43 SCS probability distributions for ^{136}Nd. Reproduced with permission and selected from Fig. 3 of Ref. [69].

The TPR describes wobbling by coupling a rigid TR with one or two excited quasiparticles. This is in contrast with the observation that the "experimental cores", i. e. the γ bands of the even-even neighbors (or the same nucleus), are γ-soft. Already Meyer-ter-Vehn was surprised how well his model applied to soft nuclei [8].

For this reason Li and advisors [71, 72] used the core-quasiparticle-coupling model [73] for studying the coupling of a high-j quasiparticle to the soft core of Ref. [74], which incorporates the fluctuations of the γ degree of freedom. The core is a simplified version of the Bohr Hamiltonian, the parameters of which were adjusted to the energies of the lowest collective states in ^{134}Ce. (See also Chapter 17.) Fig. 2.44 shows that the fitted Bohr Hamiltonian well approximates the γ-soft even-I-down pattern seen in ^{134}Ce. The resulting wobbling energies of the coupled system ^{135}Pr in Fig. 2.45 have clear LW character, which is in contrast to the experiment. The systematic study of the Pd and Rh isotopes [72] showed that γ-soft cores enhanced the increase of the wobbling energy with I as compared to a rigid $\gamma = 30°$ core. The $B(E2)$ values did not differ much while the $B(M1)$ values were larger for the soft cores.

Nomura and Petrache [75] used the IBAF approach, which couples the odd particle to a soft core described by the Interacting Boson Model, the parameters of which were obtained by mapping a mean field deformation energy surface onto the Boson space (see Chapter 11). Fig. 2.45 shows the wobbling energies calculated from the energies published in Ref. [75][5], which both for ^{135}Pr (Pr) and ^{105}Pd have clear LW character. The contrast with the observed TW behavior raises doubts that the calculations substantiate the claimed "Questioning the wobbling interpretation of low-spin bands in γ-soft nuclei".

The studies [71, 72] and [75] could not obtain the observed TW characteristics by coupling the high-j particle to a phenomenological soft core, which appears surprising at the first sight. The microscopic TPSM resolves the problem. In the series of publications [13–15, 76] the collaboration demonstrated that the energies and $E2$ and $M1$ transition matrix elements of γ-soft even-even nuclei are very well reproduced by the TPSM, provided the basis contains the quasiparticle configurations explicitly stated in Eq. (4.7, left). The softness features appear as a consequence of the mixture of basis states with different mean values of γ. Treating the wobbling nuclides in the same TPSM basis (4.7, right) incorporates the basis states that are responsible for the fluctuations of γ. Thereby it is

---

[5]Private communication of the accurate values by K. Nomura is acknowledged.

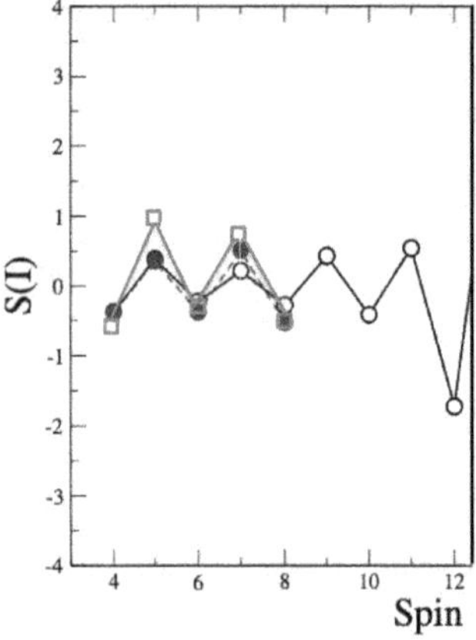

Figure 2.44 Staggering parameter for ^{134}Ce. Full circles: experiment, squares: Bohr Hamiltonian, open circles: TPSM - Private com. by G. Bhat acknowledged.

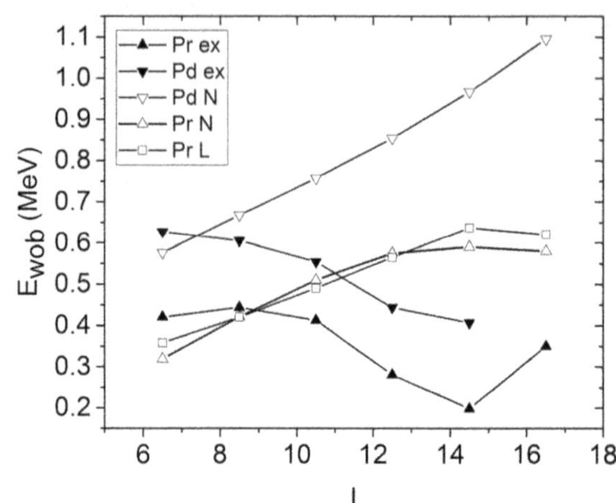

Figure 2.45 Wobbling energies from γ-soft cores for ^{135}Pr (Pr) and ^{105}Pd (pd) calculated by the model of Ref. [71] (L) and the model of Ref. [75] (N) compared with the experiment (ex).

important that the TPSM obeys the Pauli exclusion principle between the core and valence particles, which is not ensured in the case of the core-particle coupling models.

Fig. 2.44 also shows the staggering parameter $S(I)$ for ^{134}Ce, calculated by means of the TPSM with the same parameters as used in the TPSM calculation for the neighbor ^{135}Pr. It is close to the experiment and shows the even-I-down characteristics of a γ-soft nucleus. As already discussed, the TPSM well reproduces the TW features of ^{135}Pr (see Figs. 2.13–2.15).

The authors of Ref. [37] identified ^{151}Eu as another example of TW. The left panel of Fig. 2.46 displays the wobbling energies with the characteristic slight increase at low and the decrease at high spin. They are well reproduced by the TPSM calculations with the parameters $\varepsilon = 0.20$ and $\gamma = 27°$. The right panel compares the results of a TPSM calculation with the parameters $\beta = 0.20$ and $\gamma = 22°$ for the even-even neighbor ^{150}Sm [77] with the experiment. The TPSM very well accounts for the energies and $B(E2)_{in}$ values, where the staggering parameter $S(I)$ signifies γ-softness.

The two examples demonstrate that adding a high-j orbital drastically changes the behavior of the core, which may be modeled by a rigid TR although the neighbors look γ soft. Treating the core as a system of its own and taking only into account its polarization by the particle in the framework of the core-particle-coupling or IBFA approaches does not work. The success of the TPSM points to the importance of the exchange terms between the valence and core nucleons. The ratios $\frac{B(E2,I\rightarrow I-1)_{out}}{BE2,I\rightarrow I-2)_{in}} \sim 0.2$ indicate that the breaking of the axial symmetry is weaker than the breaking of the rotational symmetry. This suggest that the Pauli exclusion principle is more import for wobbling than for ordinary rotation. To investigate its role in more detail seems to be interesting.

2.7 CONCLUSIONS

The appearance of the wobbling mode in strongly and normal deformed nuclei is well established by the observation of collectively enhanced $E2$ between the bands carrying $n + 1$ and n wobbling quanta. For even-even nuclei the softness of the triaxial shape complicates the interpretation. Although the TPSM quite well accounts for the experimental energies and relevant transition rates, the relation to the simple wobbling mode of the TR remains to be clarified. The evidence for wobbling is clear in the presence of one or two high-j quasiparticles. There are a number of nuclei that

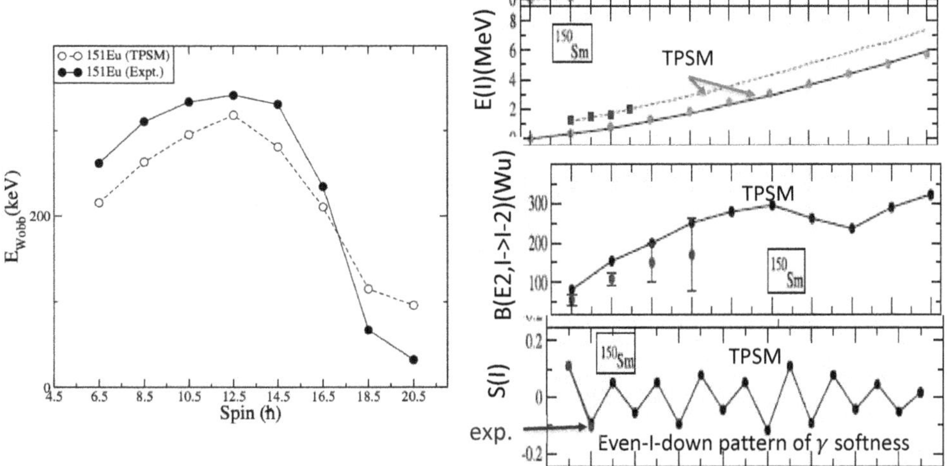

Figure 2.46 Left panel: Wobbling energies for ^{151}Eu calculated by TPSM compared with the experiment [37]. Right panel: energies, $B(E2)_{in}$ values and staggering parameter of ^{150}Sm calculated by the TPSM compared with the experiment [77]. Reproduced with permission from Refs. [37, 77]

show TW. There are less examples for LW. The expected appearance of LW for mid-shell high-j quasiparticles has not yet been demonstrated.

The PTR model has become a major tool for interpreting the data. The most successful strategy is to take the deformation parameters of the triaxial potential from mean field calculations and adjust the rotor MoI's to the observed band energies under the restriction that $\mathcal{J}_m > \mathcal{J}_s > \mathcal{J}_l$. The SCS maps of the PTR states reveal the closest proxies to the corresponding classical orbits. Their topology classifies the wobbling mode: TW orbits enclose the axes perpendicular to the axis with the largest MoI, which is enclosed by the LW orbits. The TW becomes unstable at a critical angular moment and changes into LW via the FM, where the angular momentum vector flips between four equivalent orientations in the s-m-plane. The SSS plots provide the closest proxies to the probability densities of wave functions that belong to a one-dimensional potential. Which axes the potential encloses classifies the wobbling as TW-LW-FM as well.

The microscopic RPA and TPSM approaches have only the triaxial deformations as input, which removes ambiguities of the PTR due to the choice of the MoI's. Being a small-amplitude approximation, the RPA works only for angular momenta far from the instability of the TW. Its successful application to the strongly deformed nuclei revealed the TW dynamics. It has not been applied to normal deformed nuclei yet.

The TPSM very successfully accounts for the energies of the wobbling and SP excitations as well as for the transition matrix elements between them. It resolves the apparent contradiction between the appearance of wobbling in the present of high-j quasiparticles and the features of γ softness in the even-even nuclei in their absence. The tools for extracting the physics from the results of the truncated shell model diagonalization need to be further developed.

Bibliography

1. A. Bohr and B. R. Mottelson, *Nuclear Structure* (W. A. Benjamin, New York, 1975), Vol. II.
2. J. M. Allmond and J. L. Wood, Phys. Lett. B **767**, 226 (2017).
3. S. Frauendorf, Phys. Rev. C **97**, 069801 (2018).
4. Q. B. Chen and S.Frauendorf, Euro. Phys. J. A **58**, 75 (2022).
5. E. A. Lawrie, O. Shirinda and C. M. Petrache, Phys. Rev. C **101**, 034306 (2020).

6. S. Frauendorf, Chirality: from Symmetry to dynamics, Nordita workshop on Chiral bands in Nuclei, Stockholm, 20-22 April 2015, slides from talks, https://www.nordita.org/events/workshops /list_of_workshops/index.php.

7. F. Q. Chen, *et al.*, Phys. Rev. C **96**, 051303(R) (2017).

8. J. Meyer-ter-Vehn, Nucl. Phys. A **249**, 111 (1975).

9. S. Frauendorf, Phys. Scr. **93**, 043003 (2018).

10. Evaluated Nuclear Structure Data Files.

11. Y. Toh *et al.*, Phys. Rev. C **87**, 041304(R) (2013).

12. S. Frauendorf, Int. J. Mod. Phys. E **24**, 1541001 (2015).

13. S. P. Rouoof, *et al.*, Eur. Phys. J. A, Eur. Phys. J. A **60**, 40 (2024).

14. S. Jehangir, *et al.*, Eur. Phys. J. A **57**, 308 (2021).

15. Nazira Nazir, *et al.*, Phys. Rev. C **107**, L021303 (2023).

16. J. A. Sheikh and K. Hara, Phys. Rev. Lett. **82**, 3968 (1999).

17. I. Hamamoto and B. Mottelson, Phys. Lett. B **167**, 370 (1986).

18. P. Bringel *et al.*, Eur. Phys. J. A 24, 167 (2005).

19. S. W. Ødegård, *et al.*, Phys. Rev. Lett. **86**, 5866 (2001).

20. D. R. Jensen, *et al.*, Phys. Rev. Lett. **89**, 142503 (2002).

21. G. G. Schönwasser *et al.*, Phys. Lett. B 553, 197 (2003).

22. H. Amro *et al.*, Phys. Lett. B 552, 9 (2003).

23. D. R. Hartley, *et al.*, Phys. Rev. C **80**, 041304 (2009).

24. I. Hamamoto, Phys. Rev. C **65**, 044305 (2002).

25. I. Hamamoto, G. B. Hagemann, Phys. Rev. C **67**, 014319 (2003).

26. S. Frauendorf, J. Meng, Nucl. Phys. A **617**, 131 (1997).

27. S. Frauendorf, F. Dönau, Phys. Rev. C **89**, 014322 (2014).

28. Y. R. Shimizu *et al.*, Phys. Rev. C **77**, 024319 (2008).

29. J. T. Matta *et al.*, Phys. Rev. Lett. **114**, 082501 (2015).

30. N. Sensharma *et al.*, Phys. Lett. B **792**, 170 (2019).

31. N. Sensharma *et al.*, Phys. Lett. B **820**, 136556 (2021).

32. J. Timár *et al.*, Phys. Rev. Lett. **122**, 062501 (2019).

33. J. Timár *et al.*, J. Phys.: Conf. Ser. **1555** 012025 (2020).

34. A. Karmakar *et al*, Phys. Lett. B, submitted, arXiv:2403.08235.

35. B. F. Lv *et al.*, Phys. Lett. B 824, 136840 (2022).

36. N. Sensharma *et al.* Phys. Lett B, submitted, arxiv:2403.10749.

37. A. Mukherjee *et al.* Phys. Rev. C **107**, 054310 (2023).

38. S. M. Nandi *et al.*, Phys. Rev. Lett. **125**, 132501 (2020).

39. K. Rojeeta Devi *et al.* Phys. Lett. B **823**, 136756 (2021).

40. Q. B. Chen, S. Frauendorf, C. M. Petrache, Phys. Rev. C **100**, 061301 (2019).

41. Q. B. Chen, C. M. Petrache, Phys. Rev. C **103**, 064319 (2021).

42. Q. B. Chen and S. Frauendorf, Phys. Rev. C **109**, 044304 (2024).

43. S. Frauendorf Rev. Mod. Phys. **73** 643 (2001).

44. S. Biswas *et al.*, Eur. Phys. J. A **55**, 159 (2019).

45. N. Sensharma *et al.*, Phys. Rev. Lett. **124**, 052501 (2020).

46. S. Chakaraborty *et al.*, Phys. Lett. B **811**, 135853 (2020).

47. S. Guo *et al.*, Phys. Lett. B 828, 137010 (2022).

48. Sensharma, N. (2022). In: *Wobbling Motion in Nuclei: Transverse, Longitudinal and Chiral. Springer Theses*. Springer, Cham. https://doi.org/10.1007/978-3-031-17150-5.

49. R. Budaca Phys. Rev. C **97**, 024302 (2018).

50. R. Budaca and A. I. Budaca, J. Phys. G: Nucl. Part. Phys. **50**, 125101 (2023).

51. K. Tanabe, K. Sugawara-Tanabe, Phys. Rev. C **73**, 034305 (2006); **75**, 059903(E) (2007).

52. K. Tanabe, K. Sugawara-Tanabe, Phys. Rev. C **77**, 064318 (2008).

53. G. B. Hagemann, Eur. Phys. J. A **20**, 183 (2004).
54. K. Tanabe, K. Sugawara-Tanabe, Phys. Rev. C **95**, 064315 (2017).
55. K. Tanabe, K. Sugawara-Tanabe, Phys. Rev. C **97**, 069802 (2018).
56. A. A. Raduta, R. Poenaru and L. Gr. Ixaru, Phys. Rev. C **96**, 054320 (2017).
57. A. A. Raduta, R. Poenaru and Al. H. Raduta, J. Phys. G: Nucl. Part. Phys. **45**, 105104 (2018).
58. A. A. Raduta, R. Poenaru, C. M. Raduta, Phys. Rev. C **101**, 014302 (2020).
59. A. A. Raduta, C. M. Raduta and R. Poenaru, J. Phys. G: Nucl. Part. Phys. **48**, 015106 (2021).
60. Q. B. Chen, *et al.*, Phys. Rev. C **90**, 044306 (2014).
61. Q. B. Chen, *et al.*, Phys. Rev. C **94**, 054308 (2016).
62. I. N. Michailov and D. Janssen, Phys. Lett. B **72**, 303 (1978).
63. D. Janssen and I. N. Michailov, Nucl. Phys. A **318** 390 (1979).
64. E. R. Marshalek, Nucl. Phys. A **331**, 429 (1979).
65. M. Matsuzaki *et al.*, Phys. Rev. C **65**, 041303(R) (2002).
66. M. Matsuzaki *et al.*, Phys. Rev. C **69**, 034325 (2004).
67. S. Frauendorf and F. Dönau, Phys. Rev. C **92**, 064306 (2015).
68. J. A. Sheikh and K. Hara, Phys. Rev. Lett. **82**, 3968 (1999).
69. Fang-Qi Chen and C. P. Petrache, Phys. Rev. C **103**, 064319 (2021).
70. S. P. Rouoof *et al.*, Eur. Phys. J. A **60**, 40 (2024).
71. W. Li, "Algebraic Collective Model and Its Application to Core Quasiparticle Coupling. University of Notre Dame , Thesis (2016), https://doi.org/10.7274/6682x348f37, Notre Dame Website link: Algebraic Collective Model and Its Application to Core Quasiparticle Coupling.
72. W. Li *et al.* Eur. Phys. J. A **58**, 218 (2022).
73. F. Dönau and S. Frauendorf, Phys. Lett. B **71**, 263 (1977).
74. M. Caprio, Phys. Rev. C **83**, 064309 (2011).
75. K. Nomura, C. M. Petrache, Phys. Rev. C **105**, 024320 (2022).
76. S. P. Rouoof *et al.*, in preparation, to be published in Phys. Rev. C.
77. Tabassum Naz *et al.*, Nucl. Phys. A **979**, 1 (2018).

3 The 25th anniversary for nuclear chirality

Jie Meng, Yi-PingWang
Peking University, Beijing, China

3.1 INTRODUCTION

Chirality is a subject of general interests in natural science. Since the concept of the chirality in atomic nuclei [1] was proposed 25 years ago, it has become one of the hot topics in modern nuclear physics. The history of the prediction of the nuclear chirality is quite instructive.

The gestation of the concept of the nuclear chirality can be dated back to about 30 years ago when one of the authors J. M. visited Forschungszentrum Rossendorf, now called Helmholtz Zentrum Dresden Rossendorf, in Germany. J. M. stayed there from September 1993 to August 1994. J. M. enjoyed the time with S. Frauendorf and other colleagues very much, and they published five papers.

At that time, one of the hot research topics was the superdeformed rotational band [2], while the investigation of the magnetic rotation was only at its beginning. Traditionally, it is well-known that the rotations occur only in nuclei with a stable deformation. Thus the observation of the regular rotational-like sequences in several near spherical Pb isotopes was very surprising (for a review see Ref. [3]). The explanation of such bands was first given by S. Frauendorf using the axial tilted axis cranking (TAC) approach in 1993 [4]. It was found that the angular momentum along these bands is generated by the alignment of the proton and neutron angular momenta. Unlike the normal deformed rotational bands, these rotational-like bands have strong magnetic dipole ($M1$) and very weak electric quadrupole ($E2$) transitions, thus the name "magnetic rotation" was introduced [5].

After the concept of magnetic rotation was proposed, many efforts have been devoted to verify it experimentally. However, there are big discrepancies between the predicted and observed angular momentum dependence of the $B(M1)$ values [6]. Therefore, at that time, the results of the TAC model were questioned, due to its semi-classical treatment of the angular momentum, the mean-field approximation, and the neglect of the multiparticle correlations.

After arriving in Rossendorf for one month, J. M. examined the quality of the semi-classical and mean-field approximation of the TAC model. With the recommendation of S. Frauendorf, J. M. attended the European Centre for Theoretical Studies in Nuclear Physics and Related Areas (ECT*) workshop in 1993 and presented the results, which was later published in Zeitschrift für Physik A titled "Interpretation and quality of the tilted axis cranking approximation" in 1996 [7] . In this paper, the quality of the semi-classical and mean-field approximation of the TAC model was examined by comparing with the results from the particle rotor model (PRM).

The PRM is a quantum model with the total angular momentum as a good quantum number, works in the laboratory frame, and describes properly the quantum tunneling. Surprisingly, the calculated energy spectrum as well as the $B(M1)$ and $B(E2)$ values in PRM are well reproduced by the TAC model [7]. Thus the semi-classical and mean-field approximation are not responsible for the deviation between the calculated and the observed $B(M1)$ values. Later on, with the improvement of the detector technology, the predicted decrease of the $B(M1)$ values with the angular momentum was turned out to be correct [8].

DOI: 10.1201/9781032691633-3

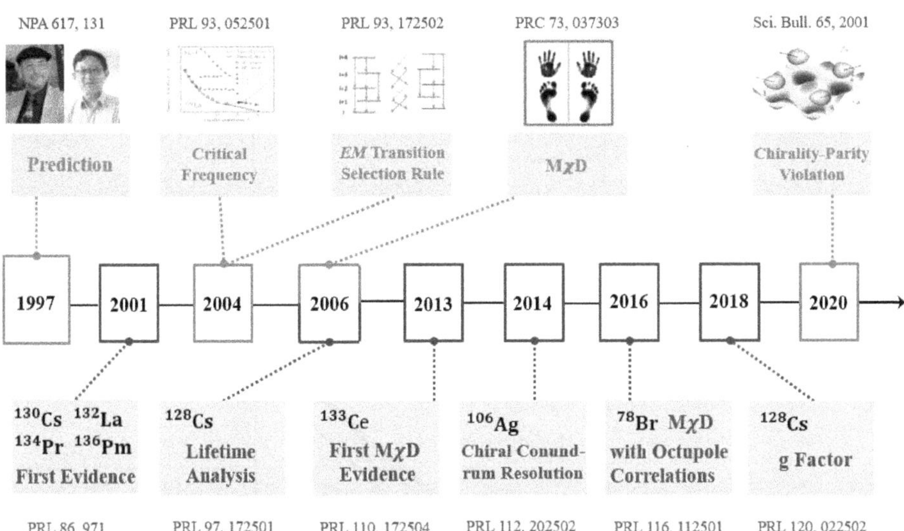

Figure 3.1 The time line for some of the significant theoretical (top) and experimental (bottom) progresses in nuclear chirality.

Following the success of the TAC model for the magnetic rotation, it is quite natural to generalize it for the triaxial nuclei. The famous paper, in which the concept of the nuclear chirality was suggested, appeared by using the triaxial TAC model and the PRM [1].

Since then, the nuclear chirality has become one of the hot topics in modern nuclear physics. Over the past two and a half decades, many efforts have been devoted to understand the nuclear chirality, and some of the significant theoretical and experimental progresses are summarized in a time line in Fig. 3.1.

In this chapter, the prediction of the nuclear chirality is presented in Sec. 14.2. The theoretical and experimental investigations of the nuclear chirality are reviewed, including the verification of chiral doublet bands in Sec. 14.3, the chiral conundrum and its resolution in Sec. 14.4, and the prediction and observation of the multiple chiral doublets (MχD) in Secs. 14.5-3.6. Some recent theoretical progresses are highlighted in Sec. 3.7 , including the chiral collective Hamiltonian, the A-plot and the K-plot, the nuclear chirality-parity (ChP) violation, the chiral rotation induced by the pairing correlations, as well as the chiral dynamics. The possibly emerging area, challenges that lie ahead, and opportunities for progress in the context of the nuclear chirality are discussed in Sec. 3.8.

3.2 PREDICTION OF NUCLEAR CHIRALITY

In Ref. [1], the nucleus is assumed to have a high-j particle and a high-j hole coupled with a triaxial rotor. The angular momenta for the particle and the hole respectively favor to align along the nuclear short (s) and long (l) axes. The angular momentum for the rotor favors to align along the intermediate (i) axis due to its largest moment of inertia. In the TAC model, the total angular momentum vector may lie outside the three principal planes of the triaxial ellipsoidal density distribution. The s, i and l principal axes of the triaxial nucleus form a screw with respect to the angular momentum vector, resulting in two systems with different intrinsic chirality, i.e., left- and right-handed systems. In the PRM, the broken chiral symmetry in the intrinsic frame would be restored, and this gives rise to the chiral doublet bands, i.e., a pair of nearly degenerate $\Delta I = 1$ bands with the same parity, as shown in Fig. 3.2.

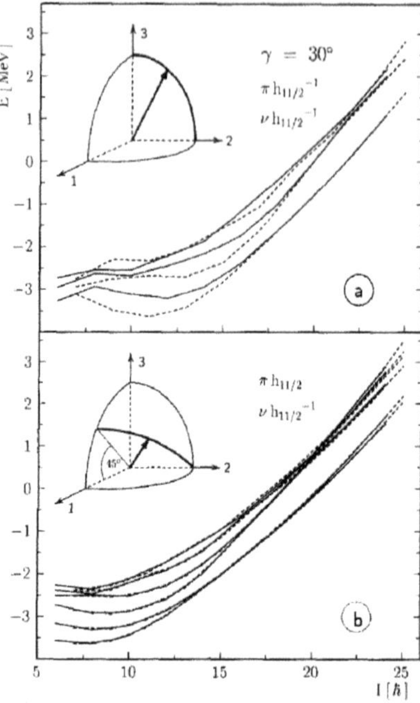

Figure 3.2 Rotational levels of $h_{11/2}$ particles and holes coupled to a triaxial rotor with $\gamma = 30°$. The upper panel shows the case of a proton hole and a neutron hole and the lower panel shows the case of a proton particle and a neutron hole. Full lines correspond to even and dashed to odd spins. Reprinted figure with permission from [1], Copyright (1997) Elsevier Science B.V.

Before the publication of Ref. [1], the examples for the pair of nearly degenerate bands had already been reported in ^{134}Pr [9]. In a conference organized by Rainer Lieder and others in the beautiful Crete island in Greece, after the first session by the talk of Frauendorf and J. M., Petrache appeared and showed his results. Therefore, the observed partner bands in ^{134}Pr were suggested as a candidate for chiral doublet bands in Ref. [1].

In the textbook *Nuclear Structure* by Bohr and Mottelson, the structure of the rotational spectra is determined by the symmetry of the nuclear deformation and the resulting rotational degrees of freedom [10]. The nuclear system with quadrupole deformation may possess the parity ($\mathcal{P}$), time-reversal ($\mathcal{T}$), spatial rotation of 180° ($\mathcal{R}$) symmetries and their combinations. The cases when the system possesses the symmetries $\mathcal{R}, \mathcal{P}, \mathcal{T}$, the symmetries $\mathcal{RP}, \mathcal{T}$, or only the symmetry $\mathcal{T}$ have been discussed in Ref. [10], as shown in Figs. 3.3 (a-c). The associated rotational spectra are schematically shown in Figs. 3.4 (a-c). The system with the symmetries $\mathcal{R}$ and $\mathcal{P}$ hasn't been discussed in Ref. [10], which corresponds to the nuclear chirality, and the corresponding symmetry as well as the associated rotational spectra are, respectively, shown in Figs. 3.3 (d) and 3.4 (d).

3.3 OBSERVATION OF NUCLEAR CHIRALITY

After the prediction in 1997, lots of experimental efforts have been devoted to search for the nuclear chirality. In 2001, four pairs of nearly degenerate bands were observed in the odd-odd $N = 75$ isotones ^{130}Cs, ^{132}La, ^{134}Pr and ^{136}Pm, respectively, and were suggested as the candidate chiral doublet bands [11]. Two years later, similar chiral doublet bands were observed in odd-A nucleus [12]. Based on the observed near-degenerate $\Delta I = 1$ doublet bands with the same parity, candidate chiral doublet bands have been proposed in a number of nuclei in the $A = 130$ region with the configuration $\pi h_{11/2} \otimes \nu h_{11/2}^{-1}$ [9, 12–23], $A = 100$ region with $\pi g_{9/2}^{-1} \otimes \nu h_{11/2}$ [24–32], $A = 190$

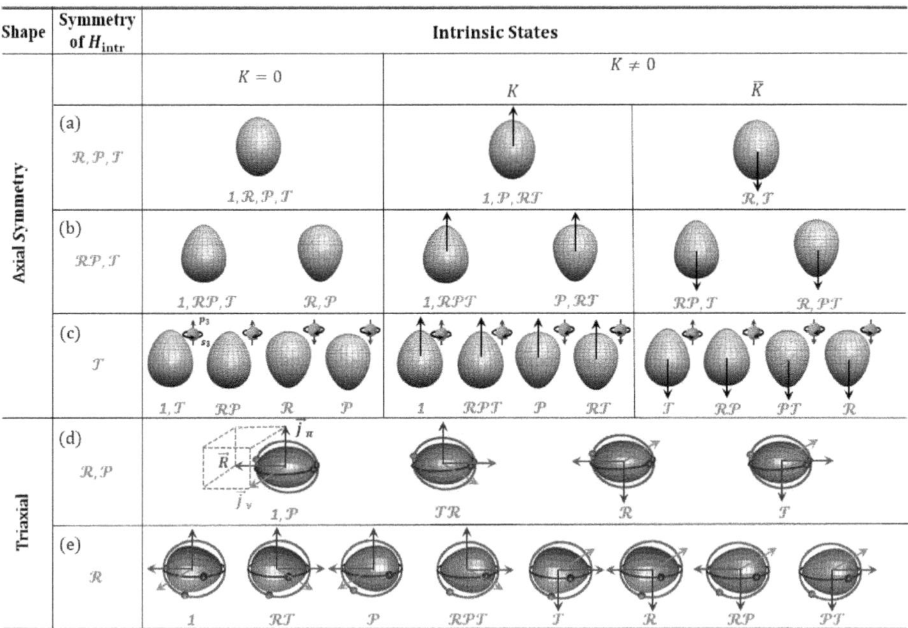

Figure 3.3 The intrinsic states for axially symmetric and triaxial systems with different combinations of $\mathcal{P}$, $\mathcal{T}$ and $\mathcal{R}$ invariance. Adapted figure from Ref. [10].

Figure 3.4 The rotational spectra for axially symmetric and triaxial systems with different combinations of $\mathcal{P}$, $\mathcal{T}$ and $\mathcal{R}$ invariance. The quantum numbers (I, π) are given to the right of the energy levels, with I the angular momentum and π the parity. Adapted figure from Ref. [10].

region with $\pi h_{9/2} \otimes \nu i_{13/2}^{-1}$ [33, 34]. In 2011, under joint efforts from scientists in China, South Africa and Hungary, the first example of chiral doublet bands based on the configuration $\pi g_{9/2} \otimes \nu g_{9/2}^{-1}$ was reported in the $A = 80$ region [35], which is the lightest known region with chiral nuclei.

In the same period, the significant progress has also been made in the study of the nuclear chirality on the theoretical side. A self-consistent three-dimensional (3D) mean-field solution has been obtained and strongly supports the possible existence of the chiral rotation in actual nuclei [36]. The selection rule for the electromagnetic transitions expected in the ideal chiral bands has been proposed, including the odd-even staggering of intraband $B(M1)/B(E2)$ ratios and interband $B(M1)$ values, as well as the vanishing of the interband $B(E2)$ transitions at high spin region [37]. The critical frequency which marks the onset of chiral rotation was suggested by the Skyrme Hartree-Fock cranking model calculations [38].

3.4   CHIRAL CONUNDRUM IN ^{134}PR

With more and more chiral doublet candidates observed, it is natural to measure other observables than the energy spectra to provide further confirmation for the chirality. In 2006, the electromagnetic transition ratios of the nearly degenerate bands in ^{134}Pr were measured [39]. The measured in-band $B(E2)$ values for the candidate partner bands show large differences, and this is in disagreement with the chiral picture. This phenomenon was known as chiral conundrum. In particular, the paper titled with "Risk of misinterpretation of nearly degenerate pair bands as chiral partners in Nuclei" [40] casted a shadow on the investigation of the nuclear chirality.

It should be noted that although the chiral partner bands have energies close to each other, it is rare to observe a crossing between them. The most famous example of such a crossing is in ^{134}Pr. The only other known case is in ^{106}Ag [31]. Thus the problem in ^{134}Pr might be a special case, and may not influence the investigation of the nuclear chirality.

Fortunately, shortly after that, lifetime measurement was performed for ^{128}Cs and the electromagnetic transition properties of the partner bands in ^{128}Cs were in good agreement with the chiral interpretation [41]. It was claimed as the best-known example revealing the chiral symmetry-breaking phenomenon. Later on, the chiral character of the bands in ^{135}Nd was also affirmed based through the lifetime measurements and it was shown that the partner bands are associated with a transition from chiral vibration to static chirality [42]. These observations renewed the enthusiasm on the study of the nuclear chirality.

3.5   PREDICTION OF MχD

It should be emphasized that caution should be taken when comparing the theoretical results with the experimental data. The nuclear chirality was predicted in an ideal model, i.e., a high-j particle and a high-j hole coupled with a triaxial rotor with $\gamma = 30°$. However, it is not specified which nuclei satisfy these conditions. In order to search for the chiral nucleus, the microscopic approaches with predictive power are demanded.

Starting from an effective nucleon-nucleon interaction with Lorentz invariance, the covariant density functional theory (CDFT) is very successful in describing many nuclear phenomena in stable and exotic nuclei of the whole nuclear chart [43–46]. It provides a powerful way for the investigation of the nuclear chirality.

In 2006, the adiabatic and configuration-fixed constrained triaxial CDFT was developed to predict chiral nuclei and guide the phenomenological models [47]. The nucleus ^{106}Rh was investigated as an example. The energy surfaces and deformation γ for ^{106}Rh are shown as functions of deformation β in Fig. 3.5. It was found that the minima, A, B, C and D, have deformations β and γ suitable for chirality. Their proton and neutron configurations were also examined in detail to see whether they correspond to the high-j particle and hole configurations required by chirality. Then a

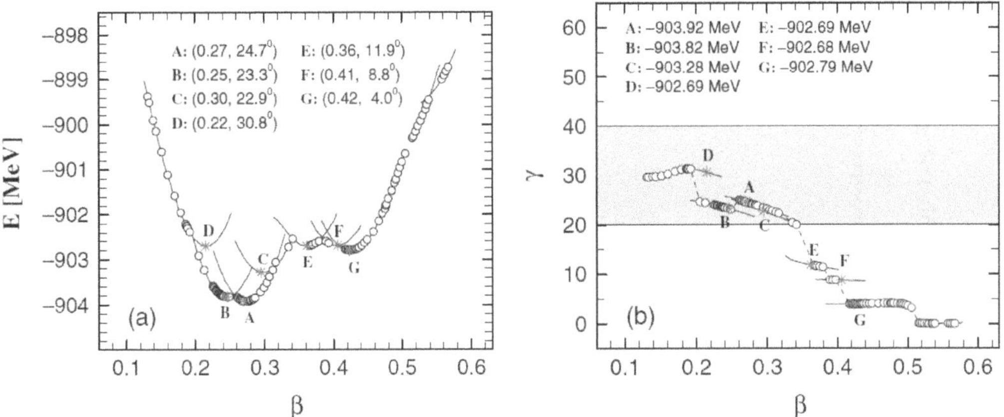

Figure 3.5 Energy surfaces (a) and deformation parameters γ (b) as functions of deformation parameter β in adiabatic (open circles) and configuration-fixed (solid lines) constrained triaxial CDFT calculations for ^{106}Rh. The minima in the energy surfaces are represented as stars and labeled A-G with their corresponding deformations β and γ (a) and energies (b). Reprinted figure with permission from [47], Copyright (2006) the American Physical Society.

new phenomenon, which was called as the multiple chiral doublets (MχD) , i.e., more than one pair of chiral doublet bands in one single nucleus, was suggested for ^{106}Rh.

3.6 OBSERVATION OF MχD

The prediction of MχD stimulated lots of experimental efforts. In 2013, two distinct sets of chiral partner bands were identified in ^{133}Ce, which was the first experimental evidence for MχD [48]. Along this line, the lifetime measurements for the chiral doublet bands in ^{106}Ag provided a resolution of the chiral conundrum [49]. Similarly, the chirality in ^{134}Pr needs further investigation. In 2014, the MχD with the identical configuration were observed in ^{103}Rh [50], showing that the chiral geometry in nuclei can be robust against the increase of the intrinsic excitation energy. In 2016, the MχD with octupole correlations were first identified in ^{78}Br [51]. This observation indicates that the nuclear chirality can be robust against the octupole correlations. It also indicates that the chirality-parity (ChP) quartet bands [52], which are a consequence of the simultaneous breaking of chiral and space-reflection symmetries, namely, ChP violation, may exist in nuclei.

Until 2019, more than 59 chiral doublet bands in 47 nuclei (including 8 nuclei with MχD) have been reported in the $A = 80$, 100, 130 and 190 regions [53]. The distribution of the observed chiral nuclei in the nuclear chart is given in Fig. 3.6.

3.7 RECENT PROGRESS

The nuclear chirality has been extensively investigated with many theoretical approaches, including the triaxial PRM [1, 37, 54–60], the 3D cranking model [1, 36, 38, 61–65], the 3D cranking model with the random phase approximation [42, 66], the 3D cranking model with the collective Hamiltonian [67–69], the interacting boson-fermion-fermion model [70], the generalized coherent state model [71] and the projected shell model [72–76]. Some of these progresses are highlighted in this section.

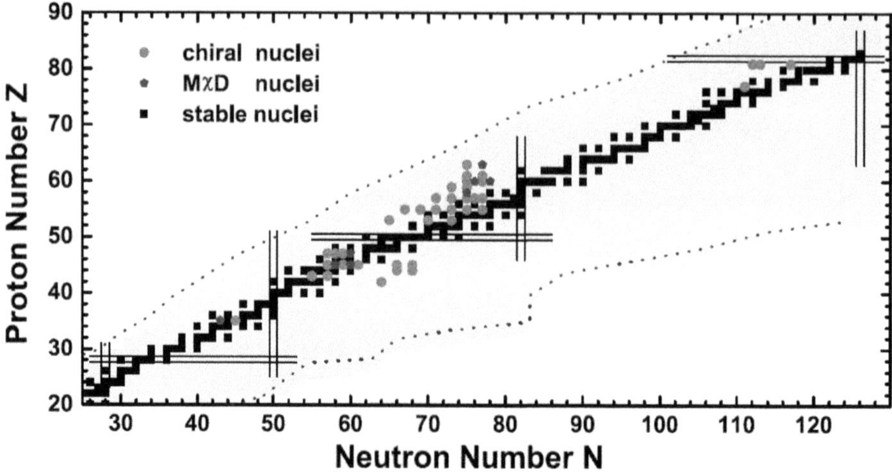

Figure 3.6 The nuclides with chiral doublet bands (circles) and MχD (pentagons) observed in the nuclear chart. The squares represent stable nuclides. Reprinted figure with permission from [53], Copyright (2018) Elsevier Inc.

3.7.1 CHIRAL COLLECTIVE HAMILTONIAN

Within the TAC mean-field approximation, the left-handed and right-handed solutions are exactly degenerate. It is not possible to calculate the energy difference between the bands due to the missing of the quantum tunneling.

To describe the energy splitting between the chiral doublet bands, a collective Hamiltonian for chiral modes was constructed in analogy to Bohr Hamiltonian [67]. Instead of the β and γ degrees of freedom in Bohr Hamiltonian, the azimuthal angle ϕ was introduced as the collective degree of freedom. The TAC with the collective Hamiltonian model reproduced well the energy spectra for the chiral partners obtained by the PRM. By considering both the azimuthal angle ϕ and the polar angle θ as the collective degrees of freedom, the collective Hamiltonian was extended to two-dimensional, and more excitation modes have been obtained [68, 77].

In the future, the collective Hamiltonian model could be combined with the microscopic 3D cranking CDFT. The collective Hamiltonian on top of the 3D cranking CDFT could become very powerful in describing and predicting chiral doublet bands.

3.7.2 A-PLOT AND K-PLOT

The projected shell model (PSM) restores the rotational symmetry broken in the mean-field approximation and combines the advantages of the TAC model and the PRM.

The attempts to understand the chiral doublet bands by the PSM have been performed in Refs. [73, 78] where the energy spectra and transitions are well reproduced. However, it is a big challenge to examine the chiral geometry of angular momentum in the PSM due to the complication that the projected basis is defined in the laboratory frame and forms a nonorthogonal set.

In Refs. [74, 75], the orientation of the angular momentum in the intrinsic frame is investigated by the distributions of its components on the three principal axes, K-plot, and those of its tilted angles, azimuthal plot (A-plot). The evolution of the chirality with spin is illustrated, and the chiral geometry is demonstrated in the PSM.

The A-plots for the chiral doublet bands A and B in ^{128}Cs are shown in Fig. 3.7. These plots clearly illustrate the evolution of the chiral mode from a chiral vibration at $I = 11\hbar$ with respect to

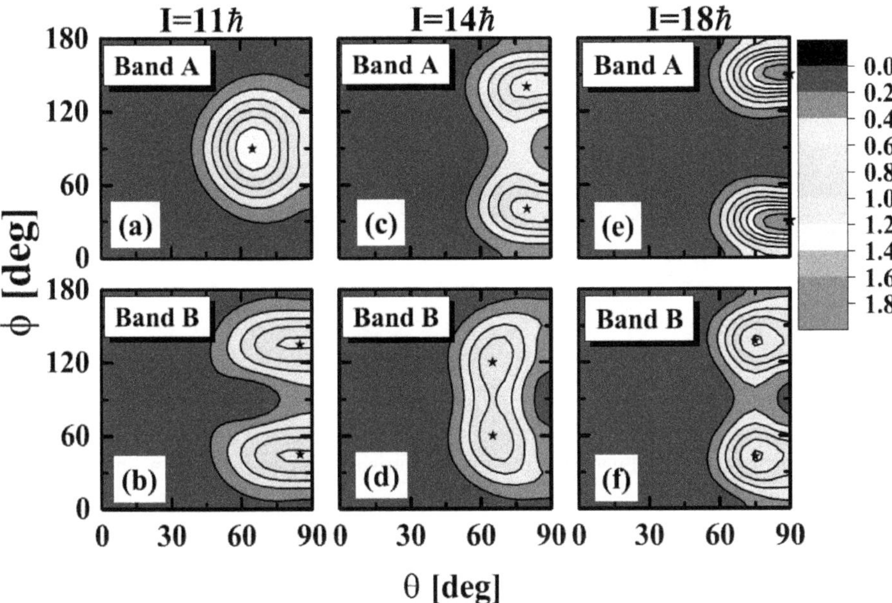

Figure 3.7 The azimuthal plots (*A*-plots), i.e., profile for the orientation of the angular momentum on the (θ, ϕ) plane, calculated at $I = 11$ [(a), (b)], 14 [(c), (d)] and $18\hbar$ [(e), (f)], respectively. Reprinted figure with permission from [74], Copyright (2017) American Physical Society.

the *l-s* plane, to a static chiral rotation at $I = 14\hbar$, and to another chiral vibration with respect to the *i-s* plane at $I = 18\hbar$.

3.7.3 NUCLEAR CHP VIOLATION

The observation in ^{78}Br, two pairs of chiral doublet bands with opposite parity connected with strong electric dipole ($E1$) transitions, provides the first evidence of MχD in octupole soft nuclei [51]. This observation indicates that the nuclear chirality can be robust against the octupole correlations, which together with the scenario in Ref. [52] encourages the exploration of the simultaneous chiral and reflection symmetry breaking, ChP violation, in a reflection asymmetric triaxial nucleus.

A schematic potential energy surface with simultaneous chiral and reflection symmetry breaking in the intrinsic (β_{30}, ϕ) plane is given in Fig. 3.8, with β_{30} the octupole deformation parameter and ϕ the azimuthal angle of the total angular momentum. The intrinsic states for ChP violation system and their transformation under different combinations of $\mathcal{P}$, $\mathcal{T}$ and $\mathcal{R}$ operators are shown in Fig. 3.3 (e). The symmetry restoration in the laboratory frame gives rise to four nearly degenerate rotational bands, i.e., ChP quartet bands, as shown in Fig. 3.4 (e).

The nuclear ChP violation is investigated with a reflection-asymmetric triaxial PRM in Ref. [60], in which a new symmetry for an ideal ChP violation system is found and the corresponding selection rules of the electromagnetic transitions are derived. By taking a two-j shell $h_{11/2}$ and $d_{5/2}$ with typical energy spacing for $A = 130$ nuclei, the fingerprints for the ChP violation including the nearly degenerate quartet bands and the selection rules of the electromagnetic transitions are discussed, it would be interesting to search for ChP quartet bands experimentally.

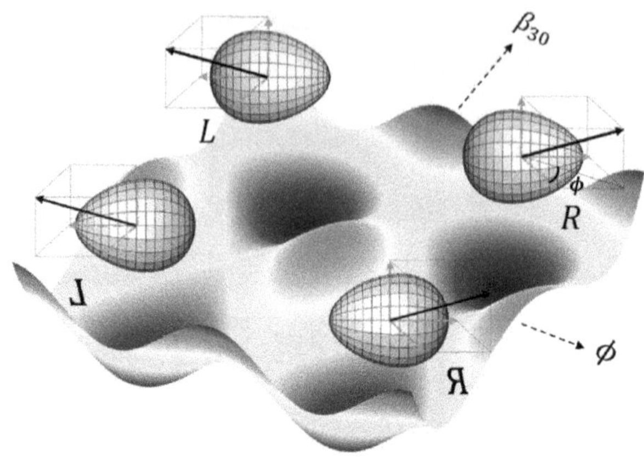

Figure 3.8 A schematic potential energy surface in the intrinsic (β_{30}, ϕ) plane for a ChP violation system, with β_{30} the octupole deformation parameter and ϕ the azimuthal angle of the total angular momentum. The sign of β_{30} stands for the orientation of the nuclear distribution parallel or antiparallel with the intrinsic axis. The sign of ϕ stands for the right- and left-handed system. Reprinted figure with permission from [60], Copyright (2020) Science China Press.

3.7.4 CHIRAL ROTATION INDUCED BY PAIRING CORRELATIONS

In the nuclear chiral rotation, the critical frequency is an important concept, which marks the onset of chiral rotation. The chiral critical frequency has been investigated by the 3D cranking models based on a Woods-Saxon potential combined with the shell correction method [36], and relativistic [62, 63] and nonrelativistic [38, 79] density functional theories. However, in the microscopic and self-consistent density functional calculations, the pairing correlations are neglected.

In Ref. [64], based on the 3D cranking CDFT, a shell-model-like approach (SLAP) with exact particle number conservation is implemented to take into account the pairing correlations and the blocking effects exactly and applied for the chiral doublet bands in ^{135}Nd [64]. The data available, including the $I - \omega$ relation, as well as the electromagnetic transition probabilities $B(M1)$ and $B(E2)$, are well reproduced. It is found that the superfluidity can reduce the critical frequency and make the chiral rotation easier.

As shown in Fig. 3.9, without pairing, the azimuthal angle is always zero. In the left panel, when the proton pairing strength is enhanced by 50%, 100% or 150%, the azimuthal angle, respectively, becomes nonzero at 0.52, 0.49 or 0.47 MeV, indicating the occurrence of the chiral rotation. Similarly, in the right panel, for the enhanced neutron pairing, the azimuthal angle, respectively, becomes nonzero at 0.54, 0.53 and 0.52 MeV. These results suggest that the pairing correlations could induce the earlier appearance of the chiral rotation.

3.7.5 CHIRAL DYNAMICS

The aforementioned studies on the nuclear chirality are based on static approaches, and the time-dependent (TD) CDFT is a dynamical extension of CDFT, which can serve as a powerful tool for simulating the dynamics of the nuclear chirality. However, to achieve this goal, the solution of the CDFT in 3D lattice space is required, which is a challenging task. Fortunately, after overcoming the longstanding problems of the variational collapse [80] and Fermion doubling [81], the solution of CDFT in 3D lattice space has been realized [81–86].

In Ref. [65], the dynamics of chiral nuclei is investigated with the TDCDFT in 3D lattice space in a microscopic and self-consistent way. The experimental energies of the two pairs of the chiral

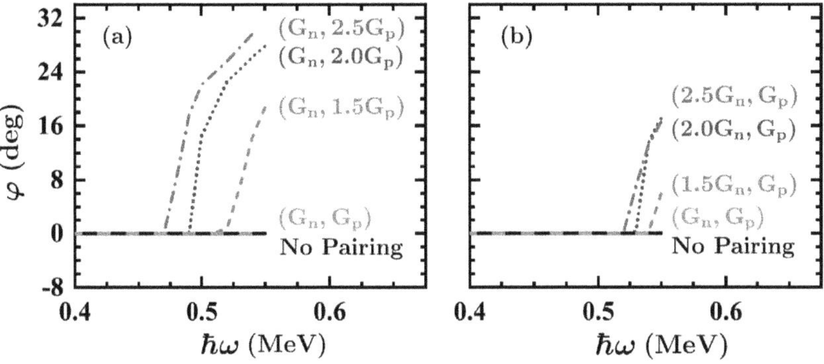

Figure 3.9 Evolution of the azimuthal angle for the total angular momentum as functions of the rotational frequency ω in the cases without pairing or with different pairing strengths for protons (left panel) and neutrons (right panel). Reprinted figure with permission from [64], Copyright (2023) The Author(s).

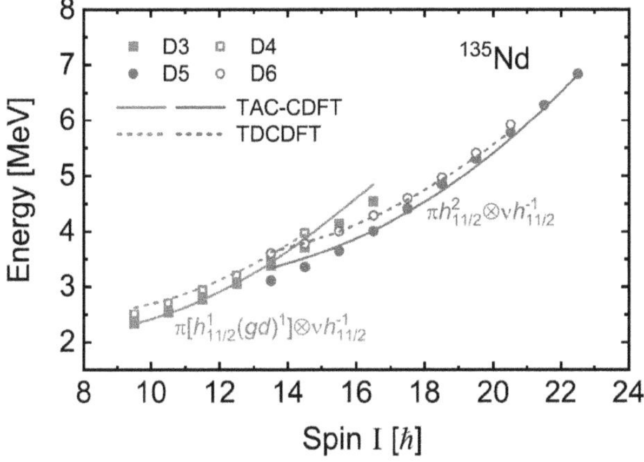

Figure 3.10 Calculated excitation energies (solid and dashed lines) for chiral doublet bands in ^{135}Nd built on the configurations $\pi[h_{11/2}^1(gd)^1] \otimes \nu h_{11/2}^{-1}$ and $\pi h_{11/2}^2 \otimes \nu h_{11/2}^{-1}$ in comparison with data (solid and open symbols) [87]. The solid lines represent the TAC-CDFT results, and the dashed ones represent the results including the chiral excitation energies obtained with the TDCDFT calculations. Reprinted figure with permission from [65], Copyright (2022) American Physical Society.

doublet bands in ^{135}Nd are well reproduced without any adjustable parameters beyond the well-defined density functional. A novel mechanism, i.e., chiral precession, is revealed from the microscopic dynamics of the total angular momentum in the body-fixed frame, whose harmonicity is associated with a transition from planar into aplanar rotations with the increasing spin.

As shown in Fig. 3.10, the calculated energies for the two pairs of chiral doublet bands in ^{135}Nd, which are built, respectively, on the configurations $\pi[h_{11/2}^2(gd)^1] \otimes \nu h_{11/2}^{-1}$ and $\pi h_{11/2}^2 \otimes \nu h_{11/2}^{-1}$, are depicted in comparison with the data [87]. The solid lines represent the TAC-CDFT results, and the dashed ones represent the results including the chiral excitation energies obtained with the TDCDFT. It can be seen that the experimental energies are well reproduced. This provides the first fully microscopic and self-consistent description for the chiral doublet bands in the framework of the density functional theory.

3.8 SUMMARY AND PERSPECTIVE

Since the concept of the nuclear chirality was proposed in 1997, the study of the nuclear chirality has been and will remain as one of the hot topics in nuclear physics. Over the past two and a half decades, many efforts have been devoted to understand chiral symmetry and its spontaneous breaking in atomic nuclei.

Both experimental and theoretical efforts are still needed in the study of the nuclear chirality. From the experimental side, the simultaneous breaking of chirality and other symmetries would be interesting, for example, the search of ChP quartet bands. Finding more observables to provide further confirmation for chirality is also important. In 2018, the g factor, which can give important information on the chiral geometry, has been measured for the bandhead state of the chiral doublet bands in ^{128}Cs [88]. The measurements of the g factors for excited states and for more nuclei are looking forward. Exploring new phenomenon is also necessary, such as the chiral wobbler in ^{74}Br, with the second and third lowest bands, respectively, suggested as the chiral partner and one-phonon wobbling excitation built on the yrast band [89]. From the theoretical side, the chirality in triaxial well-deformed nuclei has been extensively studied. In the future, it will be very interesting to further study the chirality in soft triaxial nuclei.

3.9 ACKNOWLEDGMENTS

We thank X. K. Du, T. X. Huang, F. Y. He, X. F. Jiang, T. Qu, F. F. Xu, Y. L. Yang and S. Q. Zhang for the careful reading and valuable suggestions. This work was partly supported by the National Natural Science Foundation of China (Grants No. 11935003, No. 12105004, No. 12141501), the High-performance Computing Platform of Peking University, and the State Key Laboratory of Nuclear Physics and Technology, Peking University.

Bibliography

1. S. Frauendorf and J. Meng, Nucl. Phys. A **617**, 131 (1997).
2. P. J. Twin *et al.*, Phys. Rev. Lett. **57**, 811 (1986).
3. H. Hübel and Prog. Part. Nucl. Phys. **54**, 1 (2005).
4. S. Frauendorf, Nucl. Phys. A **557**, 259 (1993).
5. S. Frauendorf, J. Meng, and J. Reif, Proceedings of the conference on physics from large γ-ray detector arrays **52** (1994).
6. M. Neffgen *et al.*, Nucl. Phys. A **595**, 499 (1995).
7. S. Frauendorf and J. Meng, Z. Phys. A: Hadrons Nucl. **356**, 263 (1996).
8. R. M. Clark *et al.*, Phys. Rev. Lett. **78**, 1868 (1997).
9. C. M. Petrache *et al.*, Nucl. Phys. A **597**, 106 (1996).
10. A. N. Bohr and B. R. Mottelson, *Nuclear Structure* (Benjamin, New York, 1975), Vol. II.
11. K. Starosta *et al.*, Phys. Rev. Lett. **86**, 971 (2001).
12. S. Zhu *et al.*, Phys. Rev. Lett. **91**, 132501 (2003).
13. T. Koike *et al.*, Phys. Rev. C **63**, 061304 (2001).
14. A. A. Hecht *et al.*, Phys. Rev. C **63**, 051302 (2001).
15. D. J. Hartley *et al.*, Phys. Rev. C **64**, 031304 (2001).
16. R. A. Bark *et al.*, Nucl. Phys. A **691**, 577 (2001).
17. E. Mergel *et al.*, Eur. Phys. J. A **15**, 417 (2002).
18. K. Starosta *et al.*, Phys. Rev. C **65**, 044328 (2002).
19. X. F. Li *et al.*, Chinese Phys. Lett. **19**, 1779 (2002).
20. T. Koike *et al.*, Phys. Rev. C **67**, 044319 (2003).
21. G. Rainovski *et al.*, Phys. Rev. C **68**, 024318 (2003).

22. A. J. Simons *et al.*, J. Phys. G: Nucl. Partic. **31**, 541 (2005).
23. S. Y. Wang *et al.*, Phys. Rev. C **74**, 017302 (2006).
24. C. Vaman *et al.*, Phys. Rev. Lett. **92**, 032501 (2004).
25. J. A. Alcántara-Núñez *et al.*, Phys. Rev. C **69**, 024317 (2004).
26. J. Timár *et al.*, Phys. Lett. B **598**, 178 (2004).
27. P. Joshi *et al.*, Eur. Phys. J. A **24**, 23 (2005).
28. S. J. Zhu *et al.*, Eur. Phys. J. A **25**, 459 (2005).
29. J. Timár *et al.*, Phys. Rev. C **73**, 011301 (2006).
30. C. Y. He *et al.*, High Ener. Phys. And Nucl. Phys. **30**, 966 (2006).
31. P. Joshi *et al.*, Phys. Rev. Lett. **98**, 102501 (2007).
32. J. Timár *et al.*, Phys. Rev. C **76**, 024307 (2007).
33. D. L. Balabanski *et al.*, Phys. Rev. C **70**, 044305 (2004).
34. E. A. Lawrie *et al.*, Phys. Rev. C **78**, 021305(R) (2008).
35. S.Y. Wang *et al.*, Phys. Lett. B **703**, 40 (2011).
36. V. I. Dimitrov, S. Frauendorf and F. Dönau, Phys. Rev. Lett. **84**, 5732 (2000).
37. T. Koike, K. Starosta and I. Hamamoto, Phys. Rev. Lett. **93**, 172502 (2004).
38. P. Olbratowski *et al.*, Phys. Rev. Lett. **93**, 052501 (2004).
39. D. Tonev *et al.*, Phys. Rev. Lett. **96**, 052501 (2006).
40. C. M. Petrache *et al.*, Phys. Rev. Lett. **96**, 112502 (2006).
41. E. Grodner *et al.*, Phys. Rev. Lett. **97**, 172501 (2006).
42. S. Mukhopadhyay *et al.*, Phys. Rev. Lett. **99**, 172501 (2007).
43. P. Ring and Prog. Part. Nucl. Phys. **37**, 193 (1996).
44. D. Vretenar *et al.*, Phys. Rep. **409**, 101 (2005).
45. T. Nikšić, D. Vretenar, P. Ring, Prog. Part. Nucl. Phys. **66**, 519 (2011).
46. J. Meng (Ed), *International Review of Nuclear Physics* (World Scientific, Singapore, 2016), Vol. 10.
47. J. Meng *et al.*, Phys. Rev. C **73**, 037303 (2006).
48. A. D. Ayangeakaa *et al.*, Phys. Rev. Lett. **110**, 172504 (2013).
49. E. O. Lieder *et al.*, Phys. Rev. Lett. **112**, 202502 (2014).
50. I. Kuti *et al.*, Phys. Rev. Lett. **113**, 032501 (2014).
51. C. Liu *et al.*, Phys. Rev. Lett. **116**, 112501 (2016).
52. S. Frauendorf, Rev. Mod. Phys. **73**, 463 (2001).
53. B. W. Xiong and Y. Y. Wang, Atom. Data Nucl. Data Tabl. **125**, 193 (2019).
54. J. Peng, J. Meng and S. Q. Zhang, Phys. Rev. C **68**, 044324 (2003).
55. S. Q. Zhang *et al.*, Phys. Rev. C **75**, 044307 (2007).
56. B. Qi *et al.*, Phys. Lett. B **675**, 175 (2009).
57. Q. B. Chen *et al.*, Phys. Lett. B **782**, 744 (2018).
58. Y. Y. Wang *et al.*, Phys. Lett. B **792**, 454 (2019).
59. Y. P. Wang, Y. Y. Wang and J. Meng, Phys. Rev. C **102**, 024313 (2020).
60. Y. Y. Wang *et al.*, Sci. Bull. **65**, 2001 (2020).
61. H. Madokoro *et al.*, Phys. Rev. C **62**, 061301(R) (2000).
62. P. W. Zhao, Phys. Lett. B **773**, 1 (2017).
63. J. Peng and Q. B. Chen, Phys. Lett. B **810**, 135795 (2020).
64. Y. P. Wang and J. Meng, Phys. Lett. B **841**, 137923 (2023).
65. Z. X. Ren, P. W. Zhao and J. Meng, Phys. Rev. C **105**, L011301 (2022).
66. D. Almehed, F. Dönau, S. Frauendorf, Phys. Rev. C **83**, 054308 (2011).
67. Q. B. Chen *et al.*, Phys. Rev. C **87**, 024314 (2013).
68. Q. B. Chen *et al.*, Phys. Rev. C **94**, 044301 (2016).
69. Q. B. Chen and J. Meng, Phys. Rev. C **98**, 031303(R) (2018).
70. S. Brant *et al.*, Phys. Rev. C **78**, 034301 (2008).

71. A. A. Raduta, Al H. Raduta and C. M. Petrache, J. Phys. G: Nucl. Part. Phys. **43**, 095107 (2016).
72. K. Hara and Y. Sun, Int. J. Mod. Phys. E **4**, 637 (1995).
73. G. H. Bhat *et al.*, Nucl. Phys. A **922**, 150 (2014).
74. F. Q. Chen *et al.*, Phys. Rev. C **96**, 051303(R) (2017).
75. F. Q. Chen, J. Meng and S. Q. Zhang, Phys. Lett. B **785**, 211 (2018).
76. Y. K. Wang *et al.*, Phys. Rev. C **99**, 054303 (2019).
77. X. H. Wu *et al.*, Phys. Rev. C **98**, 064302 (2018).
78. G. H. Bhat, J. A. Sheikh and R. Palit, Phys. Lett. B **707**, 250 (2012).
79. P. Olbratowski, J. Dobaczewski and J. Dudek, Phys. Rev. C **73**, 054308 (2006).
80. K. Hagino and Y. Tanimura, Phys. Rev. C **82**, 057301 (2010).
81. Y. Tanimura, K. Hagino, H. Z. Liang, Prog. Theor. Exp. Phys. **2015**, 073D01 (2015).
82. Z. X. Ren, S. Q. Zhang and J. Meng, Phys. Rev. C **95**, 024313 (2017).
83. Z. X. Ren *et al.*, Sci, China Phys. Mech. **62**, 112062 (2019).
84. Z. X. Ren *et al.*, Nucl. Phys. A **996**, 121696 (2020).
85. B. Li, Z. X. Ren and P. W. Zhao, Phys. Rev. C **102**, 044307 (2020).
86. F. F. Xu *et al.*, Phys. Rev. C **109**, 014311 (2024).
87. B. F. Lv *et al.*, Phys. Rev. C **100**, 024314 (2019).
88. E. Grodner *et al.*, Phys. Rev. Lett. **120**, 022502 (2018).
89. R. J. Guo *et al.*, Phys. Rev. Lett. **132**, 092501 (2024).

4 Algebraic description of wobbling motion in nuclei

Kosai Tanabe

Saitama University, Sakura-Ku, Saitama, Japan

Kazuko Sugawara-Tanabe

Otsuma Women's University, Chiyoda-Ku, Tokyo, Japan

4.1 INTRODUCTION

The purpose of this chapter is to discuss the algebraic method applied to the particle-rotor model, i.e. the top(s)-on-top model [1–5]. This model allows a practical use of the algebraic treatment based on the next-to-leading order expansion in the Holstein-Primakoff (HP) boson representation for both the total angular momentum $\vec{I}$ and the single-particle angular momentum $\vec{j}$, which keeps D_2 symmetry [1].

The wobbling mode was introduced in Ref. [6], where only the lowest order in the expansion of HP boson expansion for even nuclei was kept, and then the resulting excitation energy is like a phonon in the vibrational motion. The term "precession" is also used in Ref. [6], and originally in classical mechanics [7, 8]. According to the experimental papers [9–14], we use the terminology "wobbling" instead of "precession".

We comment on the fact that the moments of inertia (MoI) around the symmetry axis is not required to be zero. O. Klein [15] derived Euler's equations from the Heisenberg equation for $H_{\rm rot} = \sum_{i=x,y,z} I_i^2/2\mathcal{J}_i$ and the commutation relations for body-fixed components:

$$[I_x, I_y] = -iI_z, \quad (\text{for } x, y, z \text{ cyclic}). \tag{4.1}$$

Thouless and Valatin [16] derived Euler's equations based on Dirac's formalism with density matrix ρ [17]:

$$\mathcal{J}_x\frac{d\omega_x}{dt} = (\mathcal{J}_y - \mathcal{J}_z)\omega_y\omega_z, \quad \mathcal{J}_y\frac{d\omega_y}{dt} = (\mathcal{J}_z - \mathcal{J}_x)\omega_z\omega_x, \quad \mathcal{J}_z\frac{d\omega_z}{dt} = (\mathcal{J}_x - \mathcal{J}_y)\omega_x\omega_y, \tag{4.2}$$

where the expectation value of the principal axis component I_i is related to $\mathcal{J}_i$ and angular frequency ω_i by

$$\langle I_i \rangle = \mathcal{J}_i\omega_i, \quad (i = x, y, z). \tag{4.3}$$

Eq. (4.2) clearly shows that in the case of an axial symmetry about z-axis ($\mathcal{J}_x = \mathcal{J}_y \neq \mathcal{J}_z$), the third equation requires $\mathcal{J}_z d\omega_z/dt = 0$. If we consider two cases, i.e. one with $d\omega_z/dt = 0$ and $\mathcal{J}_z \neq 0$, and the other with $\mathcal{J}_z = 0$, we get the solution of $\omega_z = 0$ from the first and second equations in Eq. (4.2) for both cases. This shows that there does not exist any rotation around the symmetry axis irrelevant to the value of $\mathcal{J}_z$. Ref. [16] explained this situation by "a rotation of the coordinate system about the symmetry axis with constant ω_z does not lead to any new time-dependent solution ρ".

In Sec. 4.2, we review the HP boson expansion method in odd-A nuclei [1], where we introduce two quantum numbers n_α and n_β in the framework of the next-to-leading order in $1/(2I)$ and $1/(2j)$ together with the leading one. In order to clarify the meaning of the quantum numbers, we discuss the special case without the single-particle potential to compare the rigid (rig) MoI and hydrodynamical (hyd) MoI assumptions. Then, we discuss the general case with the single-particle Hamiltonian H_{sp}.

In Sec. 4.3, we investigate the electromagnetic transition rates for both rig MoI and hyd MoI. We discuss the results obtained by the exact diagonalization of the total Hamiltonian in comparison with the top-on-top model [1, 2]. The overlaps between the original HP bosons and the quasibosons obtained in diagonalizing the next-to-leading order Hamiltonian are very useful to explain the ratios of electromagnetic transition rates.

Because we are discussing the bands resulting from the rotational motion, the important pairing effect on MoI must be considered. In Sec. 4.4, the Coriolis-anti-pairing (CAP) effect [18] on the MoI is discussed using the formula obtained in the symmetric rotor [19].

In Sec. 4.5, this chapter is summarized and concluded.

4.2   PARTICLE-ROTOR MODEL AND TOP-ON-TOP MODEL

The Lund convention is used throughout this chapter. We start from the particle-rotor Hamiltonian given by

$$H = H_{\mathrm{rot}} + H_{\mathrm{sp}}, \quad H_{\mathrm{rot}} = \sum_{k=x,y,z} A_k (I_k - j_k)^2,$$

$$H_{\mathrm{sp}} = \frac{V}{j(j+1)} \left[\cos\gamma (3j_z^2 - \vec{j}^2) - \sqrt{3}\sin\gamma (j_x^2 - j_y^2) \right], \tag{4.4}$$

where $A_k = 1/(2\mathcal{J}_k)$, and (β_2, γ) is the deformation parameter set describing the ellipsoidal shape of the rotor. Here we remark that V is related to the γ-deformed Nilsson potential [20] through the Wigner–Eckart theorem to the 5-dimensional symmetric tensor $Y_{2\mu}$, $\mu = 0, \pm 1, \pm 2$ for a unique-parity single-j level,

$$V = \frac{1}{8}\sqrt{\frac{5}{\pi}}\hbar\omega_0\beta_2 \langle r^2 \rangle_{jm}, \tag{4.5}$$

where $\hbar\omega_0$ and $\langle r^2 \rangle_{jm}$ are defined by Eqs. (4) and (5) in Ref. [5].

We study two models for $\mathcal{J}_k$, the hydrodynamical model,

$$\mathcal{J}_k^{\mathrm{hyd}} = \frac{4}{3}\mathcal{J}_0 \sin^2\left(\gamma + \frac{2}{3}\pi k\right), \tag{4.6a}$$

and the rigid-body model,

$$\mathcal{J}_k^{\mathrm{rig}} = \frac{\mathcal{J}_0}{1 + (\frac{5}{16\pi})^{1/2}\beta_2} \left[1 - \left(\frac{5}{4\pi}\right)^{1/2}\beta_2 \cos\left(\gamma + \frac{2}{3}\pi k\right) \right]. \tag{4.6b}$$

In Eq. (4.6a), $\mathcal{J}_0$ has β_2 dependence, but abbreviated, because we are interested only in the γ-dependence of MoI. Note that the relation $\mathcal{J}_x^{\mathrm{rig}} \geq \mathcal{J}_y^{\mathrm{rig}} \geq \mathcal{J}_z^{\mathrm{rig}}$ holds for $0 \leq \gamma \leq \pi/3$, while $\mathcal{J}_y^{\mathrm{hyd}} \geq \mathcal{J}_x^{\mathrm{hyd}} \geq \mathcal{J}_z^{\mathrm{hyd}}$ holds for $0 \leq \gamma \leq \pi/6$. If we rewrite H_{sp} in Eq. (4.4) as $\sum_{k=x,y,z} B_k j_k^2$, then $1/B_x \geq 1/B_y \geq 1/B_z$ holds for $0 \leq \gamma \leq \pi/3$. In summary, A_k^{rig} and B_k have the same periodicity as a function of γ. However, A_k^{hyd} changes in the opposite direction with a different periodicity from B_k as functions of γ.

The Hamiltonian in Eq. (4.4) is invariant under a rotation through an angle π about each of three principal axes, $\exp(-i\pi R_k)$ with $R_k = I_k - j_k$ ($k = 1, 2, 3$ for x, y, z). These symmetry operations compose the D_2-symmetry group with four representations labeled by (r_1, r_2, r_3), where r_k stands for an eigenvalue of $\exp(-i\pi R_k)$, and takes the value $+1$ or -1. Bohr symmetry [21] requires that only the states belonging to the $(r_1, r_2, r_3) = (+1, +1, +1)$ representation are allowed as nuclear states, because no 1^+ rotational state is observed. We will refer $(r_1, r_2, r_3) = (+1, +1, +1)$ to A-symmetry, $(-1, -1, +1)$ to B_1-symmetry, $(-1, +1, -1)$ to B_2-symmetry and $(+1, -1, -1)$ to B_3-symmetry. A complete set of the A-symmetry basis states $(+1, +1, +1)$ is provided as

$$\left\{ \sqrt{\frac{2I+1}{16\pi^2}} \left[\mathcal{D}_{MK}^I(\theta_i)\phi_\Omega^j + (-1)^{I-j}\mathcal{D}_{M-K}^I(\theta_i)\phi_{-\Omega}^j \right]; \quad |K-\Omega| = \text{even}, \quad K, \Omega > 0 \right\}, \tag{4.7}$$

where ϕ_Ω^j stands for the single-particle state, and $\mathcal{D}_{MK}^I(\theta_i)$ for the Wigner $\mathcal{D}$-functions. We introduce two quantum numbers $n_{\alpha'}$ and $n_{\beta'}$ to relate to the HP bosons. Because the magnitude R of the rotor angular momentum $\vec{R} = \vec{I} + (-\vec{j})$ is given by $R = I - j, I - j + 1, \cdots, I + j - 1, I + j$, or an integer $n_{\beta'}$ defined by $R = I - j + n_{\beta'}$ takes the values

$$n_{\beta'} = 0, 1, 2, \cdots, 2j. \tag{4.8}$$

As R_z runs from R to 0 in this case, $R_z = I_z - j_z = K - \Omega = $ even, an integer $n_{\alpha'}$ defined by the relation $R_z = R - n_{\alpha'} \geq 0$ takes the values

$$\begin{aligned} n_{\alpha'} &= 0, 2, 4, \cdots, R, \quad \text{for} \quad R = \text{even}, \\ n_{\alpha'} &= 1, 3, 5, \cdots, R-2, \quad \text{for} \quad R = \text{odd}. \end{aligned} \tag{4.9}$$

Physical states are realized for a set of non-negative integers $n_{\alpha'}$ and $n_{\beta'}$, which are related to the magnitude of the rotor angular momentum R and its z-component R_z, and through the relation $R_z = I - n_{\alpha'} - (j - n_{\beta'})$, and the dimension of this basis is $(2I+1)(2j+1)/4$. Similarly, B_1-symmetry basis is given by

$$\left\{ \sqrt{\frac{2I+1}{16\pi^2}} \left[\mathcal{D}_{MK}^I(\theta_i)\phi_\Omega^j - (-1)^{I-j}\mathcal{D}_{M-K}^I(\theta_i)\phi_{-\Omega}^j \right]; \quad |K-\Omega| = \text{even}, \quad K, \Omega > 0 \right\}. \tag{4.10}$$

With the same definition of R and $n_{\beta'}$ in Eq. (4.8), $n_{\alpha'}(= R - R_z)$ takes the values

$$\begin{aligned} n_{\alpha'} &= R+2, R+4, \cdots, 2R, \quad \text{for} \quad R = \text{even}, \\ n_{\alpha'} &= R, R+2, \cdots, 2R-1, \quad \text{for} \quad R = \text{odd}. \end{aligned} \tag{4.11}$$

B_2-symmetry basis is given by

$$\left\{ \sqrt{\frac{2I+1}{16\pi^2}} \left[\mathcal{D}_{MK}^I(\theta_i)\phi_\Omega^j + (-1)^{I-j}\mathcal{D}_{M-K}^I(\theta_i)\phi_{-\Omega}^j \right]; \quad |K-\Omega| = \text{odd}, \quad K, \Omega > 0 \right\}. \tag{4.12}$$

With the same definition of R and $n_{\beta'}$ in Eq. (4.8), $n_{\alpha'}(= R - R_z)$ takes the values

$$\begin{aligned} n_{\alpha'} &= R+1, R+3, \cdots, 2R-1, \quad \text{for} \quad R = \text{even}, \\ n_{\alpha'} &= 0, 2, \cdots, R-1, \quad \text{for} \quad R = \text{odd}. \end{aligned} \tag{4.13}$$

B_3-symmetry basis is given by

$$\left\{ \sqrt{\frac{2I+1}{16\pi^2}} \left[\mathcal{D}_{MK}^I(\theta_i)\phi_\Omega^j - (-1)^{I-j}\mathcal{D}_{M-K}^I(\theta_i)\phi_{-\Omega}^j \right]; \quad |K-\Omega| = \text{odd}, \quad \Omega > 0 \right\}. \tag{4.14}$$

With the same definition of R and $n_{\beta'}$ in Eq. (4.8), $n_{\alpha'}(= R - R_z)$ takes the values

$$\begin{aligned} n_{\alpha'} &= 1, 3, \cdots, R-1, \quad \text{for} \quad R = \text{even}, \\ n_{\alpha'} &= R+1, R+3, \cdots, 2R, \quad \text{for} \quad R = \text{odd}. \end{aligned} \tag{4.15}$$

Each symmetry base has a dimension of $(2j+1)(2I+1)/4$. From now on we discuss the case of $I \geq j$, under A-symmetry.

Next, we explain the top-on-top model [1–5] briefly. We choose the quantization axis for the components I_x and j_x in the HP boson representation for rig MoI case, because $\mathcal{J}_x^{\text{rig}}$ and $1/B_x$ are maxima in the range $0 \leq \gamma \leq 2\pi/3$:

$$I_+ = I_-^\dagger = I_y + iI_z = -\hat{a}^\dagger\sqrt{2I-\hat{n}_a}, \quad I_x = I - \hat{n}_a, \quad \text{with} \quad \hat{n}_a = \hat{a}^\dagger\hat{a},$$
$$j_+ = j_-^\dagger = j_y + ij_z = \sqrt{2j-\hat{n}_b}\,\hat{b}, \quad j_x = j - \hat{n}_b, \quad \text{with} \quad \hat{n}_b = \hat{b}^\dagger\hat{b}. \tag{4.16}$$

These I_x, I_y and I_z satisfy the commutation relations with minus sign in Eq. (4.1), and j_x, j_y and j_z the commutation relations with plus sign [23, 24]. With Eq. (4.16), we rewrite the Hamiltonian (4.4) in terms of two kinds of boson operators, $\hat{a}$ and $\hat{b}$, by expanding the square roots $\sqrt{2I-\hat{n}_a}$ and $\sqrt{2j-\hat{n}_b}$ up to the order of $\hat{n}_a/(2I)$ and $\hat{n}_b/(2j)$ as small quantities. We call this order of approximation next-to-leading order approximation, in contrast to the leading order approximation. The latter includes only the lowest order contributions from the expansion, and the Hamiltonian includes contributions up to the terms bilinear in the boson operators. Subsequently, it cannot give the exact energy at the symmetric limit, and violates the D_2-invariance [1]. The higher order expansion is quite important to recover D_2-invariance (Eqs.(17) to (20) in Ref. [1]). We arrive at an approximate Hamiltonian written in terms of two kinds of HP bosons $H_{\text{B}} \cong H_0 + H_2 + H_4$, where H_0 denotes a constant which collects all the terms independent of boson operators, H_2 the bilinear forms of boson operators and H_4 the fourth order terms. Their explicit forms are given in Eqs. (22) to (26) in Ref. [1]. The diagonalization of H_2 is attained by the boson Bogoliubov transformation connecting the boson operators $(\hat{a}, \hat{b}, \hat{a}^\dagger, \hat{b}^\dagger)$ to new boson operators $(\alpha, \beta, \alpha^\dagger, \beta^\dagger)$ given by Eq. (25) in Ref. [5]. Their explicit formulae and the stability condition are discussed in Ref. [5] by introducing the boson numbers n_α for $\hat{n}_\alpha = \alpha^\dagger\alpha$ and n_β for $\hat{n}_\beta = \beta^\dagger\beta$ in the new boson picture. Then, H_2 is diagonalized as

$$H_2 \simeq 2\omega_\alpha(\hat{n}_\alpha + 1/2) + 2\omega_\beta(\hat{n}_\beta + 1/2), \tag{4.17}$$

where ω_α or ω_β are determined from the metric of H_2. To take account of higher order terms, we apply the boson transformation to H_4, and retain only diagonal terms which are expressed in terms of $\hat{n}_\alpha$ and $\hat{n}_\beta$. Consequently, we arrive at an approximate formula for H_{B} (see Appendix B in Ref. [1]).

To clarify the physical meaning of two quantum numbers n_α and n_β, we consider the pure rotor case, i.e., $V = 0$ in Eq. (4.4). Then, H_{B} reduces to a simple expression of the rotational energy:

$$E_{\text{rot}} \cong A_x^{\text{rig}}R(R+1) - \frac{p+q}{2}n_\alpha^2 + \left(2R\sqrt{pq} + \sqrt{pq} - \frac{p+q}{2}\right)\left(n_\alpha + \frac{1}{2}\right), \tag{4.18}$$

where

$$R = I - j + n_\beta, \quad R_x = R - n_\alpha, \quad p = A_y^{\text{rig}} - A_x^{\text{rig}}, \quad q = A_z^{\text{rig}} - A_x^{\text{rig}}. \tag{4.19}$$

The eigenvalue R can be regarded as an effective magnitude of the rotor angular momentum, and $R - n_\alpha$ as its x-component R_x. It turns out that these n_α and n_β are the same integers $n_{\alpha'}$ and $n_{\beta'}$ as defined in Eqs. (4.9) and (4.8). In the symmetric limit of $A_y^{\text{rig}} = A_x^{\text{rig}}$ ($\gamma = \pi/3$), Eq. (4.18) becomes the well-known expression, $A_z^{\text{rig}}R(R+1) - (A_z^{\text{rig}} - A_x^{\text{rig}})(R - n_\alpha)^2$. If we stay in the leading order approximation in Eq. (4.17), the coefficients of $\hat{n}_\alpha$ and $\hat{n}_\beta$ in Eq. (4.17) reduce to

$$\omega_\alpha \sim (I-j)\sqrt{pq}, \qquad \omega_\beta \sim (I-j)A_x^{\text{rig}}. \tag{4.20}$$

The wobbling energy ω_α becomes the same as Bohr and Mottelson's formula [6] with $I - j$ instead of I. This allows us to interpret the quantum number n_α as the wobbling quantum number of $\vec{R}$ with $R_x = R - n_\alpha$. The other quantum number n_β is interpreted as the precession of $\vec{j}$. For the case of rig MoI, $pq - (A_x^{\text{rig}})^2$ is always negative in the region of $0° \leq \gamma \leq \pi/3$. Thus, $\omega_\alpha (= (I - j)\sqrt{pq})$ is always smaller than $\omega_\beta (= (I - j)A_x^{\text{rig}})$, indicating that the lowest mode is the wobbling motion around the axis with the maximum MoI. As shown in Fig. 1 in Ref. [5], the yrast levels with $I - j=$ even are always $(n_\alpha, n_\beta)=(0,0)$ $(R_x = R)$, while the yrast levels with $I - j=$ odd are $(1,0)$ $(R_x = R - 1)$ except for $I = 13/2$, where $(0,1)$ $(R_x = I - j + 1)$ is the lowest and $(2,1)$ $(R_x = 0)$ becomes the second lowest. Because of A-symmetry, $R = 1$ is not allowed, and no $(1,0)$ exists for $I = 13/2$, instead $(0,1)$ and $(2,1)$ belonging to $R = 2$ appears. This is also found in the experimental data of Refs. [25, 26], where no 13/2 level has been observed in the wobbling band or Band 4.

For the system with hyd MoI and $V = 0$, where $\mathcal{J}_y^{\text{hyd}} > \mathcal{J}_x^{\text{hyd}} > \mathcal{J}_z^{\text{hyd}}$ in the region of $0° \leq \gamma \leq \pi/6$, another boson representation is preferable:

$$
\begin{aligned}
I_+ &= I_-^\dagger = I_z + iI_x = -\hat{a}^\dagger \sqrt{2I - \hat{n}_a}, \quad I_y = I - \hat{n}_a \quad \text{with} \quad \hat{n}_a = \hat{a}^\dagger \hat{a}; \\
j_+ &= j_-^\dagger = j_z + ij_x = \sqrt{2j - \hat{n}_b}\,\hat{b}, \quad j_y = j - \hat{n}_b \quad \text{with} \quad \hat{n}_b = \hat{b}^\dagger \hat{b}.
\end{aligned}
\tag{4.21}
$$

In the same approximation as Eq. (4.18), we get

$$
E_{\text{rot}} \cong A_y^{\text{hyd}} R(R+1) - \frac{p' + q'}{2} n_\alpha^2 + \left(2R\sqrt{p'q'} + \sqrt{p'q'} - \frac{p' + q'}{2} \right)\left(n_\alpha + \frac{1}{2} \right),
\tag{4.22}
$$

where

$$
R = I - j + n_\beta, \quad R_y = R - n_\alpha, \quad p' = A_z^{\text{hyd}} - A_y^{\text{hyd}}, \quad q' = A_x^{\text{hyd}} - A_y^{\text{hyd}}.
\tag{4.23}
$$

In the leading order approximation, Eq. (4.22) reduces to $\omega_\alpha \sim (I - j)\sqrt{p'q'}$, which is nothing but the wobbling mode around the axis with the maximum MoI $\mathcal{J}_y^{\text{hyd}}$, and to $\omega_\beta \sim (I - j)A_y^{\text{hyd}}$, the precession of j around the same axis. In the region of $0° < \gamma \leq 30°$, $p'q' - (A_y^{\text{hyd}})^2$ is always positive, so that $\omega_\beta = (I - j)A_y^{\text{hyd}}$ is smaller than $\omega_\alpha = (I - j)\sqrt{p'q'}$. Subsequently, the lowest mode is the precession of $\vec{j}$ around the axis with maximum MoI $\mathcal{J}_y^{\text{hyd}}$. In Fig. 2 in Ref. [5], the yrast levels with $I - j=$ even are always $(n_\alpha, n_\beta) = (0,0)$ $(R_y = R)$, the same as the rig MoI case. However, the yrast levels with $I - j=$ odd are always $(0,1)$ $(R_y = I - j + 1)$, which is the precession mode of j around the axis with the maximum MoI $\mathcal{J}_y^{\text{hyd}}$. The levels corresponding to the wobbling mode around the axis with the maximum MoI $\mathcal{J}_y^{\text{hyd}}$, i.e., the levels with $(1,0)$ $(R_y = R - 1)$ become the second lowest (yrare) levels, represented by red solid lines in $I - j=$ odd in Fig. 2 in Ref. [5]. In the hyd MoI, the lowest level is the same $(0,1)$ for $I = 13/2$ as the other $I - j =$ odd level in contradiction to the experimental data [25].

When γ is small but not 0 $(A_x \sim A_y)$, another representation is favorable where I_z and j_z are in the diagonal forms [5, 27]:

$$
\begin{aligned}
I_+ &= I_-^\dagger = I_x + iI_y = -\hat{a}^\dagger \sqrt{2I - \hat{n}_a}, \quad I_z = I - \hat{n}_a \quad \text{with} \quad \hat{n}_a = \hat{a}^\dagger \hat{a}; \\
j_+ &= j_-^\dagger = j_x + ij_y = \sqrt{2j - \hat{n}_b}\,\hat{b}, \quad j_z = j - \hat{n}_b \quad \text{with} \quad \hat{n}_b = \hat{b}^\dagger \hat{b}.
\end{aligned}
\tag{4.24}
$$

In this case with $V = 0$, we get

$$
E_{\text{rot}} \cong A_z R(R+1) - \frac{p'' + q''}{2} n_\alpha^2 - \left(2R\sqrt{p''q''} + \sqrt{p''q''} - \frac{p'' + q''}{2} \right)\left(n_\alpha + \frac{1}{2} \right),
\tag{4.25}
$$

where

$$
R = I - j + n_\beta, \quad R_z = R - n_\alpha, \quad p'' = A_z - A_x, \quad q'' = A_z - A_y.
\tag{4.26}
$$

This formula is available both for $\mathcal{J}^{\text{rig}}$ and $\mathcal{J}^{\text{hyd}}$, as seen in Fig. 3 and Fig. 4 in Ref. [5]. As seen in Fig. 4 in Ref. [5], the lowest levels for $I - j =$ even have $R_z = 0$, but those for $I - j =$ odd have $R_z = 2$ or $n_\alpha = R - 2$ except for $I = 13/2$ for $\mathcal{J}^{\text{rig}}$. On the other hand, the lowest levels are always $R_z = 0$ $(n_\alpha = R)$ for $\mathcal{J}^{\text{hyd}}$, as seen in Fig. 3 in Ref. [5].

According to Ref. [22], there is a misprint in Eq. (4-282) of Ref. [6], and the following sentences after Eq. (4-282) are misleading. In Eq. (4-282) of Ref. [6], the relation of MoI size must be interchanged for the Figure 4-33 and the definition of κ in Eq. (4.281), i.e. $A_1 \geq A_2 \geq A_3$ ($\mathcal{J}_1 \leq \mathcal{J}_2 \leq \mathcal{J}_3$). Ref. [22] solves the rotor Hamiltonian quantum mechanically without assuming the special form for non-zero MoI's. We choose $A_1 = A_z^{\mathrm{rig}}$ and $A_3 = A_x^{\mathrm{rig}}$ for $0° \leq \gamma \leq 60°$. Then, $\gamma = 0°$ corresponds to $\kappa = -1.0$, $\gamma = 60°$ to $\kappa = 1.0$ and $\gamma = 30°$ to $\kappa = 0.0$ neglecting β_2^2. The γ-dependence of A^{rig} is comparable with Figure 4-33 for $0° \leq \gamma \leq 60°$. For example, $\gamma = 18°$ corresponds to $\kappa \sim -0.448$, and $\gamma = 26°$ to $\kappa \sim -0.218$. On the other hand, the hyd MoI cannot be applicable to Figure 4-33 as a function of κ, when we choose $A_1 = A_z^{\mathrm{hyd}}$ and $A_3 = A_y^{\mathrm{hyd}}$ for $0° < \gamma \leq 30°$ due to vanishing $\mathcal{J}_z^{\mathrm{hyd}}$. For example, $\gamma = 30°$ corresponds to $\kappa = 1.0$, but $\gamma = 26°$ to $\kappa \sim 0.0446$ and $\gamma = 18°$ to $\kappa \sim -0.748$. Any virtual rotation around the symmetry axis does not require $\mathcal{J}_z^{\mathrm{hyd}} = 0$, though such a possibility is not prohibited a priori. On the other hand, $\mathcal{J}_z \neq 0$ is essential to explain the energy sequence of the $K = 2$ band, and the rotational band of odd nucleus when the other two moments of inertia are equal.

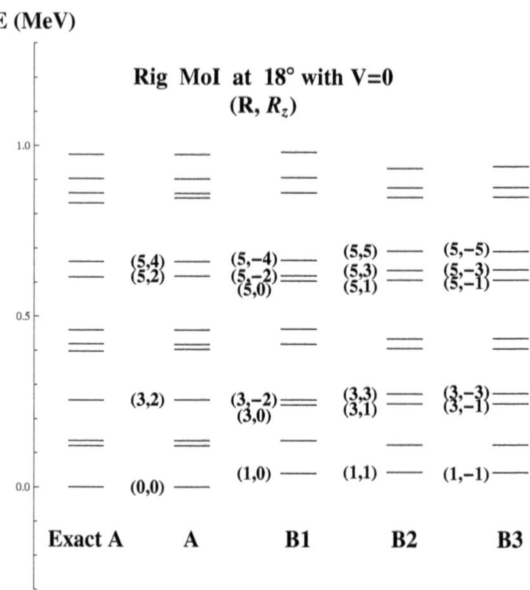

Figure 4.1 The comparison of energy levels given by H_{rot} in Eq. (4.4) between four symmetries for $R \leq 6$. "Exact A" are obtained by the exact diagonalization of H_{rot} under A-symmetry. The four symmetries "A", "B1", "B2" and "B3" are derived from top-on-top model, and explained in the text. The numerals beside each line represent (R, R_z) only for $R =$ odd. Lines between $R = 1$ and $R = 3$ are levels belonging to $R = 2$ family, those between $R = 3$ and $R = 5$ to $R = 4$ family, and those over $R = 5$ belong to $R = 6$ family. The parameter set is ($V = 0$ at $\gamma = 18°$ with $\mathcal{J}_0 = 25$ MeV^{-1}, $j = 11/2$ and $\beta_2 = 0.18$).

In Fig. 4.1, we show solutions derived from top-on-top model with $\mathcal{J}^{\mathrm{rig}}$ for 4 symmetries as "A", "B1", "B2" and "B3", together with the exact solution of A-symmetry. "Exact A" denotes the results obtained by the diagonalization with the wave function of Eq. (4.7). "A" shows results of Eq. (4.25) where n_α takes the same values of $n_{\alpha'}$ in Eq. (4.9), which agrees with "Exact A". "B1" shows results of Eq. (4.25) with n_α in Eq. (4.11), "B2" those of Eq. (4.25) with n_α in Eq. (4.13) and "B3" those of Eq. (4.25) with n_α in Eq. (4.15). The lowest energy for $I - j =$ odd, A-symmetry gives $(R, R_z) = (3,2)$ for $I = 17/2$, (5,2) for $I = 19/2$ and (7,2) for $I = 21/2$, as shown in Fig. 4 in Ref. [5]. Quite close to these levels, B_1-symmetry gives (3,-2), (5,-2) and (7,-2), respectively, for the case of $\gamma = 18°$ which is used in Ref [5]. These levels may belong to Band 2 in Ref. [26]. These characters of A-

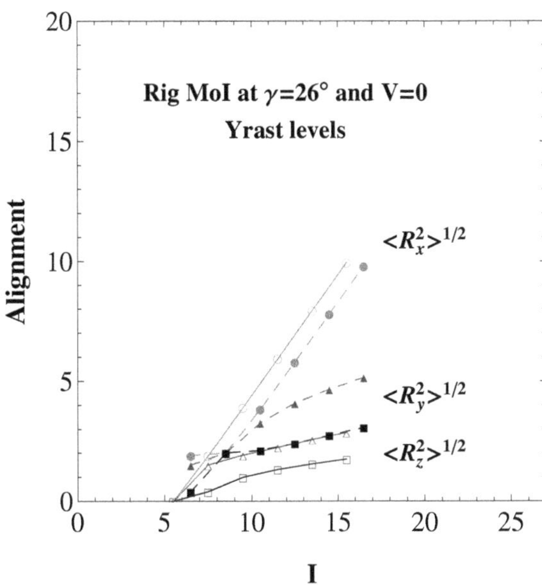

Figure 4.2 The alignments of $\langle R_x^2\rangle^{1/2}$, $\langle R_y^2\rangle^{1/2}$ and $\langle R_z^2\rangle^{1/2}$ for the rig MoI as functions of I. Circles correspond to $\langle R_x^2\rangle^{1/2}$, triangles to $\langle R_y^2\rangle^{1/2}$ and squares to $\langle R_z^2\rangle^{1/2}$. Closed symbols are for $I-j=$ odd levels, while open symbols for $I-j=$ even levels. The parameter set is ($\mathcal{J}_0=25$ MeV^{-1}, $j=11/2$, $\beta_2=0.18$).

symmetry and B_1-symmetry and also the relation to B_2-symmetry and B_3-symmetry are confirmed for $\kappa \sim -0.448$ by the calculation of Ref. [22]. Here, we mark $(R,R_z=2)$ in $I-j=$ odd level corresponds to $(R,R_x=R-1)$, i.e., $(n_\alpha,n_\beta)=(1,0)$ level. As we have not noticed the misprint in Eq. (4-282) of Ref. [6], we made mistakes in the last paragraph in the l.h.s of page 064315-6 of Ref. [5]. "Once the D_2 invariance is violated $\cdots$" should read "Once the other symmetry except for A-symmetry appears $\cdots$", "$\cdots$ $(+1,+1,+1)$ and $(+1,-1,-1)$" should be "$\cdots$ $(+1,+1,+1)$ and $(-1,-1,+1)$", and also "even when the D_2 invariance is violated (see $\cdots$)" should be "even when the other symmetry appears (see $\cdots$)". Because H in Eq. (4.4) is D_2 invariant, all four symmetry states appear equally under D_2 invariance.

The difference in $(1,0)$ and $(0,1)$ is exhibited by the alignment of core angular momentum R_x^2, R_y^2 and R_z^2 for rig MoI in Fig. 15, and for hyd MoI in Fig. 14 in Ref. [5] with $V=1.6$ MeV. Here, we show both cases with $V=0$ MeV in Fig. 4.2 (rig MoI), and Fig. 4.3 (hyd MoI). Fig. 4.2 is quite similar to Fig. 15 in Ref. [5], because $H_{\rm sp}$ supports $H_{\rm rot}$. In contrast, $H_{\rm sp}$ works against $H_{\rm rot}$ for the hyd case, as seen in Fig. 4.3 and Fig. 14 in Ref. [5], but still there remains a similarity between them. The difference between Fig. 4.2 and Fig. 4.3 is easily understood from Eq. (4.19) and Eq. (4.23). For the case of rig MoI, if $I-j=$ even, the yrast state takes $(0,0)$, and then $R_x=R=I-j\equiv R^e$. The yrast level of the neighboring $I+1$ state ($R^o \equiv I+1-j$) has $(1,0)$, and then $R_x=R^o-1=I+1-j-1=R^e$. Subsequently, the open symbols of R_x at $I-j=$ even equal to the neighboring closed symbols of R_x at $I+1-j=$ odd (see open and closed circles in Fig. 4.2). On the other hand, for the case of hyd MoI, the neighboring odd yrast $I+1$ level has $(0,1)$, and then $R^o=I+1-j+1=I+2-j$ and $R_y=R^o-0=R^o$, which is the same as $R_y=I+2-j$ for $R=I+2-j$ with $(0,0)$. Subsequently, the closed symbols equal to the neighboring open symbols (see closed and open triangles in Fig. 4.3).

For the hyd MoI, the yrare levels for $I-j=$ odd is $(1,0)$, which is shown in Fig. 2 in Ref. [5] by red lines. In Fig. 4.4, we plot $\langle R_x^2\rangle^{1/2}$, $\langle R_y^2\rangle^{1/2}$ and $\langle R_z^2\rangle^{1/2}$ between the yrast levels for $I-j=$ even (the same as in Fig. 4.3) and the yrare levels for $I-j=$ odd. We see the behavior of the alignment

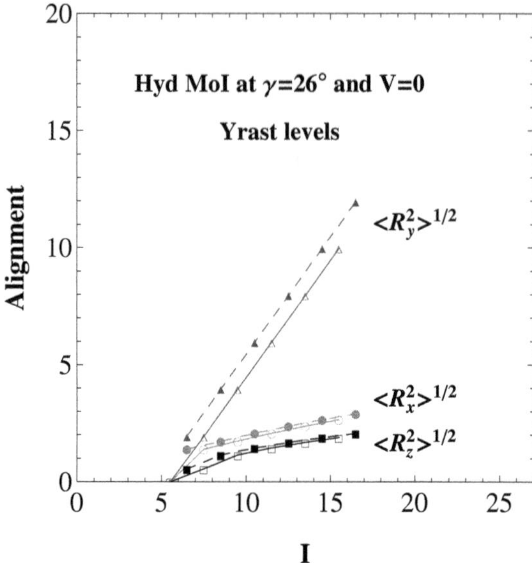

Figure 4.3   The alignments of $\langle R_x^2 \rangle^{1/2}$, $\langle R_y^2 \rangle^{1/2}$ and $\langle R_z^2 \rangle^{1/2}$ for the hyd MoI as functions of I. The definition of symbols and the parameter set are the same as in Fig. 4.2.

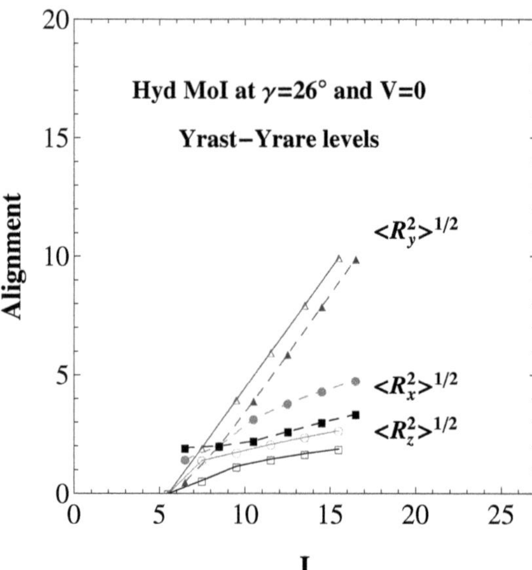

Figure 4.4   The alignments of $\langle R_x^2 \rangle^{1/2}$, $\langle R_y^2 \rangle^{1/2}$ and $\langle R_z^2 \rangle^{1/2}$ between yrast (0,0) with $I - j =$ even and yrare (1,0) with $I - j =$ odd for hyd MoI as functions of I. The definition of symbols and parameter set are the same as in Fig. 4.2.

similar to those in Fig. 4.2, indicating the wobbling around the maximum $\mathcal{J}_y^{\mathrm{hyd}}$ is also represented by $n_\alpha = 1$.

4.3   ELECTROMAGNETIC TRANSITIONS

The essential shortcoming of the hyd MoI is in the electromagnetic transitions. We have already discussed the electromagnetic transitions in Refs. [1, 2], where we have compared with the hyd MoI

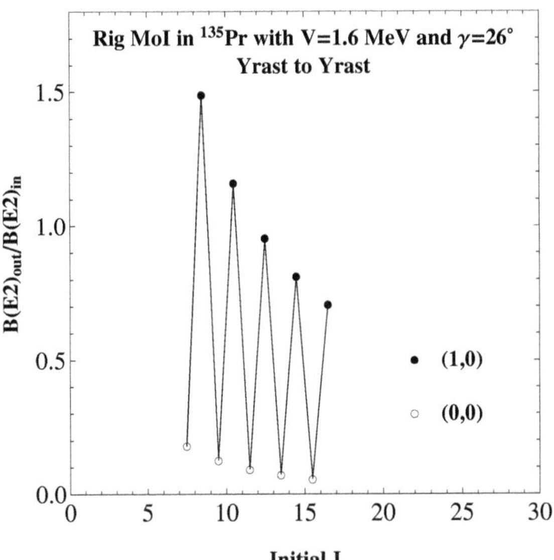

Figure 4.5 $B(E2)_{\text{out}}/B(E2)_{\text{in}}$ as functions of initial I for rig MoI. The suffix "out" denotes the transition from the initial I to the final $I-1$, while the suffix "in" the transition from the initial I to the final $I-2$. The closed circles denoted by (1,0) indicate the initial I belonging to the yrast $I-j=$ odd, and the open circles by (0,0) indicate the initial I belonging to the yrast $I-j=$ even. The parameter set is the same as in Fig. 4.2, and $V=1.6$ MeV.

which is different from the one we are discussing in this chapter. The hyd MoI in Refs. [1, 2] is the γ-reversed hyd MoI proposed in Ref. [28] to imitate the rig MoI in explaining the wobbling band in ^{163}Lu. The results obtained employing γ-reversed hyd MoI are quite different from those employing the original hyd MoI especially in the electromagnetic ratios, because H_{sp} does not work against H_{rot} with γ-reversed hyd MoI in the region of $0° \ll \gamma \leq 30°$.

In Fig. 4.5, we show the ratio $B(E2)_{\text{out}}/B(E2)_{\text{in}}$ for rig MoI obtained by the exact diagonalization of total H. The parameter set is the same as in Figs. 9, 10, 14 and 15 in Ref. [5]. The ratio $B(E2)_{\text{out}}/B(E2)_{\text{in}}$ starting from the closed circles (yrast $I-j=$ odd, (1,0)) are much larger than the one starting from the open circles (yrast $I-j=$ even, (0,0)). This behavior agrees with the experimental data in ^{135}Pr [25] and ^{163}Lu isotopes [9, 10]. The difference between $I-j=$ even and odd are simply explained with the help of the selection rule in Tables I and II in Ref. [2]. For the $I-j=$ odd level, $(n_\alpha, n_\beta) =$(1,0) to (0,0) transition, we get

$$\frac{B(E2 : I\,10 \to I-1\,00)}{B(E2 : I\,10 \to I-2\,10)} \sim \frac{6}{I}\tan^2\left(\gamma+\frac{\pi}{6}\right)\left(\frac{G_{0000}}{G_{1010}}\right)^2, \tag{4.27}$$

where $G_{n_a n_b, n_\alpha n_\beta}$ is the overlap function between the original boson state $|n_a n_b, Ij\rangle$ and the quasi-boson state $|n_\alpha n_\beta, Ij\rangle$ proposed in Refs. [27, 29] for even nuclei, extended to odd nuclei in Refs. [1–3], and to odd-odd nuclei in Ref. [4]. For the $I-j=$ even level, $(n_\alpha, n_\beta)=$ (0,0) to (1,0) transition, we get

$$\frac{B(E2 : I\,00 \to I-1\,10)}{B(E2 : I\,00 \to I-2\,00)} \sim \frac{2}{I}\left(\frac{G_{1010}}{G_{0000}}\right)^2. \tag{4.28}$$

If we take the ratio between Eq. (4.27) for $I+1$ and Eq. (4.28) for I, the ratio becomes

$$\frac{3I}{I+1}\tan^2\left(\gamma+\frac{\pi}{6}\right)\left(\frac{G_{0000}}{G_{1010}}\right)^2_I\left(\frac{G_{0000}}{G_{1010}}\right)^2_{I+1}. \tag{4.29}$$

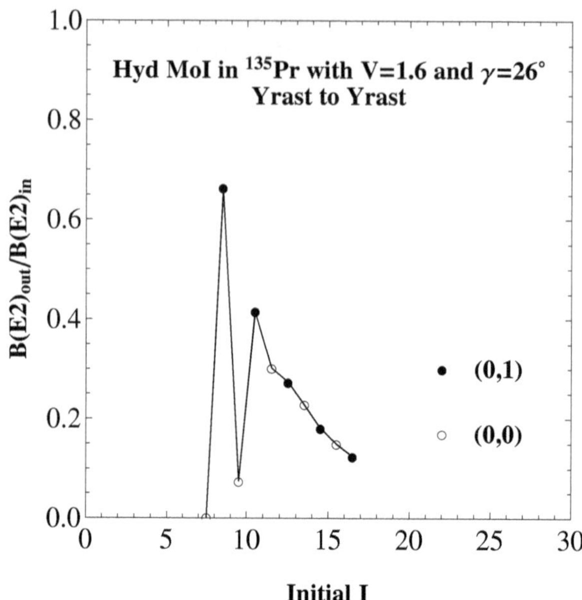

Figure 4.6 $B(E2)_{\text{out}}/B(E2)_{\text{in}}$ as functions of initial I for hyd MoI. The closed circles denoted by (0,1) indicate the initial I belonging to the yrast $I-j=$ odd, and the open circles by (0,0) the initial I belonging to the $I-j=$ even. The parameter set is the same as in Fig. 4.5.

For the case of $\gamma = 26°$, this ratio becomes $7.82I/(I+1)$, where $(G_{0000}/G_{1010})^4$ is adopted from Ref. [2] for Lu isotopes. Although it is a rough estimation without H_{sp}, it is comparable to Fig. 4.5.

In Fig. 4.6, we show the ratio $B(E2)_{\text{out}}/B(E2)_{\text{in}}$ for hyd MoI obtained by the exact diagonalization of total H. In contrast to Fig. 4.5, open circles ($I-j=$ even) are almost distributed along the same line of closed circles ($I-j=$ odd) except for $I=19/2$ and $15/2$. Note that the experimental data of $B(E2)_{\text{out}}/B(E2)_{\text{in}}$ in ^{135}Pr are from $I= 21/2$, $25/2$ and $29/2$ [25]. Associated with the change of quantization axis from z-axis to y-axis, the components of quadrupole moment Q_0 and Q_2 must be transformed to

$$Q_0'' = -Q_0 \frac{\cos(\gamma+60°)}{\cos\gamma}, \quad Q_2'' = \frac{Q_0}{\sqrt{2}} \frac{\sin(\gamma+60°)}{\cos\gamma}. \tag{4.30}$$

With the help of Tables I, II, and replacing Q_0' and Q_2' by Q_0'' and Q_2'', respectively in Eq. (4.30), we obtain the following transition rates. For $I-j=$ odd level,

$$\frac{B(E2:I01 \to I-100)}{B(E2:I01 \to I-201)} \sim \frac{6}{I} \frac{1}{\tan^2(\gamma+60°)} \left(\frac{G_{0000}G_{1001}}{G_{0101}^2}\right)^2, \tag{4.31}$$

and for $I-j=$ even level,

$$\frac{B(E2:I00 \to I-101)}{B(E2:I00 \to I-200)} \sim \frac{2}{I} \left(\frac{G_{1001}}{G_{0000}}\right)^2. \tag{4.32}$$

Both Eqs. (4.31) and (4.32) decrease with I. The ratio of Eq. (4.31) for $I+1$ to Eq. (4.32) for I at $\gamma = 26°$ becomes $0.02I/(I+1)$ almost constant, if we assume $G_{1010}^4 \sim G_{0101}^4$.

In Fig. 4.7, we show the ratio $B(M1)_{\text{out}}/B(E2)_{\text{in}}$ for rig MoI obtained by the exact diagonalization of total H. As seen in the figure, the ratio $B(M1)_{\text{out}}/B(E2)_{\text{in}}$ with open symbols ($I-j=$ even) show almost 0. From the transition rates for odd $I-j$ levels given by Tables I and IV in Ref. [2],

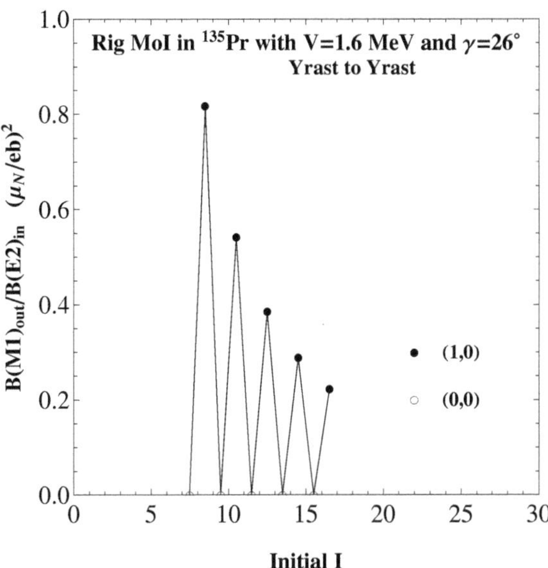

Figure 4.7 $B(M1)_{\text{out}}/B(E2)_{\text{in}}$ in units of $(\mu_N/\text{eb})^2$ as functions of initial I for rig MoI. The closed circles denoted by (1,0) indicate the initial I belonging to the yrast $I - j=$ odd, and the open circles denoted by (0,0) indicate the initial I belonging to the $I - j=$ even. The parameter set is $Q_0 = 3\cos 26°$ b, and $g_{\text{eff}}= 0.4136$. The other parameters are the same as in Fig. 4.5.

we get

$$\frac{B(M1:I10 \to I-100)}{B(E2:I10 \to I-210)} \sim \frac{4j^2}{I}\frac{8}{(\sqrt{3}-\tan\gamma)^2}\left(\frac{G_{0000}}{G_{1010}}\right)^2 F, \tag{4.33}$$

where $F = 3(\mu_N g_{\text{eff}})^2/[5(eQ_0)^2]$. For the $I - j =$ even level, the transition from $(n_\alpha,n_\beta)=(0,0)$ to $(1,0)$ is approximated as

$$\frac{B(M1:I00 \to I-110)}{B(E2:I00 \to I-200)} \sim 4j\frac{8}{(\sqrt{3}-\tan\gamma)^2}\left(\frac{G_{0110}}{G_{0000}}\right)^2 F. \tag{4.34}$$

It is easy to see that $B(M1)_{\text{out}}/B(E2)_{\text{in}}$ for $I - j =$ even is quite small, because of the small factor of G^2_{0110}.

In Fig. 4.8 we plot the ratio $B(M1)_{\text{out}}/B(E2)_{\text{in}}$ for hyd MoI obtained by the exact diagonalization of total H. The parameter set is the same as in the rig MoI case. The open circles show larger values than closed circles except for the cases of $I=$ 19/2 and 15/2, contrary to the rigid body case in Fig. 4.7. Note that the experimental $B(M1)_{\text{out}}/B(E2)_{\text{in}}$ data in ^{135}Pr and Lu isotopes are from $I - j =$ odd, and non-negligible magnitude of those from $I - j=$ even are not observed. We have chosen the same parameter set in both Figs. 4.7 and 4.8, and then it is not due to the choice of the parameter set. Whatever parameter set is chosen, such a larger $B(M1)_{\text{out}}/B(E2)_{\text{in}}$ ratio from $I - j=$ even than $I - j=$ odd happens for hyd MoI. Using Tables I and IV for Q_0'' and Q_2'' in Eq. (4.30), we get the M1 transition from (01) to (00) for $I - j =$ odd level

$$\frac{B(M1:I01 \to I-100)}{B(E2:I01 \to I-201)} \sim \frac{4j^2}{I}\frac{8}{(\sqrt{3}+\tan\gamma)^2}\left(\frac{G_{0000}G_{1001}}{G^2_{0101}}\right)^2 F. \tag{4.35}$$

For the $I - j =$ even level, the transition from (0,0) to (0,1) is approximated as

$$\frac{B(M1:I00 \to I-110)}{B(E2:I00 \to I-200)} \sim 4j\frac{8}{(\sqrt{3}+\tan\gamma)^2}\left(\frac{G_{0101}}{G_{0000}}\right)^2 F. \tag{4.36}$$

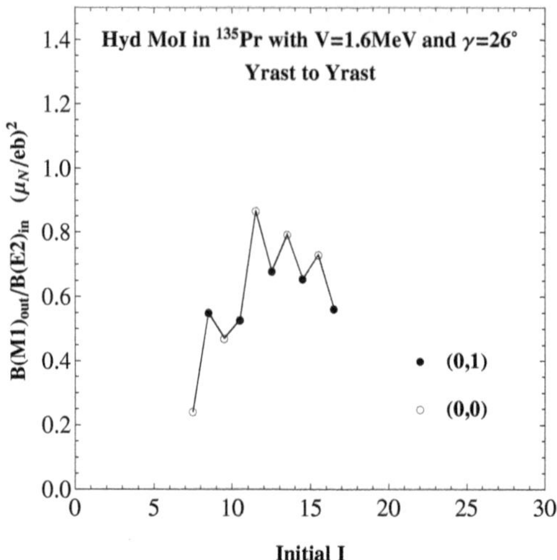

Figure 4.8 $B(M1)_{\text{out}}/B(E2)_{\text{in}}$ as functions of initial I for hyd MoI. The closed circles denoted by (0,1) indicate the initial I belonging to the yrast $I - j$= odd, and the open circles denoted by (0,0) indicate the initial I belonging to the $I - j$= even. The parameter set is the same as in Fig. 4.7.

It is easily seen that Eq. (4.35) is smaller than Eq. (4.36) due to G^2_{1001}. Usually the diagonal element of G is almost 1, i.e., $G_{mnmn} \sim 1$, but the non-diagonal element of G is quite small [2].

Not only the ratios of $B(E2)_{\text{out}}/B(E2)_{\text{in}}$ and $B(M1)_{\text{out}}/B(E2)_{\text{in}}$, but also the absolute values of $B(M1)_{\text{out}}$ and $B(E2)_{\text{in}}$ were observed in ^{135}Lu [9, 10], which have been well explained by the top-on-top model [2]. In Fig. 4.9, we compare the calculated $B(M1)_{\text{out}}$ values for rig MoI and hyd MoI. The parameter set is common for both cases. As seen in Fig. 4.9, hyd MoI gives larger $B(M1)_{\text{out}}$ than rig MoI. This difference comes from the fact that A_k^{hyd} and B_k work in an opposite way.

4.4 CORIOLIS ANTIPAIRING EFFECT

To describe physical properties of rotational states of nuclei, we must study how to include Coriolis antipairing (CAP) effect [18] on $\mathcal{J}_0$. Here, we review the essence of Ref. [19], which solves the particle number- and angular momentum-constrained Hartree-Fock-Bogoliubov (CONHFB) equation by treating the cranking term as a perturbation to the BCS solution. We start from the deformed Nilsson single-particle energy ε_α [20] measured from the chemical potential λ and the monopole-pairing interaction with strength G [30, 31];

$$H = H_0 - H_\Omega, \quad H_0 = \sum_\alpha (\varepsilon_\alpha - \lambda)c_\alpha^\dagger c_\alpha - \frac{G}{4}\sum_{\alpha,\beta} c_\alpha^\dagger c_\alpha^\dagger c_\beta c_{\tilde\beta},$$

$$H_\Omega = \Omega_x \hat{I}_x, \quad \text{with } \hat{I}_x = \sum_{\alpha,\beta}(j_x)_{\alpha\beta}c_\alpha^\dagger c_\beta. \tag{4.37}$$

The constraint on I is given by

$$\langle \hat{I}_x \rangle = I \equiv \mathcal{J}_x \Omega_x. \tag{4.38}$$

Up to the first-order perturbation, $\mathcal{J}_x$ is given by

$$\mathcal{J}_x = \sum_{\alpha,\beta\,(\varepsilon_\alpha < \varepsilon_\beta)} \frac{(j_x)^2_{\alpha\beta}}{E_\alpha + E_\beta}\left(1 - \frac{(\varepsilon_\alpha - \lambda)(\varepsilon_\beta - \lambda) + \Delta^2}{E_\alpha E_\beta}\right), \tag{4.39}$$

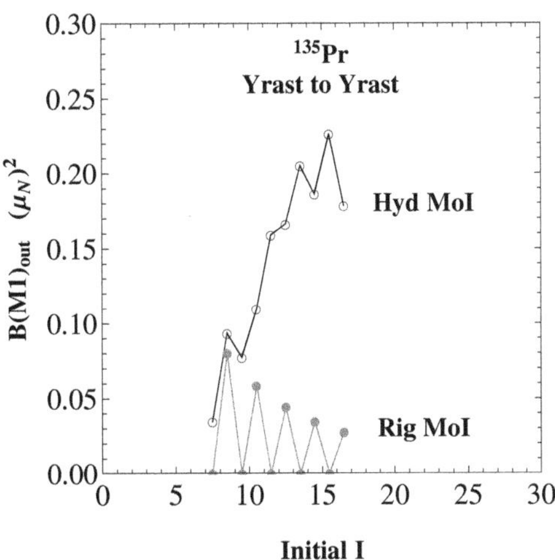

Figure 4.9 Comparison of $B(M1)_{\text{out}}$ in units of $(\mu_N)^2$ between rig MoI and hyd MoI as functions of initial I. The open circles are for hyd MoI, and the closed circles for rig MoI. The parameter set is the same as in Fig. 4.7.

where E_α is the BCS quasiparticle energy. Here, we mark that the second-order perturbation does not contribute to Eq. (4.38). The appearance of Eq. (4.39) is the same as the cranking formula, but Δ and λ deviate from the conventional BCS solution. Up to the second order perturbation, we get the gap equation,

$$= \frac{G}{4}\sum_\alpha \frac{\Delta}{E_\alpha}\left[1 - \Omega_x^2 \sum_\beta \frac{(j_x)_{\alpha\beta}^2}{E_\alpha + E_\beta}\left(\frac{E_\alpha E_\beta - (\varepsilon_\alpha - \lambda)(\varepsilon_\beta - \lambda) - \Delta^2}{E_\alpha E_\beta (E_\alpha + E_\beta)} + \frac{(\varepsilon_\alpha - \lambda)(\varepsilon_\alpha - \varepsilon_\beta)}{E_\alpha^2 E_\beta}\right)\right]. \quad (4.40)$$

The first term in Eq. (4.40) is the 0-th order contribution from the BCS solution.

Under the assumption that only large matrix elements of $(j_x)_{\alpha\beta}$ contribute to $\hat{I}_x$ with a common excitation energy $\delta(= \varepsilon_\beta - \varepsilon_\alpha)$, Eq. (4.39) is assumed to be the product of $\mathcal{J}_x^{\text{rig}}\langle g\rangle_{\text{av}}$, where $\mathcal{J}_x^{\text{rig}} \sim 2\sum_{\alpha(\varepsilon_\beta = \varepsilon_\alpha + \delta)}(j_x)_{\alpha\beta}/\delta$. The terms with E_α in Eq. (4.39) are expressed with $x = \varepsilon - \lambda$ and δ.

$$\mathcal{J}_x \cong \sum_{\alpha,\beta\,(\varepsilon_\alpha < \varepsilon_\beta)} 2\frac{(j_x)_{\alpha\beta}^2}{\delta}\langle g\rangle_{\text{av}} \equiv \sum_{\alpha,\beta\,(\varepsilon_\alpha < \varepsilon_\beta)} \frac{2(j_x)_{\alpha\beta}^2}{\delta}\left\langle \frac{\delta}{2(\sqrt{x^2 + \Delta^2} + \sqrt{(x+\delta)^2 + \Delta^2})}\right.$$
$$\left. \times\left(1 - \frac{x(x+\delta) + \Delta^2}{\sqrt{x^2 + \Delta^2}\sqrt{(x+\delta)^2 + \Delta^2}}\right)\right\rangle. \quad (4.41)$$

We integrate $\langle g(x)\rangle$ by x and obtain $\langle g\rangle_{\text{av}}$.

$$\langle g\rangle_{\text{av}} = \frac{2}{\rho\delta}\int_{-\delta/2}^{0}\rho\left[g(x) + g(x - \delta)\right]dx = 1 - \xi^2\ln\left(\frac{1 + \sqrt{1 + \xi^2}}{\xi}\right) + \frac{19}{90}\xi^2, \quad (4.42)$$

which behaves as a function of the dimensionless variable $\xi(= 2\Delta/\delta)$. As shown in Fig. 2 in Ref. [19], Eq. (4.42) comes between the results of Refs. [6, 32].

Another constraint is required for the particle-number N up to the second-order perturbation. However, the same series-expansion method for N does not change λ as is shown in Fig. 5 of Ref. [19], and we do not consider the rotational effect on λ.

We apply this method to the gap equation in Eq. (4.40).

(1) We approximate the summation $\sum_\alpha$ in Eq. (4.40) by an integral in a way similar to get Eq. (4.42). Thus, keeping terms up to the second order in ξ, replacing I by $I - I_0$, and defining $\xi = \xi_0$ at $I = I_0$, we get a relation between $I - I_0$ and Δ as

$$I - I_0 = \left[8\delta^2 \rho \, \mathcal{J}_x^{\text{rig}} \frac{\ln(\xi_0/\xi)\langle g \rangle_{\text{av}}^2}{\bar{F}} \right]^{1/2} ; \quad \bar{F} = 16(1 - \xi^2) \ln\left(\frac{1 + \sqrt{1 + \xi^2}}{\xi} \right) + \frac{319}{27}\xi^2 - \frac{371}{45}.$$

$$(4.43)$$

This relation is used to calculate the $I - I_0$ dependence of Δ.

(2) When Δ is much smaller than $d/2$ (d: the average level distance), $\ln[(1 + \sqrt{1 + \xi^2}/\xi)]$ should be replaced by $\Gamma_n - \xi^2 Z_n$, where Γ_n and Z_n are defined from the Riemann ζ function in Apendix C of Ref. [19]. With this replacement, for $\Delta \ll d/2$ we get

$$I - I_0 = \left[8\delta^2 \rho \, \mathcal{J}_x^{\text{rig}} \frac{\ln(\xi_0/\xi)\langle g \rangle_n^2}{\bar{F}_n} \right]^{1/2}, \quad (4.44)$$

where $\bar{F}_n$ and $\langle g \rangle_n$ are defined from Γ_n and Z_n. This relation is used to calculate the behavior of Δ, which is much smaller than $d/2$. As for the behavior of the moments of inertia $\mathcal{J}$ as functions of $I - I_0$ in both regions specified by $\Delta > \delta/2$ and $\Delta < d$, they are calculated from the relations given by Eqs. (4.43) and (4.44), respectively. The above method is extended also to the odd-mass case. The calculated results for both cases for even-even and odd-mass nuclei are displayed in Figs. 6 and 7 in Ref. [19].

(i) Based on Fig. 9 in Ref. [19], we apply case (1) to the yrast wobbling band in ^{135}Pr as $\mathcal{J}(I) = \mathcal{J}_0/\left(1 + \exp[-(I - a)/b]\right)$ with two parameters a and b, where $\mathcal{J}_0$ is $\mathcal{J}$ at $I = I_0$. This type of parametrization is useful for the yrast band before and after backbending, see Figs. 17, 18 and 19 in Ref. [5].

(ii) For the non-yrast wobbling bands in Lu isotopes and ^{167}Ta, we apply case (2) as $\mathcal{J}(I) = \mathcal{J}_0(I - C_1)/(I - C_2)$, with two parameters C_1 and C_2, see, Figs. 5, 6, 7, 8 in Ref. [2], Figs. 1, 2, 3 in Ref. [3] and Figs. 3, 4, 5, 6, 7 in Ref. [4].

In Fig. 5 of Ref. [2] we observe what happens without this sort of revision. As an example of case 2, we show a case of ^{163}Lu in Fig. 4.10, where $C_1 = 0.69$ and $C_2 = 23.5$. The other parameters are those used in Ref. [2].

4.5   CONCLUDING REMARKS

In this chapter we discuss the advantage of the top(s)-on-top model, which is the algebraic expression of the particle-rotor model with rigid MoI, especially for the wobbling band and related transition probabilities. We have introduced two kinds of boson $(a, a^\dagger)$ for I, and (b and $b^\dagger$) for j in the framework of Holstein-Primakoff transformation, and expand them up to the next-to-leading order. By the boson Bogoliubov transformation new bosons $\alpha, \alpha^\dagger, \beta, \beta^\dagger$ are obtained to diagonalize the total Hamiltonian $H_{\text{rot}} + H_{\text{sp}}$ within an approximation up to the boson numbers n_α^2, $n_\alpha n_\beta$ and n_β^2. In the case without H_{sp}, this model with two quantum numbers n_α and n_β perfectly reproduce the exact solution which is obtained by the diagonalization of particle-rotor model. These quantum numbers work even when H_{sp} is included.

For the particle-rotor model, MoI's are important parameters. We have compared the results obtained both from rig MoI and hyd MoI. We found rig MoI works much better than hyd MoI, not only for the energy sequences of the wobbling band and its partner rotational band, but also for their $B(E2)_{\text{out}}/B(E2)_{\text{in}}$ and $B(M1)_{\text{out}}/B(E2)_{\text{in}}$ ratios. For both reduced transition probability ratios, the

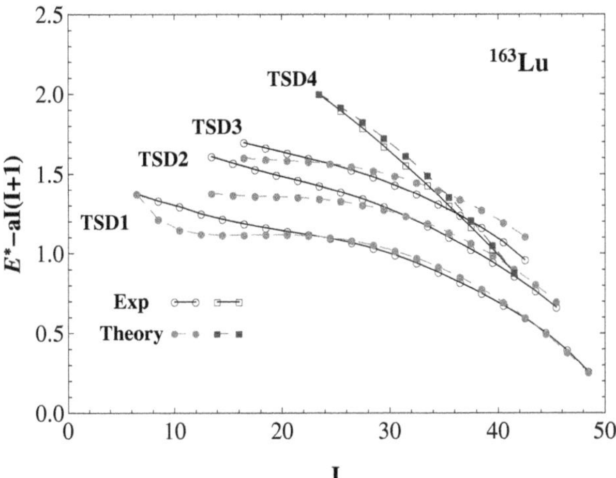

Figure 4.10 Comparison between the experimental and theoretical energy levels $E^* - aI(I+1)$ as functions of angular momentum I for ^{163}Lu with $a = 0.0075$ MeV^{-1}. The experimental data are from Refs. [9, 10].

hyd MoI is out of phase relative to the experimental data, while the rig MoI reproduces them well. This is due to the fact that H_{sp} works against hyd MoI, while cooperates with rig MoI.

The fact that no rotation exists around a symmetry axis does not require $\mathcal{J}_z = 0$ is explained in Sec. 4.1. At $\gamma = 0$, Eq. (4.26) reduces to the well-known expression, $E_{\mathrm{rot}} = A_x^{\mathrm{rig}} I(I+1) + (A_z^{\mathrm{rig}} - A_x^{\mathrm{rig}})K^2$, with the replacement of $R = I$ and $R - n_\alpha = K$. However, hyd MoI cannot give such an expression due to $\mathcal{J}_z^{\mathrm{hyd}} = 0$. Thus, the hyd MoI explains neither the wobbling band, nor the $K \geq 2$ rotational band for $\gamma \sim 0$.

The pairing interaction does not affect a unique-parity level j. However, the MoI is affected by the pairing interaction, which is included by the BCS solutions [6, 30–32], and the MoI reaches $\mathcal{J}^{\mathrm{rig}}$ in the limit of the vanishing gap Δ. We have included the CAP effect [18] on the rig MoI, which is important to reproduce the experimental data. There is no way to add the pairing effect to $\mathcal{J}^{\mathrm{hyd}}$.

We found a sentence in the bottom of the second paragraph in p. 208 in Ref. [8], "If, for example, Earth were completely fluid, then the figure axis would instantaneously adjust to the rotation axis and there could be no precession".

Bibliography

1. K. Tanabe and K. Sugawara-Tanabe, Phys. Rev. C **73**, 034305 (2006); **75**, 059903(E) (2007). In Eq. (9) $n_{\alpha'}$ should stop at R for R = even, and $R-2$ for R = odd. In Eq. (44c) sign(p-q) should read sign(q-p). In Eq. (55) $Q_2' = \frac{1}{2}(\cdots)$ should read $Q_2' = -\frac{1}{2}(\cdots)$.
2. K. Tanabe and K. Sugawara-Tanabe, Phys. Rev. C **77**, 064318 (2008). In Eq. (7) $n_{\alpha'}$ should stop at R for R = even, and $R-2$ for R = odd. In Eq. (22), sgn$(p-q)$ should read sgn$(q-p)$. In Eq. (27) $Q_2' = \frac{1}{2\sqrt{2}}(\cdots)$ should read $Q_2' = -\frac{1}{2\sqrt{2}}(\cdots)$.
3. K. Sugawara-Tanabe and K. Tanabe, Phys. Rev. C **82**, 051303(R) (2010). In Eq. (5) $n_{\alpha'}$ should stop at R for R = even, and $R-2$ for R = odd.
4. K. Sugawara-Tanabe, K. Tanabe and N. Yoshinaga, Prog. Theor. Exp. Phys. **2014**, 063D01 (2014). In Eq. (5) $n_{\alpha'}$ should stop at R for R = even, and $R-2$ for R = odd. In the second line of Eq. (16), $\frac{R}{2}$ should read $\frac{R}{4}$.
5. K. Tanabe and K. Sugawara-Tanabe, Phys. Rev. C **95**, 064315 (2017). In Eq. (10) $n_{\alpha'}$ should stop at R for R = even, and $R-2$ for R = odd. In the last paragraph in the l.h.s. of page 064315-6,

"(+1,-1,-1)" should read "(-1,-1,+1)", "Once the D_2 invariance is violated" should read "Once the other symmetry appears", and "even when the D_2 invariance is violated" should read "even when the other symmetry appears".

6. A. Bohr and B. R. Mottelson, *Nuclear Structure* (Benjamin, Reading, MA, 1975).

7. L. D. Landau and E. M. Lifshitz, *Mechanics* (Physics Mathematics Institute, Moscow, MA, 1958).

8. H. Goldstein, C. Poole and J. Safko, *Classical Mechanics* (Pearson Education International, NJ, 2002).

9. S. W. Ødegård *et al.*, Phys. Rev. Lett. **86**, 5866 (2001).

10. D. R. Jensen *et al.*, Phys. Rev. Lett. **89**, 142503 (2002).

11. A. Görgen *et al.*, Phys. Rev. C **69**, 031301(R) (2004).

12. G. Schönwaßer *et al.*, Phys. Lett. B **552**, 9 (2003).

13. H. Amro *et al.*, Phys. Lett. B **553**, 197 (2003).

14. D. J. Hartley *et al.*, Phys. Rev. C **80**, 041304(R) (2009).

15. O. Klein, Zeit. Physik **58** 730 (1929).

16. D. J. Thouless and J. G. Valatin, Nucl. Phys. **31** 211 (1962).

17. P. A. M. Dirac, Proc. Camb. Phil. Soc.**26** 376 (1930).

18. B. R. Mottelson and J. G. Valatin, Phys. Rev. Lett. **5**, 511 (1960).

19. K. Tanabe and K. Sugawara-Tanabe, Phys. Rev. C **91** 034328 (2015).

20. S. G. Nilsson, Dan. Mat. Fys. Medd. **29**, nr.16 (1955).

21. A. Bohr,Mat. Fys. Medd. Dan. Vid Selsk.**26**, no.14 (1952).

22. G. W. King, R. M. Hainer and P. C. Cross, J. Chem. Phys. **II**, 27 (1943).

23. K. Tanabe and K. Sugawara-Tanabe, Phys. Rev. C **14**, 1963 (1976). In the left-hand matrix element of Eq. (2.15), $cos\theta^1 cos\theta^2$ term in (2,2) element should read $cos\theta^1 cos\theta^3$, $cos\theta^1 sin\theta^3$ in (3,1) element should read $cos\theta^1 sin\theta^2$, and $cos\theta^3$ in (3,3) element should read $cos\theta^2$. In the left-hand matrix element of Eq. (2.17), $cos\theta^2/sin\theta^3$ term in (2,3) element should read $cos\theta^2 sin\theta^3$. In the first line of Eq. (2.22), $\frac{1}{2}$ should read $-\frac{1}{2}$.

24. K. Tanabe and K. Sugawara-Tanabe, Phys. Rev. C **19**, 552 (1979). In the left-hand side of Eq. (6c), $U j_x U^{-1}$ should read $U j_X U^{-1}$, and in the left-hand side of Eq. (8), R_X, R_Y and R_Z should read R_x, R_y and R_z. The second line of Eq. (14) should read $\sum_{J,K}(\lambda_{xJ}\lambda_{yJ\times K}I_K + \lambda_{xJ}\lambda_{yK}I_{J\times K} - \lambda_{yK}\lambda_{xK\times J}I_J)$.

25. J. T. Matta *et al.*, Phys. Rev. Lett. **114**, 082501 (2015).

26. R. Garg *et al.*, Phys. Rev. C **92**, 054325 (2015).

27. K. Tanabe and K. Sugawara-Tanabe, Phys. Lett. **B34** 575 (1971). In Eq. (4) b with $\eta_{\mp}^{*}$ should read $b^{\dagger}$. Under Eq. (6), in "$f \equiv [\cdots(1/\mathcal{J}_x - 1/\mathcal{J}_x)]$" should read "$f \equiv [\cdots(1/\mathcal{J}_z - 1/\mathcal{J}_x)]$". Under Eq. (13) "$s = -(a^{\dagger}a^{\dagger}-$" should read "$s = -\chi(a^{\dagger}a^{\dagger}-$".

28. I. Hamamoto, Phys. Rev. C **65**, 044305 (2002).

29. K. Tanabe, J. Math. Phys. **14** 618 (1973).

30. M. Sano and M. Wakai, Nucl. Phys. **67**, 481 (1965).

31. K. Sugawara, Prog. Theor. Phys. **35** 44 (1966).

32. D. Bengtsson and J. Helgessen, Lecture notes from a summer school at Oak Ridge, 1991 (unpublished).

5 Recent experimental results on chirality and wobbling motion in triaxially deformed nuclei

Song Guo and Bing Feng Lv
Institute of Modern Physics, Lanzhou, China

Excited states in nuclei carry abundant information on nuclear structure. Since such states are depopulated mainly by γ transitions, the detection of γ-rays is one of the most important approaches in the experimental studies of excited nuclear states, especially those at high spins. To study discrete states in nuclei, γ rays emitted from transitions among them are identified experimentally. The γ rays between nuclear states usually have relatively long ranges in materials. Although they carry a well defined energy, part of it can escape from the detector by Compton scattering. Therefore, the detected γ-ray spectra generally exhibit narrow peaks on top of a continuous background. A high statistics is always important in experiments involving γ-ray detection, and the peak-to-background ratios plays a vital role in the sensitivity of detector devices. For the researches on high-spin states, which rapidly decay to the ground state by a series of γ transitions, often in a sub-nanosecond range, the number of detector units is also important to determine the coincidence relations among the observed γ rays.

With the development of γ-ray detectors array technology, the highest identifiable spin values continue to increase. Many phenomena have been observed at high spins, such as band-crossing, superdeformed rotational bands, band termination, revealing the collective motions of nucleons in various quasi-particle configurations. On the other hand, as the ability to identify weak transitions improved, very weak band structures above the yrast line have been widely explored. A increased number of novel symmetries and motions have been proposed to explain the observed new phenomena in the yrare area. Among them, chirality and wobbling are two well known concepts which have been applied to explain properties of nuclei with triaxial deformation.

The spontaneous breaking of chiral symmetry and wobbling motion were predicted as two unique phenomena in nuclei with stable triaxial deformation [1, 2]. Experimental studies on the two topics both started at the beginning of 21 century [3, 4]. From then on, various experimental proofs had been reported, demonstrating that special collective excitations do exist in triaxial deformed nuclei.

In the last decade, new frontiers have emerged in the researches on chirality and wobbling. In the past, expanding the boundaries and verifying the validity of chiral candidates were the main focus in chirality studies. More recently, the studies were focused on multiple chiral doublet bands in a single nucleus, as shown in Fig. 5.1. Especially, the role of other symmetries in a multiple chiral system was discussed based on a few experimental observations. Before 2014, most of the reported wobbling bands were rotational bands at high spins in the $A = 160$ mass region. In the last ten years, wobbling bands at low and medium spins turned to be the new frontiers (see Fig. 5.1). Very recently, the first candidate of the coexistence of chiral doublet bands and wobbling band was reported in ^{74}Br [20], which will be introduced in chapter 15.

DOI: 10.1201/9781032691633-5

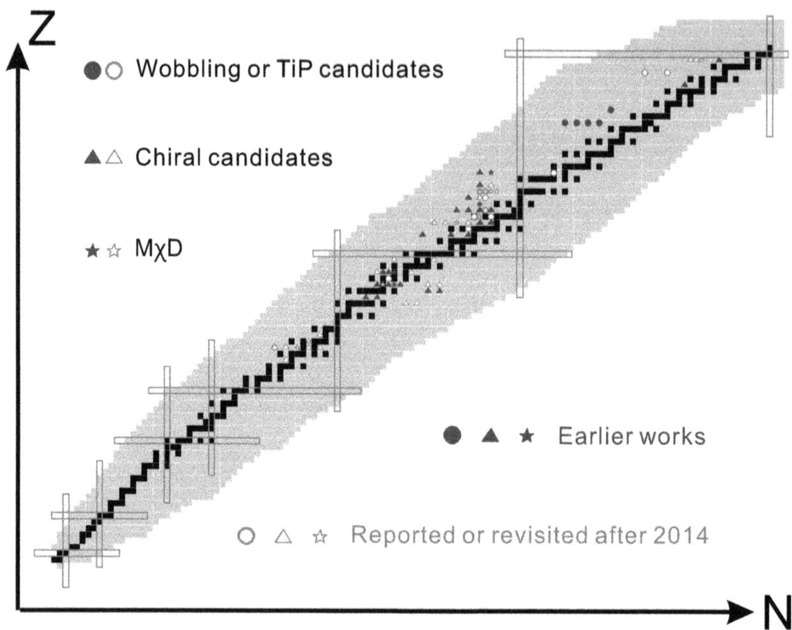

Figure 5.1 Reported candidates of chiral doublet bands and wobbling bands, among them those reported or revisited experimentally in the last decade (2014-2024) are marked with soild symbols, while the others are marked with hollow symbols. Data are collected in Ref. [5–30].

In this chapter, we intend to review the recent progress in nuclear chirality and wobbling from an experimental perspective, focusing mainly on works performed in the last decade.

5.1   RECENT EXPERIMENTAL RESULTS ON CHIRALITY

In 1997, Frauendorf and Meng introduced the concept of chiral symmetry in nuclear physics, and predicted nearly degenerate doublet bands based on different chiral geometries [1]. Two rotational bands in ^{134}Pr were suggested as the first candidate of chiral doublet bands. Up to now, more than 60 pairs of chiral doublet candidates have been reported in four areas (A$\sim$80, 100, 130, 190) in the nuclear chart [5–21], as shown in Fig. 5.1. In the last decade (2014-2024), the chiral doublet bands were established or revisited in about 30 nuclei. Among them 9 candidates of multiple chiral doublet bands in one single nucleus (MχD) were reported. This phenomenon was predict by Meng et al. in 2006 [31], and was first reported in ^{133}Ce in 2013 [32]. Two chiral doublet bands in ^{103}Rh were interpreted as MχD based on an identical configurations [33]. Later, MχD was also reported in even-even nuclei, five pairs of doublet bands were interpreted as chiral doublets based on 4- or 6-quasiparticle configurations in ^{136}Nd [34]. In 2016, two pairs of chiral bands based on two configuration with octupole correlations were reported in ^{78}Br [35]. Later, a set of pseudospin-chiral quartet bands was reported in ^{131}Ba [9], connected to another chiral doublet bands by a series of $E1$ transitions, indicating the existence of octupole correlations. Another important progress was the first measurement of the g-factor of a state belonging to a chiral candidate in ^{128}Cs [36]. This work was introduced in Chapter 7.

5.1.1 MULTIPLE CHIRAL DOUBLET BANDS

A chiral system requires three mutually orthogonal vectors. In triaxially deformed nuclei, a simple mode of chirality is expected when the angular momenta of the core and quasiparticles maintain the directions aligned along the three principal axes. In the early stage of the chirality studies, most chiral candidates were proposed in odd-odd nuclei. Generally the core is considered to rotate along the intermediate axis which corresponds to the largest irrotational-flow-like momentum of inertia. Two valence nucleons, usually one proton and one neutron, should rotate around the short and long axes. Chiral doublet bands based on such configurations have low excitation energies and low spins, and thus are easy to identify experimentally.

In 2006, Meng *et al.* predicted that multiple chiral doublet bands can exist in one single nucleus [31]. Based on the adiabatic and configuration-fixed constrained triaxial relativistic mean field theory, they found different energy minima corresponding to different configurations and deformation parameters. Three pairs of chiral doublet bands were predicted to exist in ^{106}Rh, based on configurations involving one valence proton in the $g_{9/2}$ orbital and 1-3 valence neutrons in the $h_{11/2}$ orbitals. However, the first experimental candidate of MχD was reported in ^{133}Ce [32], with two pairs of chiral doublet bands based on 3-quasiparticle configurations involving $(\pi h_{11/2})^2$ or $(\pi h_{11/2} g_{7/2})$ coupled with $\nu h_{11/2}$ orbitals. Later, similar structures have been also identified in ^{135}Nd [7] and ^{137}Nd [8]. Up to now, all the reported MχD in odd-A nuclei can be regarded as a two-quasiparticle chiral system present in the neighboring odd-odd nuclei plus an additional quasiparticle, which may either increase the angular momentum along the short/long axis or only act as a spectator. It is expected that such MχD bands generally exist in odd-A nuclei in the known chiral regions.

For odd-odd nuclei, MχD candidates were reported in three nuclei, ^{140}Eu [37], ^{78}Br [35] and ^{126}Cs [21]. Among them, the existence in ^{140}Eu of two chiral doublet bands based on the $\pi h_{11/2} \otimes \nu h_{11/2}$ and $\pi(g_{7/2}/d_{5/2}) \otimes \nu h_{11/2}$ configurations is considered an open question. The former is a typical configuration satisfying the demand of a chiral geometry, with two quasiparticles both lying on high-j orbitals. As for the latter, the quasiproton occupies medium-j pseudospin partner orbitals. Naturally, there are two questions. First, it is generally considered that a chiral geometry needs one triaxial core and two quasiparticles in high-j orbitals. Is there any quantitative requirement on the j value? Second, how to distinguish chiral doublet bands from pseudospin doublet bands? Meng *et al.* compared them and proposed one possible criterion based on the $B(M1)$ values [38]. It was found that the $B(M1)$ staggering between the partners usually show opposite odd-even phase in pseudospin doublet bands, but show the same phase in the chiral doublet bands. A similar structure was reported in ^{78}Br, with one chiral doublet bands based on $\pi g_{9/2} \otimes \nu g_{9/2}$ and the other on $\pi(f_{5/2}, p_{3/2}) \otimes \nu g_{9/2}$. In that work, the $B(M1)/B(E2)$ values were shown for the two pairs of doublet bands. However, for the band based on $\pi(f_{5/2}, p_{3/2}) \otimes \nu g_{9/2}$, there are only two experimental values which are close within error bars, and another three being only upper limits. In ^{126}Cs, two pairs of chiral doublet bands were both assigned to the $\pi h_{11/2} \otimes \nu h_{11/2}$ configuration, which will be discussed in the following.

For even-even nuclei, it was predicted that chiral geometry can be achieved with just two quasiprotons, or two quasineutrons. When a pair is broken, one nucleon can occupy the top of one set of high-j orbitals, and the other one can occupy the bottom of another set of high-j orbitals. Such candidates have been proposed in ^{104,106}Mo and ^{110,112}Ru, with a possible configuration $\nu h_{11/2}(d_{5/2}/g_{7/2})^{-1}$ [39?]. However, the assignments of chiral doublet bands were mainly based on the energy near-degeneracy and the constant signature splitting, while the interband transitions between each pair were scarce. Moreover, since the $d_{5/2}$ and $g_{7/2}$ orbitals are involved, a competing interpretation of pseudospin partner bands should be also considered.

If both protons and neutrons contribute to the chiral geometry, the chiral bands should be based on four or more quasiparticle configurations. Such multiple bands are expected to include several yrare states at high spins and high excitation energies, which are difficult to populate with relative high intensities in fusion-evaporation reactions. Powerful detector arrays and careful analyses are required to explore such structures.

An experiment to search for chiral doublet bands in even-even nuclei was conducted with the EUROBALL spectrometer array in 2002 [41]. A new dipole band and many new transitions linking it to a previously known dipole band have been discovered in ^{136}Nd. Based on the nearly degenerate energy spectra of these doublet bands as well as on the nearly maximal triaxiality of ^{136}Nd, they were proposed as first candidates for chiral doublet bands in an even-even nucleus. Subsequently, lifetimes have been measured in ^{136}Nd for the transitions in these rotational bands [42]. The experimentally extracted transition probabilities were compared with results obtained from the tilted-axis cranking and random-phase approximation calculations. It was found that the bands are built on two different quasiparticle configurations, each exhibiting significantly different transition rates. These findings contradict the previous suggestion that the bands consist of a chiral doublet bands in this even-even nucleus.

Recently, the level structure of ^{136}Nd has been re-investigated using the JUROGAM II + RITU + GREAT setup at the University of Jyväskylä [34]. The level scheme has been substantially expanded, and numerous new bands were identified spanning both low- and high-spin regimes. Among these, five nearly degenerate doublet bands (D1 and D1-C, D2 and D2-C, D3 and D3-C, D4 and D4-C, as well as D5 and D5-C, see Fig. 5.2) were identified, comforting the expectation of the multiple chiral doublet bands phenomenon. Possible configurations have been explored for the observed bands using the constrained and tilted axis cranking covariant density functional theory (TAC-CDFT). All these configurations with particle-hole excitations, which involve several different single-j shells, possess remarkable triaxial deformation which is a necessary requirement for the existence of chiral rotational bands. In three of these bands, D2, D3 and D4, a backbending was observed, which can be due to crossing with 8-, 6- and 6-qp configurations, respectively, but was beyond the reported TAC-CDFT calculations. The assigned configurations were further investigated by examining the energy spectra, angular momenta and $B(M1)/B(E2)$ rates. All these values were globally reproduced by TAC-CDFT in certain spin regions, supporting the chiral interpretation of the observed bands. As mentioned above, the unpaired nucleon configurations of the doublet bands in ^{136}Nd involve four different single-j shells. In order to describe the properties of these bands, a particle rotor model that couples nucleons in four single-j shells to a triaxial rotor core was developed [43]. The experimental energy spectra and available values were successfully reproduced, and the angular momentum geometries of the valence nucleons and the core supported the chiral rotation interpretations for all five doublet bands. The five doublet bands of ^{136}Nd have been also investigated in the framework of triaxial projected shell model, employing a new triaxial projected shell model including configurations with more than four quasiparticles in the configuration space [44]. The energy spectra, electromagnetic transition probability ratios, configuration mixing along the rotational bands and chiral geometry were examined. It was shown that an evolution from chiral vibration to static chirality is present in four pairs of partner bands. However, it was difficult to pin-down the salient chiral properties of bands D3 and D3-C due to strong configuration mixing. Various models have been developed and successfully applied to study these bands, strongly supporting ^{136}Nd as the first and best example of even-even chiral nucleus. The observation of five pairs of chiral doublet bands within a single nucleus also sets a record in the study of nuclear chirality.

5.1.2 MULTIPLE CHIRAL DOUBLET BANDS WITH IDENTICAL CONFIGURATION

After the experimental observation of the first MχD candidates based on different quasiparticle configurations and deformations, multiple chiral bands based on the same quasiparticle configuration have been further proposed in several theoretical studies [45–47].

First, the chiral systems with rigid or soft cores were studied by Droste $et\ al.$ [45]. In both cases, two pairs of nearly degenerate bands were obtained in the framework of the Core-Particle-Hole

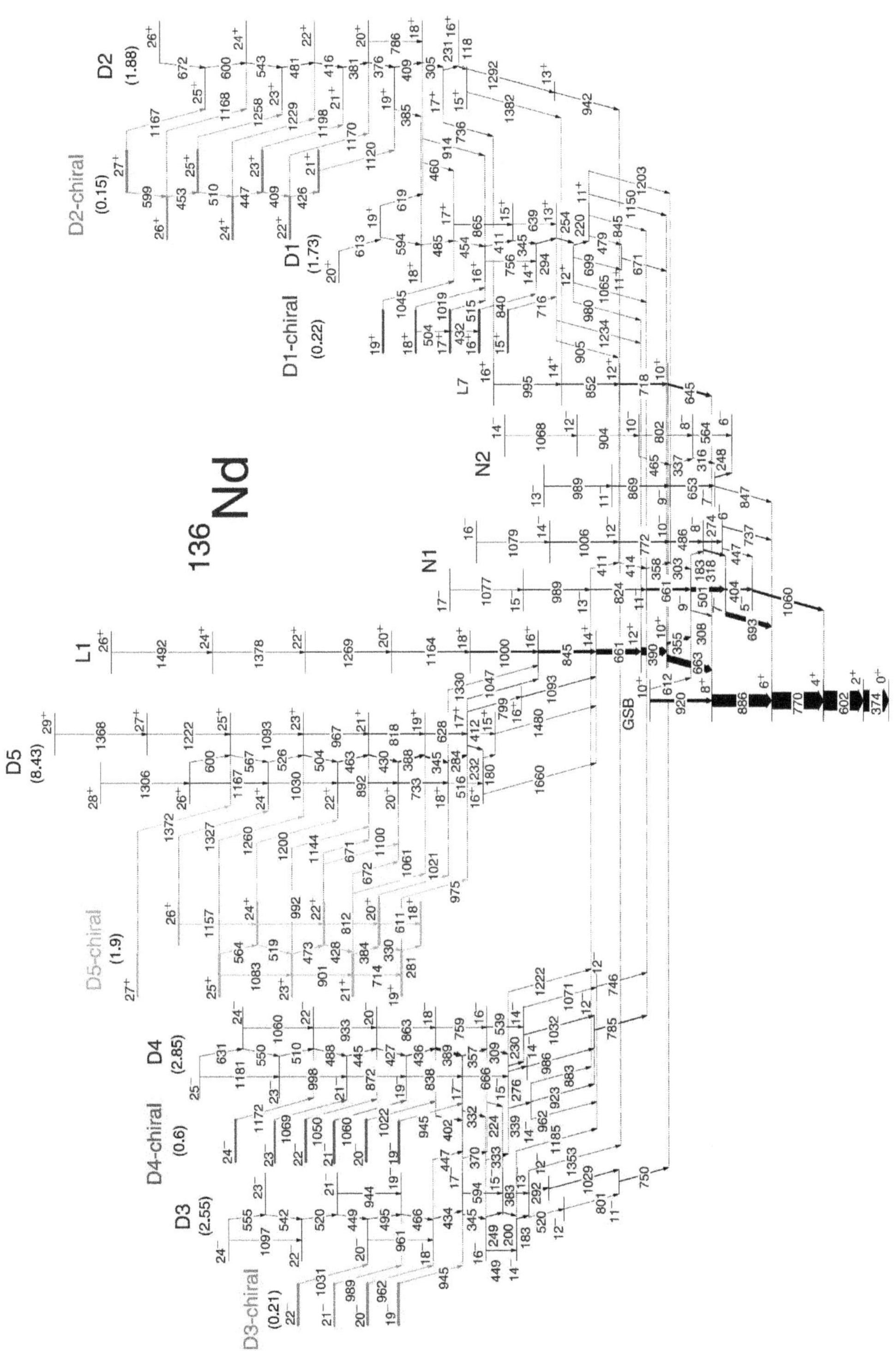

Figure 5.2 Partial level scheme of ^{136}Nd related to the newly identified five doublet bands. Adopted from Petrache *et al.*, Ref. [34].

Coupling model. However, the higher excited doublet bands were separated from the yrast pairs by 1.5-2 MeV, which render their experimental observation difficult. Later, Chen *et al.* calculated the rotational bands based on the $\pi h_{11/2}^1 \otimes \nu h_{11/2}^{-1}$ configuration in the particle-rotor model framework [46]. They found that two pairs of chiral doublet bands can be formed with both constant and spin-dependent variable moments of inertia. Hamamoto also studied the possibility of the existence of multi-chiral-pair bands with the same intrinsic configuration [47]. It was demonstrated that two pairs of chiral doublet bands can be formed in specific ranges of spin. According to the results from the latter two works, the excitation energies of the higher doublets are about several hundred keV, and thus it is possible to populate them with sufficient intensity and therefore can be observed experimentally.

The first candidate of multiple chiral doublet bands of identical configuration was reported in ^{103}Rh by Kuti *et al.* [48]. Four energy near-degenerate rotational bands were interpreted as two pairs of chiral doublet bands based on the same $\pi g_{9/2}^{-1} \otimes \nu h_{11/2}^1 g_{7/2}^1$ configuration, as shown in Fig. 5.3. The energy differences among the states with the same spin in the four bands are quite small. The energy gaps among the three yrare bands are smaller than 200 keV, while the yrast band lies about 300 keV below them. They are lower than the previous expectations from calculations. Since the two pairs of chiral doublet bands have the same quasiparticle configuration and similar chiral rotation or vibration, the yrare pair can be regarded as the core excitation of the yrast one. What is the main mechanism of this possible core excitation is matter of future investigation.

Very recently, two pairs of chiral doublet bands were reported in ^{126}Cs, with the same configuration of $\pi h_{11/2}^{1st} \otimes \nu h_{11/2}^{4th}$ [21]. As the neutron can occupy different Nilsson orbitals of the $h_{11/2}$ shell, for both pairs the internal configurations are highly mixed. Therefore, this case looks like a mixture of chiral systems based on different quasiparticle configuration, while the dominated configuration is the same for the observed two pairs of chiral doublet bands.

5.1.3   COEXISTENCE OF CHIRALITY AND OTHER SYMMETRIES

Near degeneracy between two quantum states is usually associated with fundamental symmetries and symmetry breakings in complex many-body systems like atomic nuclei. Pseudospin symmetry was introduced to describe energy degeneracy between single-particle states with quantum numbers $(n, l, j = l + 1/2)$ and $(n + 1, l + 2, j = l + 3/2)$ [49–51]. After proving itself to be critical for many phenomena such as quantized alignment [52], identical bands [53] and pseudospin partner bands [54], the pseudospin was found to be fundamental as a relativistic symmetry of the Dirac Hamiltonian [55]. Octupole interaction is derived from intrinsic reflection symmetry breaking, and reaches maximum for octupole partner orbitals (l, j) and $(l + 3, j + 3)$ [56]. For normal deformed systems, strong octupole correlations are achieved when particle numbers are around 34, 56, 88 or 134. In several nuclei, near-degenerated intrinsic energies have been found for configurations with opposite parities, leading to the observation of rotation-like bands consisting of states with both parities.

Due to pseudospin or octupole interation, two near-degenerated energy-level sequences can be generated based on two configurations different by the occupation of one quasiparticle. On the contrary, similar doublet band structures can also originate from chirality, while the difference between the two band structures is provided by diversity of the collective motion. When the spontaneous breaking of chirality is expected to exist in one structure in the presence of pseudospin partner bands or bands with octupole correlations, more complex structures consisting of four or more near-degenerate rotational bands are expected.

In 2016, Liu *et al.* reported two pairs of chiral doublet bands in ^{78}Br, connected by a series of $E1$ transitions [35]. The observed enhanced $E1$ transitions connecting the chiral doublet bands were explained as the result of octupole correlations between the $\pi g_{9/2}$ and $\pi d_{3/2}$ orbitals, while the latter is mixed with $\pi f_{5/2}$ due to pseudospin symmetry. This work will be introduced in detail later in Chapter 15.

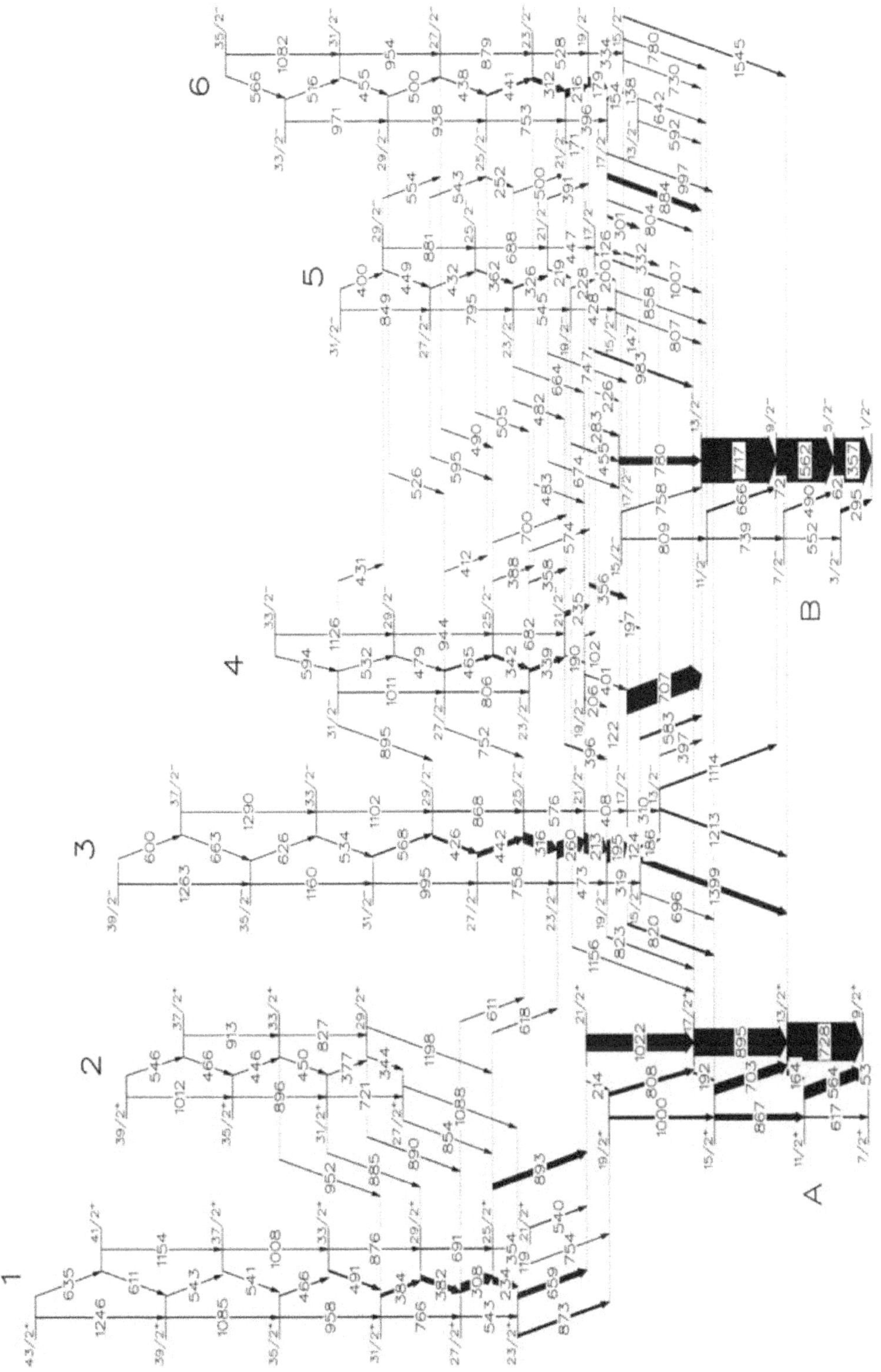

Figure 5.3 Partial level scheme of ^{103}Rh. The energies are given in keV, and the widths of the arrows are proportional to the relative transition intensities. From Kuti *et al.*, Ref. [?].

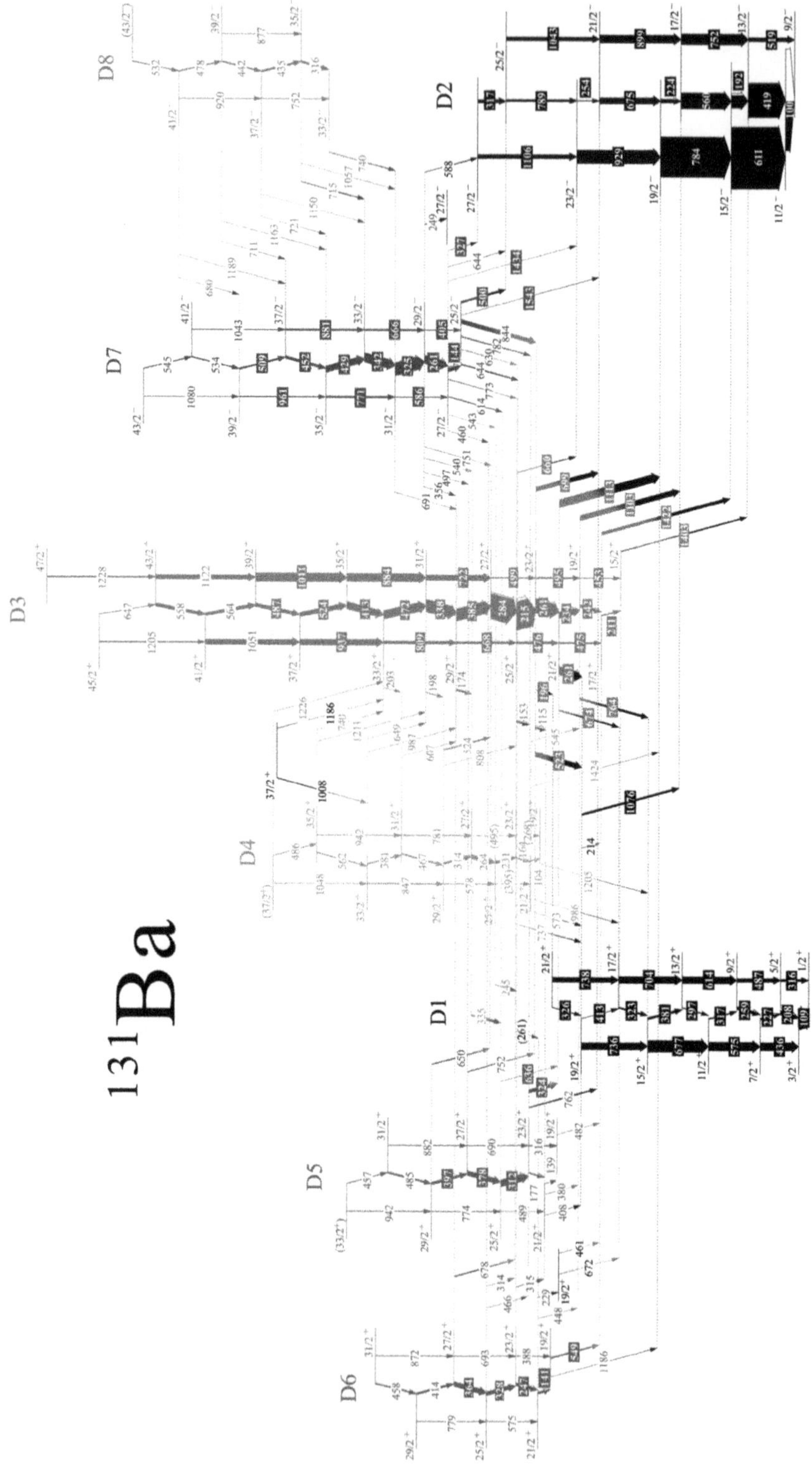

Figure 5.4 Partial level scheme of ^{131}Ba. Transition energies are given in keV and their measured relative intensities are proportional to the widths of the arrows. Levels and intraband transitions are colored in group by bands. The energies and ends of linking transitions are colored by their initial states while the tips are colored by their final states. From Guo *et al.*, Ref. [**?**].

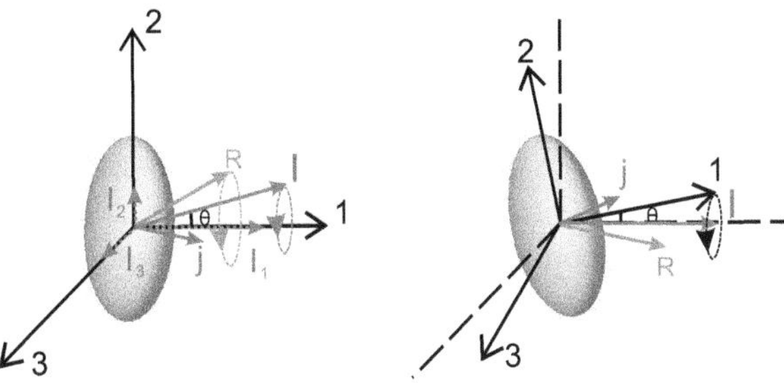

Figure 5.5   Angular momentum geometry of nuclear precession.

In 2019, four nearly degenerate positive-parity bands (bands 3-6) with configurations $\pi h_{11/2}(g_{7/2}/d_{5/2})\nu h_{11/2}$ have been observed in ^{131}Ba [9], as the first evidence of pseudospin-chiral quartet bands. A pair of negative-parity chiral bands associated with the $\pi h_{11/2}^2 \nu h_{11/2}$ configuration (bands 7, 8) was also observed, which decay to the pseudospin-chiral quartet bands via a series of $E1$ transitions, indicating the existence of octupole correlations. It was the first time when octupole correlations between two chiral systems built on 3-quasiparticle configurations was observed. This structure was later reproduced using the reflection-asymmetric triaxial particle rotor model [57].

One can compare the structure in ^{131}Ba with that in ^{103}Rh. Both structures consist of three pairs of chiral doublet bands. The proton number of ^{131}Ba is 56, to be compared to the neutron number of ^{103}Rh which is 58. Therefore, the same quasiparticle orbitals $h_{11/2}, f_{7/2}, d_{5/2}$, for protons in ^{131}Ba and for neutrons in ^{103}Rh, are close to the Fermi surface. In ^{103}Rh, two $E1$ transitions with energies of 611 and 618 keV were found to link the positive- and negative-parity chiral systems. It may also originate from the octupole correlation between $h_{11/2}$ and $d_{5/2}$ orbitals, like in the case of ^{131}Ba. As for the two pairs of chiral doublet bands with the same parity, that is the pseudospin-chiral quartet bands in ^{131}Ba, they are suggested to be chiral bands with a configuration similar to that in ^{103}Rh. Two negative-parity configurations were calculated to have triaxial deformation in ^{103}Rh, $\pi g_{9/2}\nu(h_{11/2}g_{7/2})$ and $\pi g_{9/2}\nu(h_{11/2}d_{5/2})$. The pseudospin partner orbitals $f_{7/2}$ and $d_{5/2}$ are generally regarded to be well mixed, and can be both major components in all the four bands.

5.2   RECENT EXPERIMENTAL RESULTS ON WOBBLING MOTION

Since the axial symmetry is broken in triaxial deformed nuclei, the angular momentum in the body fixed frame can have non-vanishing components on axes perpendicular to the rotation axis, inducing a precession. At high spin, such a precession can be approximately described as the coupling of a rotation and a harmonic vibration, namely wobbling [2].

As shown in Fig. 5.5, for a rotating asymmetric nucleus (with of moments of inertia $\Im_1 > \Im_2 > \Im_3$), nuclear precession occurs when the rotational angular momentum R is not oriented along the principal nuclear axis with the largest moment of inertia, but is tilted a small angle away. In the body coordinate system, the total angular momentum I, which results from the coupling of R and the single-particle angular momentum j, also precesses around the x-axis. In the laboratory reference system, the principal nuclear axis with the largest moment of inertia also precesses around the axis defined by the total angular momentum I.

For high-spin states, the angle between the total angular momentum and the x-axis can be sufficiently small, such that the precession can be approximated as the coupling of a rotation and a harmonic vibration. Such a description, namely wobbling, was first introduced by Bohr and Mottelson in the 1970s [2]. The excitation energy values are $E(\hat{n},I) = A_1 I(I+1) + (\hat{n} + \frac{1}{2})\hbar\omega$, where $\hat{n}$ is the wobbling phonon number and ω is the wobbling frequency.

Bohr and Mottelson also pointed out that the wobbling description should satisfy the approximation condition $I \gg (2\widehat{n}+1)\frac{A_2+A_3-2A_1}{2(A_3-A_1)^{1/2}(A_2-A_1)^{1/2}}$ (here $A_1 < A_2 < A_3, A_k = \frac{\hbar^2}{2\Im_k}$, k=1,2,3.). Obviously, this approximation condition is valid at high spin. Subsequently, in order to give a microscopic description for wobbling mode, several theoretical studies have been performed [58–62].

A wobbling band was experimentally observed for the first time at high-spin in ^{163}Lu [4]. Later, a series of one- and two-phonon wobbling bands have been reported in the odd-A nuclei $^{161-167}$Lu [63–66] and ^{167}Ta [67]. Several theoretical efforts were devoted to reproduce the experimental results, employing the particle-rotor model (PRM) [68] and the Random phase approximation (RPA) [69]. However, by using PRM with constant moments of inertia, the wobbling frequency was calculated to increase with increasing spin, which is in contrast with the experimental observation.

Based on the harmonic approximation and the assumption that the angular momentum of the quasiparticle is rigidly aligned with one of the principal axes (harmonic frozen approximation), Frauendorf and önau introduced the transverse wobbling and the longitudinal wobbling modes for odd-A nuclei [70]. All the reported wobbling bands in the $A = 160$ mass region have been classified as transverse wobbling, and the decrease of the wobbling frequency with spin was reproduced. In analogy to the simple wobbling mentioned above, the harmonic approximation was also discussed, indicating that the transverse and longitudinal wobbling modes also exist in the high-spin regime.

For both simple wobbling [2] and transverse/longitudinal wobbling [70], it was considered that the angular momentum should be much larger than a numerical value, which relates to the moments of inertia with respect to three principal axes of the intrinsic reference system and the number of wobbling phonons. By analyzing the ratio between the numerical values and the angular momentum, Lawrie et $al.$ [71] pointed out that the wobbling is a good approximation only for high spins. Unfortunately, a quantitative analysis of the lower boundary was absent.

Here we focus on the deviation angle θ between the axis with the largest moment of inertia and the total angular moment I, as shown in Fig. 5.5. The components of the total angular moment I_2, I_3 and the total angular moment I are related by the following relation:

$$I_2^2 + I_3^2 = I^2 sin^2\theta. \tag{5.1}$$

Following the derivation in Ref. [2], we can get

$$I = (2\widehat{n}+1)\frac{A_2+A_3-2A_1}{2(A_3-A_1)^{1/2}(A_2-A_1)^{1/2}}/sin\theta. \tag{5.2}$$

Since the arithmetic mean of two unequal numbers is larger than the geometric mean, we obtain

$$I > (2\widehat{n}+1)/sin\theta. \tag{5.3}$$

Similary, using the explicit solution in Ref. [70], one can obtain $I > \frac{3}{sin\theta}$ for one-phonon wobbling mode.

Obviously, the harmonic approximation requires small deviation angles. However, a rigorous quantitative discussion on the boundary is absent. We provide a rough estimation by analogy with a macroscopic simple pendulum which can also be regarded as small amplitude harmonic vibration. Usually it is considered that a pendulum can be regarded as a simple harmonic motion when the deviation angle is smaller than $10°$. If we consider the deviation angle $\theta \leq 10°$ for wobbling motion, we can get that $I(\widehat{n}=1) > 17$ and $I(\widehat{n}=2) > 28$ for one-phonon and two-phonon wobbling bands, respectively.

The spins of all the reported wobbling states are shown in Fig. 5.6, with the limit for one-phonon and two-phonon wobbling bands. It is clear that most states of the wobbling bands in the $A = 160$ mass region have spins above the limits. Two quasiparticle wobbling bands were recently reported in ^{130}Ba [72] and ^{136}Nd [73]. For these two bands, more than half of the states are located above the limit. However, for the recently reported wobbling bands in odd-A nuclei in the 100, 130 and 190

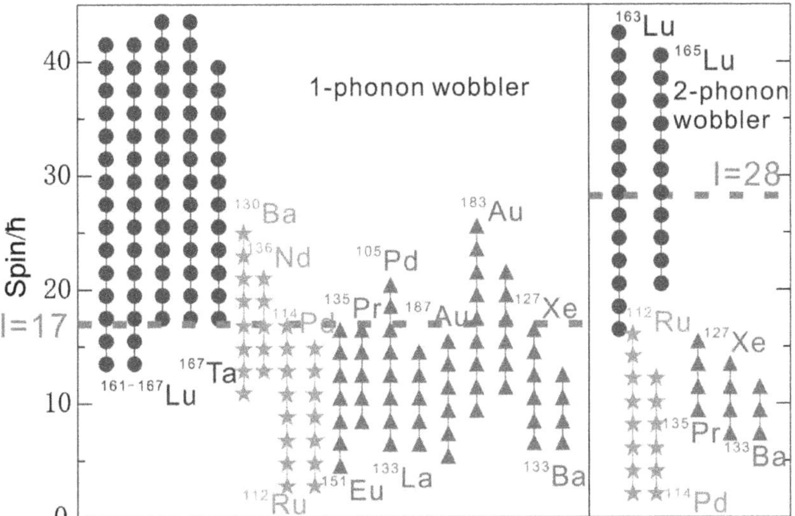

Figure 5.6 Comparing the rough estimated lowest spin limit of the one- and two-phonon wobbling mode (dash lines) with the observed level spins in the reported wobbling bands of all nuclei. The round dots and triangles correspond to the wobbling states of odd A nuclei in and out A = 160 region, respectively, and stars to those of even-even nuclei.

mass regions, most states are below the limit. The existence of wobbling at low spins is somehow not expected.

In the following sections, the recent works reporting wobbling bands at low and medium spins will be introduced.

5.2.1 WOBBLING MOTION AT LOW SPIN

The simplest mode of wobbling motion was first proposed in an even-even triaxial nucleus. Two candidates were proposed in ^{112}Ru and ^{114}Pd [74, 75]. They were regarded as γ-vibration bands previously, but interpreted as wobbling bands based on a deviation on signature splitting phase from other γ-vibration bands, and a similarity on energy spacing with the wobbling of the H_2O molecule. However, it is difficult to distinguish a wobbling band from a γ-vibration band by both experimental and theoretical approaches. Furthermore, for the wobbling assignment, it would be puzzling why the one-phonon band starts at spin 3, while the 2-phonon band starts at spin 2.

In 2015, a wobbling band out of the $A = 160$ mass region was reported by Matta *et al.* [?]. It was assigned as one-phonon transverse wobbling band in the odd-A nucleus ^{135}Pr. Different from those in the $A = 160$ mass region, this band consists of several low-spin states. Later, a number of rotational bands with low-spin states have been interpreted as transverse and longitudinal wobbling bands in the odd-A nuclei ^{135}Pr [22, 23], ^{133}La [24], ^{105}Pd [25], ^{127}Xe [26], ^{133}Ba [27], ^{183}Au [28], ^{187}Au [29] and ^{151}Eu [30]. For a few cases, the spins of the lowest observed states are only $I = j + 1$, where j is the angular momentum of the single quasi-particle (see Fig. 5.5). This means that the wobbling motion can occur for an extremely slow collective rotation (R). This seems to be in contradiction with the approximation condition for wobbling motion, which requires large angular momentum, as pointed out by Bohr and Mottelson for even-even nuclei [2] and by Frauendorf and Dönau for odd-mass nuclei [70].

The validity of the reported wobbling bands in ^{135}Pr, ^{133}La and ^{183}Au, were later criticized based on the assessment of experimental results [76–78]. The criticism focused on the $E2/M1$ mixing ratio (δ) of the linking transitions between the wobbling band and the corresponding normal band.

The extracted δ values are in direct contradiction among different works. Subsequently, the low-spin bands in ^{135}Pr [79] and ^{187}Au [80] have been re-investigated using the ^{100}Mo(^{40}Ar, 1p4n) and ^{175}Lu(^{18}O,6n) reactions, respectively. The mixing ratios were deduced from a method combining polarization and directional correlations from oriented states (P-DCO) [81], where R_{DCO} has been replaced by the two-point angular correlation ratio R_{ac}. Both new results on ^{135}Pr and ^{187}Au have shown that the inter-band transitions between the previous zero- and one-phonon wobbling bands have small mixing ratio values, suggesting predominant $M1$ character, thus excluding the wobbling nature. It is worth noting that the extracted mixing ratios in ^{187}Au are in agreement with a few earlier works [82–84].

Since wobbling motion is defined as the coupling of a rotation and a harmonic vibration, a series of rotational bands can be formed, involving different numbers of wobbling phonons. Being built on the same quasiparticle configuration, similarity and connecting transitions between them are expected. The rotational band with no wobbling phonon (normal band) should be the lowest one in energy. Several features have been proposed to identify a wobbling band. With the same quasiparticle configuration and similar deformation of the core, the wobbling bands are expected to have identical rotational properties as those of the corresponding normal band, including the dynamic moment of inertia $J^{(2)}$ and the quasi-particle angular momentum alignment i_x versus rotational frequency ω. Moreover, $\Delta I = 1$ transitions are expected to generally exist between the one-phonon wobbling band and the corresponding normal band.

However, these behaviors are quite similar with those for a pair of signature partner branches, and for other cases like bands built on different Nilsson orbitals belonging to the same j-shell. To distinguish a wobbling band from an unfavored signature branch, usually the electromagnetic characters of the linking transitions are regarded as the key experimental criterion [4]. They are expected to be $M1$ dominated for signature branches and $E2$ dominated for wobbling bands. The mixing ratio δ for $E2/M1$ transitions can be deduced from angular distribution and/or linear polarization measurements. For a given transition, both its theoretical angular distribution curve and the polarization value are determined by δ and σ/I, where σ is the variance of the projection of the angular momentum I along the beam direction. By setting σ/I to a proper value or a region, σ can be extracted from the experimental results.

It is surprising that the first experimental results on low-spin wobbling bands could not be reproduced by the subsequent works. To distinguish the correct results, and to trace the possible problems, efforts need to be made in two aspects, establishing standardized analysis methods and making experimental data public. The methods to deduce mixing ratios have been developed more than thirty years ago, and used by many researchers. However, they still need to be refined. For example, the extraction of the uncertainties is still obscure and more works are demanded to make it clear. On the other hand, when the experimental results are in direct contradiction, the simplest way is checking the possible problems by exchanging the data. An effort has to be done to prompt researchers to publicly disclose their data in a universal and clear format, which is not the present usual practice.

Despite of the debates on the experimental side, there are two consensus on this issue. First, the precession of triaxial nuclei is believed to exist in low-spin region. Second, the classic picture of wobbling motion can well describe the precession at high spins, but is not suitable for that at low spins. Different solutions are proposed by theoreticians.

After the wobbling bands have been classified as longitudinal and transverse [70], Frauendorf and Dönau reinvestigated the transverse wobbling mode in ^{163}Lu by using the quasiparticle-random-phase approximation [85]. Chen et $al.$ [86] have proposed a two-dimensional collective Hamiltonian (2DCH) on both azimuth and polar motions in triaxial nuclei to investigate the chiral and wobbling modes. Budaca [87, 88] introduced the tilted-axis wobbling in odd-mass nuclei and divided the wobbling motion into three unique phases which are determined as quantized oscillations around minima in the classical energy associated to a quantum triaxial rotor Hamiltonian with an aligned single-particle angular momentum along the first principal axis, by means of a time-dependent variational principle. Streck *et al.* [89] investigated the behavior of the collective rotor in wobbling motion within the particle-rotor model. Raduta *et al.* [90] developed a new approach for the wobbling motion in the even-odd isotopes ^{161,163,165,167}Lu within a particle-triaxial rotor semiclassical formalism. Lawrie *et al.* [71] described the rotation of a triaxial nucleus as a precession of the total angular momentum around a certain axis and at a given tilt, and introduced the tilted precession (TiP). They pointed out that TiP bands become approximately similar to wobbling bands at high spins for both zero-seniority bands in even-even nuclei, and one-quasiparticle bands with longitudinal coupling of the angular momenta in odd-mass nuclei, while the TiP and the wobbling bands have distinctly different nature for transverse coupling of the angular momenta and for one-quasiparticle configurations. Therefore, it was argued that it is not justified to use the QTR model as a theoretical description of transverse wobbling bands as this model produces TiP bands, not wobbling bands. According to this theory, the wobbling bands observed previously in ^{135}Pr [22] and ^{105}Pd [25], as well as two newly observed bands in ^{135}Pr [79] and ^{135}Nd [91] were interpreted as TiP bands.

5.2.2 WOBBLING MOTION AT MEDIUM SPIN

Recently, two yrare bands feeding the $\pi h_{11/2}^2$ bands in ^{130}Ba and ^{136}Nd were proposed to be wobbling bands [72, 92], as shown in Fig. 5.7. In Ref. [93], a large variety of band structures has been observed in the even-even nucleus ^{130}Ba. Among these, a pair of bands with even and odd spins based on the $\pi h_{11/2}^2$ configuration at medium spin were studied by Chen *et al.* and the odd-spin band was proposed to be the one-phonon wobbling band [72]. They examined this band using constrained triaxial covariant density functional theory combined with the quantum particle rotor model. Later, this structure was investigated by Wang *et al.* with the microscopic projected shell model [94]. The experimental energy spectra, energy difference between the two bands, as well as the available electromagnetic transition probabilities $B(M1)_{out}/B(E2)_{in}$ and $B(E2)_{out}/B(E2)_{in}$ were well reproduced by the two works. The investigations of the *azimuthal plots*, i.e., the probalility distribution profiles for the orientation of the angular momentum on the intrinsic (θ, ϕ) plane, the *K plot*, i.e., the K distributions of angular momenta on the three principal axes, and the root mean square angular momentum components along the three principle axes of the rotor, further demonstrated the stable transverse wobbling character of the odd-spin band. This marks the first example for transverse wobbling in an even-even nucleus. This observation in ^{130}Ba exhibits a more stable wobbling motion compared to that in odd-mass nuclei at low spin, because two quasiparticles rather than one contribute to the quasiparticle angular momentum, which is then larger, and also the triaxial shape is more stable under the polarization effect of the active quasiparticles.

Subsequently, two rotational bands in ^{136}Nd were investigated with the triaxial projected shell model, and interpreted as zero- and one-phonon wobbling bands having the same two-quasiparticle configuration $\pi h_{11/2}^2$ as those in ^{130}Ba [92]. The energy spectra and available experimental observables were reasonably reproduced. The validity of the wobbling approximation has been examined through the angular momentum geometry and the configuration components extracted from the microscopic wave functions. Additionally, the impact of the rotational alignment of the quasiparticles on the scenario of transverse wobbling was also explored. Interestingly, it showed that the $n = 0$ phonon wobbling candidate band is more affected than the $n = 1$ one, contrary to the anticipated decreasing trend in wobbling energy in the transverse case. Later, the nature of these two candidate transverse wobbling bands was investigated experimentally [95]. The critical evidence of wobbling motion, the mixing ratio of one $\Delta I = 1$ transition connecting the one-phonon and the zero-phonon wobbling bands, has been determined from a high-statistics Jurogam II γ-ray spectroscopy experiment by using the combined angular correlation and linear polarization method. The resulting wobbling excitation energy and ratios of reduced electromagnetic transition probabilities were found in good agreement with results of a new particle-rotor model which rigidly couples the total angular momentum of two quasiparticles to a triaxial core in an orthogonal geometry,

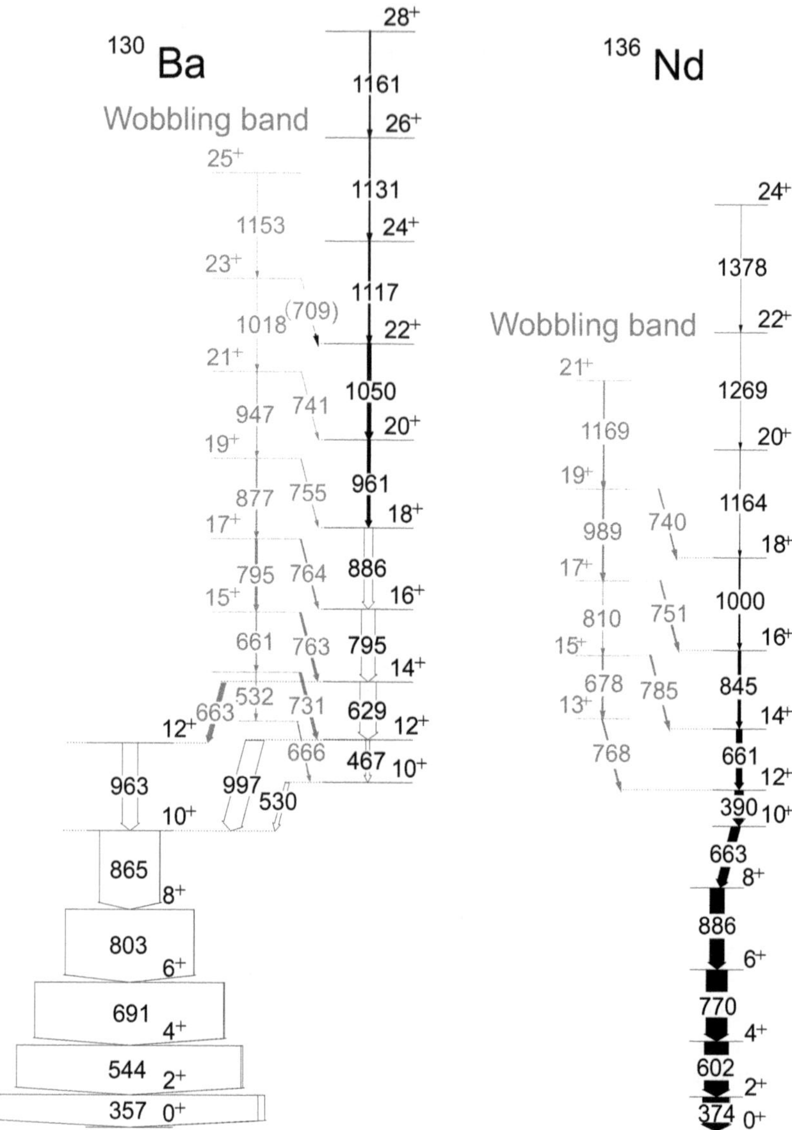

Figure 5.7 Partial level scheme of ^{130}Ba and ^{136}Nd. Transition energies are given in keV and their measured relative intensities are proportional to the widths of the arrows.

confirming thus the transverse wobbling nature of the bands. Later, the possible existence of wobbling motion in the even-even ^{130}Ba, ^{134}Ce and ^{136,138}Nd nuclei were further supported by Budaca *et al.*, using a collective Hamiltonian constructed from a semiclassical treatment applied to a triaxial rotor with a rigidly aligned pair of quasiparticles [96]. Among them the two bands in ^{134}Ce and ^{138}Nd were tentatively proposed to be wobbling bands, but the experimental proofs were not sufficient [97, 98].

In ^{130}Ba and ^{136}Nd, the $E2$ components in the linking transitions between wobbling band and normal band are weaker than the $M1$ components. Nevertheless, they are still relatively stronger than the typical $E2$ components of the transitions linking two signature partner branches. It was argued that the enhanced $E2$ components are due to the collective wobbling motion, while the $M1$ components are also enhanced due to the summed magnetic moments of the two quasiparticles.

5.3 OPEN QUESTIONS ON CHIRALITY AND WOBBLING

The experimental studies on both chirality and wobbling have a history of about 24 years. However, there are still many open questions calling for further experimental and theoretical works.

An unified description for the collective excitation of triaxial deformed nuclei, including chiral rotation, chiral vibration, wobbling, γ vibration is not yet available. What is the role of softness of the triaxial core in the nuclear motions? Will it induce any different behavior between the rigid and soft chiral or wobbling nuclei?

Several novel phenomena have been revealed in the recent years. However, most of them have only been found in a few particular nuclei, and the systematics need to be further explored.

Can a triaxial deformed core wobble without the participation of any quasiparticle? If not, what is the role of quasiparticle in the wobbling motion?

How to identify a pair of chiral doublet bands or a wobbling band experimentally? Which are the critical experimental proofs to distinguish them from other competing mechanisms?

Finally, to study the collective excitation, the validity of experimental results are of vital importance. To ensure the reliability of the extracted results from complex γ spectroscopy data, we recommend to promote standardized analyzing methods and make data public. In this way, we hope the data analysis process can be reproduced and assessed.

Bibliography

1. S. Frauendorf and J. Meng, Nucl. Phys. A **617**,131 (1997).
2. A. Bohr and B. R. Mottelson, *Nuclear Structure II*; Benjamin,W. A. Eds.; World Scientific Publishing Co. Re. Ltd.: Singapore; pp. 190 (1975).
3. K. Starosta, *et al.*, Phys. Rev. Lett. **86**, 971 (2001).
4. S. W. Ødegård, *et al.*, Phys. Rev. Lett. **86**, 5866 (2001).
5. B. W. Xiong and Y. Y. Wang, Atomic Data and Nuclear Data Tables **125**, 193 (2019).
6. C. Liu, *et al.*, Phys. Rev. C **100**, 054309 (2019).
7. B. F. Lv, *et al.*, Phys. Rev. C **103**, 019901 (2019).
8. C. M. Petrache, *et al.*, Eur. Phys. J. A **56**, 1 (2020).
9. S. Guo, *et al.*, Phys. Lett. B **807**, 135572 (2020).
10. X. C. Han, *et al.*, Phys. Rev. C **104**, 014327 (2021).
11. K. Y. Ma, *et al.*, Phys. Rev. C **103**, 024302 (2021).
12. B. M. Musangu, *et al.*; Phys. Rev. C **104**, 064318 (2021).
13. L. Mu, *et al.*, Phys. Lett. B **827**, 137006 (2022).
14. X. Xiao, *et al.*, Phys. Rev. C **106**, 064302 (2022).
15. W. Z. Xu, *et al.*, Phys. Lett. B **833**, 137287 (2022).
16. R. J. Guo, *et al.*, Phys. Lett. B **833**, 137344 (2022).
17. K. Katre, *et al.*, Phys. Rev. C **106**, 034323 (2022).
18. E. Grodner, *et al.*, Phys. Rev. C **106**, 014318 (2022).
19. W. Z. Xu, *et al.*, Phys. Lett. B **839**, 137789 (2023).
20. R.J. Guo, *et al.*, Phys. Rev. Lett. **132**, 092501 (2024).
21. T. J. Gao, *et al.*, Phys. Rev. C **109**, 024307 (2024).
22. J. T. Matta, U. Garg and W. Li, *et al.*, Phys. Rev. Lett. **114**, 082501 (2015).
23. N. Sensharma, U. Garg and S. Zhu, *et al.*, Phys. Lett. B **792**,170 (2019).
24. B. Biswas, R. Palit and S. Frauendorf, *et al.*, Eur. Phys. J. A **55**,159 (2019).
25. J. Timár, Q. B. Chen and B. Kruzsicz, *et al.*, Phys. Rev. Lett. **122**, 062501 (2019).
26. S.Chakraborty, H. P. Sharma and S. S. Tiwary, *et al.*, Phys. Lett. B **811**, 135854 (2020).
27. K. Rojeeta Devi, Kumar Suresh and Kumar Naveen, *et al.*, Phys. Lett. B **823**,136756 (2021).
28. S. Nandi, G. Mukherjee and Q. B. Chen, *et al.*, Phys. Rev. Lett. **125**, 132501 (2020).
29. N. Sensharma, U. Garg and Q. B. Chen, *et al.*, Phys. Rev. Lett. **124**, 052501 (2020).

30. A. Mukherjee, *et al.*; Phys. Rev. C **107**, 054310 (2023).

31. J. Meng, *et al.*, Phys. Rev. C **73**, 037303 (2006).

32. A. D. Ayangeakaa *et al.*, Phys. Rev. Lett. **110**, 172504 (2013).

33. I. Kuti *et al.*, Phys. Rev. Lett. **113**, 032501 (2014).

34. C. M. Petrache, B. F. Lv and A. Astier *et al.*, Phys. Rev. C **97**, 041304(R) (2018).

35. C. Liu *et al.*, Phys. Rev. Lett. **116**, 112501 (2016).

36. E. Grodner *et al.*, Phys. Rev. Lett. **120**, 022502 (2018).

37. A. A. Hecht *et al.*, Phys. Rev. C **68**, 054310 (2003).

38. J. Meng and S. Q. Zhang, J. Phys. G: Part. Phys. **37**, 064025 (2010).

39. S. J. Zhu *et al.*, Chin. Phys. C **33**(Suppl..), 145 (2009).

40. X. Y. Luo, *et al.*, *et al.*, Phys. Lett. B **607**, 370 (2009).

41. E. Mergel, C. M. Petrache and G. Lo Bianco *et al.*, Eur. Phys. J. A **15**, 417 (2002).

42. S. Mukhopadhyay, D. Almehed and U. Garg *et al.*, Phys. Rev. C **78**, 034311 (2008).

43. Q. B. Chen, B. F. Lv and C. M. Petrache, *et al.*, Phys. Lett. B **782**, 744 (2018).

44. Y. K. Wang, F. Q. Chen and P. W. Zhao *et al.*, Phys. Rev. C **99**, 054303 (2019).

45. Ch. Droste, S. G. Rohoziński and K. Starosta *et al.*, Eur. Phys. J. A **42**, 79 (2009).

46. Q. B. Chen, *et al.*, Phys. Rev. C **82**, 067302 (2010).

47. I. Hamamoto, Phys. Rev. C **88**, 024327 (2013).

48. I. Kuti *et al.*, Phys. Rev. Lett **113**, 032501 (2014).

49. K. T. Hecht and A. Adler, Nucl. Phys. A **137**, 129 (1969).

50. A. Arima, M. Harvey and K. Shimizu, Phys. Lett. B **30**, 517 (1969).

51. H. Z. Liang, J. Meng and S. G. Zhou, Phys. Rep. **570**, 1 (2015).

52. F. S. Stephens, M. A. Deleplanque and J. E. Draper, *et al.* Phys. Rev. Lett. **65**, 301 (1990).

53. J. Y. Zeng, J. Meng and C. S. Wu Phys. Rev. C **44**, R1745 (1991).

54. C. M. Petrache, G. Lo Bianco and D. Bazzacco, *et al.* Phys. Rev. C **65**, 054324 (2002).

55. J. N. Ginocchio, Phys. Rev. Lett **78**, 436 (1997).

56. P. A. Butler and W. Nazarewicz, Rev. Mod. Phys. **68**, 349 (1996).

57. Y. P. Wang, Y. Y. Wang and J. Meng, Phys. Rev. C **102**, 024313 (2020).

58. E.R. Marshalek, Nucl. Phys. A **331**, 429 (1979).

59. D. Janssen and I. N. Mikhailov, Nucl. Phys. A **318**, 390 (1979).

60. N. Onishi, Nucl. Phys. A **456**, 279 (1986).

61. K. Kaneko, Phys. Lett. B **255**,169 (1991).

62. K. Kaneko, Phys. Rev. C **45**, 2754 (1992).

63. D. R. Jensen, G. B. Hagemann and I. Hamamoto, *et al.*, Phys. Rev. Lett. **89**,142503 (2002).

64. G. Schönwaßer, H. Hübel and G. B. Hagemann, *et al.*, Phys. Lett. B **552**, 9 (2003).

65. H. Amro, W.C. Ma and G.B. Hagemann, *et al.*, Phys. Lett. B **553**, 197 (2003).

66. P. Bringel, G. B. Hagemann and H. Hübel, *et al.*, Eur. Phys. J. A **24**, 167 (2005).

67. D. J. Hartley, R. V. F. Janssens and L. L. Riedinger, *et al.*, Phys. Rev. C **80**, 041304(R) (2009).

68. I. Hamamoto, Phys. Rev. C **65**, 044305 (2002).

69. M. Matsuzaki, Y. R. Shimizu and K. Matsuyanagi, Phys. Rev. C **65**, 041303(R) (2002).

70. S. Frauendorf and F. Dönau, Phys.Rev. C **89**,014322 (2014).

71. E. A. Lawrie, O. Shirinda and C. M. Petrache, *et al.*, Phys. Rev. C **101**, 034306 (2020).

72. Q. B. Chen, S. Frauendorf and C. M. Petrache Phys. Rev. C **100**, 061301 (2020).

73. B. F. Lv, C. M. Petrache, R. Budaca, *et al.*, Phys. Rev. C **105**, 034302 (2022).

74. J. H. Hamilton *et al.*, Nucl. Phys. A **834**, 28c (2010).

75. Y. X. Luo *et al.*, in Exotic Nuclei: Exon-2012: Proceedings of the International Symposium (World Scientific, Singapore, 2013).

76. S. Guo, arXiv:2011.14364 (2020).

77. W. Hua, S. Guo and C. M. Petrache, arXiv:2011.14369 (2020).

78. S. Guo and C. M. Petrache, arXiv:2012.00343 (2020).

79. B. F. Lv, C. M. Petrache and E. A. Lawrie, *et al.*, Phys. Lett. B **824**, 136840 (2022).
80. S. Guo, X. H. Zhou and C. M. Petrache, *et al.*, Phys. Lett. B **828**,137010 (2022).
81. K. Starosta and T. Morek Ch. Droste, *et al.*, Nucl. Instrum. Methods Phys. Res., Sect. A **423**, 16 (1999).
82. D. Rupnik, E. F. Zganjar and J. L. Wood, *et al.*, Phys. Rev. C **58**, 771 (1998).
83. C. Bourgeois, P. Kilcher and J. Letessier, *et al.*, Nucl. Phys. A **295**, 424 (1978).
84. C. Bourgeois, M. G. Porquet and N. Perrin, *et al.*, Z. Phys. A **333**, 5 (1989).
85. S. Frauendorf and F. Dönau, Phys.Rev. C **92**, 064306 (2015).
86. Q. B. Chen, S. Q. Zhang and P. W. Zhao, *et al.*, Phys. Rev. C **94**, 044301 (2016.
87. R. Budaca, Phys. Rev. C **97**, 024302 (2018).
88. R. Budaca, Phys. Rev. C **103**, 044312 (2021).
89. E. Streck, Q. B. Chen and N. Kaiser, *et al.*, Phys. Rev. C **98**, 044314 (2018).
90. A. A. Raduta, R. Poenaru and C. M. Raduta, Phys. Rev. C **101**, 014302 (2020).
91. B.F. Lv, C. M. Petrache and E. A. Lawrie, *et al.*, Phys. Rev. C **103**, 044308 (2021).
92. F. Q. Chen and C. M. Petrache, Phys. Rev. C **103**, 064319 (2021).
93. C. M. Petrache, P. M. Walker and S. Guo, *et al.*, Phys. Lett. B **795**, 241247 (2019).
94. Y. K. Wang, F. Q. Chen and P. W. Zhao, Phys. Lett. B **802**, 135246 (2020).
95. B. F. Lv and C. M. Petrache, Phys. Rev. C **105**, 034302 (2022).
96. R. Budaca and C. M. Petrache, Phys. Rev. C **106**, 014313 (2022).
97. C. M. Petrache and S. Guo, arXiv:1603.08247
98. C. M. Petrache, F. Frauendorf and M. Matsuzaki *et al.*, Phys. Rev. C, **86**, 044321 (2012).

6 Tilted precession and wobbling in triaxial nuclei

Elena Atanasova Lawrie

iThemba LABS, National Research Foundation, Somerset West, and University of the Western Cape, Bellville, South Africa

6.1 EXCITED BANDS IN EVEN-EVEN TRIAXIAL NUCLEI

To date more than 3500 nuclei have been experimentally discovered. Atomic nuclei are quantum-mechanical objects, but they can also be described as drops of nuclear matter with well defined shapes. The liquid-drop model, assuming that the nuclear matter behaves similarly to a drop of liquid, predicts spherical nuclear shape. However, it is well-known that most observed nuclei are, in fact, deformed. The deformation arises as a result of the quantum mechanical nature of the building blocks of the nucleus, the nucleons. They occupy single-particle orbitals with specific shape-driving forces, that lead to an equilibrium nuclear shape that is most often deformed.

6.1.1 PARAMETRIZATION OF THE NUCLEAR SHAPE

It is common to describe the nuclear shape with an expansion of terms based on the spherical harmonics, $Y_{\lambda\mu}(\theta, \psi)$,

$$R(\theta, \psi) = R_0\left[1 + \sum_{\lambda\mu} \alpha_{\lambda\mu} Y_{\lambda\mu}(\theta, \psi)\right], \tag{6.1}$$

where R_0 is the radius of a spherical nucleus with the same volume and $\alpha_{\lambda\mu}$ are coefficients describing the magnitude of the deformation. Most deformed nuclei have quadrupole deformation with $\lambda = 2$, that can be described with five deformation coefficients, $\alpha_{2\mu}$, $\mu = -2, -1, 0, 1, 2$. Three of those refer to the orientation of the coordinate system, thus only α_{20} and α_{22} are needed. Instead of them, the quadrupole nuclear shape is often characterized by the β_2 and γ deformation parameters,

$$\alpha_{20} = \beta_2 \cos\gamma, \quad and \quad \alpha_{22} = \frac{1}{\sqrt{2}}\beta_2 \sin\gamma, \tag{6.2}$$

where $\beta_2 \geq 0$, measures the magnitude of the quadrupole deformation, while γ is an angle, that spans the range $0° \leq \gamma \leq 60°$, and indicates the asymmetry of the nuclear shape. For instance a deformed nucleus with an axial symmetry and two equal short (long) axes, has prolate (oblate) shape, and corresponds to $\gamma = 0°$ ($60°$). The asymmetric (or triaxial) nuclear shape is described by $0° < \gamma < 60°$.

6.1.2 ROTATION OF A DEFORMED EVEN-EVEN NUCLEUS

As the spherical symmetry in deformed nuclei is broken, they can rotate. An even-even nucleus with an axially-symmetric quadrupole shape can rotate only around an axis that is orthogonal to its

DOI: 10.1201/9781032691633-6

"

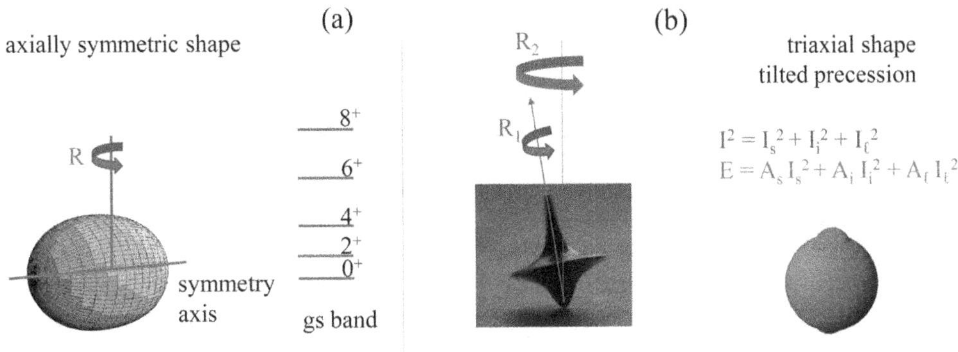

Figure 6.1 (a) The rotation of an axially symmetric nucleus proceeds around an axis that is orthogonal to its symmetry axis, and generates the ground-state rotational band. (b) The precessing top represents two simultaneous rotations, R_1 is the rotation of the top around its axis, while R_2 is the rotation of the axis of the top around the vertical direction. The precession of the total angular momentum I around the vertical direction can be presented geometrically using the classical expressions for the conservation of the total angular momentum and energy. The precession proceeds along the intersection of the angular momentum sphere and the energy ellipsoid.

symmetry axis, see Fig. 6.1(a). Thus, its excited states form a rotational band, where the excitation energy of a state with spin I is

$$E(I) = \frac{\hbar^2}{2\Im} I(I+1), \tag{6.3}$$

where $\Im$ is the moment of inertia (MoI) of the nucleus with respect to the rotational axis.

An even-even nucleus with triaxial shape can rotate simultaneously around each of its three axes, thus the rotational states are described by

$$E(I) = \frac{\hbar^2}{2\Im_s} I_s^2 + \frac{\hbar^2}{2\Im_i} I_i^2 + \frac{\hbar^2}{2\Im_\ell} I_\ell^2, \tag{6.4}$$

where I_s, I_i and I_ℓ are the projections of the total angular momentum (equal in this case to the projections of the rotational angular momentum) along the short, intermediate and long nuclear axes, respectively, and $\Im_s$, $\Im_i$ and $\Im_\ell$ are the corresponding MoIs.

The combination of a few simultaneous rotations represents a precession, as illustrated in Fig. 6.1(b). For instance the movement of a rotating top corresponds to two simultaneous rotations, R_1 where the top rotates around its axis, and R_2, where the rotational axis of the top spins around the vertical direction.

The movement can be classically described by the conservation of the angular momentum and energy equations

$$I^2 = I_s^2 + I_i^2 + I_\ell^2, \tag{6.5}$$

$$E = A_s I_s^2 + A_i I_i^2 + A_\ell I_\ell^2. \tag{6.6}$$

In a graphical representation in angular momentum space the conservation of the total angular momentum (energy) corresponds to a sphere (ellipsoid). Therefore, the simultaneous conservation of both is represented by the intersection of the sphere and ellipsoid, see Fig. 6.1(b). The intersection curve shows the precession of the total angular momentum and is equivalent to the precession of a rotating top.

The magnitudes of the projections of the rotational angular momentum around the three nuclear axes depend on the corresponding MoI. The energy required for the nucleus to rotate with angular momentum I is smallest if the rotation proceeds around the axis with largest MoI. Therefore, such rotations are energetically favored. They are equivalent to a rotation of a precessing top around its axis, R_1, see Fig. 6.1(b). The MoI of a triaxial nucleus depend on its γ deformation. It is common to describe the nuclear MoI using the irrotational-flow model, where $\mathfrak{I}_k \propto \sin^2(\gamma - \frac{2\pi k}{3})$, and $k = 1,2,3$ labels the nuclear axes. This dependence was recently supported by empirical evaluations of the three MoI in a number of triaxial nuclei [1, 2].

The MoI of an even-even triaxial nucleus (within the irrotational-flow model) is largest when the nucleus rotates around its intermediate axis. Therefore, its yrast band (the ground-state band) is formed by states with largest possible rotational angular momentum along the intermediate axis. Excited rotational bands appear when the angular momentum has an increased component along the short and (or) long axes. Such rotations tilt the total angular momentum away from the intermediate axis and make it to precess. They also cost more energy and therefore generate excited rotational bands. When the angular momentum along the short and (or) long nuclear axes is larger, the tilt of the precession increases too. The excited rotational bands that appear in triaxial even-even nuclei are commonly called γ bands and are illustrated in Fig. 6.2.

Gamma bands can be generated by other mechanisms too. For instance, it is known that (i) small γ vibrations around an average axially-symmetric shape and (ii) large γ vibrations covering the whole range of $0° < \gamma < 60°$ can also produce γ bands. In addition, it was pointed out by Bohr and Mottelson [3] that the γ bands produced by the precession in an axially asymmetric nucleus, at high spins, appear as if they were generated by excited vibrational phonons. These bands, at high spins, were called wobbling bands. Recently, questions were raised on the nature of the rotational bands in triaxial nuclei, in particular in what spin range (and whether at all) such bands can be considered as wobbling bands [4–8]. The questions were raised following a work [9] that set a precedent on the use of the term wobbling, naming rotational bands calculated with a rotational model at low spins, wobbling bands. Such use of the term wobbling for purely rotational bands, is in contrast with its known for decades meaning of describing a coupling of a simple rotation and excitation of vibrational phonons in triaxial nuclei, and creates misunderstanding and confusion.

In this work we address such issues, and propose a way forward. The distinct features of the purely rotational and the wobbling bands are studied and summarized. New terminology that is based on the nature of the excitations is proposed. Examples of how to test experimental data to extract information on the underlying nature of the observed rotational bands are discussed.

In the following, the bands generated by the rotation of a rigid triaxial nucleus are called Tilted Precession (TiP) bands, as proposed in Ref. [8]. The term highlights the fact that within the rotational model the angle of the precession of the total angular momentum (tilt angle) is different for different bands.

6.1.3 WOBBLING BANDS IN EVEN-EVEN TRIAXIAL NUCLEI

Wobbling is a coupling of a simple (one-dimensional) rotation with harmonic vibrational excitations and was introduced by Bohr and Mottelson [3]. (For other early works on wobbling see Refs. [10, 11].) The excitation energy of a state with spin I is

$$E(I,n) = \frac{\hbar^2}{2\mathfrak{I}_i}I(I+1) + (n+1/2)\hbar\omega, \qquad (6.7)$$

where the quantum number n indicates the number of excited wobbling phonons each carrying energy of $\hbar\omega$. The excited bands are illustrated in Fig. 6.2, where the ground-state band corresponds to $n = 0$, that is no excited phonons, the odd-spin states of the 2^+ γ band constitute the $n = 1$ wobbling-phonon band, the even-spin states of the 2^+ γ band belong to the $n = 2$ wobbling-phonon band, the odd-spin states of the 4^+ γ band correspond to the $n = 3$ wobbling-phonon band, etc.

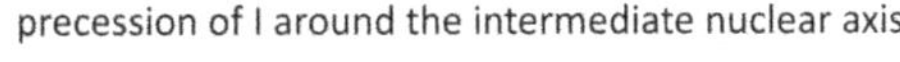
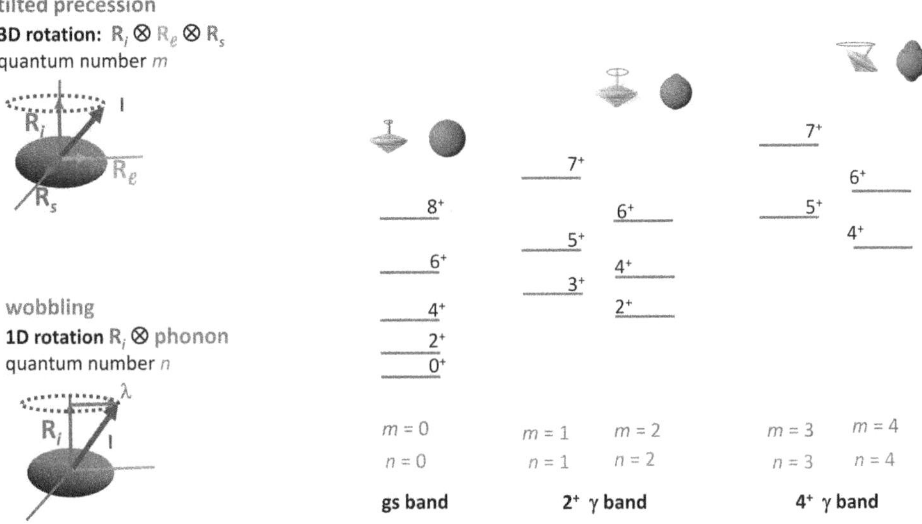

Figure 6.2 Tilted precession (TiP) bands are generated by three simultaneous rotations, where R_i is the largest, (it corresponds in Fig. 6.1(b) to the rotation of the top around its axis, R_1), while R_s and R_ℓ are small, and cause the tilt and the precession of the total angular momentum I. The tilt increases for the 2^+, and 4^+ γ bands. The bands can be described by the quantum number m. Wobbling is a coupling of a simple (one-dimensional) rotation, R_i, and one or more vibrational phonons, λ. The angular momentum of the phonon tilts the total angular momentum I away from the intermediate axis and causes it to precess. The bands are labeled by the number of wobbling phonons, n. While the two quantum numbers have the same value, $m = n$, they carry different meaning about the nature of the excitation.

Therefore, one characteristic feature of wobbling excitations is that the relative excitation energy (also called wobbling frequency),

$$E_{rel}(\Delta n = 1) = E(I,n) - E(I,n-1) = \hbar\omega, \tag{6.8}$$

is constant with n and with I.

There are also quite simple equations for the corresponding $B(E2)$ reduced transition probabilities. For instance the intra-band $B(E2)$ values are, (see e.g. Ref. [3]),

$$B(E2;n,I \to n,I-2) = \frac{5}{16\pi}e^2 Q_2^2, \tag{6.9}$$

indicating that $B(E2)_{intra}$ values are constant as a function of spin and are the same for bands with different quantum numbers n, if the quadrupole moments remain constant.

In addition, the inter-band $B(E2)$ rates are given by

$$B(E2;n,I \to n-1,I-1) = \frac{5}{16\pi}e^2 \frac{n}{I}\left(\sqrt{3}Q_{0}x - \sqrt{2}Q_{2}y\right)^2, \tag{6.10}$$

$$B(E2;n,I \to n+1,I-1) = \frac{5}{16\pi}e^2 \frac{n+1}{I}\left(\sqrt{3}Q_{0}y - \sqrt{2}Q_{2}x\right)^2, \tag{6.11}$$

where

$$x = \sqrt{\left(\frac{1}{2}\left(\frac{\alpha}{\hbar\omega}+1\right)\right)}, \quad y = \sqrt{\left(\frac{1}{2}\left(\frac{\alpha}{\hbar\omega}-1\right)\right)}, \tag{6.12}$$

and

$$\alpha = (A_s + A_\ell - 2A_i)I. \tag{6.13}$$

Therefore, for wobbling motion the $B(E2)_{inter}$ values are proportional to n, and inversely proportional to I. These $B(E2)_{inter}$ rates vanish for transitions that change n by more that 1, because such transitions infer a simultaneous creation or annihilation of more than one phonon and are thus forbidden.

These characteristic features are summarized in Table 6.1 of Section 6.4.

6.1.4 TILTED PRECESSION BANDS IN EVEN-EVEN TRIAXIAL NUCLEI

While the excitation energies and the $B(E2)$ transition probabilities for wobbling bands can be calculated using simple equations, there are no general analytical solutions for the Hamiltonian of a rotating triaxial nucleus, except for triaxial deformation of $\gamma = 30°$. In fact, nuclei with maximum triaxiality are most interesting, as they are best examples for excited bands generated by the axial asymmetry of the nuclear shape. Thus, we discuss the excitation energies and the $B(E2)$ transition probabilities for rotational, TiP, bands of a triaxial nucleus with $\gamma = 30°$, and with irrotational-flow MoI. For such nuclei $\mathfrak{I}_s = \mathfrak{I}_\ell = 1/4\,\mathfrak{I}_i$. The rotational Hamiltonian becomes

$$H = \frac{\hbar^2}{2\mathfrak{I}_i}I_i^2 + \frac{4\hbar^2}{2\mathfrak{I}_i}(I_s^2 + I_\ell^2) = \frac{\hbar^2}{2\mathfrak{I}_i}\left[I_i^2 + 4[I(I+1)-I_i^2]\right] = \frac{\hbar^2}{2\mathfrak{I}_i}[4I(I+1)-3I_i^2]. \tag{6.14}$$

where I_i is the projection of the total angular momentum on the intermediate axis. In this case I_i is a good quantum number, with possible values of $I_i = I, I - 1, I - 2, ... 0$.

One can then introduce a quantum number $m = I - I_i$. Then $m = 0$ signifies states for which $I = I_i$. These states have largest possible projection of the total angular momentum along the intermediate axis, that is, they are energetically favored and form the ground-state band. The states with $m = 1$ correspond to $I_i = I - 1$. They are the odd-spin states of the 2^+ γ band, and correspond to a small tilt of the total angular momentum away from the intermediate axis. The next band, $m = 2$, corresponds to $I_i = I - 2$, thus to a bit larger tilt of the total angular momentum away from the intermediate axis. The $m = 2$ band comprises the even-spin states of the 2^+ γ band. Therefore, while the quantum numbers m and n have different meaning, they label the ground-state and the excited bands in the same way, as illustrated in Fig. 6.2. Therefore, the TiP and the wobbling bands both correspond to a precession of the total angular momentum around the intermediate axis with a different tilt that increases for bands at higher excitation, as illustrated in Fig. 6.2. The two descriptions, however, differ in the nature of the excitation that causes the tilt; for TiP the tilt is caused by additional rotational angular momentum along the short and long nuclear axes, while for wobbling it has vibrational nature, caused by exciting vibrational phonons. Thus, the difference between TiP and wobbling stems from the different nature of their corresponding excitations, rotational and vibrational, respectively. As a consequence TiP and wobbling bands show specific differences in their behavior, which can be studied and compared with experimental data, as discussed below.

Using the quantum number m one can write the excitation energies for the TiP bands as,

$$E = \frac{\hbar^2}{2\mathfrak{I}_i}[4I(I+1)-3(I-m)^2] = \frac{\hbar^2}{2\mathfrak{I}_i}[I(I+4)+3m(2I-m)]. \tag{6.15}$$

The excitation energies of the rotational TiP states have a quadratic dependence of the quantum number m (see eq. 6.15) in contrast to the wobbling-phonon bands where the dependence of n is linear (see eq. 6.7). This causes differences in the excitation energies of the states. For instance the

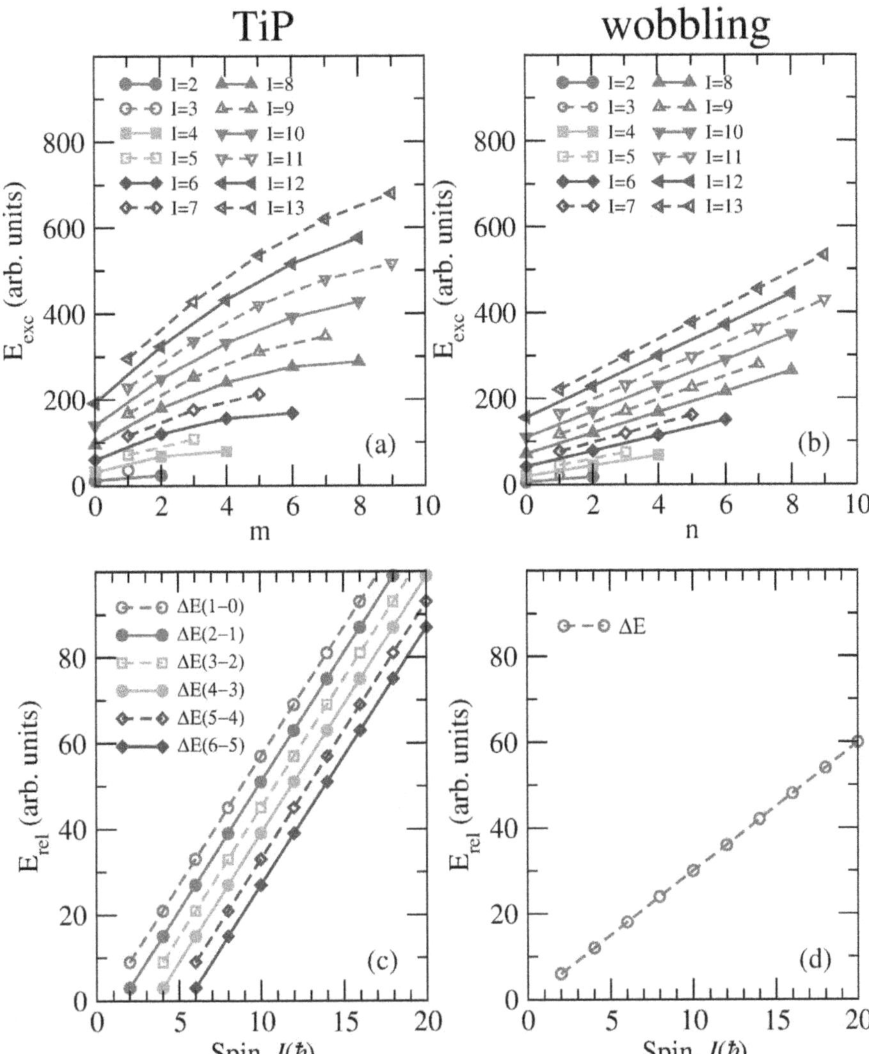

Figure 6.3 Excitation energies for given spin in even-even nuclei for TiP (a) and wobbling (b) bands, described by the corresponding quantum numbers m and n. The corresponding relative excitation energies for $\Delta m = \Delta n = 1$ are shown in (c) and (d), respectively.

excitation energies for states with given spin I are plotted in Fig. 6.3(a) and (b) as a function of the quantum number m (n) for TiP (wobbling) bands. The different patterns, which are quadratic for TiP and linear for wobbling, are obvious. In addition, there is a distinct difference in the relative excitation energies of two neighboring bands, see panels (c) and (d) of Fig. 6.3: for TiP bands $E_{rel}(\Delta m = 1)$ decreases for larger m, while for wobbling bands $E_{rel}(\Delta n = 1)$ is constant for different values of n.

Furthermore, for TiP bands the relation between the energies of some states is

$$E(3_\gamma^+) = E(2_g^+) + E(2_\gamma^+),\tag{6.16}$$

while for wobbling bands it is

$$E(3_\gamma^+) < E(2_g^+) + E(2_\gamma^+).\tag{6.17}$$

For TiP bands the $B(E2)$ reduced transition probabilities are proportional to the square of the corresponding Clebsch-Gordan coefficients, for instance the intra-band $B(E2)$ values are,

$$B(E2; m, I \rightarrow m, I-2) = Q_2^2 \; |<I(I-m)20|(I-2)(I-m)>|^2, \qquad (6.18)$$

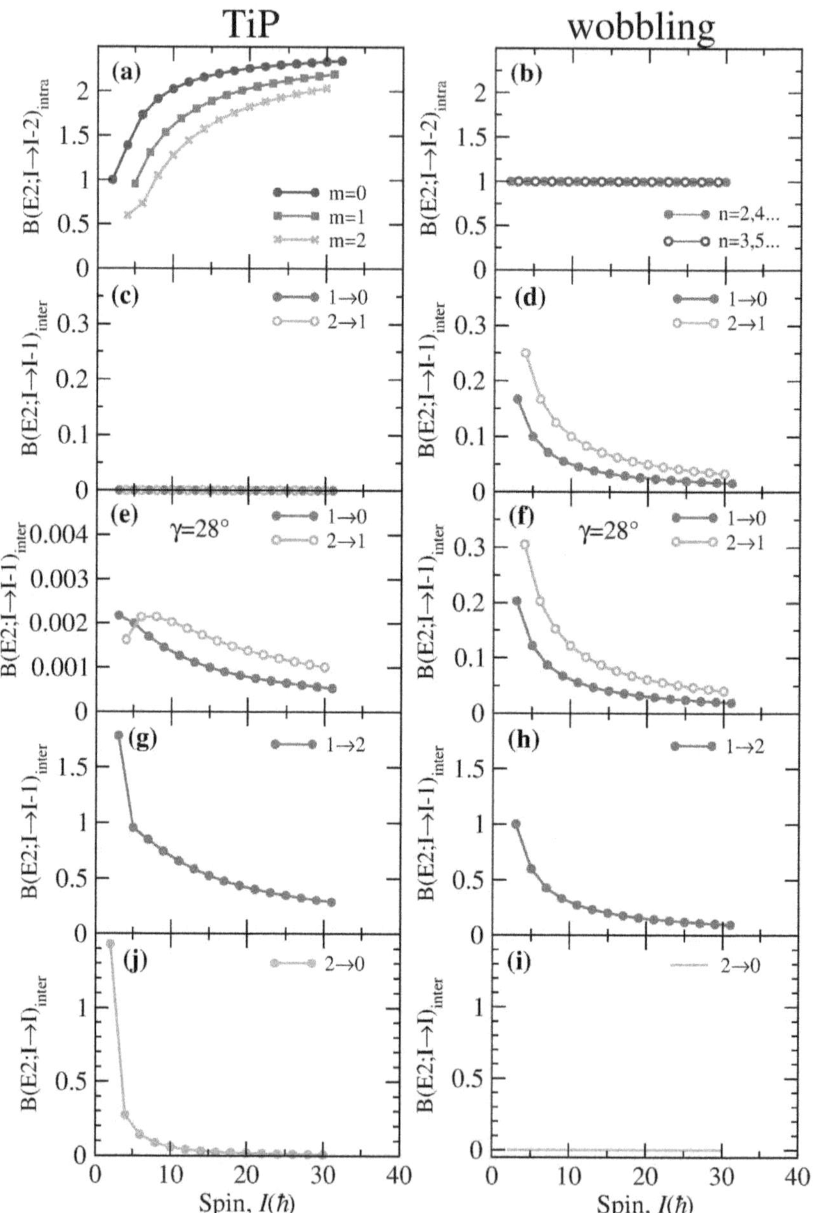

Figure 6.4 The $B(E2)$ transition probabilities for intra- and inter-band transitions in even-even triaxial nuclei normalised to $B(E2; 2_g^+ \rightarrow 0_g^+)$. The panels on the right correspond to wobbling-phonon description, while those on the left to the TiP approach. The nuclear shape has $\gamma = 30°$, except for panels (e) and (f) for which $\gamma = 28°$.

while for the inter-band transitions they are

$$B(E2; m, I \rightarrow m-2, I') = (1 + \delta_{(I-m),2}) \, Q_2^2 \; |<I(I-m)2-2|I'(I-m-2)>|^2, \qquad (6.19)$$

where the components of the intrinsic quadrupole moment with respect to the long axis,

$$Q_0 = e\frac{Z}{A}\beta_2 \cos\gamma, \quad and \quad Q_2 = \frac{1}{\sqrt{2}}e\frac{Z}{A}\beta_2 \sin\gamma, \tag{6.20}$$

should be transformed with respect to the intermediate axis,

$$Q_0' = -\frac{1}{2}Q_0 + \sqrt{\frac{3}{2}}Q_2, \quad and \quad Q_2' = \sqrt{\frac{3}{8}}Q_0 + \frac{1}{2}Q_2. \tag{6.21}$$

The intra-band $B(E2)$ values for TiP bands increase as a function of spin, in contrast to those for wobbling bands which remain constant, see Fig. 6.4 (a) and (b). In addition, the intra-band $B(E2)$s for rotational bands decrease for higher m, while for wobbling bands the $B(E2)$ values do not depend on n.

The inter-band $B(E2)$ rates also show distinct differences. For nuclear shape with $\gamma = 30°$ several inter-band $B(E2)$ probabilities for TiP bands are zero due to the vanishing quadrupole moment Q_0', for instance, $B(E2; m, I \to m - 1, I - 1) \approx 0$, indicating that such a transition cannot have an $E2$ component, see Fig. 6.4(c). Contrary to that, the inter-band $B(E2; n, I \to n - 1, I - 1)$ values for wobbling bands are large, particularly at low spins, and have characteristic dependencies of $\propto n$, and $\propto 1/I$, see Fig. 6.4(d). It should also be noted, that for nuclear shapes with $\gamma \neq 30°$ the inter-band $B(E2; m, I \to m - 1, I - 1)$ values for TiP bands are non-zero, like those shown in Fig. 6.4(e) for $\gamma = 28°$. In contrast to the $B(E2)$ probabilities for vibrational excitations, the $B(E2; m, I \to m - 1, I - 1)$ rates for TiP are not proportional to m; in fact, it is possible to have $B(E2; m + 1, I \to m, I - 1) < B(E2; m, I \to m - 1, I - 1)$, see Fig. 6.4(e). In addition, the spin dependence for TiP bands is more complex than the simple $\propto 1/I$ relation characteristic for wobbling bands.

Another important difference relates to the inter-band $B(E2; \Delta n = 2)$ transitions. For wobbling bands all inter-band transitions that link the bands of n phonons to the bands of n-2 phonons are forbidden, therefore such transitions cannot proceed. For rotational excitations, these transitions are not generally forbidden, and in some cases are very intense. For TiP bands there are also some vanishing $B(E2)$ probabilities. For instance the $B(E2)$ rates for transitions of the type $(m, I) \to (m - 2, I + 2)$ (e.g. $2_\gamma^+ \to 4_g^+$) and $(m, I) \to (m - 2, I + 1)$ vanish due to the corresponding Clebsch-Gordan coefficients. In addition, the $B(E2)$s for $(m, I) \to (m - 2, I - 2)$, (for instance $2_\gamma^+ \to 0_g^+$) and the $E2$ component for the $(m, I) \to (m - 2, I - 1)$ inter-band $\Delta m = 2$ transitions vanish if $\gamma = 30°$ due to vanishing Q_0'. They can have considerable magnitude for $\gamma \neq 30°$. Most distinct differences between bands with TiP and wobbling nature are expected for the $B(E2; \Delta m = 2)$ probabilities of transitions of the type $(m, I) \to (m - 2, I)$, particularly at low spins, e.g. for the $2_\gamma^+ \to 2_g^+$ transition. These $B(E2)$ probabilities can be very large for TiP bands, see Fig. 6.4 (i), in contrast with their vanishing values for wobbling bands.

Therefore, while for wobbling-phonon bands all transitions linking bands with $\Delta n \geq 2$ are forbidden as they correspond to a change of 2 (or more) phonons, such restrictions are not imposed for TiP bands. In particular non-zero $B(E2)$ rates are expected for transitions of the type $(m, I) \to (m - 2, I - 2)$, and $(m, I) \to (m - 2, I - 1)$ for $\gamma \neq 30°$, and particularly large $B(E2)$ values are expected for $(m, I) \to (m - 2, I)$ transitions at low spins.

Other distinct differences that can be used for testing experimental data for possible TiP or wobbling nature are:
(i) the ratio $R_{2\gamma 2g} = B(E2; 2_\gamma^+ \to 2_g^+)/B(E2; 2_g^+ \to 0_g^+)$, that spans the range between 0.4 and 1.4 for $20° \leq \gamma \leq 30°$ for TiP, while it is $R_{2\gamma 2g} = 0$ for wobbling bands; and
(ii) the ratio $R_{3\gamma 2\gamma} = B(E2; 3_\gamma^+ \to 2_\gamma^+)/B(E2; 2_g^+ \to 0_g^+) = 1.78$ for TiP (for all γ values), and $R_{3\gamma 2\gamma} = 1.0$ for wobbling bands.

The features that are distinctly different for bands with TiP and wobbling nature are summarized in Table 6.1 of Section 6.4.

6.1.5 TIP AT HIGH SPINS AND WOBBLING: HARMONIC APPROXIMATION IN EVEN-EVEN TRIAXIAL NUCLEI

Considering the $B(E2)$ transition probabilities, shown in Fig. 6.4, it is very obvious that while at low spins the two descriptions, in terms of rotational TiP bands and of wobbling-phonon excitations, are considerably different, at high spins they behave quite similarly. While at low spins the values of the intra-band $B(E2)$ probabilities decrease for higher m, see Fig. 6.4(a), as the spin increases this difference diminishes and the $B(E2)$ rates become not that different for different m. This is roughly in line with the wobbling phonon description, where the intra-band $B(E2)$ rates do not depend on n, see Fig. 6.4(b). In addition, while at low spins the intra-band $B(E2)$ rates for TiP bands increase fast as a function of spin, at high spins the dependence becomes flatter, and more similar to the constant value obtained within the wobbling description.

Similar trends can be identified for the inter-band $B(E2)$ rates. For instance, while at low spins there is a distinct difference in the inter-band $B(E2)$ rates for TiP and wobbling bands (see Fig. 6.4), this difference diminishes at high spin where the behavior of the inter-band $B(E2)$ rates becomes similar for TiP and wobbling bands.

The excitation energies for TiP and wobbling bands also show less discrepancies at high spins. While the difference caused by the quadratic and the linear dependencies in the excitation energies cannot vanish, the deviation at high spins becomes small in comparison with the rest of the excitation energy and can be considered as an anharmonic term within the wobbling-phonon framework.

Following the trend of diminishing deviations in the behavior of TiP and wobbling bands at high spin it is possible to consider them as approximately equivalent. In fact, the wobbling-phonon description was initially introduced in the late '70s by Bohr and Mottelson as an approximation of the three-dimensional rotational Hamiltonian with the coupling of a one-dimensional rotation and excitations of vibrational phonons, [3]. In that approximation the rotations around the short and long nuclear axes were approximated with the excitations of vibrational phonons. It was shown that this approximation is valid only at high spins, where

$$f(n,I) = (2n+1)\frac{A_s + A_\ell - 2A_i}{2I\sqrt{(A_s - A_i)(A_\ell - A_i)}} \ll 1. \tag{6.22}$$

Here $A_i = 2\hbar^2/\mathfrak{I}_i$ is the rotational constant with respect to the intermediate axis, the axis with largest MoI, and n is the number of excited phonons. The function $f(n,I)$ for the ground-state band of an even-even triaxial nucleus is plotted in Fig. 6.5. The approximation condition ensures that the difference between the rotational Hamiltonian, eq. 6.4 and the harmonic wobbling equation, eq. 6.7, is small and can thus be considered as small anharmonicity that does not affect the general harmonic vibrational features of the wobbling bands. Therefore, at high spins the rotational model calculations in triaxial even-even nuclei can be considered as approximately describing wobbling.

Despite all the efforts since the late '70s wobbling bands could not be discovered in the even-even triaxial nuclei. On the other hand, interpretations within the triaxial rotor model (TRM), where the motion represents precession of purely rotational nature similar to the precession of a rotating top, were often proposed for the ground-state and the γ bands of many even-even triaxial nuclei. These bands were never accepted as wobbling bands, as they were observed at low spins where the TRM rotational interpretation is not equivalent to the wobbling description based on excitations of vibrational phonons. In Section 6.4, we illustrate how experimental data can be compared with these two alternative descriptions.

6.2 EXCITED BANDS IN ODD-MASS TRIAXIAL NUCLEI

6.2.1 WOBBLING AT HIGH SPINS – THE TSD BANDS IN THE ODD-MASS LU-TA ISOTOPES

Wobbling bands have been discovered at high spins in the odd-mass $^{161-167}$Lu and ^{167}Ta isotopes [12–17]. The bands were called triaxial superdeformed (TSD) bands, as the nucleus was associated

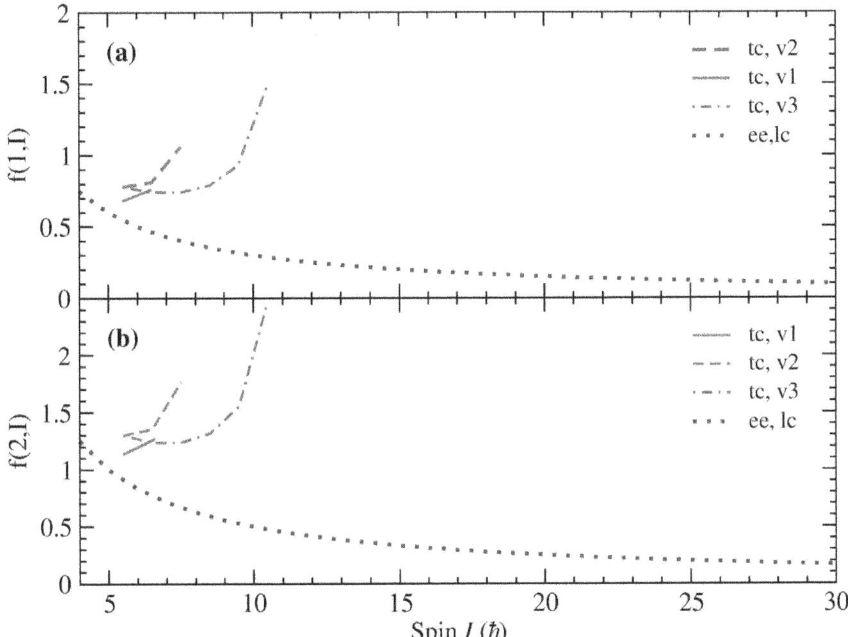

Figure 6.5   Approximation function $f(n,I)$ calculated for irrotational-flow MoI and $\gamma = 30°$ and for even-even (ee) nuclei, and odd-mass nuclei with longitudinal (lc) and transverse coupling (tc) of the single-particle and rotational angular momenta. For the odd-mass nuclei the single particle angular momentum is $j = 5.5$ and it is frozen along the intermediate (short) axis for longitudinal (transverse) coupling. Three versions for transverse coupling are considered, corresponding to MoI ratios of 1:1:4 (v1), 1:3:6 (v2) and 1:2:6 (v3).

with superdeformed shape and axial asymmetry. The bands were observed up to very high spins of $\sim 50\,\hbar$. They were interpreted as built on a $\pi i_{13/2}$ configuration, where the valence proton is aligned along the short nuclear axis. Theoretical total Routhian surface (TRS) [18], Ultimate cranker (UC) [16, 19–21] and Cranked Nilsson-Strutinsky (CNS) [22] calculations found that rotation around the short nuclear axis was favored, which led to the assumption that the MoI with respect to that axis was largest. Therefore, the rotational and the single-particle angular momenta align along the same nuclear axis (referred in the following as longitudinal coupling), in contrast to the transverse coupling case where the two angular momenta are orthogonal. The adopted description of the TSD bands is in contrast with standard assumptions that the rotational angular momentum of an even-even triaxial nucleus around the intermediate axis dominates. On the other hand the TSD bands persist up to very high spins where the pairing interactions are expected to diminish, and the nuclear rotation to begin resembling the rotation of a rigid body.

The TSD bands were interpreted using different models including the (quasi)particle-plus-triaxial-rotor (QTR) model. This is a purely rotational model that generates precession of the total angular momentum around an axis and comprises a coupling of three simultaneous rotational angular momenta with the angular momentum of a valence nucleon. In contrast to that the wobbling mode assumes that the tilt of the total angular momentum is generated by excited vibrational phonons. These two approaches (as mentioned above) can predict very similar behavior if the approximation condition $f(n,I) \ll 1$ is valid and, on the contrary, may differ substantially if it is not valid. The function $f(n,I)$ for the TSD bands and for longitudinal coupling is shown in Fig. 6.5. It is identical to $f(n,I)$ for even-even nuclei and is thus valid at high spins. Therefore, the approximation of the QTR model with the simpler wobbling-phonon approach is adequate for the TSD bands with longitudinal coupling at high spins, where QTR can be considered as similar to an anharmonic wobbling description. The TSD bands were adopted as the first observation of wobbling bands.

6.2.2 TIP AND WOBBLING BANDS IN ODD-MASS NUCLEI AT LOW SPINS

The concept of wobbling in odd-mass triaxial nuclei was revisited a few years ago [9] in a work that had an important impact including on the use of the term "wobbling". Before discussing the terminology in more details, it is important to mention that following the introduction of the TRM and QTR models in the late '70s, a large number of deformed triaxial nuclei, including even-even, and odd-mass nuclei with longitudinal or transverse coupling, have been interpreted within these two rotational models. All the calculated excited bands were understood as rotational bands, where the total angular momentum precesses around an axis, and not as wobbling bands despite the fact that wobbling has been rigorously searched for for decades. The rotational precession of the TRM and QTR descriptions was not considered equivalent to the vibrational excitation required within the wobbling approach. This fact highlights the importance of establishing the nature (vibrational or rotational) of the precession, which determines the wobbling or the TiP character of the observed bands.

As in even-even nuclei, the rotational precession in odd-mass nuclei, described by the QTR Hamiltonian can be approximated with wobbling-phonon excitations when the approximation condition $f(n,I) \ll 1$ is valid [9]. The derivation of the corresponding wobbling-phonon equations assumed that the single-particle angular momentum remains aligned (frozen) along a nuclear axis. Freezing of the single-particle angular momentum plays an important role in the derivations, as it ensures that the excitations are purely collective, as expected for a vibrational description such as wobbling.

6.2.2.1 Longitudinal coupling

The rotational Hamiltonian for a triaxial rotor coupled to a single nucleon, with the single-particle angular momentum rigidly aligned along the axis with largest MoI is

$$H = \frac{\hbar^2}{2\mathfrak{I}_i}(I_i - j)^2 + \frac{\hbar^2}{2\mathfrak{I}_s}I_s^2 + \frac{\hbar^2}{2\mathfrak{I}_\ell}I_\ell^2, \tag{6.23}$$

where the largest MoI is assumed to be around the intermediate axis.

This Hamiltonian describes a precession of the total angular momentum around the intermediate axis which is very similar to that in even-even nuclei. It can be illustrated geometrically using the conservation of the total angular momentum and energy. The single-particle angular momentum simply shifts the center of the energy ellipsoid along the intermediate axis, as sketched in Fig. 6.6, while the total angular momentum moves along the intersection of the angular momentum sphere and the energy ellipsoid, as illustrated in Fig. 6.6. For higher excited bands the tilt of the precessing total angular momentum increases, while it decreases for the higher-spin states of a band.

For irrotational-flow MoI and $\gamma = 30°$ the excitation energy of the rotational states has the simple analytic expression

$$E = \frac{\hbar^2}{2\mathfrak{I}_i}(I_i - j)^2 + \frac{4\hbar^2}{2\mathfrak{I}_i}(I_s^2 + I_\ell^2) = \frac{\hbar^2}{2\mathfrak{I}_i}\left[4I(I+1) + (I_i - j)^2 - 4I_i^2\right]. \tag{6.24}$$

In this case the projection I_i is a good quantum number, and the rotational bands correspond to $I_i = I, I-1, \dots j$. As above, one can introduce quantum number m that represents the difference of I and I_i,

$$m = I - I_i. \tag{6.25}$$

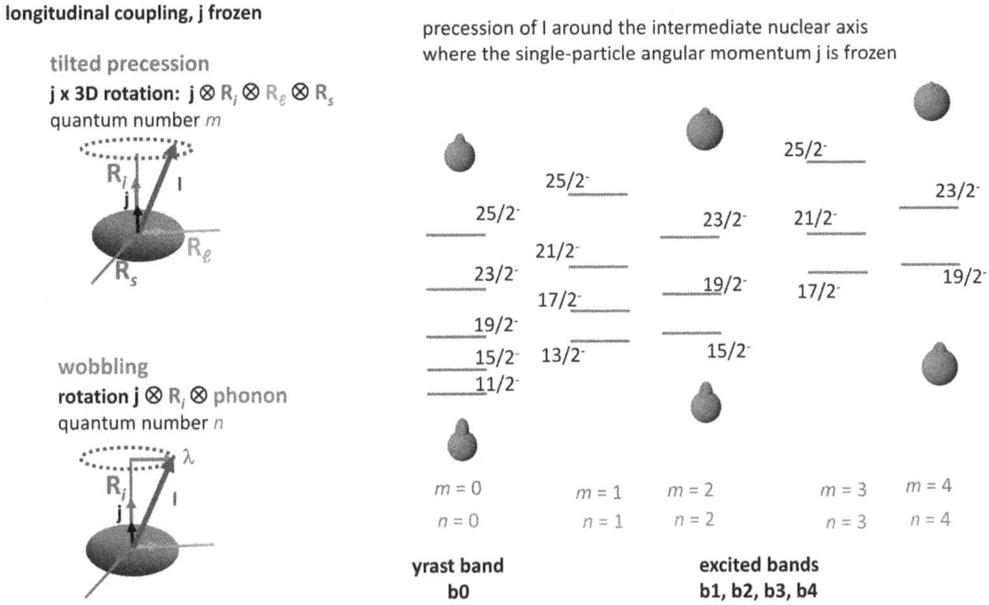

Figure 6.6 For longitudinal coupling the TiP bands are generated by the coupling of the single-particle angular momentum j, (considered as frozen along the intermediate axis), and the three simultaneous rotations of the core, where R_i is the largest. The precession occurs around the intermediate axis. The tilt of the precessing total angular momentum increases for the excited bands and decreases at higher spins. Wobbling is a coupling of the frozen single-particle angular momentum j with a simple rotation R_i of the core, and one or more vibrational phonons, λ. The angular momentum of the phonons tilts the total angular momentum I away from the intermediate axis and causes it to precess. The wobbling bands are labeled by the number of wobbling phonons, n, while the TiP bands by the rotational quantum number m.

Then the yrast band corresponds to $m = 0$, the first excited band to $m = 1$, etc. The excitation energy can be expressed as a function of m,

$$
\begin{aligned}
E(m,I) &= \frac{\hbar^2}{2\Im_i}\left[4I(I+1)+(I-m-j)^2-4(I-m)^2\right] \\
&= \frac{\hbar^2}{2\Im_i}\left[4I(I+1)+(-I+m-j)(3I-3m-j)\right] \\
&= \frac{\hbar^2}{2\Im_i}\left[(I-j)^2+4I+m(6I+2j-3m)\right].
\end{aligned}
\tag{6.26}
$$

The dependence of $E(m,I)$ with respect to m is quadratic. In contrast, the wobbling energy for longitudinal coupling of the angular momenta are

$$
E(n,I) = A_i(I-j)(I-j+1)+\hbar\omega(n+1/2),
\tag{6.27}
$$

where the intermediate axis is the axis with largest MoI. The excitation energies have a linear dependence on the number of excited wobbling phonons n, in contrast with the quadratic dependence on m in eq. 6.26 valid for TiP bands. Each phonon tilts the total angular momentum away from the intermediate axis, and causes precession as sketched in Fig. 6.6.

The $B(E2)$ transition probabilities for wobbling bands with longitudinal coupling are the same as those for even-even nuclei, see eqs. 6.18 and 6.19.

The approximation function is also the same as that for even-even nuclei, eq. 6.22, plotted in Fig. 6.5. Therefore, while the wobbling and the QTR equations produce similar descriptions of the rotational bands at high spins, at low spins the predictions of the two models are different and one can test the experimental data to deduce whether the vibrational or the rotational approach represents a more adequate interpretation. Examples of such evaluations are discussed in Section 6.4.

6.2.2.2 Transverse coupling

The rotational Hamiltonian for transverse coupling (where the single-particle angular momentum is aligned along an axis orthogonal to the axis with largest MoI, for instance along the short axis) and within the frozen single-particle assumption, reads

$$H = \frac{\hbar^2}{2\mathfrak{I}_s}(I_s - j)^2 + \frac{\hbar^2}{2\mathfrak{I}_i}I_i^2 + \frac{\hbar^2}{2\mathfrak{I}_\ell}I_\ell^2. \tag{6.28}$$

At low spins this Hamiltonian describes a precession of the total angular momentum around the short axis which changes to a precession around a tilted axis at higher spins. This motion is illustrated geometrically as shown in Fig. 6.7. The single-particle angular momentum shifts the center of the energy ellipsoid along the short axis. At low spins the rotational angular momentum is small, thus the total angular momentum (mostly of single-particle nature), precesses around the short axis following the alignment of the single-particle angular momentum. At higher spins the rotational angular momentum becomes larger. This collective angular momentum has dominant component along the intermediate axis and causes the total angular momentum to gradually re-align toward the intermediate axis, while still precessing. The process is illustrated in Fig. 6.7, where the precession at low and at high spin is sketched. It should be noted that this re-alignment proceeds even faster when a realistic description is considered, where the single-particle angular momentum is not frozen but free to align toward the intermediate axis at higher spins.

The wobbling approximation for transverse angular momentum coupling was derived in the same way as those for even-even nuclei, under the assumption that the single-particle angular momentum j remained fixed (frozen) along the nuclear axis (in this case the short axis) [9]. The excitation energy is

$$E(n,I) = A_s(I-j)(I-j+1) + \hbar\omega(n+1/2), \tag{6.29}$$

where

$$\hbar\omega = \sqrt{\alpha^2 - \beta^2} = 2I\sqrt{(A_i - A_s')(A_\ell - A_s')}, \tag{6.30}$$

and

$$\alpha = (A_i + A_\ell - 2A_s')I \quad and \quad \beta = (A_\ell - A_i)I, \tag{6.31}$$

and

$$A_s' = A_s\left(1 - \frac{j}{I}\right). \tag{6.32}$$

It should be noted that the wobbling description for transverse coupling [9] is limited by a maximum spin, I_{max}, beyond which it fails, as the term $(A_i - A_s')(A_\ell - A_s')$ in eq. 6.30 becomes negative.

The $B(E2)$ reduced transition probabilities are also similar to previous cases, they are described by eqs. 6.18 and 6.19, except that the quadrupole moments are defined with respect to the short axis [9].

The $B(M1)$ transition probabilities [9] are given by

$$B(M1,n,I \to n-1,I-1) = \frac{3}{4\pi}\frac{n}{I}\left[j(g_j - g_R)x\right]^2 \tag{6.33}$$

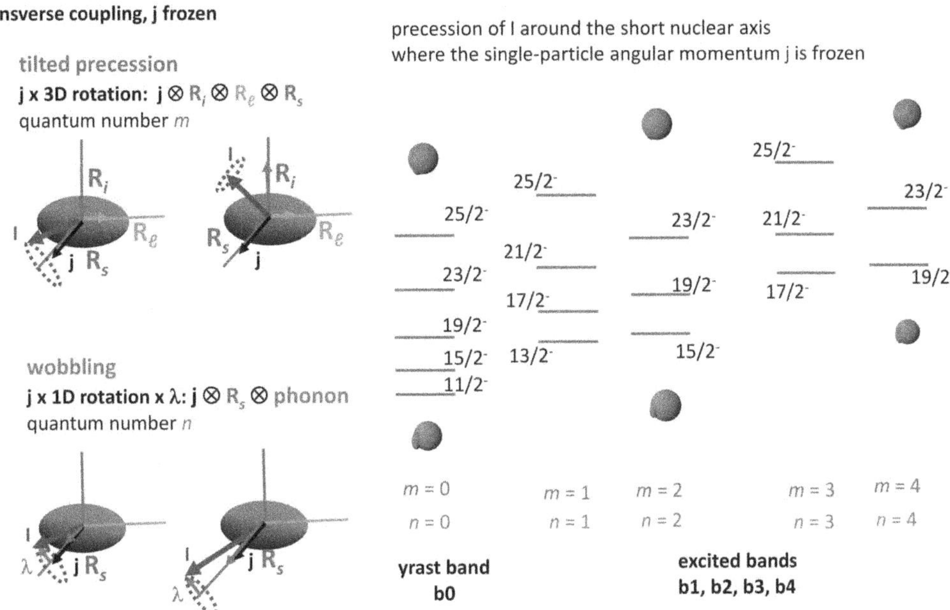

Figure 6.7 For transverse coupling the TiP bands represent a coupling of the single-particle angular momentum j (considered here as frozen along the short axis) and the three simultaneous rotations of the core, where R_i is the largest. At low spins the total angular momentum I is mostly due to j, and thus it precesses around the short axis. At high spins the rotational angular momentum, in particular R_i, becomes substantial and tilts the axis of precession toward the intermediate nuclear axis. In general the tilt is larger for the excited bands. Wobbling is a coupling of the frozen single-particle angular momentum j with a simple rotation R_s of the core, and one or more vibrational phonons, λ. The angular momentum of the phonons tilts the total angular momentum I away from the short axis and causes it to precess. The excited wobbling bands are labeled by the number of excited wobbling phonons, n.

$$B(M1, n, I \rightarrow n+1, I-1) = \frac{3}{4\pi} \frac{n+1}{I} [j(g_j - g_R)y]^2. \tag{6.34}$$

In principle, as in previous cases, there might be a spin range where the rotational- and the vibrational-excitation descriptions become similar. This occurs when the approximation condition for transverse coupling,

$$f(n, I) = (2n+1) \frac{A_i + A_\ell - 2A'_s}{2I\sqrt{(A_i - A'_s)(A_\ell - A'_s)}} \ll 1, \tag{6.35}$$

is valid.

The approximation function $f(n, I)$ is shown in Fig. 6.5 for bands with transverse angular momentum coupling, and for $j = 11/2$ and rotational constants of $A_1 : A_2 : A_3 = 1 : 4 : 4$ (labeled as v1), $A_1 : A_2 : A_3 = 1 : 3 : 6$ (labeled as v2) and $A_1 : A_2 : A_3 = 1 : 2 : 6$ (labeled as v3). In all cases I_{max} is quite low. In contrast to the previous two cases (of even-even nuclei and odd-mass nuclei with longitudinal coupling) there is no spin range where the values of the $f(n, I)$ function become negligible, see the examples shown in Fig. 6.5 for which $f(n, I) > 0.6$ for $n=1$, and $f(n, I) > 1.1$ for all spins. More cases with other rotational constants are considered in [8], where it was shown that for transverse coupling the approximation $f(n, I) \ll 1$ is generally not fulfilled. Therefore, in this case the rotational (TiP) and the vibrational (wobbling) descriptions show distinct differences in the observed spin range of the rotational bands. The divergence stems from their different underlying nature, rotational and vibrational, respectively. It also indicates that one can test the experimental data against the two models to establish whether the rotational or the vibrational scenario is favored.

To illustrate the differences in the features of TiP and wobbling bands, transverse coupling was considered for a frozen $j = 11/2$ nucleon. In contrast with the previous cases where the rotational model reduces to simple analytical expressions for the excitation energies and transition probabilities, for transverse coupling there are no naturally good quantum numbers and the solutions have to be calculated numerically. Thus, QTR calculations were done with standard parameters for ^{135}Pr, and for $\gamma = 30°$. The calculated excitation energies, and the $B(E2)$ and $B(M1)$ reduced transition probabilities are shown in Figs. 6.8 and 6.9.

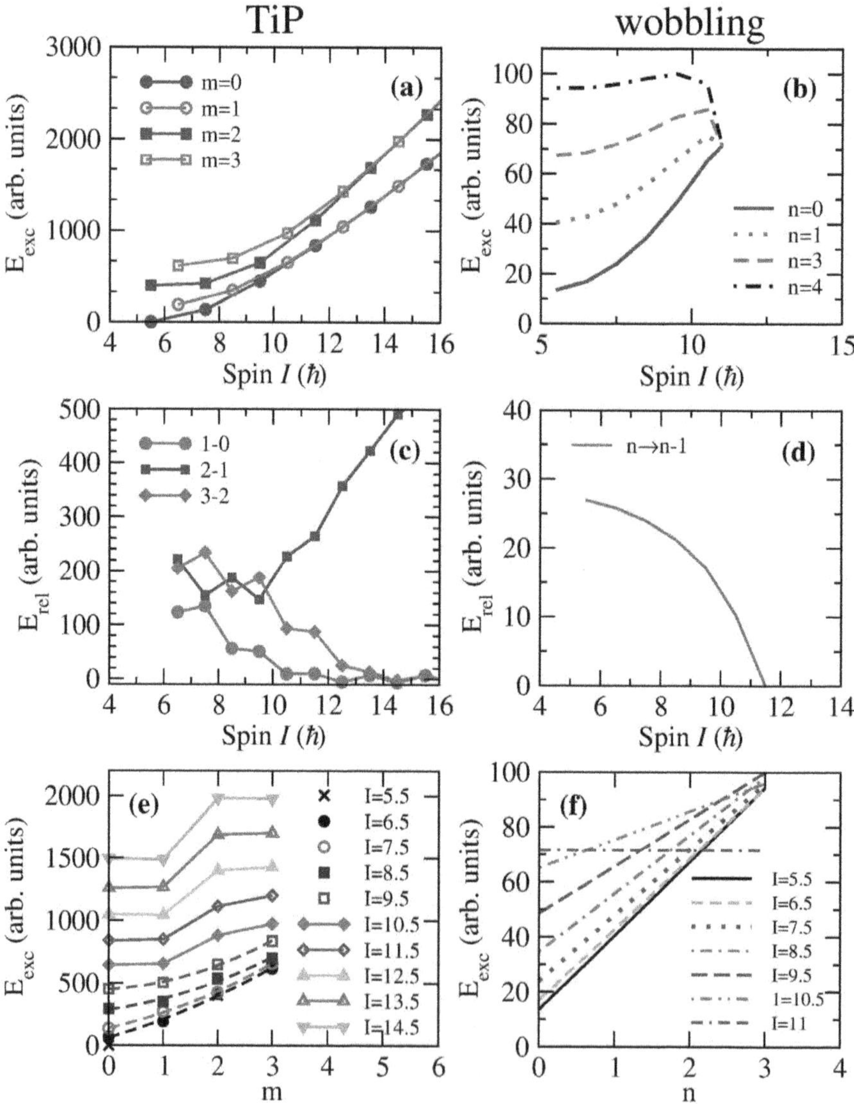

Figure 6.8 Excitation energies and relative excitation energies for transverse coupling in odd-mass nuclei within rotational (a), and vibrational (b) excitation nature. The corresponding relative excitation energies are shown in (c) and (d), while the corresponding excitation energies for given spin are plotted in (e) and (f). The dashed lines represent quadratic fits to the data to guide the eye.

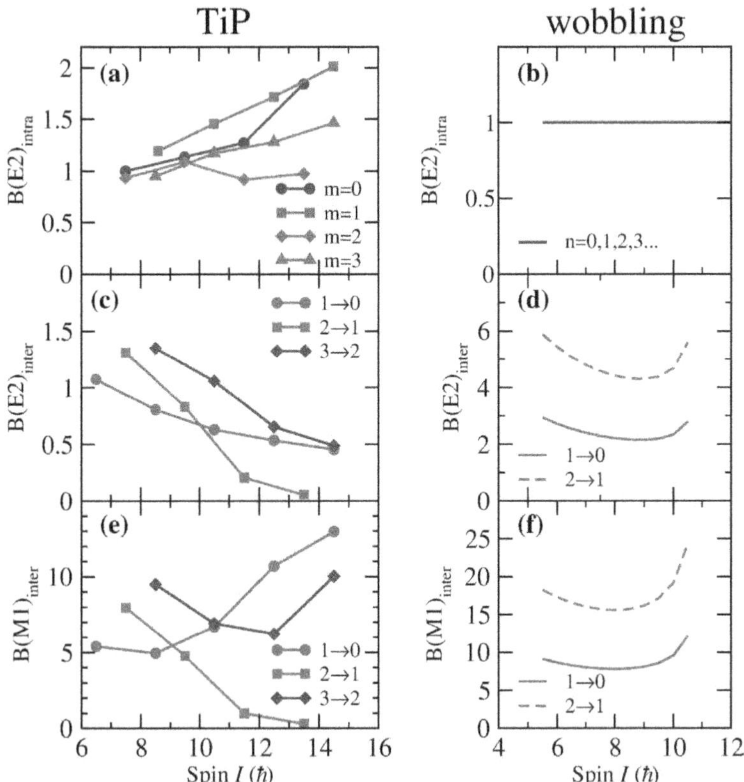

Figure 6.9 Intra- and inter-band $B(E2)$ and $B(M1)$ reduced transition probabilities normalized to the intra-band $B(E2; 15/2^- \to 11/2^-)$ value for transverse coupling in odd-mass nuclei within the rotational (left panels) and vibrational (right panels) excitation models.

The excitation energies and the corresponding $B(E2)$ values calculated using the wobbling equations are illustrated in Figs. 6.8 and 6.9. It should be noted that these equations are applicable only up to a maximum spin I_{max}. Irrotational-flow MoI with rotational constants of $A_s = A_\ell = 4A_i$ and $j = 11/2$ yield $I_{max} = 7.5$. The alternative versions of $3A_s = A_\ell = 6A_i$ and $2A_s = A_\ell = 6A_i$ yield a bit higher value for I_{max} of 8.25 and 11. The latter version was adopted as it allows largest I_{max}. The excitation energies for the states of the wobbling bands with $n = 0$, 1, 2 and 3 are shown in Fig. 6.8(b). It is clear that the relative excitation energies decrease as a function of spin, until near $I = I_{max}$ where the excitation energies of all bands become equal. In contrast, the excitation energies for the TiP bands show a different trend. These bands group in pairs, and the excitation energies of the bands in a pair become equal at a given spin I_c, see Fig. 6.8(a). In fact, the two bands of the pair can be considered as showing large signature splitting for $I < I_c$, while at higher spins, where $I > I_c$, the signature splitting vanishes. It should also be noted that the value of I_c is different for each pair, and that I_c increases for pairs at higher excitation energies.

The different trends in the excitation energies are also visible in the plots of the relative excitation energies, $E_{rel}(\Delta n = 1)$, shown in panels (c) and (d) of Fig. 6.8. Panel (d) highlights the characteristic feature of the vibrational wobbling approach that the relative excitation energy (also called wobbling frequency) is independent on the number of exited phonons n, and is therefore the same for all bands. In contrast, the relative excitation energies for TiP bands are different as a function of the rotational quantum number m. For instance, for two bands that belong to a pair, (e.g. $E_{rel}(1-0)$), $E_{rel}(I)$ decreases as a function of spin, while for two bands that do not belong to a pair (e.g. $E_{rel}(2-1)$),

$E_{rel}(I)$ increases for $I > I_c$. In addition, for bands that belong to pairs E_{rel} increases for excited bands (e.g. $E_{rel}(3-2) > E_{rel}(1-0)$).

Panels (e) and (f) show the excitation energies of states in different bands for a constant I. For the wobbling-phonon model the excitation energies are a linear function of the number of excited phonons n, thus the excitation energies $E_{exc}(I = const)$ are straight lines, as shown in Fig. 6.8(f), (only states below I_{max} are shown). Note also the decreasing slope of the lines for larger I. The rotational model does expand to states at higher spins, including beyond I_c. For states with $I > I_c$ the signature splitting in the bands of a pair vanishes, which is reflected by the equal excitation energy of the two signature partners, see for instance the data connected with solid lines in panel (e). The excitation energies of the states with $I < I_c$, (e.g. the states with $I \leq 9.5$) are in good agreement with a quadratic trend, illustrated with the dashed lines in panel (e). The quadratic dependence is in contrast with the linear trend for wobbling-phonon excitations of panel (f).

Important differences are also apparent in the $B(E2)$ and $B(M1)$ reduced transition probabilities shown in Fig. 6.9. While the intra-band $B(E2)$ transition probabilities for wobbling bands are the same for all bands as they do not depend on n, and are also constant as a function of I, see panel (b), there are no such rules for the corresponding intra-band $B(E2)$ rates for TiP bands, as shown in panel (a). The latter are not the same for all bands, and may be approximately constant or increase with I.

Different behaviors for TiP and wobbling bands are also expected for the inter-band $B(E2)$ probabilities, see panels (c) and (d) of Fig. 6.9. While the inter-band $B(E2; n, I \rightarrow n-1, I-1)$ probabilities are proportional to n for wobbling bands, there is no such dependence for TiP bands. In fact, it may happen that a TiP band with higher quantum number m has lower values of $B(E2; m, I \rightarrow m-1, I-1)$ than a band with lower value of m, as is the case for $B(E2; 1 \rightarrow 0)$ and $B(E2; 2 \rightarrow 1)$ for $I > 11$, see Fig. 6.9(c).

Similar differences are found for the $B(M1, n, I \rightarrow n-1, I-1)$ transition probabilities. These probabilities are proportional to the number of excited phonons n for wobbling bands, while no such dependence is in place for TiP bands, see panels (e) and (f) of Fig. 6.9.

In addition, for wobbling bands both the $B(E2)$ and the $B(M1)$ probabilities vanish for transitions linking states with $\Delta n > 1$, because such transitions are forbidden. In contrast, some transitions of this type do proceed within the rotational model.

A summary of the features of TiP and wobbling bands for transverse coupling that are distinct and can be used to test the experimental data is included in Table 6.1 of Section 6.4.

6.3 CONFLICTING TERMINOLOGY

The wobbling motion is a precessional motion caused by harmonic vibrational excitations, quantized as wobbling phonons. On the other hand the TRM and QTR models also describe precessional motion, but with purely rotational nature. In fact, both approaches depict a precession of the total angular momentum of the nucleus around an axis, similar to the precession of a rotating top. Both approaches find that the excited bands correspond to larger tilted angles of the precessing total angular momentum. The difference in the two descriptions stems from the underlying nature that causes the tilt of the total angular momentum, and is vibrational for wobbling and rotational for TRM and QTR.

It was noted that the features of the calculated vibrational and rotational precession bands may look very similar if an approximation condition $f(n, I) \ll 1$ is valid. In such cases, where the approximation condition is met, the rotational Hamiltonian is almost equivalent to the Hamiltonian of a harmonic wobbling excitation, and the difference between them can be considered as small anharmonicity. One can, in these cases, refer to the solution of a rotational model as anharmonic wobbling. On the other hand, if the approximation condition is not valid the bands produced by the two approaches show significant differences stemming from the different nature of the underlying excitations which are rotational (TiP) and vibrational (wobbling), respectively.

The wobbling was first introduced as a harmonic vibrational approximation of the TRM model applicable at high spins where the approximation condition, $f(n,I) \ll 1$, was met [3]. It does not mean that wobbling cannot exist at low spins, but simply that at low spins it should be described not by the TRM calculations, but by equations that preserve its vibrational character. Such descriptions can be based on the equations listed above, or can be modeled involving RPA or other similar approaches.

For many decades after the introduction of wobbling it was described either by models involving vibrational excitations, or by rotational models where the approximation condition was valid. However, in 2014, the wobbling motion was re-visited and in particular the transverse coupling in odd-mass nuclei was investigated [9]. The equations in Section 6.1.3, expressing the harmonic vibrational nature of the wobbling motion were derived and the approximation condition of eq. 6.35, indicating where the rotational QTR description looks like vibrational wobbling motion, was presented. But in contrast with the previous works on wobbling, the rotational-model solutions were referred to as wobbling, despite the fact that they were used in a regime where the approximation condition was not valid, and therefore the calculations were representing bands with rotational rather than vibrational excitation nature.

This approach was followed by a number of more recent experimental and theoretical works. For instance, the observation of wobbling bands of transverse and longitudinal coupling character were reported in ^{105}Pd, ^{127}Xe, ^{133}La, ^{133}Ba, ^{135}Pr, ^{183,187}Au, [23–29]. The proposed wobbling interpretation was based on measured large mixing ratios of the linking transitions and on an agreement with rotational-model calculations. It should be noted that more recent experimental data found small values for the mixing ratios of these transitions in ^{135}Pr and ^{187}Au [30, 31], in conflict with previous experimental results. Such small values ruled out the wobbling interpretation and suggested that these bands are strongly impacted by single-particle excitations. Among the theoretical studies, some of the works questioned the transverse wobbling presented in Ref. [9], for instance [4, 6, 8, 32, 33], others accepted the idea that the solutions of the rotational models can be considered as wobbling even when the approximation condition was not valid and proceeded to investigate details of the behavior of the bands [34–43], while a few other works treated the wobbling excitations as based on vibrational phonons [7, 32, 41, 44, 45].

The precedent of adopting wobbling character for rotational bands even when the approximation condition is not valid, infers severe conflicts in the meaning of the term wobbling. In its original meaning wobbling is caused by harmonic vibrational excitations, while possible anharmonicity should remain small (limited by the approximation condition $f(n,I) << 1$). In line with that definition wobbling bands were labeled by the number of excited phonons. These bands were also expected to have vibrational characteristics, (for instance the $B(E2; nI \rightarrow n-1, I-1) \propto n$, as well as vanishing $B(E2; \Delta n > 1)$ rates). The collective solutions of the rotational TRM and QTR models could be considered as wobbling only at high spins, where the approximation condition is valid and wobbling remains a vibrational excitation. Therefore, using the term wobbling at low spins, where the bands correspond to purely rotational mode of excitation, can inflict a severe conflict about the meaning of the term wobbling, creating misunderstandings and confusion.

For instance, if the rotational TiP bands are to be called wobbling disregarding the approximation condition, then all γ bands in the even-even triaxial nuclei, generated by TRM would qualify as wobbling bands. Within this description they all represent a precession of the total angular momentum around the intermediate axis, similar to the precession of a rotating top, and have purely rotational nature. Then it becomes incomprehensible, why despite the rigorous search for wobbling bands at the end of the last century, and despite the fact that TRM calculations have been successfully applied to the γ bands of many triaxial nuclei at low spin, these bands were not adopted as wobbling.

The failure for decades to discover wobbling bands led to viewing wobbling as a very rare and very special phenomenon. And even when the search for wobbling bands finally succeeded with the identification of the TSD bands, the phenomenon remained very rare, restricted to only a few Lu, and Ta isotopes. Through that time many low-spin bands in even-even and in other odd-mass triaxial nuclei have been successfully interpreted within the TRM and QTR models, respectively, but they were not considered as wobbling bands.

Therefore, applying the term wobbling to the TRM and QTR solutions when the approximation condition is not met implies a change in the definition of wobbling, and inflicts contradictions with past research, for instance

(i) all γ bands predicted by the TRM models would, within this changed definition of wobbling, be re-named to wobbling bands. This would diminish the efforts of many colleagues in the past who have been searching for wobbling, but disregarded the well-known low-spin γ bands.

(ii) many rotational bands in odd-mass triaxial nuclei interpreted within the QTR model where the excitation is dominated by collective rotation (producing large $B(E2)$ components in the linking transitions) would, within the changed definition of wobbling, be understood as wobbling bands. This is in contrast with the originally-proposed wobbling which still remains a rare and very special phenomenon.

(iii) wobbling bands are customary labeled by a quantum number n, that refers to the number of excited vibrational phonons. Within the changed definition of wobbling, the calculated bands (of purely rotational nature) would be labeled as vibrational phonon excitations, which is clearly contradictory.

(iv) the term wobbling that was reserved in the past for precessional motion caused by vibrational excitations, within the changed definition, would describe precession of purely rotational nature.

The problems listed above are severe. A clear and correct terminology is needed, that is consistent with the past, and that would address the similarities and differences of the precessional motion caused by rotational and vibrational excitations. To address the misunderstanding and the conflicts with the terminology we propose the following:

(i) to keep the original term "wobbling" as highlighting the vibrational nature of the excitation in agreement with the way wobbling was introduced and used for several decades. As in the past, the term can also be used for the TRM and QPR descriptions if the approximation condition is valid, as in that case the rotational models can be considered as describing vibrational excitations with small anharmonicity.

(ii) to refer to the general TRM and QPR descriptions as precession of rotational nature, for instance by labeling the bands tilted precession (TiP) bands, as proposed in [8].

(iii) in that case one can call "rotational TiP" the special cases where the excitations are predominantly collective rotation, highlighting that the precession of the total angular momentum has rotational nature.

(iv) to use "single-particle TiP" for the cases of TiP where the excitations are predominantly of single-particle nature highlighting that the precession of the total angular momentum is due to single-particle excitations.

Within this terminology, the QTR descriptions of the TSD bands are wobbling, as they appear at high spins, where the approximation condition is valid. In addition the excitations are of collective nature as indicated by the observed large mixing ratios of the linking transitions. It should be noted that other features of the TSD bands were also found in agreement with a vibrational type of the excitations, in particular the $B(E2)$ values. In contrast, the term wobbling should not be used for the low-spin TRM γ bands in even-even nuclei and for the low-spin QTR one-quasiparticle bands in odd-mass nuclei where the approximation condition is not valid. These TiP bands, if generated by collective excitations can be called rotational TiP, and if the single-particle component is dominant can be regarded as TiP of single-particle nature. Such terminology would underline the precessional motion of the total angular momentum, that might be caused by vibrational, rotational and single-particle excitations. On the other hand, low-spin bands can be tested for possible wobbling nature by comparing their features with calculations based on vibrational excitations, such as models based on RPA, or the wobbling equations in the previous sections.

We believe that this approach will clarify the definitions, while underlying the nature of the excitations and remaining consistent with past research works. In the following sections, we illustrate how experimental data can be examined in order to determine the underlying nature of the excited bands using the proposed terminology.

6.4 APPLICATION TO EXPERIMENTAL DATA: ESTABLISHING THE NATURE OF THE PRECESSION IN TRIAXIAL NUCLEI

When examining the excited rotational bands in a triaxial nucleus in order to test whether they may represent vibrational, rotational or single-particle excitations, first of all one can test the significance of the single-particle component. The collectivity of the excitation can be tested by measuring the mixing ratios of the linking transitions and (if possible) the lifetimes of the nuclear states. Large mixing ratios, and in particular large $B(E2)$ values for the linking transitions, suggest that the excitation has collective nature. The linking transitions would have collective nature if the excited bands are generated for instance by wobbling, rotational TiP or other collective type of excitation, such as small γ vibration around an axially symmetric nuclear shape, large gamma vibration of a γ-soft nucleus, etc.

One can test whether bands showing collective nature of the excitations are generated by vibrational (wobbling) or rotational (TiP) modes, by simply comparing the experimental data with the predictions of the two approaches. Table 6.1 presents a summary of the expected differences in the features of the observed bands generated by rotational TiP and vibrational (wobbling) type of excitations at low spins for $\gamma = 30°$, where the approximation condition is not valid.

In the following sections, we test the two models with experimental data to illustrate how information on the nature of the excited bands can be obtained.

6.4.1 TESTING FOR ROTATIONAL (TIP) AND VIBRATIONAL (WOBBLING) NATURE IN EVEN-EVEN TRIAXIAL NUCLEI: ^{192}OS

In this case we examine the low-spin states of ^{192}Os, including the ground-state and the 2^+ and 4^+ γ bands. Since in this spin-range there are no single-particle excitations, the excited bands are collective, thus we can test whether they can be associated with a precession with vibrational (wobbling) or rotational (TiP) nature.

We have chosen the ^{192}Os nucleus because diagonal and transitional matrix elements of the $E2$ operator were measured in a Coulomb excitation experiment, and the triaxiality of the nuclear shape, γ, was deduced in a model-independent analysis using the Kumar-Cline sum rule [46]. A value of $\gamma \sim 26(4)°$ was deduced for the states of the ground-state band and a similar value of $\gamma \sim 30(8)°$ for the 2^+ and 4^+ γ bands. Therefore, the ^{192}Os nucleus has a shape with near maximal triaxiality. Furthermore, lifetime measurements have been carried out for a number of states in the collective bands. Experimental data from the NNDC data base are used [47].

The excitation energies of the states in the ground-state and the γ bands, and the corresponding measured values of the $B(E2)$ transition probabilities are plotted in Fig. 6.10. The bands are labeled with a quantum number k, where $k = m = n$, but k does not carry a meaning about the nature of the bands, as the rotational m and the vibrational n quantum numbers do. Therefore, k is 0 for the ground-state band, 1 for the odd-spin members of the 2^+ γ band, 2 for the even-spin members of the 2^+ γ band, 3 for the odd-spin members of the 4^+ γ band and 4 for the even-spin members of the 4^+

Table 6.1

Differences in the excitation energy and the intra- and inter-band $B(E2)$ transition probabilities for vibrational (wobbling) and rotational (TiP) at low spins where the approximation condition is not valid. The comparison is for even-even (ee) and odd-mass triaxial nuclei with longitudinal (lc) and transverse (tc) coupling, and for $\gamma = 30°$ (unless stated otherwise). Entries marked with stars (*) correspond to $\gamma \neq 30°$ and vanish for $\gamma{=}30°$. The entries marked with & refer to bands in pairs. The entry marked with hash (#) relates to $(m,I) \rightarrow (m-2,I-1)$ and $(m,I) \rightarrow (m-2,I-2)$ transitions for $\gamma \neq 30°$, and to $(m,I) \rightarrow (m-2,I)$ transitions for any γ.

Item	For	Property	Vibrational	Rotational
1	ee, lc, tc	quantum number k	n	m
2	ee, lc, tc	$E_{exc}(I = const)$	$\propto n$	$\propto m^2$
3	ee	$E(2_g^+) + E(2_\gamma^+)$	$> E(3_\gamma^+)$	$= E(3_\gamma^+)$
4	ee, lc	E_{rel}	const with n	decreasing with m
5	ee, lc	$B(E2)_{intra}$	const with n	decreasing with m
6	ee, lc	$B(E2)_{intra}$	const with I	increasing with I
7	ee, lc, tc	$B(E2; k, I \rightarrow k-1, I-1)$	$\propto n$	not proportional to m *
8	ee, lc	$B(E2; k, I \rightarrow k-1, I-1)$	$\propto 1/I$	not proportional to $1/I$ *
9	ee, lc, tc	$B(E2)_{inter}$ for $\Delta n > 1$	0	> 0 (allowed)#
10	ee	$R_{2\gamma 2g} = \dfrac{B(E2; 2_\gamma^+ \rightarrow 2_g^+)}{B(E2; 2_g^+ \rightarrow 0_g^+)}$	0	1.4
11	ee	$R_{3\gamma 2\gamma} = \dfrac{B(E2; 3_\gamma^+ \rightarrow 2_\gamma^+)}{B(E2; 2_g^+ \rightarrow 0_g^+)}$	1	1.8
12	tc	E_{rel}	const with n	increasing with m &
13	tc	$E_{rel}(I)$	decreasing for $I < I_{max}$	decreasing for $I < I_c$ &
14	tc	$B(E2)_{intra}$	const with n	not const with m
15	tc	$B(E2)_{intra}$	const with I	not const with I

γ band. Excitation energies for all observed states, as well as excitation energies for the intermediate states of all bands, for instance,

$$E(k, I-1) = \frac{1}{2}\left[E(k,I) + E(k, I-2)\right], \tag{6.36}$$

are shown in panel (a) of Fig. 6.10. Fits with polynomials of second order are added to guide the eye. It is clear that the excitation energies of the states of a given spin I are consistent with a quadratic dependence of k, in agreement with TiP excitations and in contrast with the wobbling approach requiring a linear dependence, see item 2 in Table 6.1.

The relative excitation energies were calculated as

$$E_{rel}(k, k-1, I) = E(k,I) - \frac{1}{2}\left[E(k-1, I+1) + E(k-1, I-1)\right], \tag{6.37}$$

and are shown in panel (b) of Fig. 6.10. They are not well described by either of the models, however, the trend of $E_{rel}(k, k-1)$ is in better agreement with rotational nature, for which the values should decrease for higher values of k, than for vibrational excitations for which the values should be the same for all k, see item 4 in Table 6.1.

There is an excellent agreement between the excitation energy of the 3^+ state, $E(3_\gamma^+) = 690.4$ keV and the sum, $E(2_g^+) + E(2_\gamma^+) = 694.9$ keV, which is in line with the expectations for rotational excitation (see item 3 in Table 6.1). Within the wobbling model, $E(3_\gamma^+)$ should be considerably smaller.

The behavior of the $B(E2)$ rates was also examined. Within the wobbling-phonon excitations all bands should have constant intra-band $B(E2)$ values, that do not depend on I and on n, see items 5 and 6 in Table 6.1. Within the rotational model the intra-band $B(E2)$ rates increase with spin and

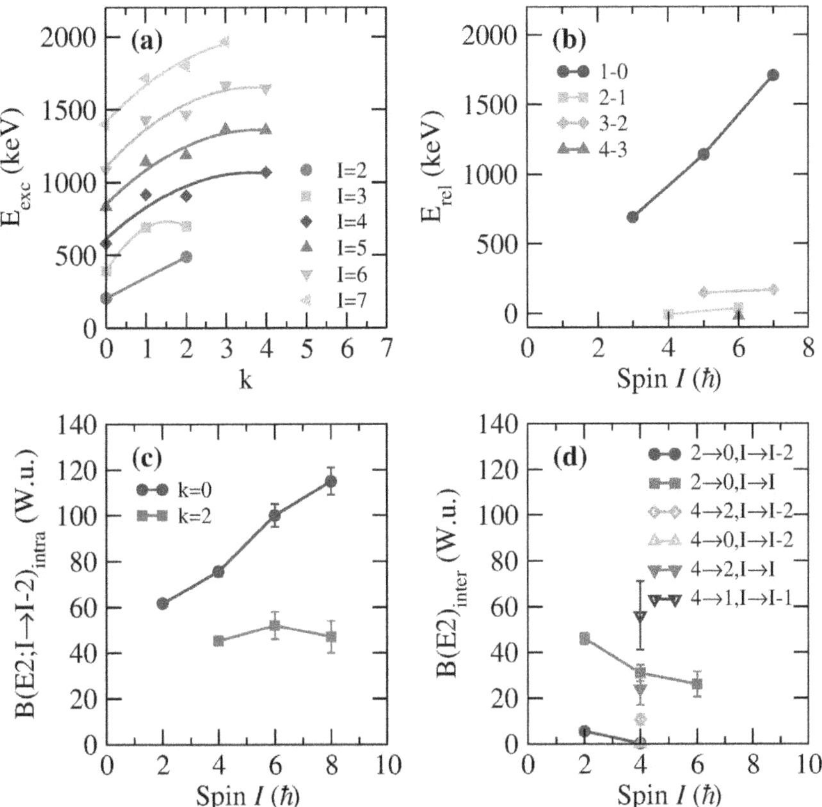

Figure 6.10 Experimental data for the ground-state and the 2^+ and 4^+ γ bands of ^{192}Os. (a): Excitation energies, (second-order polynomials are shown to guide the eye); (b): relative excitation energies, $E_{rel}(k,k-1)$; (c): intra-band and (d): inter-band $B(E2)$ transition probabilities in W.u. Experimental data are from the NNDC data base, [47].

decrease for larger k. The experimental data shown in Fig. 6.10(c) favors the rotational excitations based on (i) the observed difference in the $B(E2)$ values for the two bands and (ii) the increasing with spin $B(E2)$ values for the ground state band. In contrast, the almost constant values of the $B(E2)$ rates for the even-spins of the γ band are more in line with vibrational-type of excitation.

The available experimental data on the inter-band $B(E2)$ rates are plotted in panel (d) of Fig. 6.10. The rates correspond to transitions of the type $\Delta k > 1$, which are forbidden within the wobbling model, see item 9 in Table 6.1. Note that the disagreement is significant, as the majority of the measured inter-band $B(E2)$ values are with a magnitude of tens of W.u., establishing collective nature for these transitions. This is in stark contrast with the wobbling excitations model, where such transitions are forbidden. Within the rotational model the transitions of the type $(k,I \rightarrow k-2,I)$ can be large, in particular for large γ deformation as the one measured in ^{192}Os. Thus the large values for the $I \rightarrow I$ transitions are in agreement with the expectations of the rotational model. The small $B(E2)$ values of the transitions $(k,I \rightarrow k-2,I-2)$ and $(k,I \rightarrow k-4,I-2)$ are also in rough agreement with the expectations of the rotational model. These values are vanishing for $\gamma = 30°$, and can have a small non-zero value for γ near $30°$. The ratio of $R = B(E2;2_\gamma^+ \rightarrow 2_g^+)/B(E2;2_g^+ \rightarrow 0_g^+) = 0.75(5)$ is large, in contrast to the vibrational wobbling model for a vanishing value, see item 10 in Table 6.1, but in agreement with rotational TiP nature.

To summarize, the rotational TiP and the vibrational wobbling models were tested with the excited bands of the triaxial ^{192}Os nucleus. The rotational TiP description is in very good (although not perfect) agreement with the experimental data, while the vibrational wobbling approach is unsatisfactory. This conclusion is in line with the failure to discover wobbling bands, (within the original meaning of the term wobbling as indicating vibrational phonon excitations), at low spins in even-even triaxial nuclei, in contrast with the successful application of the triaxial-rotor model for the ground-state and γ bands of such nuclei.

6.4.2 TESTING FOR ROTATIONAL (TIP) AND VIBRATIONAL (WOBBLING) NATURE IN ODD-MASS TRIAXIAL NUCLEI WITH LONGITUDINAL COUPLING AT HIGH SPINS: TSD IN ^{163}LU

As another example we will examine the experimental data for three TSD bands in ^{163}Lu, (TSD1, TSD2 and TSD3), for which the excitation energies and the spins were established. In addition, experimental data on the mixing ratios of several linking transitions and on the lifetimes of a number of excited states are available [47].

The TSD bands were initially interpreted as corresponding to longitudinal coupling of the single-particle and rotational angular momenta, and here we will adopt the same approach. For longitudinal coupling and at high spins, the approximation condition is valid. Therefore both models, assuming rotational TiP and vibrational wobbling nature of the excitations produce similar features for the rotational bands. Therefore, by examining the properties of the bands one cannot give preference to one of these models. In the following, we will compare the bands with the vibrational wobbling approach as the simpler model.

The experimental excitation energies and the $B(E2)$ transition probabilities for three TSD bands in ^{163}Lu are shown in Fig. 6.11.

Consider first the excitation energies plotted for a given spin in Fig. 6.11(a). For vibrational wobbling these energies are proportional to the number of the excited phonons n, see item 2 in Table 6.1. Indeed, the excitation energies of the TSD bands in ^{163}Lu are in excellent agreement with this requirement for linearity.

Consider next the relative excitation energies shown in Fig. 6.11 (b). According to the wobbling description the relative excitation energies should be independent of the number of excited phonons n, see item 4 in Table 6.1. The experimental data show a considerable difference, which, however becomes small at spins around $40\hbar$, which indicates that the wobbling interpretation works better at high spins.

The experimental data for the intra-band $B(E2)$ probabilities are in very good agreement with the expectation of the wobbling model, requiring that these rates are independent of n, see item 5 of Table 6.1. Indeed, the experimental intra-band $B(E2)$ values for TSD1 ($k=0$) and TSD2 ($k=1$) are the same, see panel (c) of Fig. 6.11.

There is also a reasonable agreement with item 6 of Table 6.1, that the intra-band $B(E2)$ values remain constant as a function of spin. The experimental data might be showing a slow and gradual reduction in the intra-band $B(E2)$ rates, although the uncertainties on the experimental data do not exclude an almost constant trend.

The inter-band $B(E2; kI \to k-1, I-1)$ transition probabilities for the transitions linking TSD2 and TSD1 are shown in panel (c) of Fig. 6.11. The values are large, with magnitudes of more than 100 W.u., indicating a collective excitation for TSD2 with respect to TSD1. Before we proceed further we will stress the importance of the lifetime measurements, as they (together with the mixing ratios) yield the magnitude of the $B(E2)$ reduced transition rates of the transitions linking the excited bands and provide important evidence whether the excitation, like in the present case, is collective. The experimental data on the inter-band $B(E2)$ transition rates, see Fig. 6.11(c), are also consistent with the $B(E2) \propto 1/I$ dependence expected by the wobbling equations, (shown with dashed line in the figure), see also item 8 in Table 6.1.

As lifetime measurements could not be performed for states of TSD3, instead of the absolute inter-band $B(E2)$ values the ratios of inter- and intra-band rates, $B(E2; kI \to k-1, I-1)/B(E2; kI \to$

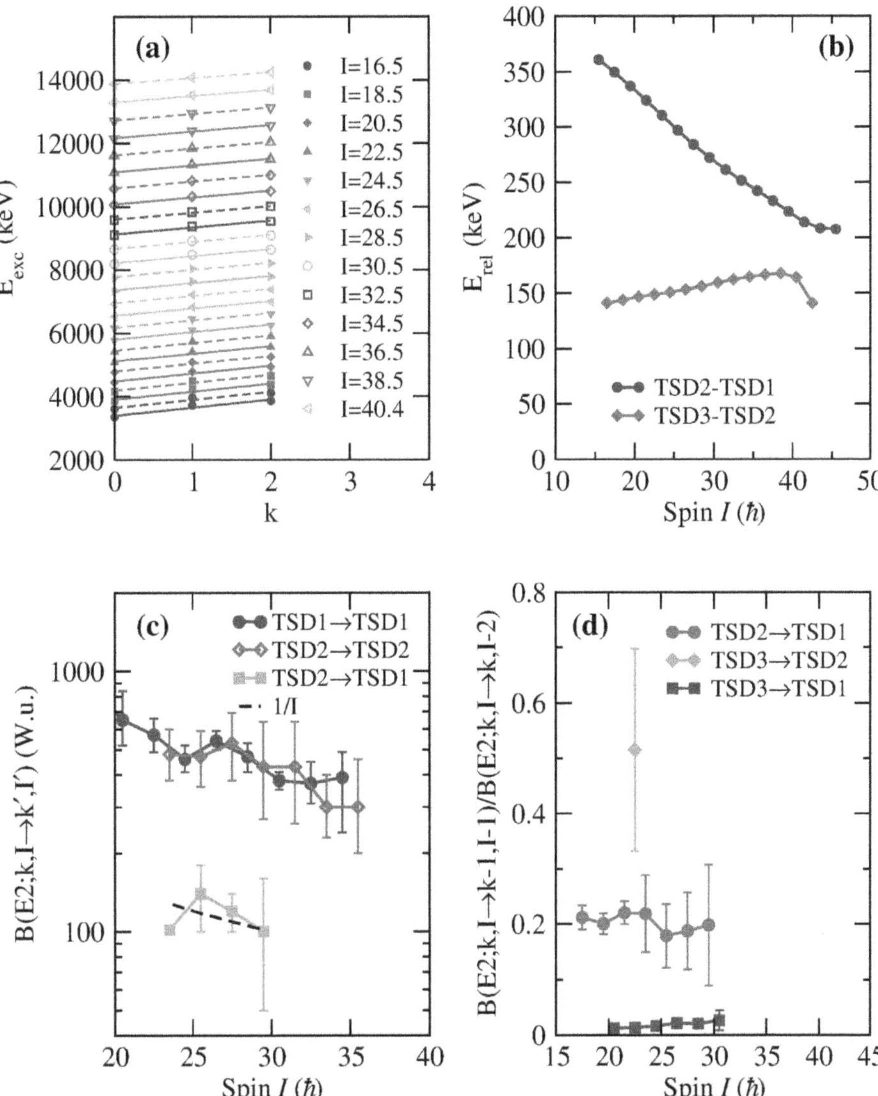

Figure 6.11 Experimental data for three TSD bands of ^{163}Lu. (a): The excitation energy for a given spin; linear fits to the data are shown to guide the eye. The filled and open symbols show data for spin I and $I+1$, respectively. (b): The relative excitation energies. (c): The intra-band $B(E2; k, I \rightarrow k, I-2)$ for TSD1 and TSD2, and the inter-band $B(E2; k, I \rightarrow k-1, I-1)$ for TSD2 $\rightarrow$ TSD1 reduced transition probabilities in W.u. (d): The ratio of the inter-band $B(E2; k, I \rightarrow k-1, I-1)$ to the intra-band $B(E2; k, I \rightarrow k, I-2)$ rates. Experimental data are from the NNDC data base, [47].

$k, I-2$), are shown in panel (d) of Fig. 6.11. We assume that the intra-band $B(E2)$ values for TSD3 are approximately constant (similarly to those of TSD1 and TSD2), therefore, using these $B(E2)$ ratios we can test the behavior of the linking transitions decaying out of TSD3. An inter-band $B(E2)$ ratio is measured for only one transition linking TSD3 and TSD2, see Fig. 6.11(d). This ratio is in excellent agreement with the expectation of the wobbling model that the inter-band $B(E2)$ rates are proportional to n, see item 7 of Table 6.1, as it is approximately twice as large as the corresponding inter-band $B(E2)$ ratios for TSD2 $\rightarrow$ TSD1 linking transitions.

Experimental data on the $B(E2)$ ratios for six inter-band TSD3 $\rightarrow$ TSD1 linking transitions are also shown in Fig. 6.11(d). Within the wobbling model such decays are forbidden because they correspond to simultaneous destruction of two phonons. The experimental data is in excellent agreement with this expectation, as the corresponding $B(E2)$ ratios are indeed very small.

In summary, the available experimental data for three TSD bands in ^{163}Lu are in excellent agreement with the wobbling interpretation of these bands, based particularly on the features of the transition probabilities of the relevant intra- and inter-band transitions and excitation energies.

6.4.3 TESTING FOR ROTATIONAL (TIP) AND VIBRATIONAL (WOBBLING) NATURE IN ODD-MASS TRIAXIAL NUCLEI WITH TRANSVERSE COUPLING AT LOW SPINS: ^{135}PR

As another example we will test the yrast band and the two excited bands in ^{135}Pr that were initially interpreted as zero-, one- and two-phonon wobbling bands, associated with a $\pi h_{11/2}$ configuration [9, 26, 27]. Recent experimental data contradicted that interpretation based on established dominant $B(M1)$ component of the linking transitions, suggesting that the single-particle excitations are very important for these bands [30]. In the following, we will test the nature of the yrast and two excited $\pi h_{11/2}$ bands taking into consideration the available experimental data on the mixing ratios and on the lifetimes of the yrast-band states.

The excitation energies of these three bands, labeled with quantum number k, are shown in Fig. 6.12(a). For bands with transverse coupling of the angular momenta the relative excitation energy decreases with spin, a feature that is similar for the vibrational wobbling and the rotational TiP descriptions. However, for wobbling the relative excitation energy does not depend on the quantum number n, that is, it should be the same for $E_{rel}(1-0)$ and $E_{rel}(2-1)$, see Fig. 6.8(d). This is in disagreement with the experimental observations, and more in line with rotational TiP, where E_{rel} is different for bands with different m. The excitation energy for given spin, shown in Fig. 6.12(c), is consistent with a quadratic dependence on k, in agreement with the expectations for TiP nature, and in contrast with the linear dependence of a wobbling excitation, see item 2 in Table 6.1.

Some experimental data on the lifetime of the states of the yrast band in ^{135}Pr are available. The deduced $B(E2)$ values are shown in Fig. 6.12(d). In general the data show collective nature of the intra-band transitions in the yrast band, but its trend to decrease with spin is in contrast with both the rotational TiP, where $B(E2)$ is expected to increase, and the vibrational wobbling, where $B(E2)$ is expected to be constant with spin, see item 15 in Table 6.1. In general, the decreasing trend indicates decreasing collectivity in the band, underlying that toward higher spins the single-particle component becomes more and more important. This suggests that a description that includes a considerable single-particle component would be more appropriate than both the rotational TiP and the wobbling models which infer collective excitations. Such a more general description can be for instance the general TiP that combines the three-dimensional rotation of the core with a free single-particle able to align under the influence of the Coriolis interaction, or alternatively a model that treats a coupling of vibrational (wobbling or γ vibrations) and single-particle degrees of freedom.

As mentioned above, limited lifetime data are available for ^{135}Pr. Therefore, where the lifetimes of relevant states remain unknown, it is still possible to analyze the $B(E2)_{inter}/B(E2)_{intra}$ ratios provided that the mixing ratios of transitions linking $(k, I \rightarrow k-1, I-1)$ states are measured. Two conflicting values were published for the $21/2^- \rightarrow 19/2^-$ transition linking the $k=1$ and $k=0$ bands in ^{135}Pr, yielding $B(E2)_{inter}/B(E2)_{intra} = 0.843(32)$ and $0.12(8)$, respectively. (The conflict arises from adopting different solutions for the measured mixing ratios, for more details see Refs. [30, 31]). The former $B(E2)$ ratio suggests large $B(E2)_{inter}$ values of similar order as the collective intra-band $B(E2)$ rates. That supports dominant collective excitation, which can be both of vibrational wobbling or of rotational TiP nature. The data on the mixing ratio alone does not allow to favor one of these options. In contrast, the later smaller value for the $B(E2)$ ratio suggests small collectivity for the $B(E2)_{inter}$ probability, supporting a considerable single-particle contribution. It would, similar to the decreasing trend of the $B(E2)_{intra}$ values shown in Fig. 6.12(d), indicate a major impact from the single-particle degree of freedom.

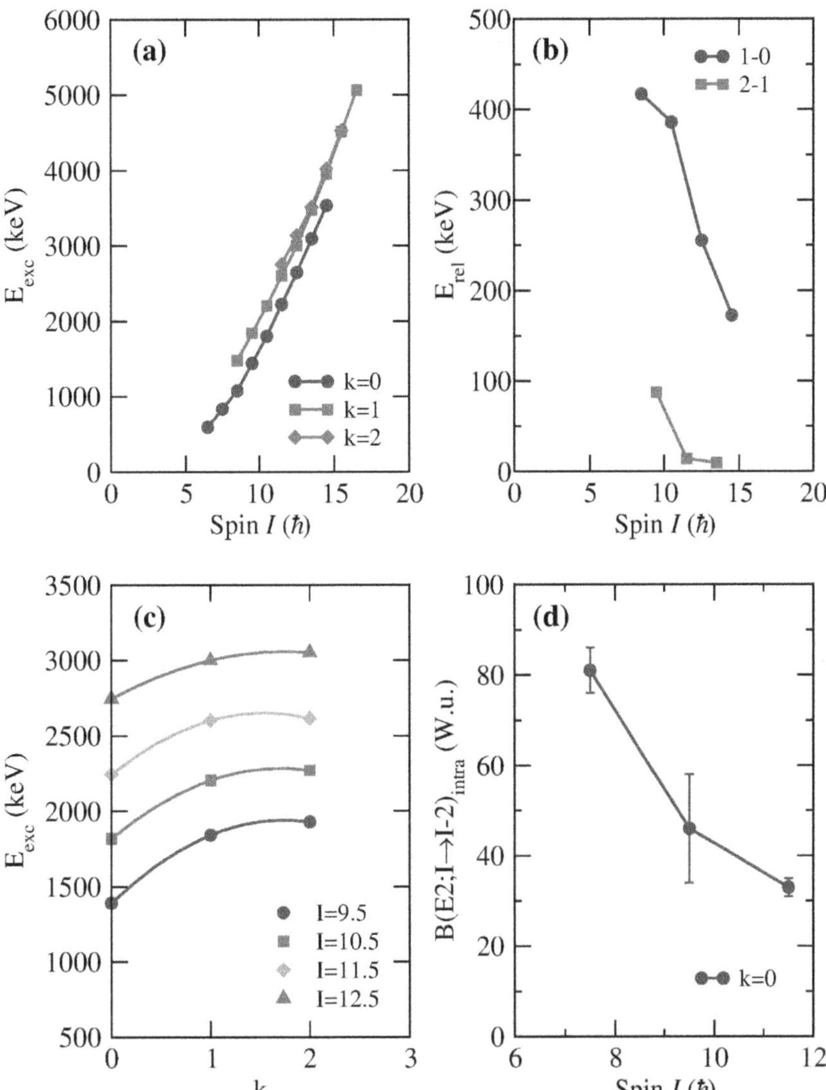

Figure 6.12 Experimental data for the $h_{11/2}$ bands of ^{135}Pr. (a): Excitation energies, and (b): relative excitation energies, as a function of spin. (c): Excitation energies for given spin. The solid lines are polynomials of second order to guide the eye. (d): Intra-band $B(E2;k,I \rightarrow k,I-2)$ transition probabilities in W.u. Experimental data are from the NNDC data base, [47].

In summary, the available experimental data for three $\pi h_{11/2}$ bands of ^{135}Pr do not support wobbling phonon interpretation of these bands, based on the features of the excitation energies and the available transition probabilities of the intra-band transitions. The data are in better agreement with TiP nature, comprising both single-particle and rotational excitations. Additional experimental data on the mixing ratios and on the lifetimes of the states are crucial in order to confidently establish the underlying nature of these bands.

6.5 SUMMARY

In this work the current controversy in the use of the term "wobbling" is discussed. Wobbling was introduced in triaxial nuclei to model the precession of the total angular momentum as a coupling of a simple one-dimensional rotation and vibrational excitations. The vibrational phonons cause the tilt of the precessing total angular momentum. Despite its vibrational nature wobbling can be modeled with rotational models, provided that an approximation condition is valid. The condition is typically met at high spins for even-even nuclei and for longitudinal coupling in odd-mass nuclei. It requires that the discrepancies between the rotational and the wobbling descriptions are small and thus allows to view the rotational-model solutions as wobbling with small anharmonicity. In line with the harmonic nature of the wobbling excitations the bands are labeled by the number of the excited phonons n. They also have characteristic vibrational features, such as vanishing probabilities for transitions linking two bands with $\Delta n > 1$, inter-band $B(E2; n, I \to n - 1, I - 1) \propto n$, etc. This definition of wobbling was consistently used for several decades.

Following a precedent in 2014, the solutions of standard rotational models began to be associated with wobbling in conditions where the approximation condition is not valid. In such conditions the calculated bands are purely rotational and do not have the vibrational-excitation character required by the original definition of wobbling. The discrepancies create confusion and misunderstandings.

In this work the similarities and the differences in the nature of the wobbling (in its original definition) and the rotational excitations of the standard rotational models are described. A way forward to resolve the controversy is proposed. It involves the following terminology:

(i) the term "wobbling" is used in its original meaning, expressed in the vibrational wobbling equations of Bohr and Mottelson for even-even nuclei [3], and of S. Frauendorf for odd-mass nuclei [9]. This preserves the vibrational nature of the wobbling bands and ensures consistency in the common labeling of wobbling bands by the number of the excited wobbling phonons n;

(ii) tilted precession (TiP) bands refer to the solutions of standard rotational models for triaxial nuclei. The calculated bands correspond to a precession of the total angular momentum, but have purely rotational nature. Such bands do not qualify as wobbling bands because they do not exhibit the vibrational-excitation features expected for the wobbling motion, for instance the γ bands in even-even triaxial nuclei are not adopted as wobbling bands. TiP bands in odd-mass triaxial nuclei represent a coupling of the three-dimensional rotation of the core with the excitations of the valence nucleon;

(iii) rotational TiP can be used for TiP bands that are mostly of collective rotational nature, such as the γ bands in a nucleus with rigid triaxial shape or the bands in odd-mass triaxial nucleus with longitudinal coupling. The precession has purely rotational nature. When the approximation condition $f(n, I) \ll 1$ is met these bands show very similar features to wobbling bands and can be interpreted as such, as for example the TSD bands in the Lu and Ta isotopes.

The proposed terminology was then applied to examine experimental data and deduced the underlying nature of the observed bands. First, the similarities and the distinct features of bands produced by the rotational TiP and the vibrational wobbling approaches were presented and discussed. Then the predictions were tested against available experimental data. In particular, the trends of the excitation energies for a given spin and of the relative excitation energies, as well as the absolute values and the trends of the intra- and inter-band $B(E2)$ reduced transition probabilities were examined for a number of bands of ^{192}Os, ^{163}Lu and ^{135}Pr. The features of the bands at low spins in the even-even triaxial ^{192}Os isotope were found consistent with rotational TiP character, the three TSD bands of ^{163}Lu were associated with wobbling nature, while the three $\pi h_{11/2}$ bands in ^{135}Pr were best described as TiP bands, where the single-particle excitations play a significant role in addition to the three-dimensional rotation of the core. It is also important to highlight that such comparisons necessitate detailed experimental data, in particular reliable results on the mixing ratios of the $M1 + E2$ transitions and on the lifetimes of the relevant states.

6.6 ACKNOWLEDGMENT

This work is based on research supported in part by the National Research Foundation of South Africa (Grants Number 150650).

Bibliography

1. J. Allmond and J. Wood, Phys. Lett. B **767**, 226 (2017).
2. J. Allmond, presentation at the CWNA'23 conference, 10-14 June 2023, Huizhou, China, to be published.
3. A. Bohr and B. R. Mottelson, *Nuclear Structure* (Benjamin, New York, 1975), Vol. II. (1975).
4. K. Tanabe and K. Sugawara-Tanabe, Phys. Rev. C **95**, 064315 (2017).
5. S. Frauendorf, Phys. Rev. C, **97**, 069801 (2018).
6. K. Tanabe and K. Sugawara-Tanabe, Phys. Rev. C **97**, 069802 (2018).
7. A A Raduta, C M Raduta and R Poenaru, J. Phys. G: Nucl. Part. Phys., **48**, 015106 (2020).
8. E. A. Lawrie, O. Shirinda and C. M. Petrache, Phys Rev. C **101**, 034306 (2020).
9. S. Frauendorf and F. Dönau, Phys. Rev. C **89**, 014322 (2014).
10. R.J. Liotta and R.A. Sorensenfs, Nucl. Phys. A, **297**, 136-162 (1978).
11. E.R. Marshalek, Nucl. Phys. A, **331**, 429-463 (1979).
12. S. W. Ødegård, *et al.*, Phys. Rev. Lett. **86**, 5866 (2001).
13. D. R. Jensen, *et al.*, Phys. Rev. Lett. **89**, 142503 (2002).
14. G. Schönwer, *et al.*, Phys. Lett. B **552**, 9 (2003).
15. H. Amro, *et al.*, Phys. Lett. B **553**, 197 (2003).
16. P. Bringel, *et al.*, Eur. Phys. J. A **24**, 167 (2005).
17. D. J. Hartley, *et al.*, Phys. Rev. C **80**, 041304(R) (2009).
18. D. Jensen, *et al.*, Nucl. Phys. A, **703**, 3 (2002).
19. H. Schnack-Petersen, *et al.*, Nucl. Phys. A, **594**, 175 (1995).
20. G. Schönwaßer, *et al.*, Phys. Lett. B **552**, 9 (2003).
21. G. B. Hagemann, Eur. Phys. J. A **20**, 183 (2003).
22. Yoshifumi R. Shimizu, Takuya Shoji and Masayuki Matsuzaki, Phys. Rev. C **77**, 024319 (2008).
23. J. Timár, *et al.*, Phys. Rev. Lett. **122**, 062501 (2019).
24. S. Chakraborty, *et al.*, Phys. Lett. B **811**, 135854 (2020).
25. S. Biswas, *et al.*, Eur. Phys. J. A **55**, 159 (2019).
26. J. T. Matta, *et al.*, Phys. Rev. Lett. **114**, 082501 (2015).
27. N. Sensharma, *et al.*, Phys. Lett. B **792**, 170 (2019).
28. S. Nandi, *et al.*, Phys. Rev. Lett. **125**, 132501 (2020).
29. N. Sensharma, *et al.*, Phys. Rev. Lett. **124**, 052501 (2020).
30. B. F. Lv *et al.*, Phys. Rev. C **98**, 044304 (2018).
31. S. Guo *et al.*, Phys. Lett. B, **828**, 137010 (2022).
32. A. A. Raduta, R. Poenaru and C. M. Raduta, Phys. Rev. C **101**, 014302 (2020).
33. K. Nomura and C. M. Petrache, Phys. Rev. C **105**, 024320 (2022).
34. Q. B. Chen, S. Q. Zhang, P. W. Zhao, R. V. Jolos, and J. Meng, Phys. Rev. C, **94**, 044301 (2016).
35. Fang-Qi Chen and C. M. Petrache, Phys. Rev. C, **103**, 064319 (2021).
36. B. Qi, H. Zhang, S Y Wang and Q.B. Chen, J. Phys. G: Nucl. Part. Phys., **48**, 055102 (2021).
37. R. Budaca, Phys. Lett. B, **817**, 136308 (2021).
38. C. Broocks, Q. B. Chen, N. Kaiser and Ilf-G. Meißner, Eur. Phys. J. A **57**, 161 (2021).
39. Q. B. Chen and S. Frauendorf, Eur. Phys, J. A **58**, 75 (2022).
40. H. Jia, S.Y. Wang, B. Qi, C. Liu and L. Zhu, Phys. Lett. B, **833**, 137303 (2022).

41. R. Budaca and C. M. Petrache, Phys. Rev. C **106**, 014313 (2022).
42. R. Budaca and A. I. Budaca, J. Phys. G: Nucl. Part. Phys., **50**, 125101 (2023).
43. Q. B. Chen and S. Frauendorf, Phys. Rev. C **109**, 044304 (2024).
44. R. Budaca, Phys. Rev. C **97**, 024302 (2018).
45. R. Poenaru and A. A. Raduta, Int. J. Mod. Phys. E, **30**, 2150033 (2021).
46. C.Y. Wu, *et al.*, Nucl. Phys. A, **607**, 178 (2002).
47. NNDC Online Data Service, ENSDF database, http://www.nndc.bnl.gov/ensdf/.

7 Experimental search for non-chiral to chiral configuration transition in ^{126,128}Cs isotopes

Ernest Grodner

National Centre for Nuclear Research, Otwock, Poland

7.1 INTRODUCTION

Nuclear chirality has initially been predicted for odd-odd triaxial nuclei in mass region A=130. Around 10 nuclei have then been identified as chiral candidates among which the Cs isotopes have been extensively studied experimentally. Various spectroscopic techniques have been used in chirality research, from identification of chiral rotational bands, through lifetime measurements of states belonging to these bands up to the g-factor investigations performed recently. Numerous chiral-characteristic signatures and phenomena have been predicted and observed. Detection of multi-chiral doubled bands $M\chi D$ expanded the set of nuclei in which chirality occurs to odd-even and even-even ones which numbers more than 60 isotopes today. Chirality became a significant part of nuclear spectroscopy due to the increasing number of chiral nuclei as well as identification of many chiral partner bands per single nucleus. Thus, the phenomenon of nuclear chirality needs to be studied in detail, especially in the transition region between non-chiral and chiral configuration in low spin range close to the chiral bandheads. Various signatures of chirality have been identified so far [1] which are mostly observed in the chiral rotational bands. These signatures and their relations are investigated here using a simple formalism and experimental data already available in Cs isotopes.

7.2 CHIRALITY, 2×2 MIXING PICTURE

The phenomenon of chirality is an interesting example of nuclear symmetry, which has both: known, simple features of the abelian Z_2 symmetry group and lesser known subtleties in the field of low energy nuclear excitations related to the involvement of the time reversal operator T. A symmetry operator O belonging to the Z_2 symmetry group has the known property: acting twice restores the original state, i.e. $O^2 = 1$. The eigenstates of the O-invariant Hamiltonian can therefore be classified to symmetric (eigenvalue $+1$) or antisymmetric (eigenvalue -1) states often called doublets. Well known examples of such symmetry operators are the charge conjugation C or the space inversion P. In case of the latter one, the invariance of the nuclear Hamiltonian $PHP^+ = H$ leads to the classification of nuclear states to positive or negative parity. Similar doublets are expected for the nuclear chirality symmetry operator $O = R_Y T$ since also in this case $O^2 = 1$ holds. Unlike parity, where the main role is played by the spatial, and especially static, features of the state in the context of space inversion operations, the nuclear chirality is based on dynamic properties of the state since time reversal operator is involved. This dynamics – in the form of nuclear rotation and odd nucleons motion with respect to the deformed nuclear shape – depends on several factors. Therefore, the nuclear chirality arises when specific physical conditions are reached. How the nuclear chirality depends on changing of these conditions has not yet been studied experimentally although presence of nuclear chirality has been found in more than 60 nuclei placing the phenomenon as an essential part of the low excitations nuclear spectroscopy.

Chirality concept in general means the existence of two states $|\sigma\rangle$ with opposite handedness – the left-handed $|\sigma\rangle = |L\rangle$ and the right-handed $|\sigma\rangle = |R\rangle$, which describe a given phenomenon or a structure of an object. The right-handed state $|R\rangle$ is transformed to the left handed $|L\rangle$ one (and vice-versa) by acting of the corresponding symmetry operator and not just by a rotation. In the nuclear chirality the symmetry operator is $R_Y T$. Due to the presence of T the $|\sigma\rangle$ states with specified handedness are defined by dynamical, motion-like properties of the nucleus. These properties are represented by angular momentum vectors (axial vectors). In the remaining part of this paper, the shorter term "spins" will be used interchangeably. The spin-based handedness in the most simple case can be defined in the odd-odd nuclei that can be described as a composition of three elements: an unpaired proton, an unpaired neutron and an even-even core each with its own angular momentum vector j_p, j_n and j_R, respectively. These vectors may form a reference frame with specified handedness as long as they do not lie in one plane – the aplanar configuration. The handedness parameter can be defined as a mixed spin product $\sigma = \frac{\langle(j_p \times j_n)\cdot j_R\rangle}{\langle j_p\rangle\cdot\langle j_n\rangle\cdot\langle j_R\rangle}$. The handedness is unable for experimental observation thus the left-handed $|\sigma = L\rangle$ and right-handed $|\sigma = R\rangle$ states are often called "intrinsic states". Like in the Schrödinger's cat problem, only the states superimposed of $|\sigma\rangle$ are accessible for experimental investigations. Such states are formally the projections of $|\sigma\rangle$ onto chirality χ being symmetric $|\chi = +\rangle$ or antisymmetric $|\chi = -\rangle$ eigenstates of $R_Y T$ operator. A projection operator in the spirit of Generator Coordinate Method (GCM), Section 11.4.6 in Ref. [2] deals with integration over a continuous symmetry parameters – coordinates. The handedness σ, as a mixed spin product, also is a continuous parameter. A detailed wave function composition in terms of σ is only accessible in advanced nuclear models and out of the scope of the present paper which is focused on the most general and simple properties of chirality expected for experimental observation. Therefore, a simplified formalism is used here where integration is limited to summation of opposite handed states. This formalism is further called 2×2 mixing picture.

The general description of nuclear chirality is greatly simplified by the following commutation relations of the $R_Y T$ operator, eq.(1-39) of Ref. [3]

$$
\begin{aligned}
\left[R_Y T, I_z\right] &= 0 \\
\left[R_Y T, I^2\right] &= 0,
\end{aligned}
\tag{7.1}
$$

These relations cause the projection order not to be significant. The rotational symmetry operators I, I_z does not change the handedness and vice versa the chiral symmetry operator $R_Y T$ does not change spin and magnetic quantum numbers IM. Thus the intrinsic $|\sigma\rangle$ states can be projected on the $|IM\rangle$ basis first

$$
P_I^M|\sigma\rangle = |IM,\sigma\rangle.
\tag{7.2}
$$

And then the handedness σ can be projected on chirality χ without change of IM values, what in the 2×2 mixing picture gives

$$
\begin{aligned}
|IM,\chi = +\rangle &= \frac{1}{\sqrt{2}N_+}(|IM,L\rangle + |IM,R\rangle) \\
|IM,\chi = -\rangle &= \frac{i}{\sqrt{2}N_-}(|IM,L\rangle - |IM,R\rangle)
\end{aligned}
\tag{7.3}
$$

with $N_\pm$ being the normalization factors. Now all matrix elements can be expressed by $|L\rangle$ and $|R\rangle$ handed states and experimentally expected features can be deduced as a function of these handed states properties. This especially helps when investigating the experimental signatures in the transition region between the chiral configuration (large aplanarity with well defined handed states) and not chiral one (near planar spins configuration). Conclusions drawn from such an analysis are presented in the following sections where the 2×2 mixing picture limitation is assumed to hold. Before diving into this interesting topic, one more feature related to the presence of time-reversal operator T needs to be mentioned. Calculating the Hamiltonian matrix element on a handed state $|IM,L\rangle$ gives

$$\langle IM,L|H|IM,L\rangle = \langle IM,L|R_y T^+ H R_y T|IM,L\rangle^* = \langle IM,R|H|IM,R\rangle^*. \tag{7.4}$$

since the chiral symmetry operator reverses the handedness, i.e. $R_y T|L\rangle = |R\rangle$. The time-reversal operator T is not a unitary one, thus the complex conjugation star appears in the right hand-side of the equation. This means that reversing the handedness in the matrix element needs a complex conjugation to be applied. This holds for Hamiltonian and also for any other operator commuting with $R_Y T$.

7.3 CHIRAL DOUBLETS DEGENERATION

It is a known fact that stable chirality requires small energy splitting between the $|IM,\chi=+\rangle$ and $|IM,\chi=-\rangle$ doublet members. The energy splitting is often linked to the possible tunneling process $\langle IM,L|H|IM,R\rangle$ between the $|IM,L\rangle$ and $|IM,R\rangle$ intrinsic states. In the 2×2 mixing picture, however, the energy splitting may also appear when tunneling process becomes negligible what may explain the observed nonzero energy splitting in nuclei identified to have stable chiral rotation regime. To see this, let us examine the energy expectation values for the $|IM,\chi=+\rangle$ and $|IM,\chi=-\rangle$ doublet members

$$\begin{aligned} E_+ &= \langle IM,+|H|IM,+\rangle \\ E_- &= \langle IM,-|H|IM,-\rangle \end{aligned} \tag{7.5}$$

Substituting the $|IM,\chi\rangle$ from eqs.(7.3) leads to

$$\begin{aligned} E_+ &= \frac{\langle IM,L|H||IM,L\rangle + \langle IM,R|H|IM,R\rangle}{2N_+^2} \\ &\quad + \frac{\langle IM,L|H|IM,R\rangle + \langle IM,R|H|IM,L\rangle}{2N_+^2} \\ E_- &= \frac{\langle IM,L|H||IM,L\rangle + \langle IM,R|H|IM,R\rangle}{2N_-^2} \\ &\quad - \frac{\langle IM,L|H|IM,R\rangle + \langle IM,R|H|IM,L\rangle}{2N_-^2} \end{aligned} \tag{7.6}$$

The above formulae can be simplified owing to the fact that reversing handedness means a complex conjugation. So there are summations of complex conjugated values leaving only the real component of each of them

$$\begin{aligned} E_+ &= \frac{Re\langle IM,L|H||IM,L\rangle + Re\langle IM,L|H|IM,R\rangle}{N_+^2} \\ E_- &= \frac{Re\langle IM,L|H||IM,L\rangle - Re\langle IM,L|H|IM,R\rangle}{N_-^2} \end{aligned} \tag{7.7}$$

Now, two reasons why the doublet members may split are clearly visible: significant tunneling process making the $Re\langle IM,L|H|IM,R\rangle$ a non-zero value and different normalization factors N_+, N_- of the $|IM,\chi=+\rangle$, $|IM,\chi=-\rangle$ states. For the tunneling effects being negligible the $Re\langle IM,L|H|IM,R\rangle$ component must approach zero value but still a significant energy splitting may exist due to difference in the normalization factors. To see when such a situation may happen one needs to express the left-handed and right-handed states $|IM, \sigma = L\rangle$, $|IM, \sigma = R\rangle$ by the chirality basis elements $|IM, \chi = +\rangle$, $|IM, \chi = -\rangle$ and substitute them into the tunneling matrix element $\langle IM, L|H|IM, R\rangle$

$$\langle IM, L|H|IM, R\rangle = \frac{1}{2}\left(N_+^2 \langle IM, +|H|IM, +\rangle - N_-^2 \langle IM, -|H|IM, -\rangle\right) \tag{7.8}$$

The above formula has been obtained with the condition that Hamiltonian does not mix the eigenstates basis elements, i.e $\langle IM, +|H|IM, -\rangle = \langle IM, -|H|IM, +\rangle = 0$. For the tunneling to vanish at nonzero energy splitting of the doublet members one needs the following condition to be met

$$N_+^2 \langle IM, +|H|IM, +\rangle = N_-^2 \langle IM, -|H|IM, -\rangle \tag{7.9}$$

meaning that the normalization factors must be inverse proportional to the energies of the doublet members. In principle a situation may occur when an initially non-chiral nucleus undergoes a transformation to the chiral configuration with increasing rotational frequency. In the transition region the tunneling process may be significant giving large doublet members energy splitting. Finally a nucleus may reach a stable chiral configuration with reduced tunneling, however, the energy splitting may remain large due to dominating effects of different normalization factors. The last element in the puzzle of the energy splitting analysis is the physical reason for the normalization factors to be different. This is explained by substituting eqs.(7.3) into the normalization conditions $\langle IM, +|IM, +\rangle = \langle IM, -|IM, -\rangle = 1$

$$\langle IM, +|IM, +\rangle = 1 \quad = \quad \frac{1}{N_+^2}\left(Re\langle L|L\rangle + Re\langle L|R\rangle\right) \tag{7.10}$$

$$\langle IM, -|IM, -\rangle = 1 \quad = \quad \frac{1}{N_-^2}\left(Re\langle L|L\rangle - Re\langle L|R\rangle\right)$$

where the subtlety of complex conjugation when reversing handedness in matrix elements has been used again. The reason for normalization factors difference is now clearly seen in the mutual overlap $\langle L|R\rangle$ of the opposite handed states. Getting an estimate of the overlap value is out of the scope of this paper and out of range of the 2×2 mixing picture since it requires a detailed knowledge of the intrinsic state distribution over the continuous handedness parameter σ. Nevertheless both the tunneling and the intrinsic states overlap plays a role in the doublet members energy splitting. For an ideal chiral configuration the handedness of the intrinsic states is well defined meaning the overlap value being close to zero. The simplest odd-odd chiral nuclei can be close to such an ideal configuration since three spins j_p, j_n and j_R form a reference frame in which the handedness may clearly be defined. For nuclear chirality in single-odd or even-even nuclei where more unpaired nucleons are involved the definition of $|L\rangle$ and $|R\rangle$ handedness is more complicated leaving room for the opposite handed states to have common components and therefore larger values of the overlap in question. Decreasing the energy splitting as a function of spin is therefore the first indication of the non-chiral to chiral transition region and should be best visible in simple cases of odd-odd nuclei. Search for such a decrease is presented in [4] and shown in Fig.7.1. No clear behavior of the energy splitting with increasing spin is observed showing a difficulty for this condition to be used i.e. the requirement for doublets to be observed for wide spin region. Often this is not the case. Below a certain spin value only one doublet member is often observed (the one forming the yrast rotational band) while the other not (the one forming the side band low spin levels). An energy splitting should be observed for low (close to the bandhead) spins which should decrease and then stabilize for higher spins when stable chirality is attained. Such a behavior is not observed. Either the energy splitting does not systematically change and represent rather stable value (like in Cs isotopes) or a systematic decrease is observed without the stabilization behavior at high spin levels indicating a different moment of inertia of both bands rather than effects related to the nuclear

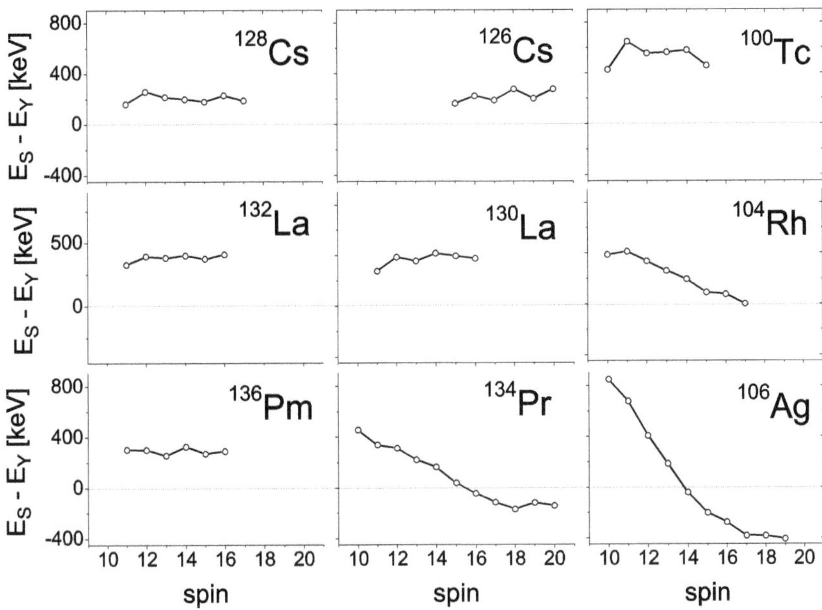

Figure 7.1 Energy difference of corresponding chiral doublet members in rotational bands as a function of spin. The expected behavior of large energy splitting that gets smaller and stabilizes with increasing spin is not observed. Either the energy difference is nearly constant (indicating absence of non-chiral to chiral transition) or systematic decrease is observed without stabilization region (possibly different moments of inertia in both bands).

chirality. Unlike to experimental data, the energy splitting values in the non-chiral to chiral transition region are available in nuclear theoretical models. A model reproducing a large set of experimental observables has recently been used in ref. [5] for the description of ^{126}Cs chiral properties. Still, no striking decrease of the doublet members energy splitting as a function of spin is predicted.

7.4 ELECTROMAGNETIC TRANSITION PROBABILITIES

Study of the chiral doublets energy behavior in the chiral rotational bands does not provide sufficient information on nuclear chirality mechanisms occurring in nature. Another observables are the electromagnetic (EM) transition probabilities linking states in those rotational bands. Unlike diagonal matrix elements in case of Hamiltonian, the non-diagonal matrix elements can be investigated with EM transition measurements. In nuclear chirality phenomenon the two rotational bands $|IM, \chi = +\rangle$ and $|IM, \chi = -\rangle$ have the same parity of rotational states, thus the transition types of M1, E2 are only observed. This is a favorable situation since for these transition types the commutation relation (eqs. (1A-73) and (3C-10) in Ref. [3]) of EM transition operator $[EM, R_Y T] = 0$ is the same as for Hamiltonian $[H, R_Y T] = 0$ and the same formalism as for Hamiltonian can be used again. For electromagnetic transitions from spin I_1 to spin I_2 within one chirality band (the inband transitions linking either $\chi = +$ states only or $\chi = -$ states only) we therefore obtain

$$\langle I_2, \chi = + ||EM||I_1, \chi = +\rangle = \frac{Re\langle I_2, L||EM||I_1, L\rangle + Re\langle I_2, L||EM||I_1, R\rangle}{N_{I_2+}N_{I_1+}}$$

$$\langle I_2, \chi = - ||EM||I_1, \chi = -\rangle = \frac{Re\langle I_2, L||EM||I_1, L\rangle - Re\langle I_2, L||EM||I_1, R\rangle}{N_{I_2-}N_{I_1-}}. \qquad (7.11)$$

The probabilities for corresponding inband transitions in both bands are expected to be similar when the $Re\langle I_2,L||EM||I_1,R\rangle$ part has a negligible value and the normalization factors N for corresponding levels are similar. This is expected in the ideal chiral configuration with significant aplanarity in odd-odd nuclei. In ideal case the right- and left-handed states are well defined and separated having $\langle IM,L|IM,R\rangle \approx 0$ overlap and thus similar normalization factors. Significant aplanarity means that all three spins j_p, j_n and j_R are almost perpendicular to each other and have similar lengths. The unpaired nucleons in the A=130 region have spins j_p and j_n around $5\hbar$. Assuming similar spin value for j_R leads to the expected small value of $Re\langle I_2,L||EM||I_1,R\rangle$ component. Otherwise the M1 or E2 electromagnetic transition could swap the left-handed state to the right-handed one. This effectively means reversing one of the three spins, i.e changing its spin value about $10\hbar$ units. To do this, a significant reorientation of the other two spins would be required for the total angular momentum vector I to change not more then $1\hbar$ or $2\hbar$ units. Such dramatic reorientation is rather unlikely. In this context an interesting experimental data are published in [11, 12] for ^{126}Cs and ^{128}Cs isotopes. In these isotopes the reduced B(M1) and B(E2) transition probabilities for inband transitions in both bands become similar above $14\hbar$ spin value suggesting that a stable chirality is attained with similar normalization factors N of corresponding levels. Thus the non-zero energy splitting of the doublet states can be attributed to domination of the tunneling effect (between the right- and left-handed states) over the effects of different normalization factors. Similar conclusions hold for interband EM transitions (linking states with opposite chirality). Again using the same formalism as for Hamiltonian one gets for the interband transitions

$$\langle I_2,\chi = +||EM||I_1,\chi = -\rangle \;\; = \;\; -\frac{Im\langle I_2,L||EM||I_1,L\rangle + Im\langle I_2,L||EM||I_1,R\rangle}{2N_{I_2+}N_{I_1+}} \qquad (7.12)$$

$$\langle I_2,\chi = -||EM||I_1,\chi = +\rangle \;\; = \;\; -\frac{Im\langle I_2,L||EM||I_1,L\rangle - Im\langle I_2,L||EM||I_1,R\rangle}{2N_{I_2-}N_{I_1-}}.$$

Because of the same reasons as for inband transitions, the probabilities of interband ones between corresponding levels are expected to be similar in the ideal chiral configurations. So the EM transition probabilities in both directions, from band build of $|IM,\chi = +\rangle$ states to band build of $|IM,\chi = -\rangle$ states and vice versa, should be similar. Thus, there should be symmetric behavior of transition probabilities for all EM transitions in both $\chi = +$ and $\chi = -$ bands for stable nuclear chirality case. The condition of similar normalization parameters may not be fulfilled rigorously, especially in nuclei with more than two unpaired nucleons like in ^{136}Nd [6] or [7–10] since more that 3 spins are involved in definition of the handedness, thus the left- and right-handed configuration may posses common components giving non-zero $\langle L|R\rangle$ overlap. For chirality built of two unpaired nucleons and an even-even core like Cs isotopes, the picture is cleaner and some properties for the non-chiral to chiral transition region can be deduced. In this case both unpaired nucleons occupy $h_{11/2}$ states where proton has particle-like character while neutron the hole-like one. The mutual orientation of the j_p and j_n angular momentum vectors is a result of interaction energy with the deformed even-even core and the Coriolis effects. For low spins (close to the head of rotational band) the Coriolis effects become negligible leaving the deformation of the core as a main factor governing the j_p and j_n mutual orientation. Here the β deformation parameter plays the major role. This parameter does not change much along the chiral rotational bands indicating that mutual orientation of j_p and j_n does not change dramatically in the non-chiral to chiral transition region. More complicated situation relates to the spin of the core j_R. Unlike spins of unpaired nucleons the length of j_R increases with increasing spin of rotational states. Moreover its direction may strongly depend on the triaxial deformation parameter γ. For little values of j_R in the transition region the EM transitions may connect the opposite handed states resulting in non-zero $Re\langle I_2,L||EM||I_1,R\rangle$ parameter. Thus an asymmetry in electromagnetic behavior of both bands should increase when going from chiral to non-chiral configuration close to the bandhead. These effects have been investigated via lifetime measurements of states in chiral bands [9, 11–18]. Similar values of B(M1) and B(E2) in both bands have been observed in Cs [11, 12] isotopes together with characteristic B(M1) staggering. In case of ^{128}Cs isotope an asymmetry in EM behavior of both bands is observed for spin lower than $14\hbar$. However, the data are scarce and to draw a conclusion on chirality buildup close to the $I = 9^+$ bandhead a lifetime measurements in spin region 9-14$\hbar$ are required. Similar to the energy splitting between the chiral doublets members, an asymmetry in electromagnetic behavior of both chiral bands may exist even for stable chiral configuration due to differences in N normalization factors.

7.5 ELECTROMAGNETIC SELECTION RULES

In year 2004 a letter article [19] was published predicting a set of EM selection rules expected for experimental observation in two chiral bands – the B(M1) staggering. A rigid even-even core with maximum triaxiality has been assumed with two unpaired nucleons (one particle-like, the other hole-like) occupying a single-j shell. Such a configuration is expected in a set of odd-odd isotopes in mass region A=130. Roughly the same year the electromagnetic transition probabilities have been measured in ^{132}La and ^{128}Cs isotopes belonging to this isotope set. The expected chiral EM selection rules, B(M1) staggering, are observed in ^{128}Cs [11] and in the later measured ^{126}Cs [12]. The isotopes around A=130 belonging to the set in question posses a triaxially-soft deformation contradicting to a rigid deformation assumed in [19]. In 2011, the theoretical description of chiral selection rules has been revisited and generalized to triaxially-soft deformation [20]. The rigid deformation condition has been lifted in the generalized description where a new S-symmetry has been introduced as responsible for the B(M1) staggering existence. The S-symmetry deals with α-parity of the nucleus deformation (reversal of five α parameters of the quadrupole deformation) and with exchange of unpaired nucleons states occupation, i.e proton states are exchanged with neutron ones. Some isotopes are invariant with respect to the S-symmetry and the characteristic B(M1) staggering appears in their chiral bands. The exchange of the unpaired nucleons states occupation means that these selection rules are expected in isotopes with same number of odd protons and odd neutrons both occupying the same j-shell (like odd-odd Cs isotopes) and like ^{136}Nd isotope (chirality based on four quasiparticle states). For chirality based on odd number of unpaired nucleons or based on different orbitals occupied by neutrons and protons – a nucleus is not invariant to the S-symmetry and characteristic B(M1) staggering needs not to be present. The Cs isotopes are unique ones since the EM selection rules in the chiral bands are clearly observed for spin $I \geq 14\hbar$ in ^{128}Cs as well as in ^{126}Cs, see Fig.7.2. In ^{126}Cs the B(M1) staggering is observed in the entire measured spin range $14\hbar \leq I \leq 21\hbar$. In case of ^{128}Cs the B(M1) staggering vanishes in one band for spin less than $15\hbar$, while in the other band is still observed with increased extent. This might be a particularly important observation suggesting that for decreasing spin value the S-symmetry may still exist while the nuclear chirality begins to vanish. To see this one needs to examine the formulae (7.11) once again. As stated in the previous section, for stable chiral configuration the $Re\langle I_2,L||EM||I_1,R\rangle$ component is expected to vanish. Also the normalization factors N are expected to be similar. In such a case, the entire electromagnetic behavior is governed by the $Re\langle I_2,L||EM||I_1,L\rangle$ component in both bands and must reproduce the B(M1) staggering too. Thus, the value of this component must have alternate small and large values as a function of spin for the S-invariant nucleus. With decreasing spin one enters the non-chiral to chiral transitional region close to the bandhead. In this case the $Re\langle I_2,L||EM||I_1,R\rangle$ becomes a non-zero value. In order for the B(M1) staggering in one of the bands, say $\chi = +$, to vanish the alternate values of $Re\langle I_2,L||EM||I_1,L\rangle$ must be summed with opposite alternate values of $Re\langle I_2,L||EM||I_1,R\rangle$ representing an "anti-staggering" of B(M1). The "anti-staggering" component in the $\chi = -$ band (7.11) has opposite sign thus an opposite behavior is expected: increasing the observed B(M1) staggering in the $\chi = -$ band instead of vanishing. This leads to asymmetry in the EM behavior of both bands, the staggering vanishes in one of the band while it increases in the other chiral band. As explained above such an observation may suggest that the S-symmetry is still valid while the nucleus looses its chiral configuration. Lifetime

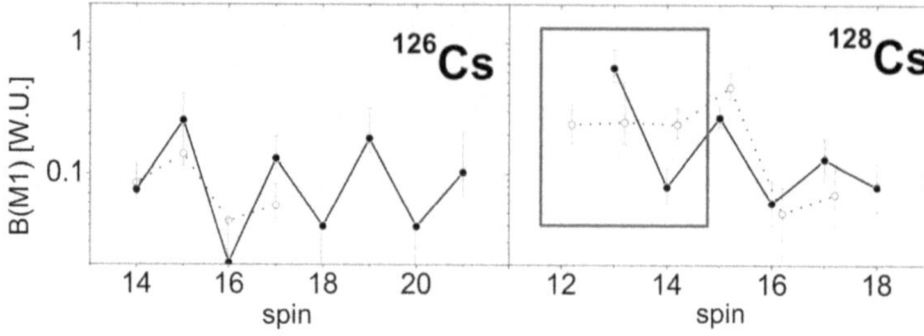

Figure 7.2   Chiral electromagnetic selection rules (resulting from S-symmetry invariance) seen in chiral rotational bands of ^{126,128}Cs isotopes as characteristic B(M1) staggering. The B(M1) staggering exists in both chiral bands in ^{126}Cs. Red rectangle indicates the spin region below $15\hbar$ in ^{128}Cs where the staggering vanishes in one band while it exists in the other band. Such a behavior may indicate that the S-symmetry is still valid while the nucleus looses its chiral character for low spins close to the bandhead (see text for explanation).

measurements of the states belonging to chiral bands in Cs isotopes in spin region $9\hbar \leq I \leq 13\hbar$ close to the bandhead are extremely important for investigating the relation of the S-symmetry with chirality. These lifetime measurements are now conducted by National Centre for Nuclear Research, Swierk, Otwock and Heavy Ion Laboratory, University of Warsaw spectroscopy groups.

7.6   RECENT PLUNGER MEASUREMENTS IN ^{128}CS

First of the lifetime measurements in the non-chiral to chiral transitional region was already conducted in July 2022. The goal of that experiment was to measure lifetimes of states in the spin region $10\hbar \leq I \leq 12\hbar$ of the yrast band in ^{128}Cs. That should answer the question whether the B(M1) staggering exists in both bands or only in one band or does not exists at all in the transition spin region. These experimental data would help find the relation between the S-symmetry and chirality in ^{128}Cs isotope. The Recoil Distance Method (RDM) uses plunger device to set specific distances between target and stopper. Due to physical properties of the target, it can't be set lower than a certain value. This limits the shortest lifetimes accessible by this method. Extrapolation of the B(M1) staggering observed for spin $I \geq 14\hbar$ to the transitional region shows that the lifetime of the 12^+ state should be over 5 ps, and for the 11^+ state – below 1.5 ps. Preliminary results shows those lifetimes might be around 5 and 2.5 ps (see Fig. 7.3), but they are below the methods limit. The 10^+ lifetime longer than 60 ps is expected from the B(M1) staggering extrapolation. Preliminary results for the state is 9 ± 5 ps, therefore the staggering must vanish at some point before 10^+ state. The experiment showed also that it should be possible to use another reaction to get to the limit 2-3 times shorter lifetimes. Such experiment was proposed and accepted for late 2024 and should measure lifetimes of both 11^+ and 12^+ states, so the whole spin region $9\hbar \leq I \leq 18\hbar$ would be known.

7.7   GYROMAGNETIC FACTOR IN CHIRAL BANDS

Unlike the previously described observables where the nuclear chirality was traced through indirect effects like energy degeneracy or B(M1) staggering the gyromagnetic factor measures the chiral configuration of a nucleus directly. The principle for the direct chirality determination is based on two features of the g-factor measurement: i) magnetic moments of the system components (unpaired proton, unpaired neutron and even-even core) are involved, ii) the total magnetic moment does not

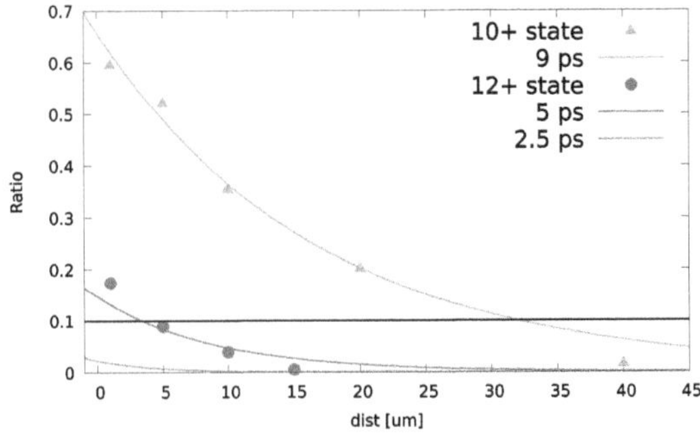

Figure 7.3 Preliminary results from plunger lifetime measurements in low spin states of chiral bands in ^{128}Cs. Ratios should decrease in the exponential rate. Uncertainty for each point is 0.1, so values below the dotted line are indistinguishable from 0, so for 12^+ state only 1 point is valid and for 11^+ state none.

distinguish the handedness σ in which the system actually may exist. The first feature means that the relative values of the three spins j_p, j_n and j_R are different then the relative values of corresponding magnetic moments or g-factors g_p, g_n and g_R since the components in question have different electric charges. While different mutual orientations of j_p,j_n and j_R form the same total spin I of an excited state, the total gyromagnetic factor g of this state built of g_p, g_n and g_R changes when mutual orientation of the three spins varies due to different relative values of the components g-factors with respect the relative values of their spins. The g-factor of an excited state is therefore sensitive to mutual orientation of the spins involved in the σ handedness parameter definition. The second feature relates to the Schrödinger cat problem in form of unobserved $|L\rangle$ and $|R\rangle$ intrinsic states. As explained in sec. 3 only symmetric or antisymmetric superpositions of these states are observed experimentally. The handedness parameter, according to its definition $\sigma = \frac{(j_p \times j_n)\cdot j_R}{|j_p|\cdot|j_n|\cdot|j_R|}$ has opposite sign for opposite handedness, thus the expectation value of handedness and any operator sensitive to it always vanish in the superimposed states. As explained in refs.[1, 21, 22] the gyromagnetic factor is not sensitive to handedness itself but rather to the volume spanned by the three spins forming the handed state thus the expectation value is the same for corresponding $|L\rangle$ and $|R\rangle$ configurations and does not vanish in the superposed $|\chi = +\rangle$ and $\chi = -\rangle$ states. The g-factor measurement in the chiral states has been performed for the first time in $I = 9^+$ chiral bandhead in ^{128}Cs. The obtained result proves that the bandhead $I = 9^+$ spin cannot be formed chiefly by the spins of unpaired proton j_p and unpaired neutron j_n. Significant contribution of the core rotation with spin j_R comparable to the ones of odd nucleons is required to reproduce the obtained g-factor value. The bandhead angular momentum is formed by three spins (of proton, neutron and core) that may form a handed state thus being chiral (aplanar configuration) on non-chiral (planar configuration). Moreover, the three spins must have near planar configuration in order for the expected g-factor value to match the experimentally measured one, thus the $I = 9^+$ bandhead has been measured to be non-chiral. The g-factor has therefore proven that the non-chiral to chiral transition must exist in ^{128}Cs in the spin region $9\hbar \leq I \leq 14\hbar$ near the bandhead. The phenomena described in the previous sections should be observed in this transitional region. The 56 ns half-life of the isomeric $I = 9^+$ bandhead state allowed precise g-factor measurement using traditional spin-precession method. Similar measurements for other states in chiral bands of ^{128}Cs are difficult due to their much shorter picoseconds half-lives. In case of ^{126}Cs nucleus neither the chiral bandhead half-life nor its g-factor are known. The experimental investigation of these properties represent the closest opportunity to study the nuclear chirality using currently available techniques.

7.8 CONCLUSIONS

The nuclear chirality has initially been predicted for 2-qp configurations in triaxial odd-odd nuclei around mass A=130. At the beginning of the present century a pair of rotational bands resembling the chiral ones have been identified in some isotopes in this mass region. Around 10 isotopes represented the first chiral candidates. The next two decades were characterized by a rapid development of chirality research, both theoretical and experimental. A multi-chiral doublet bands $M\chi D$ phenomenon has been predicted in which multiple chiral bands form on various multi-quasiparticle configurations within a single isotope. Not only the number of isotopes identified as chiral has been extended from odd-odd to od-even [7–10] and even-even [6] nuclei but also the number of chiral bands in a single nucleus has been multiplied by its various quasiparticle configurations. Multiple chiral bands are even expected for two quasiparticles located in one shell [5]. Therefore, a set of nuclear collective states with chiral character is often significant with respect to the number of non-chiral excited states in the level scheme of a single nucleus. Recently [23] the nuclear chirality has been predicted for octupole deformation of a nucleus as chirality-parity χP effect where not two but four chiral bands exist per one quasiparticle configuration. The number of chiral nuclei (today more than 60) is expected to increase as well as the contribution of chirality to the nuclear level schemes making this phenomenon a significant part of nuclear spectroscopy in general. Thus, the mechanisms of chirality needs to be investigated in details and observed effects interpreted correctly. As shown in this paper, some features, like small energy splitting of the doublets or the same EM behavior may not strictly be considered as a necessary condition for the rotational bands to be chiral. The chiral EM selection rules, the B(M1) staggering, may exist for some quasiparticle configurations that are invariant to additional S-symmetry. The B(M1) staggering may exists only in one band while in the other band vanishes indicating that the S-symmetry still holds while a nucleus looses its chiral character. All these effects should be investigated in the non-chiral to chiral transitional spin region close to the chiral bandhead state. Major part of chiral properties are observed in ^{128,126}Cs making them good candidates for chirality mechanisms investigations in this transitional spin region.

7.9 ACKNOWLEDGMENTS

I would like to thank Adam Nałęcz-Jawecki for performing plunger experiments and data analysis resulting in preliminary lifetime results for low spin chiral states in ^{128}Cs. These results are part of Adam Nałęcz-Jawecki PhD study at National Centre for Nuclear Research, Otwock, Poland.

Bibliography

1. E. Grodner *et al.*, Frontiers of Physics, Topical Review, Volume 19, Issue 3, 34202 (2024).
2. P. Ring and P. Schuck, Nuclear Many-Body Problem Springer-Verlag, (1975).
3. A. Bohr and B. R. Mottelson, Nuclear Structure, Vol. I, Benjamin Inc., (1969).
4. E. Grodner, Acta Physica Polonica, B **38**, 3769 (2007).
5. T. J. Gao *et al.*, Physical Review C **109**, 024307 (2024).
6. C. M. Petrache *et al.*, Physical Review C **97**, 041304 (2018).
7. B. F. Lv *et al.*, Physical Review C **100**, 024314 (2019).
8. S. Zhu *et al.*, Physical Review Letters **91**, 132501 (2003).
9. S. Mukhopadhyay *et al.*, Physical Review Letters **99**, 172501 (2007).
10. S. Guo *et al.*, Physics Letters B **807**, 135572 (2020).
11. E. Grodner *et al.*, Physical Review Letters **97**, 172501 (2006).
12. E. Grodner *et al.*, Physics Letters B **703**, 46 (2011).
13. P. L. Masiteng *et al.*, European Physical Journal A **52**, 28 (2016).
14. R. J. Guo *et al.*, Physics Letters B **833**, 137344 (2022).

15. W. Z. Xu *et al.*, Physics Letters B **833**, 137287 (2022).
16. Wu Xiaoguang *et al.*, Plasma Science and Technology, Vol. 14, No.6, Jun. (2012).
17. K. Selvakumar *et al.*, Physical Review C **92**, 064307 (2015).
18. T. Marchlewski *et al.* Acta Physica Polonica B **46**, 689 (2015).
19. T. Koike *et al.*, Physical Review Letters **93**, 172502 (2004).
20. L. Prochniak *et al.*, Acta Physica Polonica B **42**, 465 (2011).
21. E. Grodner *et al.*, Physical Review Letters **120**, 022502 (2018).
22. E. Grodner *et al.*, Physical Review C **106**, 014318 (2022).
23. C. Liu *et al.*, Physical Review Letters **116**, 112501 (2016).

8 Semiclassical feature of wobbling and chiral properties in atomic nuclei

Apolodor Aristotel Raduta

National Institute of Nuclear Physics and Engineering, Magurele, and Academy of Romanian Scientists, Bucharest, Romania

Cristian Mircea Raduta

National Institute of Nuclear Physics and Engineering, Magurele, Romania

Robert Poenaru

UNESCO International Centre for Advanced Training and Research in physics, Magurele, Romania

8.1 WOBBLING MOTION IN EVEN-EVEN NUCLEI

We suppose that some properties of triaxial nuclei can be quantitatively described by a triaxial rigid rotor. Therefore, we consider a triaxial rigid rotor with the moments of inertia $\mathcal{I}_k$, k=1, 2, 3, corresponding to the axes of the intrinsic frame, described by the Hamiltonian:

$$\hat{H}_R = \frac{\hat{I}_1^2}{2\mathcal{R}_1} + \frac{\hat{R}_2^2}{2\mathcal{I}_2} + \frac{\hat{R}_3^2}{2\mathcal{I}_3}.$$

(8.1)

In principle it is easy to find the eigenvalues of H_R by using a diagonalization procedure within a basis exhibiting the D_2 symmetry. However, when we restrict the considerations to the yrast band it is by far more convenient to use a closed expression for the excitation energies.

We suppose that a certain class of properties of the Hamiltonian H_R can be obtained by solving the time dependent equations provided by the variational principle:

$$\delta \int_0^t \langle \psi(z)|H - i\frac{\partial}{\partial t'}|\psi(z)\rangle dt' = 0.$$

(8.2)

If the trial function $|\psi(z)\rangle$ spans the whole Hilbert space of the wave functions describing the system, solving the equations provided by the variational principle is equivalent to solving the Schrödinger equation associated to H_R. Here we restrict the Hilbert space to the subspace spanned by the the variational state:

$$|\psi(z)\rangle = \mathcal{N}e^{z\hat{R}_-}|IMK\rangle,$$

(8.3)

where z is a complex number depending on time and $|IMK\rangle$ denotes the eigenstates of the angular momentum operators $\hat{R}^2$, R_z and $\hat{R}_3$ with R_z denoting the angular momentum projection on the z axis of the laboratory frame. $\mathcal{N}$ is a factor which assures that the function $|\psi\rangle$ is normalized to unity:

$$\mathcal{N} = (1 + |z|^2)^{-I}.$$

(8.4)

DOI: 10.1201/9781032691633-8

$\hat{R}_-$ denotes the lowering operator which for the intrinsic components is:

$$\hat{R}_- = \hat{R}_1 + i\hat{R}_2. \tag{8.5}$$

The function (8.3) is a coherent state for the group $SU(2)$ [1], generated by the angular momentum components, and is suitable for the description of the classical features of the rotational degrees of freedom. Due to the supercompletness property, the variational state comprises all basis vectors spanning the Hilbert space. Actually this is the feature which assures a good approach to the eigenfunctions of H. As a matter of fact this will be concretely checked out within the present section.

The average of H_R and the time derivative operator with the function (2.3) have the expressions:

$$\langle \hat{H} \rangle = \frac{I}{4}\left(\frac{1}{\mathcal{I}_1} + \frac{1}{\mathcal{I}_2}\right) + \frac{I^2}{2\mathcal{I}_3} + \frac{I(2I-1)}{2(1+zz^*)^2}\left[\frac{(z+z^*)^2}{2\mathcal{I}_1} - \frac{(z-z^*)^2}{2\mathcal{I}_2} - \frac{2zz^*}{\mathcal{I}_3}\right],$$

$$\langle \frac{\partial}{\partial t} \rangle = \frac{I(\dot{z}z^* - z\dot{z}^*)}{1+zz^*}. \tag{8.6}$$

Denoting the average of H_R by $\mathcal{H}$, the time dependent variational equation yields:

$$\frac{\partial \mathcal{H}}{\partial z} = -\frac{2iI\dot{z}^*}{(1+zz^*)^2}, \quad \frac{\partial \mathcal{H}}{\partial z^*} = \frac{2iI\dot{z}}{(1+zz^*)^2}. \tag{8.7}$$

By a suitable change of variables the classical equations acquire the canonical Hamilton form. Indeed, taking the polar form of the complex variable $z(=\rho e^{i\varphi})$ and

$$r = \frac{2I}{1+\rho^2}, \quad 0 \leq r \leq 2I. \tag{8.8}$$

the new variables (φ, r) are canonical conjugate and satisfy the equations:

$$\frac{\partial \mathcal{H}}{\partial r} = \dot{\varphi}, \quad \frac{\partial \mathcal{H}}{\partial \varphi} = -\dot{r}. \tag{8.9}$$

Accordingly, φ and r play the role of generalized coordinate and momentum, respectively. In the new coordinates, the classical energy function acquires the expression:

$$\mathcal{H}(r,\varphi) = \frac{I}{4}\left(\frac{1}{\mathcal{I}_1} + \frac{1}{\mathcal{I}_2}\right) + \frac{I^2}{2\mathcal{I}_3} + \frac{(2I-1)r(2I-r)}{4I}\left[\frac{\cos^2\varphi}{\mathcal{I}_1} + \frac{\sin^2\varphi}{\mathcal{I}_2} - \frac{1}{\mathcal{I}_3}\right]. \tag{8.10}$$

Averaging the angular momentum components with the function $|\psi(z)\rangle$ one obtains:

$$\langle I_1 \rangle = \frac{2I\rho}{1+\rho^2}\cos\varphi, \quad \langle I_2 \rangle = \frac{2I\rho}{1+\rho^2}\sin\varphi, \quad \langle I_3 \rangle = I\frac{1-\rho^2}{1+\rho^2}. \tag{8.11}$$

Another pair of canonically conjugate coordinates is:

$$\xi = I\frac{1-\rho^2}{1+\rho^2} = \langle I_3 \rangle \text{ and } \phi = -\varphi, \tag{8.12}$$

Indeed, their equations of motion are:

$$\frac{\partial \mathcal{H}}{\partial \xi} = -\dot{\phi}, \quad \frac{\partial \mathcal{H}}{\partial \phi} = \dot{\xi}. \tag{8.13}$$

Taking the Poisson bracket defined in terms of the new conjugate coordinates one finds:

$$\{\langle I_1 \rangle, \langle I_2 \rangle\} = \langle I_3 \rangle, \quad \{\langle I_2 \rangle, \langle I_3 \rangle\} = \langle I_1 \rangle, \quad \{\langle I_3 \rangle, \langle I_1 \rangle\} = \langle I_2 \rangle \tag{8.14}$$

Therefore, the angular momentum components form a classical algebra, $SU(2)_{cl}$, with the inner product $\{,\}$. The correspondence

$$\{\langle I_k \rangle, \{,\} i\} \longrightarrow \{I_k, [,]\} \tag{8.15}$$

is an isomorphism of $SU(2)$ algebras, which accomplishes the quantization of classical angular momentum.

Solving the classical equations of motion (8.9), the classical trajectories given by $\varphi = \varphi(t)$, $r = r(t)$ are found. Due to Eq. (8.9), one finds that the time derivative of $\mathcal{H}$ is vanishing. This means that the system energy is a constant of motion and, therefore, the trajectory lies on the surface $\mathcal{H} = const$. Another restriction for trajectory consists in the fact that the classical angular momentum squared is equal to $I(I+1)$. This restriction is automatically fulfilled by the classical angular momentum. The intersection of the two surfaces, defined by the two constants of motion, determines the manifold to which the system trajectory belongs.

Studying the sign of the Hessian associated to $\mathcal{H}$, one obtains the points where $\mathcal{H}$ acquires extremal values. Here we consider only the case $\mathcal{I}_1 > \mathcal{I}_3 > \mathcal{I}_2$, when $(0, I)$ is a minimum point for energy, while $(\frac{\pi}{2}, I)$ a maximum.

The second order expansion for $\mathcal{H}(r, \varphi)$ around the minimum point yields:

$$\tilde{\mathcal{H}}(r, \varphi) = \frac{I}{4}\left(\frac{1}{\mathcal{I}_2} + \frac{1}{\mathcal{I}_3}\right) + \frac{I^2}{2\mathcal{I}_1} + \frac{2I-1}{4I}\left(\frac{1}{\mathcal{I}_3} - \frac{1}{\mathcal{I}_1}\right)r'^2 + \frac{(2I-1)I}{4}\left(\frac{1}{\mathcal{I}_2} - \frac{1}{\mathcal{I}_1}\right)\varphi'^2. \tag{8.16}$$

This equation describes an oscillator with the frequency:

$$\omega_I = \left(I - \frac{1}{2}\right)\sqrt{\left(\frac{1}{\mathcal{I}_3} - \frac{1}{\mathcal{I}_1}\right)\left(\frac{1}{\mathcal{I}_2} - \frac{1}{\mathcal{I}_1}\right)}. \tag{8.17}$$

This frequency is associated to the precession motion of the angular momentum around the x axis. In our description, the yrast band energies are therefore given by:

$$E_I = \frac{I}{4}\left(\frac{1}{\mathcal{I}_2} + \frac{1}{\mathcal{I}_3}\right) + \frac{I^2}{2\mathcal{I}_1} + \frac{\omega_I}{2}. \tag{8.18}$$

The classical energy function can be further quantized with the following scheme. One first expresses $\mathcal{H}$ in terms of the complex conjugate variable C_1 and B_1^*

$$C_1 = \sqrt{2I}\sqrt{\frac{2I-r}{r}}e^{-i\varphi}, \quad B_1^* = \frac{1}{\sqrt{2I}}\sqrt{r(2I-r)}e^{i\varphi}. \tag{8.19}$$

and then replacing these by the bosons b and b^+, one obtains the Dyson boson representation of H_R denoted hereafter by H_D. Although this boson operator is not Hermitian it has real eigenvalues [2]. We searched for the eigenvalues of $H_D^{\dagger}$ by using the Bargmann representation of the boson operators [3–5]:

$$b^\dagger \to x, \quad b \to \frac{d}{dx} \tag{8.20}$$

In this way the eigenvalue equation of $H_D^{\dagger}$ is transformed into a differential equation:

$$\left[\left(-\frac{k}{4I}x^4 + x^2 - kI\right)\frac{d^2}{dx^2} + (2I-1)\left(\frac{k}{2I}x^3 - x\right)\frac{d}{dx} - k\left(I - \frac{1}{2}\right)x^2\right]G = E'G. \tag{8.21}$$

where

$$k = \frac{\frac{1}{\mathcal{I}_1} - \frac{1}{\mathcal{I}_2}}{\frac{1}{\mathcal{I}_1} + \frac{1}{\mathcal{I}_2} - \frac{2}{\mathcal{I}_3}}. \tag{8.22}$$

It can be easily proved that this equation can be brought to the algebraic form of the Lamé equation [6, 7]. Performing now the change of function and variable:

$$G = \left(\frac{k}{4I}x^4 - x^2 + kI\right)^{1/2} F, \quad t = \int_{\sqrt{2I}}^{x} \frac{dy}{\sqrt{\frac{k}{4I}y^4 - y^2 + kI}}, \tag{8.23}$$

Eq. (8.21) is transformed into a second order differential Schrödinger equation:

$$-\frac{d^2 F}{dt^2} + V(t)F = E'F, \tag{8.24}$$

with

$$V(t) = \frac{I(I+1)}{4} \frac{\left(\frac{k}{I}x^3 - 2x\right)^2}{\frac{k}{4I}x^4 - x^2 + kI} - k(I+1)x^2 + I. \tag{8.25}$$

The considered ordering for the moments of inertia is such that $k > 1$. Under this circumstance the potential $V(t)$ has two minima for $x = \pm\sqrt{2I}$, and a maximum for $x = 0$.

The minimum value for the potential energy is:

$$V_{min} = -kI(I+1) - I^2. \tag{8.26}$$

Note that the potential is symmetric in the variable x. Due to this feature the potential behavior around the two minima are identical. To illustrate the potential behavior around its minima we make the option for the minimum $x = \sqrt{2I}$. To this value of x it corresponds, $t = 0$. Expanding $V(t)$ around $t = 0$ and truncating the expansion at second order we obtain:

$$V(t) = -kI(I+1) - I^2 + 2k(k+1)I(I+1)t^2. \tag{8.27}$$

Inserting this expansion in Eq. (8.24), one arrives at a Schrödinger equation for an oscillator. The eigenvalues are

$$E'_n = -kI(I+1) - I^2 + [2k(k+1)I(I+1)]^{1/2}(2n+1). \tag{8.28}$$

The quantized Hamiltonian associated to $\mathcal{H}$, i.e. $H^{\dagger}_D$, has an eigenvalue which is obtained from the above expression. The final result is:

$$E_n = \frac{I(I+1)}{2\mathcal{I}_1} + \hbar\omega_I(n + \frac{1}{2}). \tag{8.29}$$

where

$$\omega_I = \left[\left(\frac{1}{\mathcal{I}_2} - \frac{1}{\mathcal{I}_1}\right)\left(\frac{1}{\mathcal{I}_3} - \frac{1}{\mathcal{I}_1}\right)I(I+1)\right]^{1/2} \tag{8.30}$$

defines the wobbling frequency of the angular momentum.

The Bargmann representation of the angular momentum components is obtained by inserting the correspondence (8.20) into the Dyson boson expansion. The result is:

$$I_+ = \sqrt{2I}x, \quad I_- = \sqrt{2I}(\frac{d}{dx} - \frac{x}{2I}\frac{d^2}{dx^2}), \quad I_0 = I - x\frac{d}{dx}. \tag{8.31}$$

From these expressions one may derive the angular momentum component I_1, which may be further averaged with the wave function provided by the Schrödinger equation for a given value of I. As a result one obtains a maximal value ($= I$) which, in fact, confirms the result we got at the classical level.

It is instructive to compare the K-amplitudes of the yrast states obtained through diagonalization, A_K^{diag}, and those corresponding to the coherent state (2.3) considered in the minimum point $(\varphi, r) = (0, I)$ for $I = 20$. The latter function can be written in a different form:

$$|\Phi_{IM}\rangle = |\Psi_{IM}\rangle|_{0,I} = \sum_K \frac{1}{2^I} \binom{2I}{I-K}^{1/2} |IMK\rangle \equiv \sum_K A_K^{coh} |IMK\rangle. \tag{8.32}$$

The two sets of amplitudes were plotted in Fig. 8.1, from where we see that the two functions have a similar K dependence. The small difference is caused by the fact that the diagonalization provides non-vanishing amplitudes only for $K = even$, while the coherent state comprises all K-components. The former state is degenerate with the second yrast state which has only $K = odd$ components. Combining the two degenerate functions to a normalized function, the K-distribution of the new function is almost identical to that of $|\Phi_{IM}\rangle$.

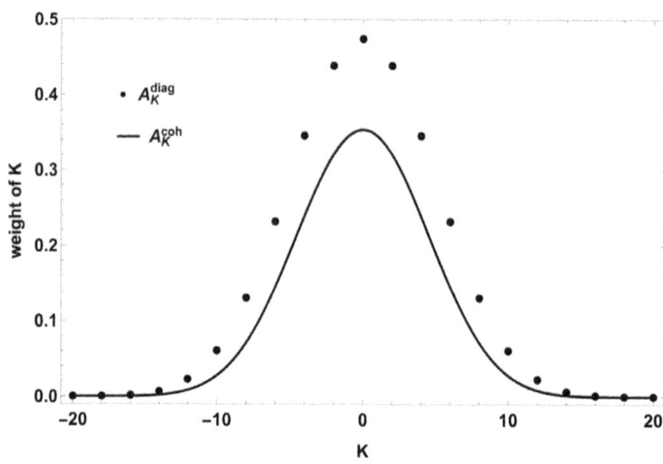

Figure 8.1 The K-amplitudes supplied by the diagonalization procedure and the coherent state, respectively.

Having the wave functions for the states I, one can calculate the $E2$ and $M1$ transition probabilities using the electric quadrupole and magnetic dipole transition operators:

$$\mathcal{M}(E2; \mu) = \frac{3}{4\pi} Z e_{eff} R_0^2 \left(D_{\mu 0}^2 \beta \cos\gamma + (D_{\mu 2}^2 + D_{\mu, -2}^2) \beta \sin\gamma / \sqrt{2} \right),$$

$$\mathcal{M}(M1; \mu) = \sqrt{\frac{3}{4\pi}} g_R D_{\mu\nu}^1 R_\nu. \tag{8.33}$$

Here Z and R_0 denote the nuclear charge and radius, respectively, while D_{MK}^I stands for the Wigner function describing the rotation matrix, μ_N is the nuclear magneton and e_{eff} the effective charge. β is the nuclear deformation and γ represents the nuclear shape deviation from the axial symmetry. For the case of ^{158}Er the nuclear quadrupole deformation is taken equal to 0.203 [8], while γ is determined using for the moments of inertia the expressions given by the hydrodynamic model on one hand and the values provided by the adopted fitting procedure for energies, on the other hand. Thus, one obtains:

$$\cot\gamma = \left(-1 + 2(\mathcal{J}_1^{hyd} / \mathcal{J}_3^{hyd})^{1/2} \right) / \sqrt{3}. \tag{8.34}$$

where $\mathcal{J}_k^{hyd}$ are the moments of inertia corresponding to the k-th axis, respectively, in the liquid drop model [9]. Using the results for the moments of inertia yielded by the fitting procedure one obtains:

$$\cos\gamma = 0.5772, \quad \sin\gamma = 0.8166 \tag{8.35}$$

The value extracted from the equations (8.35) is about $55°$, which indicates a shape close to the prolate-oblate transition, i.e. a γ-soft picture. This is reflected in the low part of the spectrum by a even-odd staggering which for large rotation frequency is washed out. Note that the moments of inertia considered in this section are rigid and consequently the corresponding γ is not depending on spin. This is at variance with the microscopic description which studied the shape dependence on angular momentum [10–12]. For example Ref. [12] pointed out that in ^{158}Er, approaching the band termination the state's collectivity decreases, which might be caused by the fact that the corresponding shape is almost oblate. Also, the coupling with two or four quasiparticle states may lead to discontinuities in the yrast spectrum [10, 13]. Combination of the single-particle level crossing effect and the prolate-oblate competition was accounted for in Ref. [11] and a prolate-oblate shape coexistence was pointed out for high spins. Similar result was obtained in Ref. [14] when the rotation axis coincides with one principal axis. However, when the rotational axis changes the direction, the higher energy minimum becomes a saddle point. Since the approach adopted for ^{158}Er yields an energy level sequence which agrees well with the corresponding experimental data as well as with the exact results obtained through diagonalization, it seems that the rigid soft γ regime is able to describe the global properties of the wobbling band in ^{158}Er.

When the intra-band transition is concerned, the initial and final states of the yrast band are described by the function defined in Eq. (8.58). Since the one phonon operator of the yrast band is associated with the quantas in the parameter space, it commutes with the transition operator and moreover gives zero when acts on the final yrast state, unless this deviates from (8.58) due to parameter fluctuations. The normalized first order expansion of Ψ around Φ is:

$$|\Psi_{IM}\rangle = N_I \frac{1}{2^I} \sum_{K=-I}^{K=I} \left[1 + \frac{i}{\sqrt{2}}\left(\frac{\alpha_I K}{I} + \frac{I-K}{\alpha_I}\right)\right]\left(\begin{matrix}2I\\I-K\end{matrix}\right)^{1/2} |IMK\rangle a_I^\dagger |0\rangle_I, \ I \neq 0,$$

$$(N_I)^{-2} = \frac{1}{2^{2I}} \sum_{K=-I}^{K=I} \left[1 + \frac{1}{2}\left(\frac{\alpha_I K}{I} + \frac{I-K}{\alpha_I}\right)^2\right]\left(\begin{matrix}2I\\I-K\end{matrix}\right). \tag{8.36}$$

Here $a_I^\dagger$ denotes the creation operator for a wobbling quanta on the top of the yrast state of angular momentum I. The corresponding vacuum state is $|0\rangle_I$. The canonical transformation relating the conjugate coordinate and momentum with the creation and annihilation operators depend on the parameter α_I having the expression

$$\alpha_I = \left(I^2 \frac{\frac{1}{\mathcal{J}_2} - \frac{1}{\mathcal{J}_1}}{\frac{1}{\mathcal{J}_3} - \frac{1}{\mathcal{J}_1}}\right)^{1/4}. \tag{8.37}$$

For what follows we introduce the following notation for a wobbling multi-phonon state:

$$|\Phi_{IM};n_w\rangle = |\Phi_{IM}\rangle \frac{\left(a_I^\dagger\right)^{n_w}}{\sqrt{n_w!}}|0\rangle_I. \tag{8.38}$$

The reduced transition probability is readily obtained:

$$B(E2;In_w \to I'n_w') = \left|\langle\Phi_I;n_w||\mathcal{M}(E2)||\Phi_{I'};n_w'\rangle\right|^2. \tag{8.39}$$

The transition amplitude can be also used to calculate the quadrupole moment of an yrast state of angular momentum I:

$$Q_I = \sqrt{\frac{16\pi}{5}}C_{I01}^{I2I}\langle\Phi_{II}|\mathcal{M}(E2)|\Phi_{II}\rangle. \tag{8.40}$$

The magnetic properties were studied with the dipole transition operator defined by Eq. (8.33). The result for the magnetic dipole moment for an yrast state I is:

$$\mu_I \equiv \sqrt{\frac{4\pi}{3}} \langle \Phi_{II} | I_0 | \Phi_{II} \rangle = \sqrt{\frac{4\pi}{3}} g_R C_{I\,0\,I}^{I\,1\,I} \sqrt{R(R+1)} \mu_N. \tag{8.41}$$

where the standard notation μ_N for the nuclear magneton has been used.

These expressions for the electric and magnetic transition operator matrix elements will be used in the next subsection to calculate the corresponding observables for the case of ^{158}Er.

8.1.1 NUMERICAL ANALYSIS

Here we address the issue of how do the results obtained through diagonalization, by solving the Schrödinger equation and by the harmonic approximation leading to the wobbling motion of the angular momentum, respectively, compare with each other. The moments of inertia used here are those obtained in Ref. [13] by fitting the experimental excitation energies with the wobbling energy formula, i.e. $\mathcal{I}_1 = 125\ \hbar^2 MeV^{-1}$, $\mathcal{I}_2 = 31.4\ \hbar^2 MeV^{-1}$, $\mathcal{I}_3 = 42\ \hbar^2 MeV^{-1}$. The variable x from the Bargmann representation of the rotor Hamiltonian is defined in the interval $(-\infty, +\infty)$, while the current variable t entering the Schrödinger equation is restricted in a finite interval which is close to $[-1.5, +1.5]$.

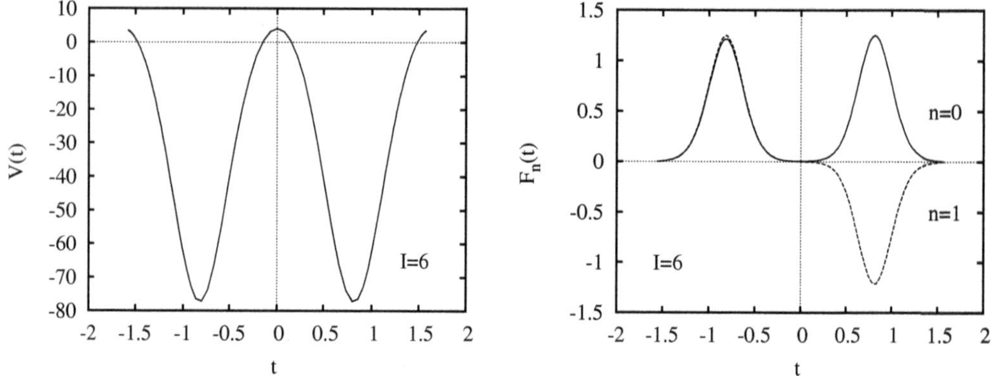

Figure 8.2 Left panel: The dependence of the potential on the variable t, defined by Eq. (2.38). Right panel: Eigenfunction F_n as function of t, for $n = 0, 1$.

In Ref. [15], the potential energy was considered as function of x, while here its dependence on the variable t is represented in Fig. 8.2 for I=6. If one calculates the average of R_1 with trial function $|\psi(z)\rangle$ and the result is considered in the two potential minima one obtains that $\langle \hat{R}_1 \rangle = \pm I$. This shows that in one minimum the system rotates around the OX axis, while in the other minimum the rotation is performed around OX. A useful insight to the system behavior, for a given solution of the Schrödinger equation, is obtained by plotting the wave function $F_n(t)$ for $n = 0, 1$ and $n = 3, 4, 5, 6$ in Figs. 8.2 and 8.3, respectively, for $I = 6$. The pair of states represented in each of the mentioned figures are degenerate.

The probability distributions $|F_n|^2$ for the degenerate states are identical. Note that if the states corresponding to F_n and F_{n+1} are degenerate, then the states described by $F_n + F_{n+1}$ and $F_n - F_{n+1}$ are also degenerate and localized each in a separate well. For I running from zero to I_{max}, the lowest two eigenstates of the Schrödinger equation for each I form two degenerate bands, one localized inside the well corresponding to the positive minimum and one in the well associated to the negative minimum. The same is also true for the next two degenerate bands and so on.

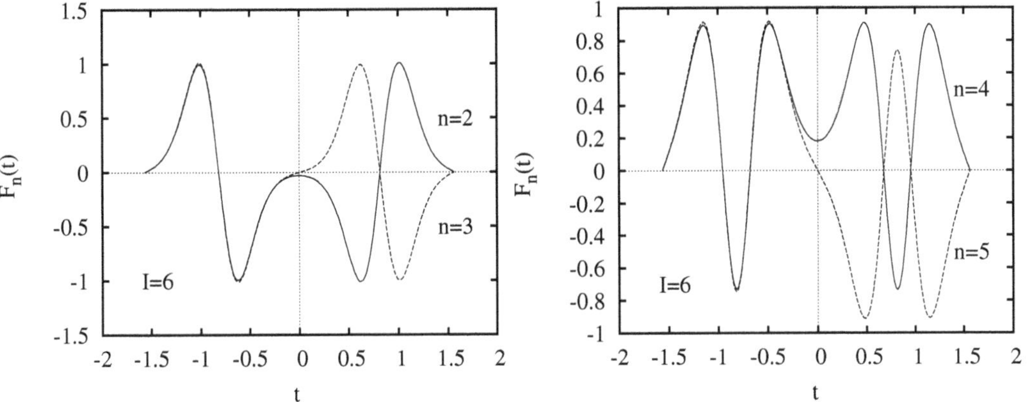

Figure 8.3 Left panel: The F_n as function of t for n=2,3. Right panel: The F_n as function of t for n=4,5.

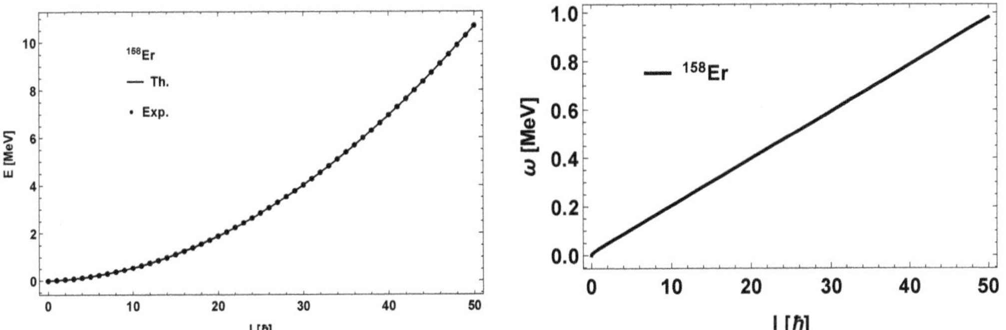

Figure 8.4 Left panel: Results (Th.) obtained with Eq. (2.29) are compared with experimental data (Exp.) taken from [16]. Right panel: The wobbling frequency for ^{158}Er as function of the angular momentum.

The parameters $\mathcal{I}_k$ used in our numerical analysis are those obtained by fitting the experimental yrast energies for ^{158}Er with the relation (8.18).

We calculated the yrast energies provided by three methods: diagonalization, solving the Schrödinger equation and by the wobbling energy formula (2.29). The results is that the three sets of energies are almost equal to each other, the maximal deviations being less that 5 keV. Results are illustrated in the left panel of Fig. 8.4. It is known that for triaxial nuclei, i.e. when the three moments of inertia are all different, the projection of $\mathbf{R}$ on the z axis is not a good quantum number. The present result shows that the coherent states we used as trial function are an optimal mixture of the K components to approximate the exact wavefunction. Also, the wobbling approximation of the yrast energies describes very well the exact solution of the Schrödinger equation. The agreement of the calculated yrast wobbling energies and the corresponding experimental data is very good, as shown in the left panel of Fig. 8.4. Theoretical results were obtained with the wobbling formula (2.29). We notice that the wobbling frequency depends almost linearly on the angular momentum I. This dependence can be seen in the right panel of Fig. 8.4.

Results for transition probabilities and moments are obtained with an effective charge $e_{eff} = 1.4e$ and collected in Table 8.1. The agreement with the corresponding experimental data for the B(E2) values is good. Concluding, the semi-classical formalism presented in this section describes the experimental data in a realistic fashion. We showed that the variational equations corresponding to a coherent state as trial function provides, after re-quantization, a very good description of the quantal system.

Table 8.1

The calculated intra-band B(E2) values are compared with the available experimental data, taken from Ref. [16]. Also, the calculated B(M1) values connecting the $\Delta I = 1$ yrast states.

I	$B(E2;In_w \to (I-2)n'_w)$ [W.u]		$B(E2;In_w \to (I-2)n'_w)$ [W.u]	$B(E2;In_w \to I-1n'_w)$ [W.u]	$B(M1;In_w \to I-1n'_w)$ $[10^{-4}\mu_N^2]$
	$n_w = n'_w = 0$		$n_w = 1\,n'_w = 0$	$n_w = 1\,n'_w = 0$	$n_w = 1\,n'_w = 0$
	Th.	Exp. [16]	Th.	Th.	Th.
2	61.287	129±9	0	0.916	0.05
4	170.243	186± 6	75.961	0.453	0.250
6	212.149	246± 8	130.908	0.258	0.420
8	234.334	298± 10	165.756	0.165	0.511
10	248.068	250± 4	189.322	0.114	0.558
12	257.407	260±3	206.216	0.084	0.585

8.2　DESCRIPTION OF THE WOBBLING MOTION IN EVEN-ODD NUCLEI

We suppose that the odd-mass nuclear system consists of an even-even core described by a triaxial rotor Hamiltonian and a single j-shell particle moving in a quadrupole deformed mean-field described by:

$$H_{sp} = \frac{V}{j(j+1)}\left[\cos\gamma(3j_3^2 - \mathbf{j}^2) - \sqrt{3}\sin\gamma(j_1^2 - j_2^2)\right]. \tag{8.42}$$

It is convenient to express the rotor Hamiltonian in terms of the total angular momentum $\mathbf{I}$ and the angular momentum carried by the odd particle:

$$H_{rot} = \sum_{k=1,2,3} A_k(I_k - j_k)^2. \tag{8.43}$$

Where A_k are expressed in terms of the moments of inertia associated to the principal axes of the inertia ellipsoid as $A_k = \frac{1}{2\mathcal{I}_k}$.

In what follows, the moments of inertia are taken as given by rigid-body model in the Lund convention:

$$\mathcal{I}_k^{rig} = \frac{\mathcal{I}_0}{1+(\frac{5}{16\pi})^{1/2}\beta}\left[1 - \left(\frac{5}{4\pi}\right)^{1/2}\beta\cos\left(\gamma+\frac{2}{3}\pi k\right)\right] + \varepsilon_j, \quad k = 1,2,3 \tag{8.44}$$

where ε_j denote the single particle energy of the odd nucleon. To the total Hamiltonian, $H(= H_{rot} + H_{sp})$, we associate the time dependent variational equation

$$\delta\int_0^t \langle\Psi|H - i\frac{\partial}{\partial t'}|\Psi\rangle dt' = 0, \tag{8.45}$$

where the trial function is chosen as:

$$|\Psi\rangle = \mathbf{N}e^{z\hat{I}_-}e^{s\hat{j}_-}|IMK\rangle|jj\rangle, \tag{8.46}$$

with $\hat{I}_-$ and $\hat{j}_-$ denoting the lowering operators for the intrinsic angular momenta $\mathbf{I}$ and $\mathbf{j}$, respectively, while $\mathbf{N}$ is the normalization factor having the expression:

$$\mathbf{N}^{-2} = (1+|z|^2)^{2I}(1+|s|^2)^{2j}. \tag{8.47}$$

The variables z and s are complex functions of time and play the role of classical phase space coordinates describing the motion of the core and the odd particle, respectively:

$$z = \rho e^{i\varphi}, \quad s = f e^{i\psi}, \tag{8.48}$$

The variables (φ, r) and (ψ, t) with r and t defined as:

$$r = \frac{2I}{1+\rho^2}, \quad 0 \le r \le 2I, \quad t = \frac{2j}{1+f^2}, \quad 0 \le t \le 2j, \tag{8.49}$$

bring the classical equations provided by the variational principle to the canonical form:

$$\frac{\partial \mathcal{H}}{\partial r} = \dot{\varphi}; \quad \frac{\partial \mathcal{H}}{\partial \varphi} = -\dot{r}; \quad \frac{\partial \mathcal{H}}{\partial t} = \dot{\psi}; \quad \frac{\partial \mathcal{H}}{\partial \psi} = -\dot{t}, \tag{8.50}$$

where $\mathcal{H}$ denotes the average of H (Eq.3.5) with the function $|\Psi\rangle$ and has the expression:

$$
\begin{aligned}
\mathcal{H} \;=\;& \frac{I}{2}(A_1+A_2)+A_3 I^2 + \frac{2I-1}{2I} r(2I-r)\left(A_1\cos^2\varphi + A_2\sin^2\varphi - A_3\right) \\
+\;& \frac{j}{2}(A_1+A_2)+A_3 j^2 + \frac{2j-1}{2j} t(2j-t)\left(A_1\cos^2\psi + A_2\sin^2\psi - A_3\right) \\
-\;& \sqrt{r(2I-r)t(2j-t)}\left(A_1\cos\varphi\cos\psi + A_2\sin\varphi\sin\psi\right) + A_3\left(r(2j-t)+t(2I-r)\right) \\
+\;& V\frac{2j-1}{j+1}\left[\cos\gamma - \frac{t(2j-t)}{2j^2}\sqrt{3}\left(\sqrt{3}\cos\gamma - \sin\gamma\cos 2\psi\right)\right].
\end{aligned}
\tag{8.51}
$$

From Eq. (8.50) we see that the angles φ and ψ play the role of generalized coordinates, while r and t are the corresponding conjugate momenta. Looking for the extremal points of the energy surface $\mathcal{H} = const$, one finds out that the point $(\varphi, r; \psi, t) = (0, I; 0, j)$ is a minimum point for the classical energy function. Linearizing the equations of motion around the mentioned minimum point, one arrive at a homogeneous and linear system for the generalized conjugate coordinates. The compatibility restriction leads to an equation defining the wobbling frequency:

$$\Omega^4 + B\Omega^2 + C = 0, \tag{8.52}$$

where the coefficients B and C have the expressions:

$$
\begin{aligned}
-B \;=\;& \left[(2I-1)(A_3-A_1)+2jA_1\right]\left[(2I-1)(A_2-A_1)+2jA_1\right]+8A_2A_3 Ij \\
+\;& \left[(2j-1)(A_3-A_1)+2IA_1 + V\frac{2j-1}{j(j+1)}\sqrt{3}(\sqrt{3}\cos\gamma+\sin\gamma)\right] \\
\times\;& \left[(2j-1)(A_2-A_1)+2IA_1 + V\frac{2j-1}{j(j+1)}2\sqrt{3}\sin\gamma\right],
\end{aligned}
\tag{8.53}
$$

$$
\begin{aligned}
C \;=\;& \left\{\left[(2I-1)(A_3-A_1)+2jA_1\right]\left[(2j-1)(A_3-A_1)+2IA_1 + V\frac{2j-1}{j(j+1)}\sqrt{3}(\sqrt{3}\cos\gamma+\sin\gamma)\right] - 4Ij A_3^2\right\} \\
\times\;& \left\{\left[(2I-1)(A_2-A_1)+2jA_1\right]\left[(2j-1)(A_2-A_1)+2IA_1 + V\frac{2j-1}{j(j+1)}2\sqrt{3}\sin\gamma\right] - 4Ij A_2^2\right\}.
\end{aligned}
\tag{8.54}
$$

Under certain restrictions for the moments of inertia (MoI), the dispersion equation (8.52) admits two real and positive solutions. Hereafter, these will be denoted by Ω_1^I and Ω_2^I for $j = i_{13/2}$, with $\Omega_1^I < \Omega_2^I$.

In Refs. [17, 18], the wobbling frequencies have been used in order to define the rotational bands in the even-odd isotopes ^{161,163,165,167}Lu. Most experimental available data are in ^{163}Lu, where four bands denoted by TSD1, TSD2, TSD3 and TSD4 are known. The name is the achronym for "triaxial strongly deformed", which reflects the fact that the nuclear deformation and the deviation from the axial symmetric shape are large ($\beta \approx 0.38$ and $\gamma \approx 20°$). In the quoted paper the variational principle is used only for the states of TSD1, while the other three bands are interpreted as one, two and three boson excitations of the ground band. The ground band excitation energies are just the zero point energies associated to each member of the band. The band TSD4 is known only for ^{163}Lu and has a negative parity in contrast with the other bands which are of positive parity. The single-particle orbital for the odd nucleon was taken as $i_{13/2}$ in the first three bands, while for TSD4 it is $h_{9/2}$. Due to the core polarization caused by the particle-core interactions the core moments of inertia for TSD1,TSD2 and TSD3 are the same but different from those characterizing TSD4. A good description of the data was obtained although due to the situation mentioned in the above statement the fitting procedure is somewhat tedious.

Here we give the results of Ref. [19, 20] where the variational principle is applied to the states of TSD1, TSD2 and TSD4, while TSD3 is considered to be an one phonon excitation of TSD2. Moreover, for all the four bands the odd nucleon is placed in the $i_{13/2}$ orbital. The core states are given by the triaxial rotor and are of positive parity for TSD1, TSD2 and TSD3 and of negative parity for TSD4. Note the fact that the pair of bands TSD1, TSD3 and TSD2, TSD4 have the same signature, $+1/2$ and $-1/2$, respectively. In this context the bands TSD2 and TSD4 may be called *parity partner bands*.

Further, to the $TSD_{1,2,3,4}$ bands we associate the energies:

$$E_I^{\text{TSD1}} = \varepsilon_j + \mathcal{H}_{\text{min}}^{(I,j)} + \mathcal{F}_{00}^I, \quad I = R+j, R = 0,2,4,\ldots,$$

$$E_I^{\text{TSD2}} = \varepsilon_{j,1} + \mathcal{H}_{\text{min}}^{(I,j)} + \mathcal{F}_{00}^I, \quad I = R+j, R = 1,3,5,\ldots$$

$$E_I^{\text{TSD3}} = \varepsilon_j + \mathcal{H}_{\text{min}}^{(I,j)} + \mathcal{F}_{10}^I, \quad I = R+j, R = 0,2,4,\ldots$$

$$E_I^{\text{TSD4}} = \varepsilon_{j,2} + \mathcal{H}_{\text{min}}^{(I,j)} + \mathcal{F}_{00}^I, \quad I = R+j, \ R = 1,3,5,\ldots.$$

$$\tag{8.55}$$

where $\mathcal{F}_{n_{w_1} n_{w_2}}$ is function of the wobbling frequencies

$$F_{n_{w_1} n_{w_1}}^I = (n_{w_1} + \frac{1}{2})\Omega_1^I + (n_{w_2} + \frac{1}{2})\Omega_2^I. \tag{8.56}$$

while $\mathcal{H}_{\text{min}}^{(I,j)}$ is the minimal classical energy. We considered different re-normalizations for the single-particle mean field in the signature unfavored as well as in the negative parity states, which result in two distinct energy shifts for the excitation energies of the TSD2 and TSD4 bands, respectively. These two quantities will be adjusted throughout the numerical calculations such that the energy spectrum is best reproduced. In order to save the space, only the results for ^{163}Lu are presented here. The fitting procedure yields the set $\mathcal{P} = (\mathcal{I}_1, \mathcal{I}_2, \mathcal{I}_3)$ shown in Table 8.2.

Table 8.2

The parameter set $\mathcal{P}$ that was determined by a fitting procedure of the excitation energies of ^{163}Lu.

$\mathcal{I}_1$ [$\hbar^2$/MeV]	$\mathcal{I}_2$ [$\hbar^2$/MeV]	$\mathcal{I}_3$ [$\hbar^2$/MeV]	γ [deg.]	V [MeV]
72	15	7	22	2.1

Using these parameters, the excitation energies in the four bands were calculated and compared with he corresponding experimental data, as shown in Fig. 8.5. In terms of the stability of the wobbling motion with respect to the total angular momentum, several contour plots were plotted, using the obtained parameter set $\mathcal{P}$ with the help of Eq. (8.51). For each band, a spin close to the band head of each sequence was chosen. Due to the obtained MOI ordering, the surfaces have minimum points indicated by the red dots for each panel. Results can be seen in Fig. 8.6. The four figures have many similarities suggesting common collective properties, but also differences caused

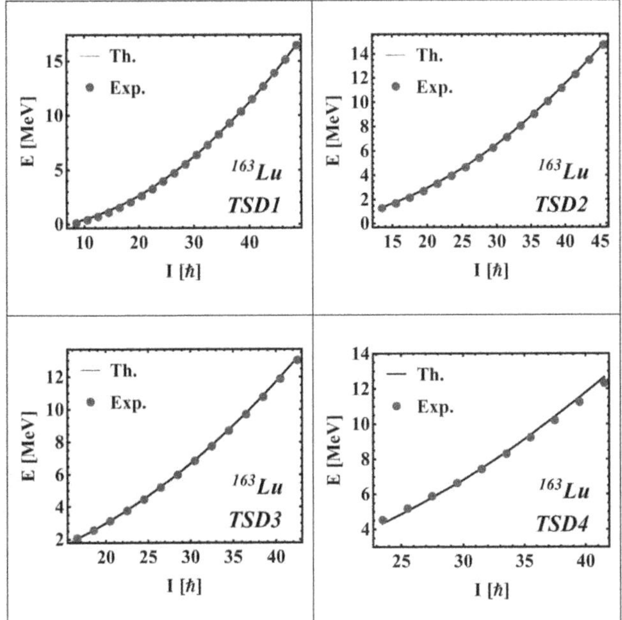

Figure 8.5 The excitation energies for the bands TSD1, TSD2, TSD3 and TSD4.

by the fact that minima have different depths. The common feature consists of that the equi-energy curves surround a sole minimum for low energy, while for higher energies the trajectories go around all minima, the lack of localization indicating an unstable picture.

Finally we are interested in finding out the dependence of the classical trajectories on angular momenta as well as on energies. When the model Hamiltonian is diagonalized for a given I, a set of $2I + 1$ energies are obtained. Therefore, it makes sense to study the trajectory change with increasing energy. Trajectories are represented in Fig. 8.7 as the manifold given by intersecting the surfaces corresponding to the two constants of motion, the energy and the angular momentum. The first energy in each column corresponds to the real excitation energy for that particular spin state, the second one represents the point at which the ellipsoid touches the sphere at the equator, which marks a nuclear phase transition, while the third one is the trajectory of the system at energies sufficiently large that the system changes its wobbling regime. For low energies, one notices two distinct trajectories having as rotation axes the 1-axis and -1-axis, respectively. As energy increases the two trajectories approaches each other which results in a tilted rotation axis for each of trajectories, the rotation axes being misaligned. Note that this picture is fully consistent with that of Ref. [21]. When the two trajectories intersect each other, the trajectories surround both minima. Increasing the energy even more one arrives again at a two trajectories regime but with different rotation axes which become close to the 3-axis. This reflects another phase transition for the system.

8.2.1   OTHER SCENARIOS

An alternative way to interpret the four bands in ^{163}Lu is to keep the definition for TSD1, TSD2 and TSD3 as before, but as concerns TSD4 the odd nucleon is moving in the single shell $h_{9/2}$, otherwise the variational principle is formulated for each state of the negative-parity band. The wobbling frequencies will be denoted by Ω_1^I and $\Omega_{1'}^I$ for $i_{13/2}$, and by Ω_2^I and $\Omega_{2'}^I$ for $h_{9/2}$. They are ordered

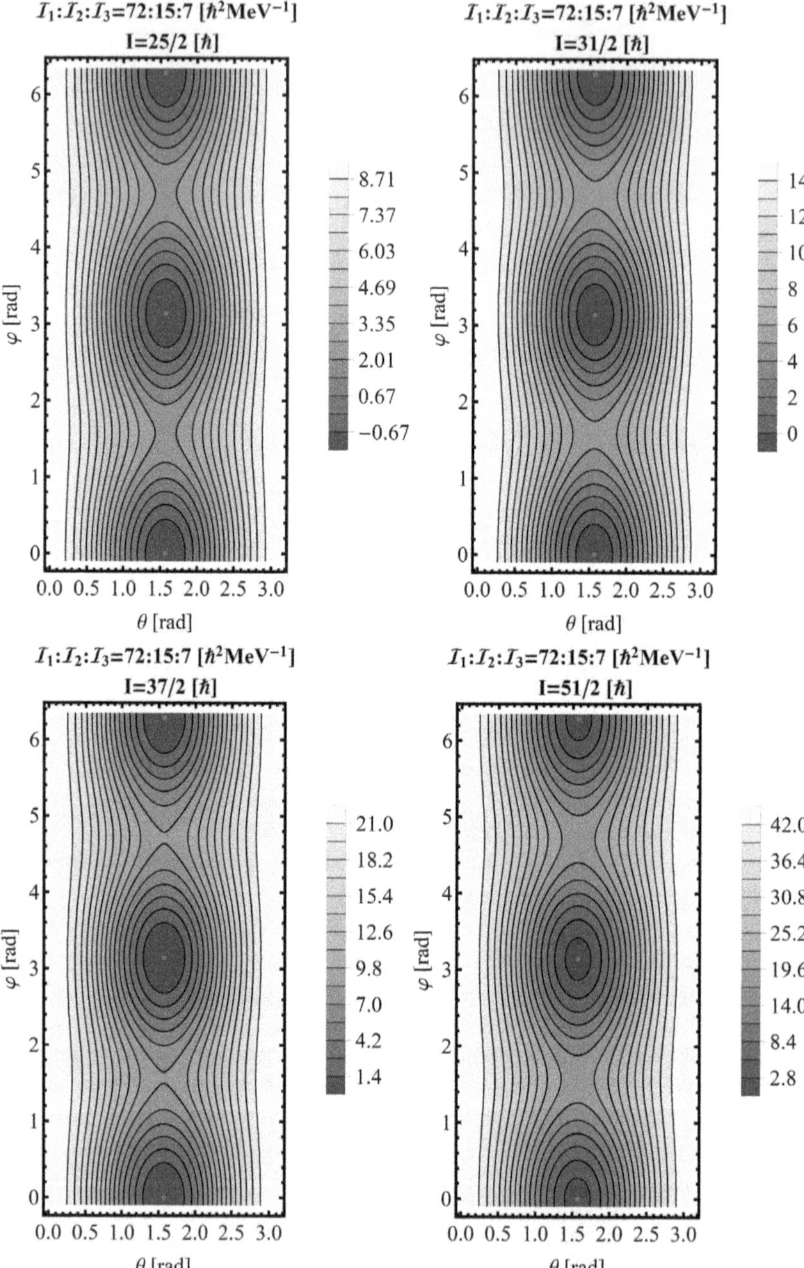

Figure 8.6 A contour plot with the energy function $\mathcal{H}$ for TSD1, TSD2, TSD3 and TSD4. The parameter set $\mathcal{P}$ was used for the numerical calculations.

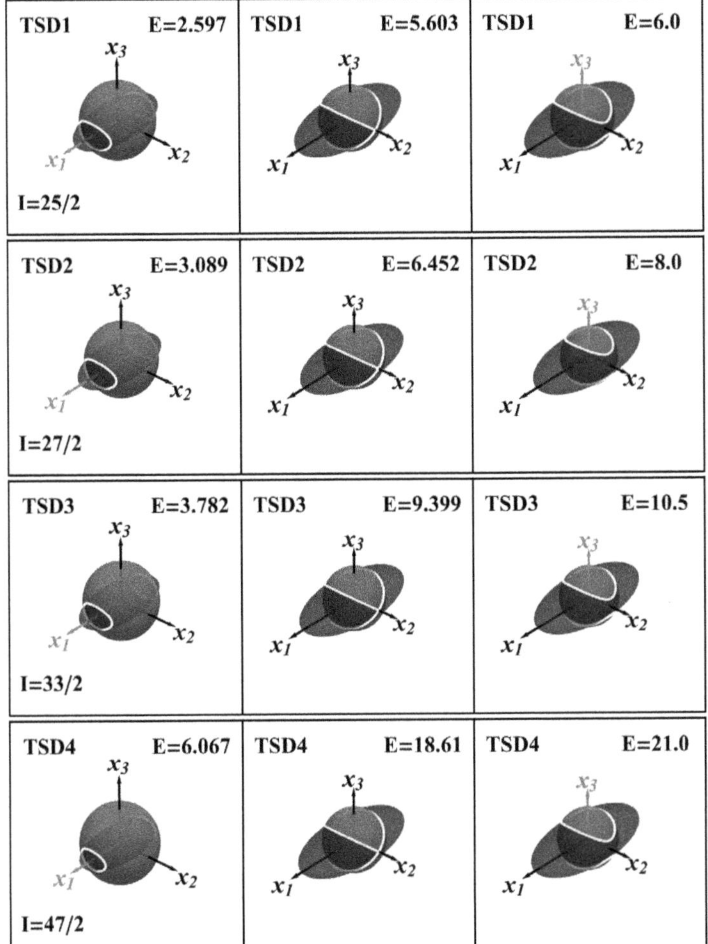

Figure 8.7 The nuclear trajectory of the system for a spin state belonging to each of the four TSD bands of ^{163}Lu. Intersection line represents the actual orbits.

as: $\Omega_1^I < \Omega_{1'}^I$ and $\Omega_2^I < \Omega_{2'}^I$. Energies of the states in the four bands are defined as:

$$
\begin{aligned}
E_I^{TSD1} &= \varepsilon_{13/2} + \mathcal{H}_{I,min}(13/2) + \frac{1}{2}\left(\Omega_1^I + \Omega_{1'}^I\right), \\
&\qquad I = 13/2, 17/2, 21/2, ... \\
E_I^{TSD2} &= \varepsilon_{13/2} + \mathcal{H}_{I,min}(13/2) + \frac{1}{2}\left(\Omega_1^I + \Omega_{1'}^I\right), \\
&\qquad I = 27/2, 31/2, 35/2, ... \\
E_I^{TSD3} &= \varepsilon_{13/2} + \mathcal{H}_{I-1,min}(13/2) + \frac{1}{2}\left(3\Omega_1^{I-1} + \Omega_{1'}^{I-1}\right), \\
&\qquad I = 33/2, 37/2, 41/2, ... \\
E_I^{TSD4} &= \varepsilon_{9/2} + \mathcal{H}_{I,min}(9/2) + \frac{1}{2}\left(\Omega_2^I + \Omega_{2'}^I\right), \\
&\qquad I = 47/2, 51/2, 55/2, ...
\end{aligned}
\tag{8.57}
$$

The excitation energies are obtained by subtracting $E_{13/2}^{TSD1}$ from the above expressions. The excitation energies for the TSD4 states contain the constant term $\varepsilon_{9/2} - \varepsilon_{13/2} = -0.334$ MeV. The involved parameters are fitted through the least mean square procedure with the result shown in Table 8.3.

Table 8.3

The MoI's in $\hbar^2$/MeV, the strength of the single particle potential V in MeV, and the triaxial parameter (γ) in degrees as provided by the adopted fitting procedure.

isotope	j	bands	$\mathcal{I}_1$	$\mathcal{I}_2$	$\mathcal{I}_3$	V	γ	nr. states	r.m.s.[MeV]
^{163}Lu	13/2	TSD1,TSD2,TSD3	63.2	20	10	3.1	17	52	0.264
	9/2	TSD4	67	34.5	50	0.7	17	10	0.057

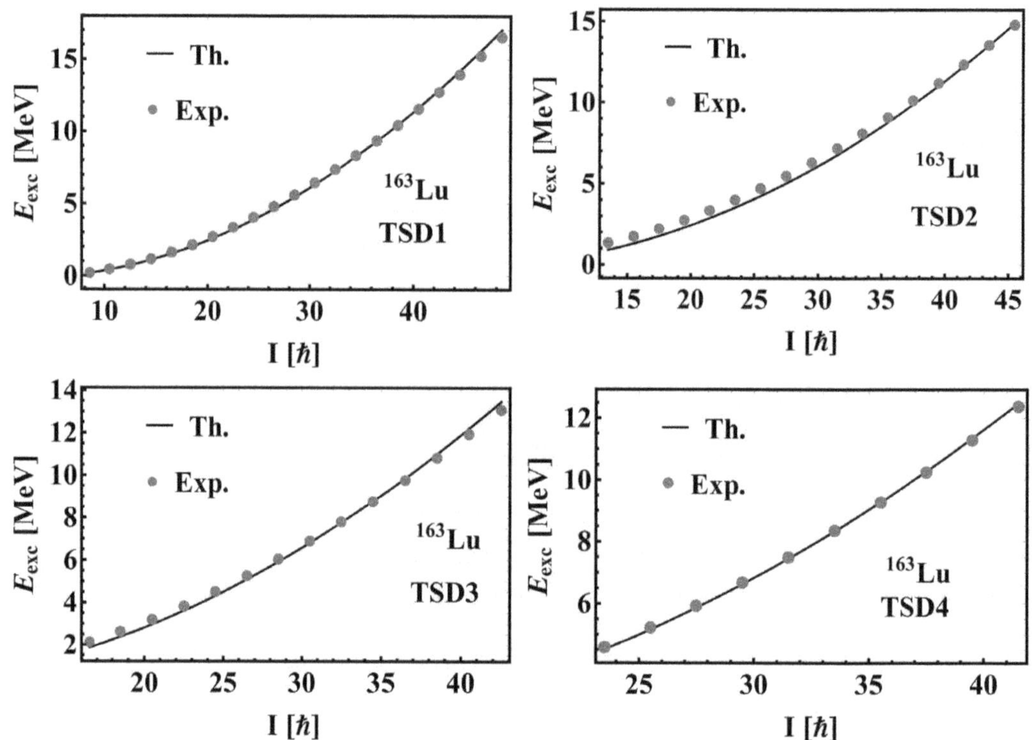

Figure 8.8   Calculated energies for the bands TSD1, TSD2, TSD3 and TSD4 are compared with the corresponding experimental data [22, 23] for ^{163}Lu.

As seen from Fig. 8.8 the fitted parameters provide excitation energies which agree very well with the corresponding experimental data. The adopted option of fitting the MoI's is naturally required by the observation that the experiment indicates that they are neither irrotational nor rigid but satisfy the relation $\mathcal{I}_1^{irr} < \mathcal{I}_1 < \mathcal{I}_1^{rig}$. Note that the MoI's provided by the fitting procedure take care of the particle-core interaction. In that respect the fact that the maximal MoI is $\mathcal{I}_2$ does not necessarily imply that the system motion is of transversal character. Indeed, even though for the non-interacting core the ordering is $\mathcal{I}_2 > \mathcal{I}_1 > \mathcal{I}_3$, as the microscopic studies show, the particle-core interaction renormalizes the bare MoI's which results in a strong increasing of $\mathcal{I}_1$ (due to the alignment) and only a moderate decreasing of $\mathcal{I}_2$ (caused by the pairing interaction), ending with the dominance of the new $\mathcal{I}_1$, characterizing the whole system [24]. Then, one can assert that the interaction with the odd proton stabilizes the system into a large deformed shape and moreover drives it to a longitudinal-like motion where the maximal MoI is the normalized $\mathcal{I}_1$. This change in the rotation regime is caused by both the angular momentum alignment and the pairing interaction. The transition from a transversal to a longitudinal wobbling motion is not abruptly achieved, but only at a certain critical angular momentum I_{cr}. Note that the MoI's were fixed such that the best agreement with the corresponding experimental data is obtained for energies of the whole spectrum and therefore no angular momentum dependence can be inferred. Furthermore, the study of the phase transition, transversal-longitudinal, cannot be performed with the present formalism. However, due to this feature one may say that the present formalism does not exclude the possible transversal wobbling motion in the low lying spectrum where the alignment is small, but that part of bands cannot be explored because the adopted fitting procedure does not use any I dependence for MoI's.

The fitted parameters yield the excitation energies in the four bands, which are compared with the corresponding experimental energies in Fig. 8.8. In Figs. 8.9 and 8.10, the aligned angular momenta

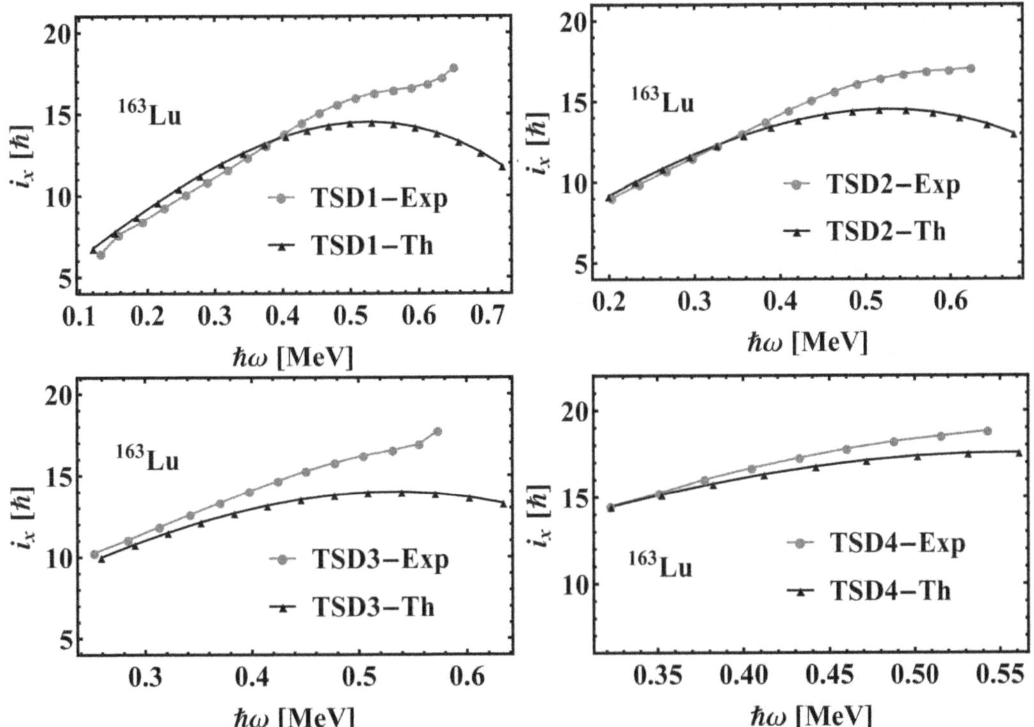

Figure 8.9 Results for the aligned angular momenta, i_x, relative to a reference $I_{ref} = \mathcal{J}_0\omega + \mathcal{J}_1\omega^3$ with $\mathcal{J}_0 = 30\hbar^2 MeV^{-1}$ and $\mathcal{J}_1 = 40\hbar^4 MeV^{-3}$, in ^{163}Lu, are compared with the corresponding experimental data [22, 23].

i_x and the excitation energies normalized to the energy of a rigid rotor, respectively, are compared with the corresponding experimental data [22, 23]. The calculated and experimental dynamic moments inertia are compared in Fig. 8.11.

To calculate the quadrupole electric transition probabilities we need the expression of the wave functions describing the states involved in the given transition and the quadrupole transition operator. The available experimental data concern the states of TSD1 and TSD2. As we saw before, the level energies from these bands account for the wobbling motion through the zero point energy. Therefore, the wave function is considered to be the one corresponding to the classical minimum energy, corrected by the first order expansion term, with the coordinates deviations from the minimum

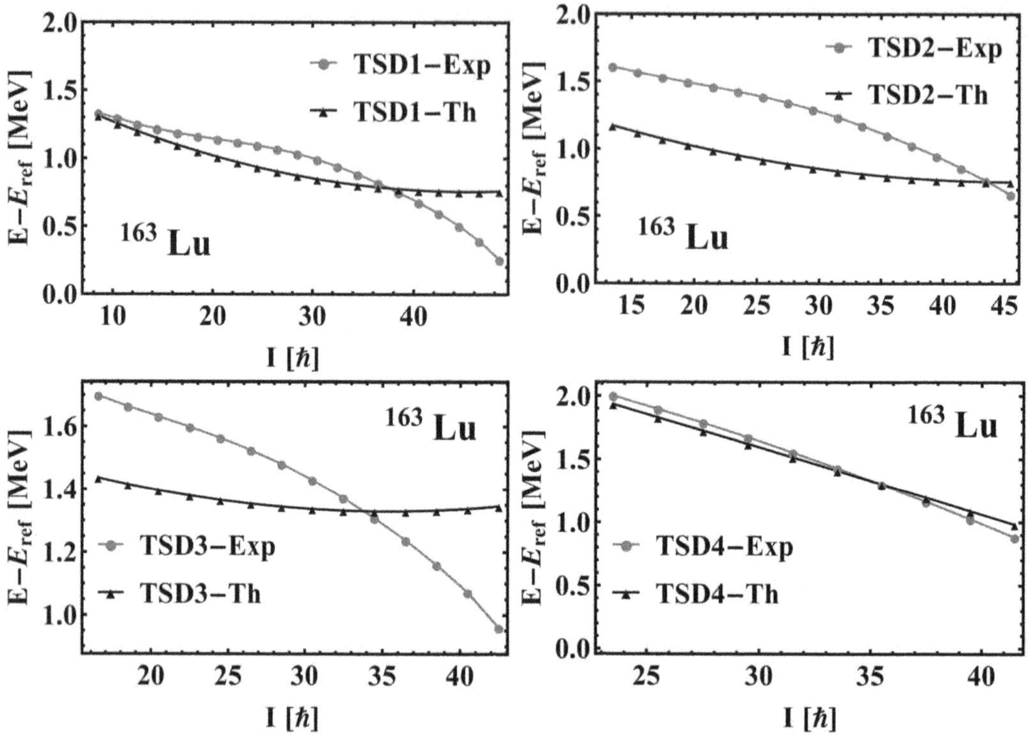

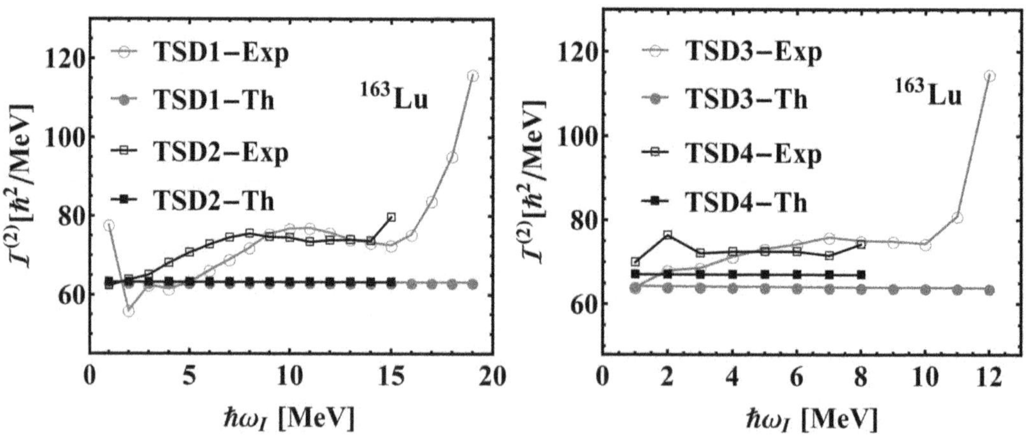

Figure 8.10 Theoretical and experimental excitation energies for TSD1 in ^{161}Lu normalized to the energy of a rigid rotor with an effective moment of inertia, i.e. $E_{REF} = 0.0075I(I+1)(MeV)$, are plotted as function of the angular momentum for the bands TSD1, TSD2, TSD3 and TSD4 of ^{163}Lu.

Figure 8.11 Results for the dynamic MoI's in the bands TSD1, TSD2, TSD3 and TSD4 of ^{163}Lu, are compared with the corresponding experimental data taken from Ref. [22, 25].

quantized. Thus, one arrived at the following wave function:

$$\Phi^{(1)}_{IjM} = \mathbf{N}_{Ij} \sum_{K,\Omega} C_{IK}C_{j\Omega}|IMK\rangle|j\Omega\rangle \left\{ 1 + \frac{i}{\sqrt{2}} \left[\left(\frac{K}{I}k + \frac{I-K}{k} \right) a^\dagger + \left(\frac{\Omega}{j}k' + \frac{j-\Omega}{k'} \right) b^\dagger \right] \right\} |0\rangle_I,$$

with $\mathbf{N}_{Ij}$ standing for the normalization factor, and $|0\rangle_I$ for the vacuum state of the bosons $a^\dagger$ and $b^\dagger$ determined by the classical coordinates φ, ψ and the corresponding conjugate momenta r and t through the canonical parameters k and k', which are analytically expressed in Appendix A of Ref. [20]. Expansion coefficients of the trial function corresponding to the minimum point, in terms of the normalized Wigner function, C_{IK}, were analytically expressed in Ref. [18].

The electric quadrupole transition operator is defined by:

$$\mathcal{M}(E2,\mu) = \left[Q_0 D_{\mu 0}^2 - Q_2 (D_{\mu 2}^2 + D_{\mu -2}^2) \right] + e \sum_{v=-2}^{2} D_{\mu v}^2 Y_{2v} r^2, \tag{8.58}$$

with Q_0 and Q_2 taken as free parameters which are to be fixed by fitting two intra-band transitions for TSD1 and TSD2. Note that MoI's are free parameters, that is, no option for their nature, rigid or hydrodynamic, is adopted. To be consistent with this picture, the strengths Q_0 and Q_2 were also considered as free parameters. However, this is not consistent with the structure of the single-particle potential, which considers the collective quadrupole operator as emerging from the hydrodynamic model. These are fixed by fitting the $B(E2)$ values for one intra-band (TSD1) and one inter-band ($TSD2 \rightarrow TSD1$) transition. The remaining $B(E2)$ transitions and the quadrupole transition moments are free of any adjustable parameter. Results for the $B(E2)$ values are compared with the corresponding data in Tables 8.4–8.6.

The magnetic transition operator used in our calculations is:

$$\mathcal{M}(M1,\mu) = \sqrt{\frac{3}{4\pi}}\mu_N \sum_{v=0,\pm 1} [g_R R_v + q g_j j_v] D_{\mu v}^1, \tag{8.59}$$

with R_v denoting the components of the core's angular momentum with the corresponding gyromagnetic factor, $g_R = Z/A$, while g_j is the free gyromagnetic factor for the proton angular momentum $j(=13/2)$, which was quenched by a factor $q = 0.43$ in order to account for the polarization effects not included in g_j. This factor takes care of the interaction of the odd-proton with the currents distributed inside the core, as well as of the internal structure of the proton, which may also influence its magnetic moment [27].

To evaluate the transition matrix elements of the core's angular momentum, the involved states are written in the form:

$$\Psi_{IM} = \frac{1}{\sqrt{2j+1}} \sum_{M_R \Omega K} C_{M_R \Omega M}^{RjI} C_{RK} |RM_R K\rangle |j\Omega\rangle. \tag{8.60}$$

Again, the expansion coefficients of the core's wave function in the basis of the normalized Wigner function are denoted by C_{RK}. Results for the relevant $B(E2)$ and $B(M1)$ values of the inter-band transitions $I \rightarrow I - 1$ as well as for the mixing rations are collected in Table 8.4. Here the coordinate fluctuations around their minima are ignored since their contribution is negligible. Results on the branching ratios and transition quadrupole moments are collected in Tables 8.5 and 8.6.

The alignment, the excitation energy relative to the energy of a rigid rotor with an effective moment of inertia, and the dynamic moment of inertia are plotted in Figs. 8.9–8.11 as function of the rotational frequency, angular momentum and again on the rotational frequency, respectively. One notices a reasonable agreement of our results with the corresponding experimental data. One specific feature for the wobbling motion consists of a strong $E2$ transition from the TSD2 to the TSD1 bands. This is reflected by the relative large values of the branching ratios characterizing the states from TSD2. This is confirmed by the results shown in Table 8.4, where the calculated branching ratios are compared with the corresponding experimental data. Also, the computed ratios $B(M1)/B(E2)_{in}$ are in good agreement with the experimental data in ^{163}Lu. Another specific wobbling feature is the large transition quadrupole moment, as shown in Table 8.5. From there one can see a very good agreement of our calculated results for ^{163}Lu and the corresponding data. Concluding the application

Table 8.4

The $B(E2)$ and $B(M1)$ values for the transitions from TSD2 to TSD1. Mixing ratios are also mentioned. Theoretical results (Th.) are compared with the corresponding experimental (Exp.) data taken from Ref. [26, 28]. Data labeled by [a] are from Ref. [29].

	I^π	$B(E2)[e^2b^2]$ $I^+ \to (I-1)^+$ Th.	Exp.	$B(M1)[\mu_N^2]$ $I^+ \to (I-1)^+$ Th.	Exp.	$\delta_{I\to(I-1)}$ $[MeV.fm]$ Th.	Exp.
^{163}Lu	$\frac{47}{2}^+$	0.54	$0.54^{+0.13}_{-0.11}$	0.017	$0.017^{+0.006}_{-0.005}$	-1.55	$-3.1^{+0.36}_{-0.44}$
	$\frac{51}{2}^+$	0.49	$0.54^{+0.09}_{-0.08}$	0.018	$0.017^{+0.005}_{-0.005}$	-1.58	$-3.1\pm0.4^{a)}$
	$\frac{55}{2}^+$	0.44	$0.70^{+0.18}_{-0.15}$	0.019	$0.024^{+0.008}_{-0.007}$	-1.61	$-3.1\pm0.4^{a)}$
	$\frac{59}{2}^+$	0.34	$0.65^{+0.34}_{-0.26}$	0.019	$0.023^{+0.013}_{-0.011}$	-1.64	$-3.1\pm0.4^{a)}$
	$\frac{63}{2}^+$	0.36	$0.66^{+0.29}_{-0.24}$	0.020	$0.024^{+0.012}_{-0.010}$	-1.66	

Table 8.5

The $E2$ intra-band transitions $I \to (I-2)$ for TSD1 and TSD2 bands. Also, the transition quadrupole moments, defined as in Ref. [30], are given. Theoretical results (Th.) are compared with the corresponding experimental data (Exp.) taken from Ref. [26]. $B(E2)$ values are given in units of e^2b^2, while the quadrupole transition moment in b.

^{163}Lu TSD1 I^π	$B(E2;I^+ \to (I-2)^+)[e^2b^2]$ Th.	Exp.	Q_I $[b]$ Th.	Exp.	^{163}Lu TSD2 I^π	$B(E2;I^+ \to (I-2)^+)[e^2b^2]$ Th.	Exp.	Q_I $[b]$ Th.	Exp.
$\frac{41}{2}^+$	2.80	$3.45^{+0.80}_{-0.69}$	8.89	$9.93^{+1.14}_{-0.99}$	$\frac{47}{2}^+$	2.71	$2.56^{+0.57}_{-0.44}$	8.71	$8.51^{+0.95}_{-0.73}$
$\frac{45}{2}^+$	2.74	$3.07^{+0.48}_{-0.43}$	8.77	$9.34^{+0.72}_{-0.65}$	$\frac{51}{2}^+$	2.66	$2.67^{+0.41}_{-0.33}$	8.62	$8.67^{+0.66}_{-0.53}$
$\frac{49}{2}^+$	2.69	$2.45^{+0.28}_{-0.25}$	8.66	$8.32^{+0.47}_{-0.42}$	$\frac{55}{2}^+$	2.62	$2.81^{+0.53}_{-0.41}$	8.53	$8.88^{+0.83}_{-0.64}$
$\frac{53}{2}^+$	2.64	$2.84^{+0.24}_{-0.22}$	8.57	$8.93^{+0.38}_{-0.35}$	$\frac{59}{2}^+$	2.58	$2.19^{+0.94}_{-0.65}$	8.46	$7.82^{+1.66}_{-1.15}$
$\frac{57}{2}^+$	2.60	$2.50^{+0.32}_{-0.29}$	8.50	$8.37^{+0.54}_{-0.49}$	$\frac{63}{2}^+$	2.54	$2.25^{+0.75}_{-0.48}$	8.39	$7.91^{+1.32}_{-0.84}$
$\frac{61}{2}^+$	2.56	$1.99^{+0.26}_{-0.23}$	8.43	$7.45^{+0.49}_{-0.43}$	$\frac{67}{2}^+$	2.51	$1.60^{+0.52}_{-0.37}$	8.34	$6.66^{+1.09}_{-0.76}$
$\frac{65}{2}^+$	2.53	$1.95^{+0.44}_{-0.30}$	8.36	$7.37^{+0.82}_{-0.57}$	$\frac{71}{2}^+$	2.49	$1.61^{+0.82}_{-0.49}$	8.28	$6.68^{+1.70}_{-1.02}$
$\frac{69}{2}^+$	2.50	$2.10^{+0.80}_{-0.48}$	8.31	$7.63^{+1.46}_{-0.88}$					

Table 8.6

Branching ratios $B(E2)_{out}/B(E2)_{in}$ and $B(M1)/B(E2)_{in}$ of some states from the band TSD2. Experimental data are from Refs. [23, 31–34] and [31], respectively.

	I^π	$B(E2)_{out}/B(E2)_{in}$ Th.	Exp.	$B(M1)/B(E2)_{in}[10^2 \frac{\mu_N^2}{e^2b^2}]$ Th.	Exp.
^{163}Lu	$\frac{31}{2}$	0.29	0.21 ± 0.11		
	$\frac{35}{2}$	0.26	0.22 ± 0.02	0.502	$0.439^{+0.082}_{-0.076}$
	$\frac{39}{2}$	0.24	0.21 ± 0.02	0.560	$0.447^{+0.077}_{-0.078}$
	$\frac{43}{2}$	0.22	0.22 ± 0.02	0.608	$0.509^{+0.088}_{-0.086}$
	$\frac{47}{2}$	0.20	0.21 ± 0.03	0.650	$0.498^{+0.091}_{-0.084}$
	$\frac{51}{2}$	0.18	0.21 ± 0.02	0.685	$0.709^{+0.182}_{-0.196}$
	$\frac{55}{2}$	0.17	0.26 ± 0.05		
	$\frac{59}{2}$	0.15	0.30 ± 0.09		
	$\frac{63}{2}$	0.14	0.30 ± 0.11		

part of the present paper, one may say that the proposed semi-phenomenological approach seems to be an efficient tool to account for the main features of electromagnetic properties of the even-odd Lu isotopes.

The portrait of the stationary points characterizing the energy function $\mathcal{H}$ split the parameter space in several regions associated to distinct nuclear phases. These are bordered by separatrices defined by the following equation:

$$\det\left(\frac{\partial^2 \mathcal{H}}{\partial (q_i)^k \partial (p_j)^l}\right) = 0; \ i,j,=1,2; k,l=0,1,2; k+l=2. \tag{8.61}$$

where the canonical conjugate coordinates are suggestively denoted as:

$$q_1 = \varphi, \ q_2 = \psi, \ p_1 = r, \ p_2 = t. \tag{8.62}$$

After some algebraic manipulations, the above equation leads to:

$$C = 0, \tag{8.63}$$

where C has the expression from Eq. (8.54). Eq. (8.63) splits to the following two equations:

$$z = f_1(x), \ z = f_2(x,y), \tag{8.64}$$

with

$$\begin{aligned}
f_1(x) &= \left[(1-4(I-j)^2)x^2 + (4I^2+4j^2-8Ij+2j+2I-2)x \right.\\
&\quad - \left. (2I+2j-1)\right]\left[G_1\left((2I-2j-1)x-(2I-1)\right)\right]^{-1},\\
f_2(x,y) &= \left[(1-4(I-j)^2)x^2 + (1-2(I+j))y^2 + 2\left(2(I-j)^2\right.\right.\\
&\quad + \left.\left. (I+j)+1\right)xy\right]\left[G_2\left((2I-2j-1)x-(2I-1)y\right)\right]^{-1}.
\end{aligned} \tag{8.65}$$

The separatrices bordering the nuclear phases are plotted in the phase diagram given by Fig. 8.12. Here the coordinates x, y and z are defined as:

$$x = \frac{A_1}{A_3}, \ y = \frac{A_2}{A_3}, \ z = \frac{V}{A_3}. \tag{8.66}$$

For a fixed $\gamma(=17°)$, the equations (8.64) represent two singular surfaces, having the asymptotic planes:

$$x = \frac{2I-1}{2I-2j-1}, \ y = \frac{2I-2j-1}{2I-1}x. \tag{8.67}$$

For the fixed MoI's and V shown in Table 8.1, the coordinates (x,y,z) corresponding to the isotopes ^{161,163,165,167}Lu, respectively, are represented by small circles of different colors: purple (^{161}Lu), white (^{163}Lu), yellow (^{165}Lu) and black (^{167}Lu). Here only the set of MoI's corresponding to ^{163}Lu are given. In the case of ^{161}Lu, the coordinate y is too large and therefore drops out the range shown in Fig. 8.12. In order to keep it inside the figure we modified y to $y-7$. Even so, the purple circle falls in an adjacent phase, which suggests that this isotope belongs to a different nuclear phase. Inside a given phase the classical Hamiltonian has specific stationary points. If one of these is a minimum, then the classical trajectories surround it with a certain time period. If the point in the parameter space approaches the separatrices, the period tends to infinity [35]. When $V > 0$, $\mathbf{j}$ is always oriented along the short axis, that is the 1-axis, and the region where $\mathcal{I}_2 > \mathcal{I}_1 > \mathcal{I}_3$ is the phase where the transversal wobbling may take place. More specifically, this region is bounded by four planes, one being the diagonal plane, one is given by the second Eq. (8.67), one is the plane $x=1$, and the fourth one is the plane $y=1$. There are other two planes bordering

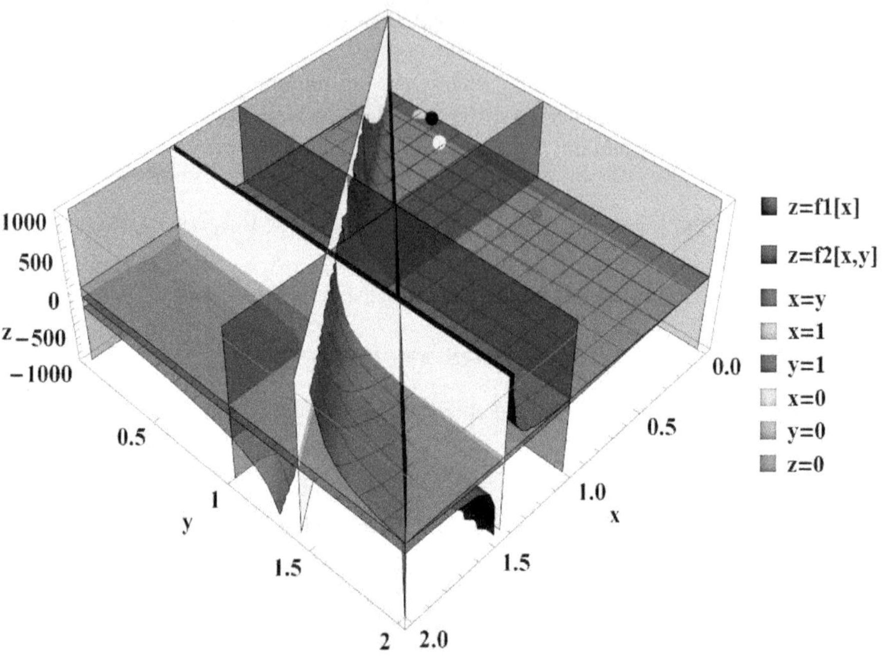

Figure 8.12 The phase diagram for a j-particle-triaxial rotor coupling Hamiltonian with $j = 13/2$ and $I = 45/2$. The coordinates x, y and z are a-dimensional.

the phase of interest defined in Appendix B of Ref. [20]. For this region we have to depict the minimum of $\mathcal{H}$, if that exists. If $\mathcal{H}$ exhibits, indeed, a minimum in the considered sector for γ, i.e. $[0°, 60°]$, further, the frequencies describing the small oscillations around the found minimum are to be determined. *The results of our investigations [20] confirm the existence of a transversal mode but for ideal restrictions [43], while within the Holstein-Primakoff description, the minimum for energy surface reflecting a transversal wobbling regime does not exist [44], if one keeps all energy terms. Therefore, there is no contradiction between the two formalisms [45, 46], since they deal with different Hamiltonians. Moreover, if in each of the two formalisms as well as in the present one, one keeps all terms from the starting Hamiltonian, it seems that no solution for the transversal wobbling exists.* Although in our case the transversal mode is determined by a part of the starting Hamiltonian, for the time being we cannot say whether this ideal picture is preserved when the remaining interaction is accounted for or is totally spoiled by the Coriolis interaction. An indirect answer to this question is, actually, provided by the wobbling energy behavior as function of spin, this being considered as a signature for the wobbling character. Contrary to what is stated in Ref. [43], in the present approach the monotony of the experimental and theoretical curves are the same, namely both are increasing functions of the angular momentum. This, in fact, confirms that the considered isotopes are longitudinal wobblers.

8.3 SIMULTANEOUS DESCRIPTION OF THE WOBBLING AND CHIRAL MOTION IN NUCLEI

Here we study an odd-mass system consisting of an even-even core described by a triaxial rotor Hamiltonian H_{rot} and a single j-shell proton moving in a quadrapole deformed mean-field:

$$H_{sp} = \varepsilon_j + \frac{V}{j(j+1)} \left[\cos\gamma(3j_3^2 - \mathbf{j}^2) - \sqrt{3}\sin\gamma(j_1^2 - j_2^2) \right]. \tag{8.68}$$

Here ε_j is the single-particle energy and γ the deviation from the axial symmetric picture. In terms of the total angular momentum $\mathbf{I}(=\mathbf{R}+\mathbf{j})$ and the angular momentum carried by the odd particle, $\mathbf{j}$, the rotor Hamiltonian is written as:

$$H_{rot} = \sum_{k=1,2,3} A_k (I_k - j_k)^2, \tag{8.69}$$

where A_k are half of the reciprocal moments of inertia associated to the principal axes of the inertia ellipsoid, i.e. $A_k = 1/(2\mathcal{I}_k)$, which are considered as free parameters.

The expressions for the single-particle coupling potential, H_{sp}, and the triaxial rotor term, H_{rot}, have been previously used by many authors, the first being Davydov [36]. In the context of rigid coupling of the single particle to the core, the term H_{sp} does not contribute to the equations of motion for the angular momentum components I_k, k=1, 2, 3.

We consider that the maximal moment of inertia (MoI) is $\mathcal{I}_1$. For a rigid coupling of the odd proton to the triaxial core, we suppose that $\mathbf{j}$ stays out of any principal plane: $\mathbf{j} = (j\cos\theta_0, j\sin\theta_0\cos\varphi_0, j\sin\theta_0\sin\varphi_0)$.

We recall that within the liquid drop model (LDM) the odd nucleon may be coupled either to the deformation or to the core angular momentum. Correspondingly, the single-particle angular momentum is oriented either to the symmetry axis or to the core's angular momentum [9]. These scenarios are reached for weak and strong coupling regimes, respectively. For an intermediate coupling one may meet the situation when $\mathbf{j}$ lays outside the principal planes. Within a microscopic picture, the orientation of the single-particle angular momentum depends on the location of the Fermi level. Thus, when the Fermi level of valence nucleon is located in the lower/upper part of a high-j sub-shell, its angular momentum is oriented along the short/long axis of the triaxial core, and in the middle part with its angular momentum easily aligned with the intermediate axis with the maximum MoI. When the Fermi level is different from these special cases, the angular momentum of the odd proton might be oriented along a line which is different from the three mentioned axes. In these phenomenological and microscopic contests, it seems reasonable to fix the single-particle angular momentum outside any principal plane of the triaxial core.

Note that the linear term in $\hat{I}$ from (8.69) looks like the cranking term in the microscopic cranking formalism. According to the pioneering paper of Bengtsson [37] the equations for a general orientation of the cranking term admit a real solution.

In this context we ask ourselves, whether the phenomenological Hamiltonian (8.69) admits a harmonic solution within a classical treatment. To this aim we dequantized the model Hamiltonian by replacing the operators $\hat{I}_k$, k=1, 2, 3 with the classical components of the angular momentum hereafter denoted by x_k, k=1, 2, 3, respectively, and the commutators with the Poisson brackets:

$$\hat{I}_k \to x_k,$$
$$[,] \to i\{,\}. \tag{8.70}$$

with "i" denoting the imaginary unity and $\{,\}$ the Poisson bracket.

According to these rules, the classical energy can be written as:

$$\mathcal{H}_{rot} = AH' + A_1 I^2 + \sum_{k=1,2,3} A_k j_k^2, \tag{8.71}$$

with

$$
\begin{aligned}
H' &= x_2^2 + u x_3^2 + 2v_1 x_1 + 2v_2 x_2 + 2v_3 x_3, \\
u &= \frac{A_3 - A_1}{A_2 - A_1}, \ v_k = -\frac{j_k A_k}{A_2 - A_1}, k = 1,2,3., A = A_2 - A_1.
\end{aligned}
\tag{8.72}
$$

Also, the angular momentum components obey:

$$
\{x_i, x_j\} = -\varepsilon_{i,j,k} x_k.
\tag{8.73}
$$

where $\varepsilon_{i,j,k}$ denotes the three dimensional unity tensor. The classical counterpart of the Heisenberg quantal equation for the classical image o, of an operator $\hat{O}$, is:

$$
\{o, H\} = i \frac{\partial o}{\partial t}.
\tag{8.74}
$$

with o denoting the classical image of $\hat{O}$.

Since $\mathcal{H}_{rot}$ and H' differ from each other by one multiplicative and one additive constant, the motion described by $\mathcal{H}_{rot}$ is readily known once that corresponding to H' is given. Due to this feature, it is convenient to deal first with H'. The equations of motion for H' are non-linear. Linearizing them one finds that the stationary points coordinates satisfy:

$$
\frac{v_1}{x_1} - \frac{v_2}{x_2} = 1; \quad \frac{v_1}{x_1} - \frac{v_3}{x_3} = u, \quad \frac{v_2}{x_2} - \frac{v_3}{x_3} = u - 1.
\tag{8.75}
$$

From these relations we can express x_2 and x_3 in terms of x_1 and then insert the result in the angular momentum conservation equation:

$$
x_1^2 + x_2^2 + x_3^2 = I^2.
\tag{8.76}
$$

It results an algebraic equation for the component x_1:

$$
F(x_1) \equiv \sum_{k=0}^{6} B_k x_1^k = 0,
\tag{8.77}
$$

with the coefficients B_k depending on u, v_k and I.

We note that, since $B_0 \neq 0$, the equation (8.77) does not admit vanishing solutions, which as a matter of fact is a specific feature for the chiral motion. The solution for x_1 corroborated with Eqs.(8.75) leads to the stationary points $(\overset{\circ}{x}_1, \overset{\circ}{x}_2, \overset{\circ}{x}_3)$ for the surface of constant energy, $H' = E$.

8.3.1 SMALL OSCILLATIONS AROUND THE DEEPEST MINIMUM

The equations of motion for the components x_k are non-linear. However, these can be linearized by replacing one factor of the quadratic terms with the coordinates of the deepest minimum point.

A solution of the linear system of equations may be found by searching for the linear combination:

$$
C^* = X_1 x_1 + X_2 x_2 + X_3 x_3,
\tag{8.78}
$$

such that the following equation is fulfilled:

$$
\{C^*, H'\} = \omega C^*.
\tag{8.79}
$$

This restriction leads to a homogeneous system of linear equations for the coefficients X_1, X_2, X_3. The compatibility condition yields an equation for the frequency ω:

$$
\omega^3 + 3S\omega - 2T = 0.
\tag{8.80}
$$

with the coefficients given by:

$$3S = -\left(2v_1 - u\overset{\circ}{x}_1\right)\left(\overset{\circ}{x}_1 - 2v_1\right) + \left(u\overset{\circ}{x}_3 + 2v_3\right)\left(2v_3 - (1-u)\overset{\circ}{x}_3\right) + \left(\overset{\circ}{x}_2 + 2v_2\right)\left(2v_2 + (1-u)\overset{\circ}{x}_2\right),$$

$$2T = \left(2v_3 + u\overset{\circ}{x}_3\right)\left(2v_2 + (1-u)\overset{\circ}{x}_2\right)\left(\overset{\circ}{x}_1 - 2v_1\right) - \left(\overset{\circ}{x}_2 + 2v_2\right)\left((1-u)\overset{\circ}{x}_3 - 2v_3\right)\left(2v_1 - u\overset{\circ}{x}_1\right). \tag{8.81}$$

The solutions of Eq.(8.80) are analytically given by the Cardano formula.

Note that the system under consideration exhibits two constants of motion, namely the energy and the angular momentum squared. Furthermore, there is only one independent angular momentum component; adding to this the corresponding conjugate momentum one arrives at a two dimensional phase space, which is conventionally called *the reduced space*. To define this space, it is convenient to use the polar coordinates which define the two independent variables θ, ϕ associated to the angular momentum **I**. These depend, of course, on the choice for the quantization axis:

$$axis\ 1: \quad x_1 = I\cos\theta_1,\ x_2 = I\sin\theta_1\cos\varphi_1,\ x_3 = I\sin\theta_1\sin\varphi_1,$$

$$axis\ 2: \quad x_2 = I\cos\theta_2,\ ^\circ\ x_3 = I\sin\theta_2\cos\varphi_2,\ x_1 = I\sin\theta_2\sin\varphi_2,$$

$$axis\ 3: \quad x_3 = I\cos\theta_3,\ x_1 = I\sin\theta_3\cos\varphi_3,\ x_1 = I\sin\theta_3\sin\varphi_3. \tag{8.82}$$

For each of the above choices, the Hamiltonian H' becomes a function of the conjugate coordinates (x_k, ϕ_k) which admits a minimum. Expanding H' in the first order around the minimum point. The linear equations in the deviations describe a harmonic motion of frequencies:

$$\omega^{(1)} = 2\left[\left(\cos^2\overset{\circ}{\varphi}_1 + u\sin^2\overset{\circ}{\varphi}_1 + \frac{1}{I}(v_2\cos\overset{\circ}{\varphi}_1 + v_3\sin\overset{\circ}{\varphi}_1)\right)\right.$$
$$\left.\left((I^2 - \overset{\circ}{x}_1^2)(1-u)\cos 2\overset{\circ}{\varphi}_1 + \frac{1}{I}(I^2 - \frac{\overset{\circ}{x}_1^2}{2})(v_2\cos\overset{\circ}{\varphi}_1 + v_3\sin\overset{\circ}{\varphi}_1)\right)\right]^{1/2}$$

$$\omega^{(2)} = 2\left[\left(1 - u\cos^2\overset{\circ}{\varphi}_2 - \frac{1}{I}\left(v_1\sin\overset{\circ}{\varphi}_2 + v_3\cos\overset{\circ}{\varphi}_2\right)\right)\right.$$
$$\left.\left((\overset{\circ}{x}_2^2 - I^2)u\cos 2\overset{\circ}{\varphi}_2 + (\frac{\overset{\circ}{x}_2^2}{2I} - I)\left(v_1\sin\overset{\circ}{\varphi}_2 + v_3\cos\overset{\circ}{\varphi}_2\right)\right)\right]^{/2}$$

$$\omega^{(3)} = 2\left[\left(-\sin^2\overset{\circ}{\varphi}_3 + u - \frac{1}{I}(v_1\cos\overset{\circ}{\varphi}_3 + v_2\sin\overset{\circ}{\varphi}_3))\right)\right.$$
$$\left.\left((I^2 - \overset{\circ}{x}_3^2)\cos 2\overset{\circ}{\varphi}_3 + (\frac{\overset{\circ}{x}_3^2}{2I} - I)(v_1\cos\overset{\circ}{\varphi}_3 + v_2\sin\overset{\circ}{\varphi}_3)\right)\right]^{1/2}. \tag{8.83}$$

If these frequencies are all real, then they describe the wobbling frequencies for the motion along the axes 1, 2, 3, respectively. It is interesting to see what is the relations between the frequencies given above and the solutions of the cubic equation (8.80). This issue will be pointed out in what follows.

In the space of angular momentum, a chiral transformation is equivalent to the space inversion operation, i.e. $C = \mathbf{I} \rightarrow -\mathbf{I}$. Due to the linear terms in angular momentum components, the Hamiltonian H' is not invariant to chiral transformations. On the other hand, if there exists an operator O which satisfies the relation,

$$\{H, O\} = 0, \tag{8.84}$$

then, if Ψ is an eigenfunction of H corresponding to the eigenvalue λ, it results that $O\Psi$ is also an eigenfunction of H with the eigenvalue $-\lambda$. Therefore, the eigenvalues λ and $-\lambda$ are mirror images of one another. In our case H' is a sum of two terms, one invariant, H_1, and another non-invariant, H_2, to chiral transformations C. The non-invariant term H_2 and the transformation C satisfy Eq. (8.84). Due to this feature the eigenvalues of H' are mirror images of those for $CH'C^{-1}$. The two sets of energies define the so called chiral bands. We note that $CH'C^{-1}$ is obtained from H' by changing $v_k \to -v_k$, which results that the wobbling frequencies, $\omega_{ch}^{(k)}$, built up with $CH'C^{-1}$ are obtained from those obtained with H' with the transformation $v_k \to -v_k$.

The notations $H'^{(k)}_{min}$ and $H'^{(k)}_{ch,min}$ are used for minimal energy when the axis "k" is the quantization axis. The energies, defined with the above mentioned frequencies are angular momentum dependent. For a given n and $I = \alpha + 2n$ with α being the signature, the set of energies $E^{(k)}_{I,n}$ defines a wobbling band, while $E^{(k)}_{ch,I,n}$ the chiral partner band. In this way we found out a set of states which are simultaneously of wobbling and chiral character. The wobbling motion is conciliated with the ingredient that the rotation axis is outside the principal planes.

Here we present an illustrative example. Thus, we consider an odd particle from the single particle orbit $i_{13/2}$ moving around a triaxial rigid rotor core with the moments of inertia (MoI):

$$(\mathcal{J}_1, \mathcal{J}_2, \mathcal{J}_3) = (60, 20, 30)\hbar^2 MeV^{-1}. \tag{8.85}$$

The composite system moves in a state of angular momentum $I = 35/2\hbar$. The odd particle is rigidly coupled to the core : $\mathbf{j}$ are: $\mathbf{j} = (j, \theta_0, \varphi_0) = (13/2, \pi/4, \pi/4)$. The stationary points for the equations of motion for the classical angular momentum components x_k, k=1 , 2, 3, obey a set of equations which leads to an algebraic sixth-order equation for x_1, i.e $F(x_1, I) = 0$. This equation admits four real solutions for x_1: $-13.062; -8.811; -1.81; 16.185 [\hbar]$. Making use of relations expressing x_2 and x_3 in terms of x_1 one arrives at the final result:

$$(\overset{\circ}{x}_1, \overset{\circ}{x}_2, \overset{\circ}{x}_3) = \begin{pmatrix} -13.062 & 5.916 & 10.029 \\ -8.811 & 6.595 & 13.588 \\ -1.810 & 16.886 & -4.223 \\ 16.185 & 4.269 & 5.062 \end{pmatrix} \hbar. \tag{8.86}$$

To the four stationary vectors the following classical energies correspond:

$$H_{rot} = \begin{pmatrix} axis-1 & axis-2 & axis-3 \\ 3.542 & 3.027 & 2.887 \\ 3.559 & 3.093 & 2.839 \\ 5.921 & 4.920 & 5.793 \\ 1.200 & 1.452 & 1.424 \end{pmatrix} MeV. \tag{8.87}$$

For example, for the stationary angular momenta components from the row 1 of Eq. (8.86), the energies of the row 1 from Eq. (8.87) correspond, for the situations when the quantization axes are the axis-1, axis-2 and axis-3, respectively.

The frequencies characterizing the linearized equations of motion satisfy a third order algebraic equation. The minimum value for the energy H_{rot} when $\mathbf{I} \parallel \mathbf{j}$, i.e. when the two angular momenta are aligned, is 1.765 MeV. With this data there exists a real solution for the wobbling frequency: $\omega = 0.362$ MeV. The chiral partner state has the frequency equal to 3.651 MeV.

The coordinates and spins of all minima are collected in Table 8.7; these minima are taken from the contour plots shown in Fig. 8.13, respectively. Furthermore, we studied the equations of motion for H' in the reduced space of generalized phase space coordinates. The coordinates and spins of all minima are collected in Table 8.8.

A major conclusion of this analysis is that irrespective of the chosen quantization axis, the deepest minimum of H_{rot} is met for an angular momentum lying outside any principal plane of the inertia ellipsoid which is a prerequisite of a chiral motion.

Table 8.7

Coordinates of the minima points for H_{rot} and the corresponding values of the spin components.

Quantization axis	θ_{min}[rad]	φ_{min}[rad]	$I_1[\hbar]$	$I_2[\hbar]$	$I_3[\hbar]$	$H_{rot,min}$[MeV]
axis-1	0.388	0.8703	16.198	4.269	5.063	1.203
axis-2	1.206	1.236	15.443	6.238	5.370	1.381
axis-2	1.104	-0.983	- 13.003	7.879	8.666	2.960
axis-3	1.124	0.283	15.152	4.403	7.569	1.361

Table 8.8

Coordinates of the minima points for the chirally transformed Hamiltonian, H_{rot}^{ch}, and the corresponding values of the spin components.

Quantization axis	θ_{min}	φ_{min}	$I_1[\hbar$	$I_2[\hbar]$	$I_3[\hbar]$	$H_{rot,min}^{ch}$[MeV]
axis-1	2.753	- 2.27	- 16.198	- 4.269	- 5.063	1.202
axis-1	2.894	3.141	- 16.965	- 4.293	≈ 0.0	1.478
axis-2	1.935	- 1.905	- 15.443	- 6.238	- 5.370	1.381
axis-2	2.148	3.141	≈ 0	- 9.553	- 14.662	2.873
axis-3	2.018	- 2.859	- 15.152	- 4.403	- 7.568	1.361
axis-3	2.021	3.142	- 15.754	≈ 0	- 7.620	1.719

The transformed Hamiltonian H_{rot}^{ch} has the contour plots graphically represented in Fig. 8.14 for the quantization axes 1, 2 and 3, respectively, with the minima coordinates collected in Table 8.8.

The wobbling frequencies corresponding to the deepest minima when the quantization axis is the axis 1, 2 and 3, respectively, have the values: 0.245, 0.209 and 0.210 MeV, respectively. One notes that when axis-1 is the quantization axis, the Hamiltonians H_{rot} and H_{rot}^{ch} have the same wobbling frequencies, while for axis-2 as quantization axis the wobbling frequencies are very close to each other. Moreover the two Hamiltonians have the same/almost the same values in the respective minima. This is a reflection of the fact that for the two cases the chiral invariant part of H_{rot} prevails over the non-invariant part. Consequently, for the cases when the quantization axis is the axis-1 or 2, one can build up two wobbling bands with a similar structure which results in having a twin pair of bands.

Note that contrary to the previous publications, where the odd nucleon was rigidly fixed either to an axis [38] or to a principal plane [39] of the inertia ellipsoid, here the rigid coupling is achieved to a direction which does not belong to a principal plane. In this way one conciliates between the two signatures of triaxial nuclei, the wobbling and chiral motion, these being simultaneously considered.

8.4 BOSON DESCRIPTION OF THE WOBBLING MOTION

Assuming a rigid coupling of an odd nucleon to a triaxial core, the Hamiltonian for the even-odd system may be approximated as:

$$\hat{H}_{rot} = \sum_{k=1,2,3} A_k(\hat{I}_k - \hat{j}_k)^2, \tag{8.88}$$

with $A_k = \frac{1}{2\mathcal{J}_k}$ and I standing for the total angular momentum.

For a rigid coupling of the odd proton to the triaxial core, we suppose that $\mathbf{j}$ stays in the principal plane (1,2). Also, we consider that the maximal moment of inertia is $\mathcal{J}_2$; furthermore, we expand

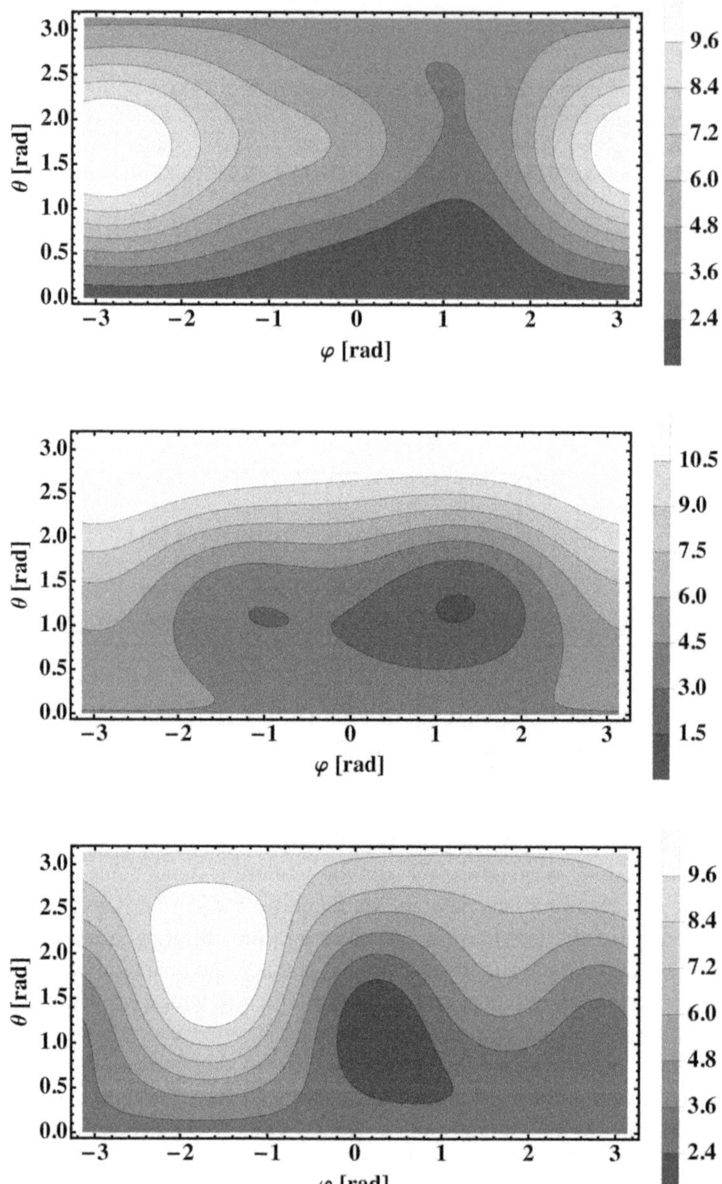

Figure 8.13 The energy function is H_{rot} given by Eq. (2.2) as a function of the polar coordinates. Contour plot when the axis-1, axis-2 and axis-3 are the quantization axes, respectively.

the linear term in the first order of approximation:

$$\hat{I}_2 = I\left(1 - \frac{1}{2}\frac{\hat{I}_1^2 + \hat{I}_3^2}{I^2}\right). \tag{8.89}$$

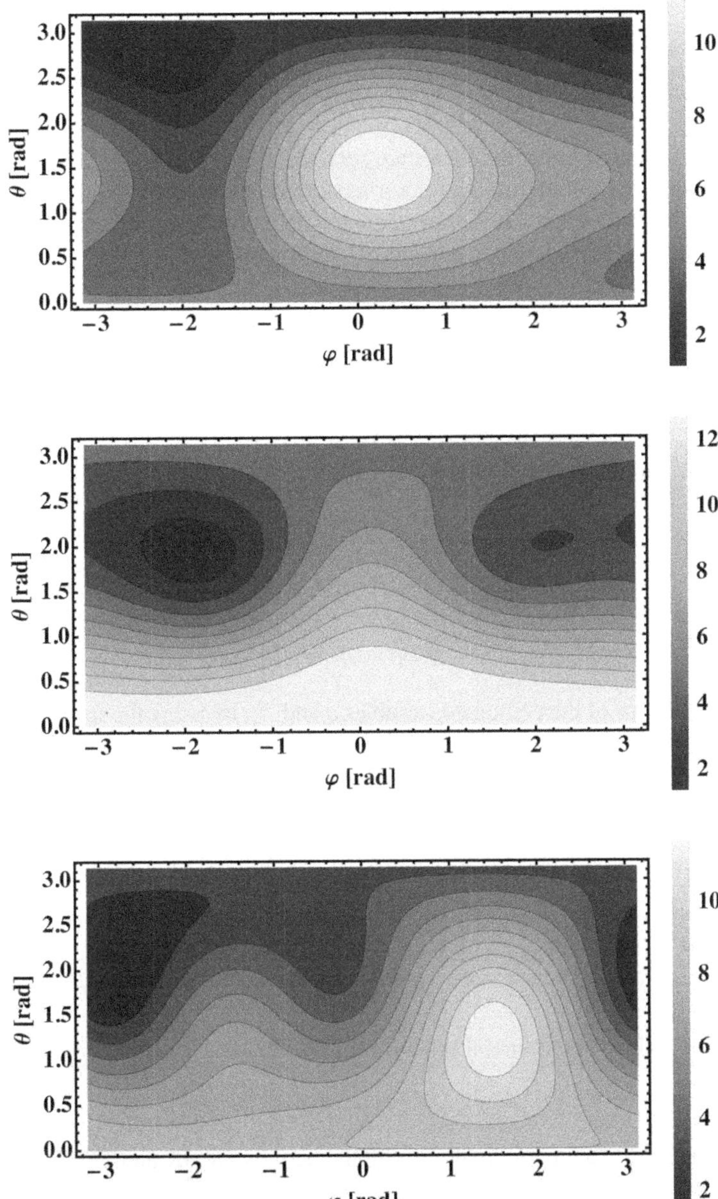

Figure 8.14 Contour plot when the axis-1, axis-2 and axis-3 are the quantization axes, respectively. The energy function is H_{rot}^{ch}, the chiral image of the classical energy.

Thus, the Hamiltonian acquires the form:

$$\hat{H}_{rot} = A\hat{H}' + (A_1 I^2 - A_2 j_2 I) + \sum_{k=1,2} A_k \hat{j}_k^2, \tag{8.90}$$

where the following notations have been used:

$$\hat{H}' = \hat{I}_2^2 + u\hat{I}_3^2 + 2v_0\hat{I}_1, \quad \text{with}$$

$$A = A_2(1 - \frac{j_2}{I}) - A_1, \quad u = \frac{A_3 - A_1}{A}, \quad v_0 = \frac{-A_1 j_1}{A}. \tag{8.91}$$

For what follows, we suppose that the MoI's are such that $1 > u > 0$.

Note that $\hat{H}'$ looks like a Hamiltonian for a triaxial rotor amended with a term, which cranks the system to rotate around the 1-axis. It is convenient to choose the cranking axis as quantization axis. Moreover, it is useful to express the considered Hamiltonian in terms of the raising and lowering angular momenta operators:

$$\hat{I}_\pm = \hat{I}_2 \pm i\hat{I}_3, \quad \hat{I}_0 = \hat{I}_1. \tag{8.92}$$

In the intrinsic frame of reference, the angular momentum components satisfy the commutation relations:

$$\left[\hat{I}_-, \hat{I}_+\right] = 2\hat{I}_0, \quad \left[\hat{I}_\mp, \hat{I}_0\right] = \mp\hat{I}_\mp. \tag{8.93}$$

In terms of the new variables, one obtains:

$$\hat{H}' = \frac{1-u}{4}\left(\hat{I}_+^2 + \hat{I}_-^2\right) + \frac{1+u}{4}\left(\hat{I}_+\hat{I}_- + \hat{I}_-\hat{I}_+\right) + 2v_0\hat{I}_0. \tag{8.94}$$

The Schrödinger equation associated to $\hat{H}'$,

$$\hat{H}'|\Psi\rangle = E|\Psi\rangle, \tag{8.95}$$

is further written in terms of the conjugate variables q and $\frac{d}{dq}$, by using the following representation for the angular momentum components:

$$\hat{I}_\mp = i\frac{c \pm d}{k's}\left(I \mp \hat{I}_0\right), \quad \hat{I}_0 = Icd - s\frac{d}{dq} \equiv \hat{I}_1, \tag{8.96}$$

where s, c and d denote the Jacobi elliptic functions:

$$s = sn(q,k), \quad c = cn(q,k), \quad d = dn(q,k), \quad \text{with}$$

$$k = \sqrt{u}, \quad k' = \sqrt{1-k^2}, \quad q = \int_0^\varphi \left(1 - k^2\sin^2(t)\right)^{-1/2} dt \equiv F(\varphi,k). \tag{8.97}$$

Their connection with the trigonometric function is given by:

$$s = \sin\varphi, \quad c = \cos\varphi, \quad d = \sqrt{1 - k^2 s^2}. \tag{8.98}$$

The functions s, c and d are periodic in φ, with the periods $4K, 4K$ and $2K$, respectively, where:

$$K = F(\frac{\beta}{2}, k) = \frac{\pi}{2}{}_2F_1(\frac{1}{2}, \frac{1}{2}, 1; k^2). \tag{8.99}$$

The standard notation for the hyper-geometric function ${}_2F_1(\alpha, \beta, \gamma; \varepsilon)$, has been used. It is worth mentioning that the transformation (8.96) preserves the commutation relations obeyed by the angular momentum components (2.6).

In terms of the newly introduced conjugate coordinates, $\hat{H}'$ becomes:

$$\hat{H}' = -\frac{d^2}{dq^2} - 2v_0 s\frac{d}{dq} + I(I+1)s^2 k^2 + 2v_0 cdI. \tag{8.100}$$

Changing the wave-function by the transformation:

$$|\Psi\rangle = (d - kc)^{-\frac{v_0}{k}}|\Phi\rangle, \tag{8.101}$$

the Schrödinger equation acquires a new form, where the kinetic and potential energies are separated:

$$\left[-\frac{d^2}{dq^2} + V(q) \right] |\Phi\rangle = E|\Phi\rangle. \tag{8.102}$$

The potential energy term has the expression:

$$V(q) = \left[I(I+1)k^2 + v_0^2 \right] s^2 + (2I+1)v_0 cd. \tag{8.103}$$

Note that V(q) is invariant with respect to the transformation $q \to -q$. This leads to the fact that in the interval, for example of [-4K,4K], the potential exhibits two deep symmetric wells with degenerate minima, and three local minima in $q = 0, \pm 4K$. States inside the local minima are metastable since they are tunneling to the adjacent deep minima. The states in the deepest wells are degenerate. The shape of the potential $V(q)$ in the interval of [0,4K] is shown in Fig. 8.15, for a few angular momenta I. To visualize the symmetry mentioned above, we plotted $V(q)$ in a larger interval, namely [-4K,4K]. Denoting by ψ_+ the wave function of a state in the right deepest well, and by ψ_- the function corresponding to the same energy as the former state, but the left deepest minimum, they are both spread over the whole interval of [-4K,+4K]. However, the sum $\psi_+ + \psi_-$ is mainly located in the right deepest well, while the difference $\psi_+ - \psi_-$ is mainly spread inside the left deepest well.

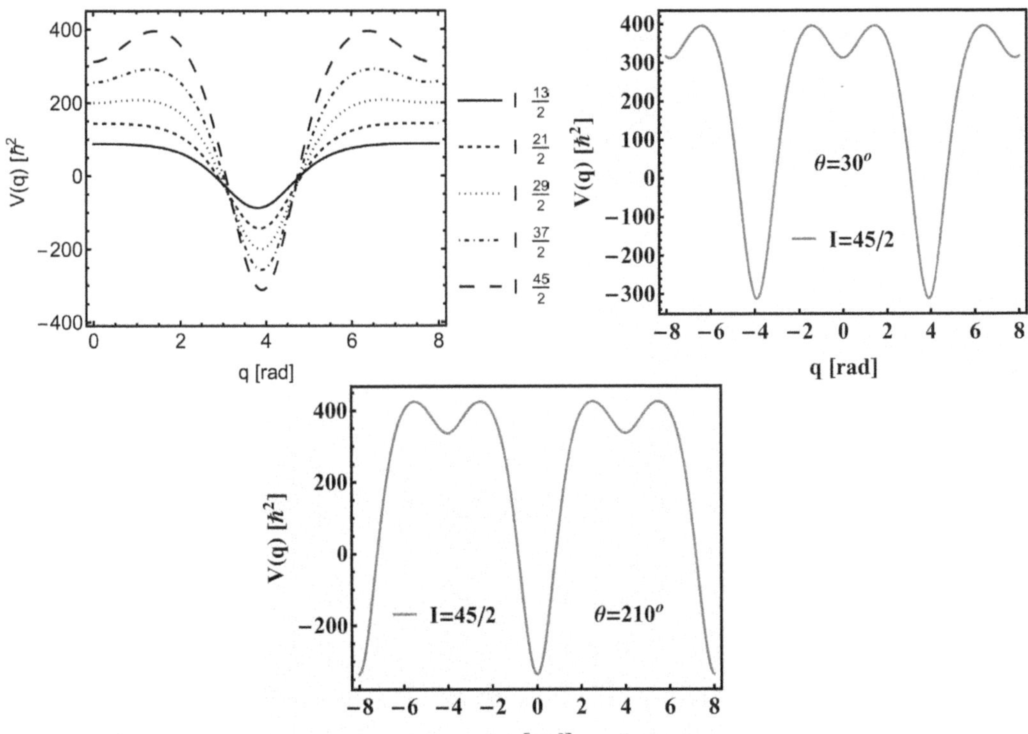

Figure 8.15 The potential energy is plotted as function of q for a particular set values for the moment of inertia $(MoI): \mathcal{J}_2 : \mathcal{J}_3 : \mathcal{J}_1 = 100 : 40 : 20\ \hbar^2 MeV^{-1}$, the odd particle angular momentum j=13/2 and $\theta = \pi/6$. The middle and the right figures correspond to $\theta = \pi/6$ and $\theta = 7\pi/6$, respectively. In both cases the total angular momentum is I=45/2.

The formalism described above was applied to ^{135}Pr. The excitation energies in three bands, conventionally called band 1 (B1), band 2 (B2) and band 3 (B3), and the electromagnetic properties of the states have been described by a simple Hamiltonian (5.1), associated to the even-even core and the odd proton, which stays in the orbital $h_{11/2}$. The core properties are simulated by a triaxial core with the moments of inertia $\mathcal{J}_k$ (k=1, 2, 3), considered to be free parameters, while the odd proton is rigidly coupled to the core and having the angular momentum $j = 11/2$, placed in the inertial plane (1,2), and having the polar angle θ. Thus, the approach involves four free parameters $\mathcal{J}_k$ (k=1, 2, 3), and θ, which were fixed by a least mean square procedure, fitting the excitation energies for the three bands.

The parameters yielded by the fitting procedure are listed in Table 8.9. With the parameters thus determined, the excitation energies are readily obtained:

$$E_I^{exc;1} = A_1 I^2 + (2I+1)A_1 j_1 - IA_2 j_2 + \omega_I/2 - E_{11/2}, \quad I = R+j, \quad R = 0,2,4,...,$$

$$E_I^{exc;2} = A_1 I^2 + (2I+1)A_1 j_1 - IA_2 j_2 + \omega_I/2 - E_{11/2}, \quad I = R+j, \quad R = 1,3,5,...,$$

$$E_{I+1}^{exc;3} = A_1 I^2 + (2I+1)A_1 j_1 - IA_2 j_2 + 3\omega_I/2 - E_{11/2}, \quad I = R+j, \quad R = 1,3,5,....$$

Table 8.9

The MoI's, and the parameter θ as provided by the adopted fitting procedure.

$\mathcal{I}_1$ $[\hbar^2/MeV]$	$\mathcal{I}_2$ $[\hbar^2/MeV]$	$\mathcal{I}_3$ $[\hbar^2/MeV]$	θ [degrees]	nr. of states	r.m.s. [MeV]
89	12	48	-71	20	0.174

They are visualized in Fig. 8.16 and compared with the corresponding experimental data taken from Refs. [40, 41]. From there, one sees the quality of the agreement with the data, that might be appraised by the r.m.s. of the deviation which is also given in Table 8.9. We may conclude that the agreement between theoretical, and experimental results is good.

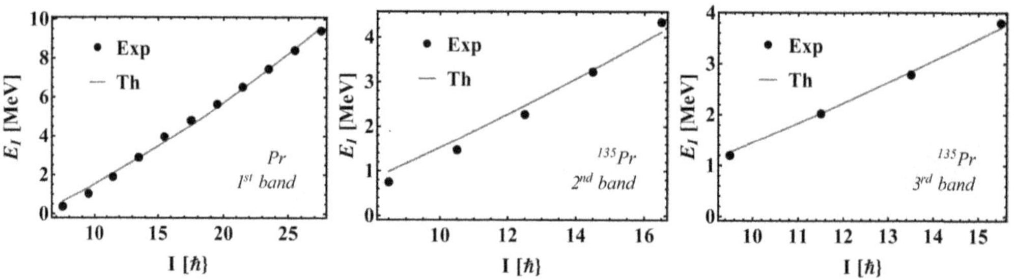

Figure 8.16   The excitation energies yielded by our calculations for the yrast band of ^{135}Pr, using the MoI's, and θ determined by a fitting procedure.

From Table 8.9 we see that the MoI's ordering predicted by our calculations is: $\mathcal{J}_1 > \mathcal{J}_3 > \mathcal{J}_2$. However, due to the adopted fitting procedure this, however, is a global result. In order to check whether this ordering holds also by using another fitting procedure, we fixed the MoI's by equating the calculated excitation energies for the lowest two states of band 1 and the second state of band 2, otherwise fixing θ to obtain a global best fit. In this way we found a set of MoI's which reclaims a transverse wobbling regime for the odd system under consideration. Indeed, for $\theta = 140°$, the result is $\mathcal{J}_1 = 13.5285 \, \hbar^2/MeV$, $\mathcal{J}_2 = 101.759 \, \hbar^2/MeV$, $\mathcal{J}_3 = 52.9364 \, \hbar^2/MeV$. However, the overall agreement is of a poor quality. We also calculated the potential determined by the fitted parameters and $I = 19/2$ which results in having the deepest minimum at $q = 0$, where $I_1 = I$, which means that the angular momentum is oriented along the axis one. Therefore, $\mathcal{J}_2 =$ maximal, does not necessarily mean that the rotation axis is the 2-axis. The fact that the maximal MoI established by a global fit is $\mathcal{J}_1$, indicates that for a larger angular momentum a change of MoI hierarchy takes place, and a new nuclear phase begins.

The classical energy function exhibits several stationary points. The character of the stationary point to be minimum, maximum or saddle point is decided by the signs of the diagonal matrix elements of the Hessian: a) if all diagonal elements are positive, the stationary point is minimum; b) if all diagonal m.e. are negative, then we deal with a maximum, while; c) is a saddle point if one m.e. is positive and the other is negative. Equating the Hessian to zero, one obtains the parameters u and v for which the critical points are degenerate. The resulting equations may be unified in a single formulae:

$$(1-u)(1-v^2)(v^2-u^2)(v^2-u) = 0. \tag{8.104}$$

The last factor in the above equation is obtained by equating the critical energies E_M and E_s. Each factor generates a curve, called separatrices, in the parameter space spanned by (u,v). As shown in Fig. 8.17, the separatrices are bordering manifolds defining unique nuclear phases characterized by a specific portrait of the stationary points. Indeed, among the factors involved in Eq. (8.104), we recognize those defining the two wobbling frequencies. On the other hand a vanishing energy defines a Goldstone mode [47] which, as a matter of fact, render evidently a phase transition.

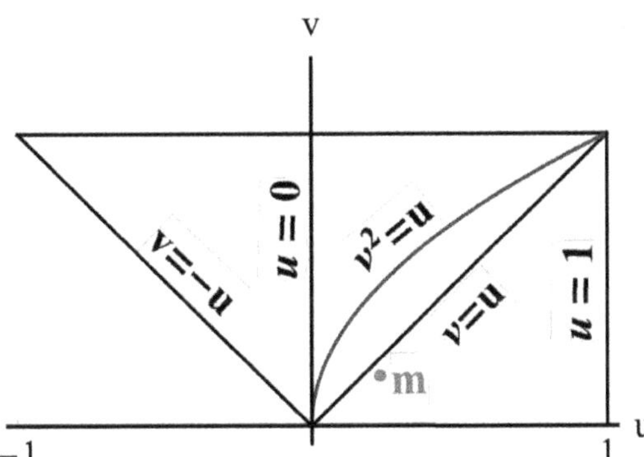

Figure 8.17 The phase diagram for $I = 15/2$ of ^{135}Pr. Using the MoI's, and θ determined by the adopted fitting procedure, and $\theta = 150°$, we mentioned the minimum point by a full circle having a lowercase m.

8.4.1 COMMENT ON THE CHIRAL FEATURES OF THE WOBBLING MOTION

In the angular momentum space, the change of sign of the a.m. defines a chiral transformation. A system is invariant to a chiral transformation if its rotational energy is preserved when the sense of rotation around an axis is changed. Note that our starting Hamiltonian is a sum of two terms, one being symmetric to chiral transformations, $\hat{H}_s$, and one antisymmetric to chiral transformations, $\hat{H}_a$.

If $|\psi\rangle$ is an eigenstate for $\hat{H}_s$, and C is a chiral transformation, then $C|\psi\rangle$ is also eigenstate for $\hat{H}_s$, and corresponds to the same energy. In this case, the function $|\psi\rangle$ has the chirality equal to one, since $C|\psi\rangle = |\psi\rangle$. For $\hat{H}_a$, the above mentioned property changes to : If $|\psi\rangle$ is an eigenstate of $\hat{H}_a$ corresponding to the eigenvalue E, then $C|\psi\rangle$ is also eigenstate, but corresponding to the energy $-E$. Therefore, the eigenvalues of $\hat{H}_a$ split in two sets, one being the mirror image of the other one. This property is of a chiral nature. The eigenstates of $\hat{H}_a$ have the chirality -1 since $C|\psi\rangle = -|\psi\rangle$. The eigenstates of $\hat{H}_{rot}$ are mixtures of the two chiralities. When there are two sets of energies that are one the mirror image of the other, one says that a definite chirality is projected out [42]. In our calculation, the change of $\mathbf{I} \rightarrow -\mathbf{I}$ is achieved by changing θ to $\theta + \pi$. The a.m. dependence of the wobbling frequencies corresponding to $\theta = -71°$, and $\theta = 109°$, respectively, is shown in Fig. 8.18, left panel. The two curves are parallel to each other, which suggests that the corresponding states have similar properties. The look of the potentials V, and CVC^{-1}, are shown in Fig. 8.18, middle and right panels, respectively, for $\theta = \pi/6$, and $\theta = 7\pi/6$, respectively, and $\mathcal{J}_1 : \mathcal{J}_2 : \mathcal{J}_3 = 100 : 40 : 20$ $\hbar^2 MeV^{-1}$. From these two potentials we may extract the symmetric and antisymmetric parts of V:

$$V_s = \frac{1}{2}V(\pi/6) + V(7\pi/6), \quad V_a = \frac{1}{2}V(\pi/6) - V(7\pi/6). \tag{8.105}$$

The two potential of definite chirality, are visualized in Fig. 8.18 middle panel, and Fig. 8.18 right panel, respectively. A similar analysis can be performed also for the excitation energies. Indeed, Eq. (5.19) expresses explicitly the dependence of the wobbling frequency on the angle θ, which fixes the orientation of $\mathbf{j}$. Therefore, it is easy to calculate $\omega_I(\theta + \pi)$, with $\theta = -71°$. The frequencies $\omega_I(\theta)$, and $\omega_I(\theta + \pi)$, with one being the chiral image of the other one, are plotted in Fig. 8.18 for the yrast band. The two curves are parallel to each other, which suggests that the corresponding states have similar properties. However, they do not correspond to states of definite chirality. However, one can extract the symmetric and antisymmetric terms of the excitation energy. Here we give the result for the yrast states:

$$\begin{aligned} E_{I,s}^{exc;1} &= A_1 I^2 + (\omega_I(\theta) + \omega_I(\theta + \pi))/2 - E_{11/2}, \; E_{I,a}^{exc;1} \\ &= (2I+1)A_1 j_1 - IA_2 j_2 + (\omega_I(\theta) - \omega_I(\theta + \pi))/2, \; I \\ &= R + j, \; R = 0,2,4,..,\theta = -71^0. \end{aligned} \tag{8.106}$$

Since for asymmetric states, the energies $-E_{I,a}^{exc;1}$ are also eigenvalues of the antisymmetric Hamiltonian, we interpret the two sets of energies $E_{I,s}^{exc;1}$, and $-E_{I,a}^{exc;1}$ as defining two bands of chirality $+1$, and -1, respectively. Although we don't have enough data to conclude that the two bands are indeed of real chiral type, however, due to the above mentioned features, they might be considered as germinos of chiral bands. Indeed, there are properties unanimously accepted, which prevent us to make a decisive statement on this matter. For a chiral band the system rotates around an axis which doesn't belong to any of principal planes, while here the rotation axis is a principal axis. However, since in our case the Hamiltonian involves linear terms in the total angular momentum, the wobbling motion and chiral properties seem not to be disconnected.

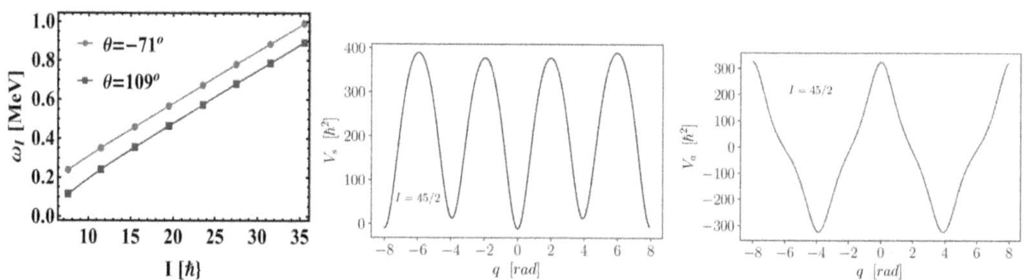

Figure 8.18 The wobbling frequencies corresponding to the angles $\theta°$, and $\theta = -71°$, respectively (left panel). The symmetric potential, with respect to the chiral transformations, as function of q (middle panel). The antisymmetric potential, with respect to the chiral transformations, as function of q (right panel).

Within this formalism we described the experimental data for the electric quadrupole intra- and inter-band transitions as well as the magnetic dipole transitions. As for the $E2$ transitions, we need the wave functions describing the involved states and the transition quadrupole operator. The wave function for an I-state is the solution of the Schrödinger equation for the given total angular momentum, I. Note that the wave function is degenerate with respect to "M", the projection of I on the x-axis, in the laboratory frame. Since the ground state is the vacuum state for the wobbling phonon operator, and moreover, in the minimum point of the constant energy surface, the angular momentum projection on the 1-axis of the intrinsic frame is equal to $-I$, it results that the K quantum number is equal to $-I$. Therefore, the solution of the Schrödinger equation must be labeled by the mentioned quantum numbers, i.e.

$$\Psi_{IM} = \Phi_{I,-I}|IM,-I\rangle, \text{ with } |IMK\rangle = \sqrt{\frac{2I+1}{8\pi^2}}D^I_{M,K}. \tag{8.107}$$

Here we consider the first two wobbling bands as signature partner bands, the arguments being in detail given in Ref. [19]. More specifically, the spin sequence of the first band is $j+R$, for $R = 0,2,4,\dots$, while for the second band the spin succession is $j+R$, with $R = 1,3,5,\dots$. Note that the quadrupole inter-band transitions are forbidden since, for the states mentioned above have $\Delta K = 1$. In this case, considering the component $K = -I+1$ in one of the involved states is necessary. The quadrupole transition operator is taken as:

$$\mathcal{M}(E2;\mu) = \sqrt{\frac{5}{16\pi}}e\left(Q_0 D^2_{\mu 0} + Q_2\left(D^2_{\mu 2} + D^2_{\mu -2}\right)\right), \tag{8.108}$$

where Q_0 and Q_2 denote the $K = 0$ and $K = \pm 2$ components of the quadrupole transition operator, respectively. Note that since the intrinsic component of the wave function depends on one of the conjugate variables q and d/dq, that is q, we must express the quadrupole operators in terms of the q variable. This will be achieved by writing Q in the space of angular momentum and then use the Bargmann representation of the angular momentum components. Thus we have:

$$Q_0 = -\frac{1}{4}\sqrt{\frac{2}{3}}\left(\hat{I}_+\hat{I}_- + \hat{I}_-\hat{I}_+\right) + \sqrt{\frac{2}{3}}\hat{I}_1^2, \quad Q_2 = \frac{1}{2}\left(\hat{I}_+^2 + \hat{I}_-^2\right). \tag{8.109}$$

The reduced $E2$ transition probability is expressed as the square of the reduced matrix element of the quadrupole transition operator. The magnetic dipole transition operator is:

$$\mathcal{M}(M1;\mu) = \sqrt{\frac{3}{4\pi}}\mu_N \sum_v \left(g_R\hat{R}_v + g_j\hat{j}_v\right)D^1_{\mu v} \equiv M^{coll}_{1\mu} + M^{sp}_{1\mu}, \tag{8.110}$$

where R_v and j_v are the spherical components of the core and the odd nucleon angular momenta, respectively. g_R and g_j stand for the gyromagnetic factors of the core, and the coupled odd nucleon, respectively. Also, the standard notation for the Wigner function, D^J_{MK}, and for the nuclear magneton, μ_N, are used. To calculate the collective part of the transition matrix element, we need to express the wave function describing the odd system as a Kronecker product of the core and the odd particle wave functions:

$$|IMK\rangle = \frac{1}{2j+1}\sum_{M_R,\Omega,R}' C^{R\,j\,I}_{M_R\,\Omega\,M}|RM_R K\rangle\,\psi_{j\Omega}. \tag{8.111}$$

To calculate the reduced m.e. of the single-particle $M1$ operator we need the wave function describing the odd proton whose angular momentum is placed in the plane (x,y) making the angle θ with the x axis. This function is obtained by rotating around the axis 3, the function $\psi_{j,j}$ associated to the odd proton having the angular momentum along the 1-axis, i.e. $\psi'_j = R_3(\theta)\psi_{jj}$. Finally, the magnetic dipole reduced transition probability is given by:

$$B(M1;I \to I') = [\langle\Psi_I||\mathcal{M}(M1)||\Psi_{I'}\rangle]^2. \tag{8.112}$$

Results for the branching ratios of $E2$ and $M1$ transitions are collected in Table 8.10.

Table 8.10

The calculated branching ratios $B(E2)_{out}/B(E2)_{in}$, and $B(M1)_{out}/B(E2)_{in}$ as well as the mixing ratios δ are compared with the corresponding experimental data taken from Ref. [40].

I^π	$\dfrac{B(E2;I^-\to(I-1)^-)}{B(E2;I^-\to(I-2)^-)}$		$\dfrac{B(M1;I^-\to(I-1)^-)}{B(E2;I^-\to(I-2)^-)}\left[\dfrac{\mu_N^2}{e^2b^2}\right]$		$\delta_{I^-\to(I-1)^-}$ [MeV.fm]	
	Exp.	Th.	Exp.	Th.	Exp.	Th.
$\frac{21}{2}^-$	0.843 ± 0.032	0.510	0.164 ± 0.014	0.164	-1.54 ± 0.09	-0.542
$\frac{25}{2}^-$	0.500 ± 0.025	0.500	0.035 ± 0.009	0.066	-2.384 ± 0.37	-0.703
$\frac{29}{2}^-$	$\geq0.261\pm0.014$	0.487	$\leq0.016\pm0.004$	0.033	-	-0.873
$\frac{33}{2}^-$	-	0.473	-	0.019	-	-1.052

8.5   SUMMARY AND CONCLUSIONS

Here we presented a selected issues of our contribution to the field of wobbling and chiral motion in atomic nuclei. Since one deals with high spin state we chose a semi-classical approach. Moreover the trial function involved in the variational equation is a coherent state for SU(2) group generated by the angular momenta determining a rigid rotor Hamiltonian, for the even-even system and a product of SU(2) coherent states the factors being associated to the core and the odd nucleon for an even-odd system.For the even-even system a new quantization method for the classical orbitals is proposed which results in obtaining a Schödinger equation characterized by that the kinetic and potential energies are separated. In the harmonic approximation for the potential energy term, a compact formulae for the wobbling frequency, different from those proposed by other authors is obtained. An excellent agreement its experimental data is obtained for both excitation energies and $B(E2)$ values measured for ^{158}Er.

As for the even-odd nuclei to wobbling frequencies are obtained which reflects the fact that the angular momentum of the odd nucleon is frozen. The formalism is was applied to ^{163}Lu where plenty of experimental data are available. Two scenarios of defining the bands TSD1, TSD2, TSD3 and TSD4 are proposed. In one case each state member of the four bands is treated variationally with $j=13/2$ and the core states of positive parity for the first three bands and negative parity for TSD4. In the other scenarios the first three bands are defined in a similar way but the odd nucleon for TSD4, moves in the orbital $h_{9/2}$. Both variants yield results for excitation energies, transition probabilities, mixing ratios, branching ratios, angular momentum alignment, excitation energies relative to an rigid rotor effective energy, dynamic moment of inertia which are in good agreement with the corresponding data. The classical orbital dependence on energy and angular momentum is studied by intersecting the surfaces of constant energy and constant angular momentum, respectively.The contour plots for all the four bands as well as the phase diagram are given.

A simultaneous description of wobbling and chiral motion is presented in Section 8.4. If in the particle-triaxial rigid rotor Hamiltonian the coupling potential is neglected, due to the freezing of the particle angular momentum, one arrives to a cranked rotor Hamiltonian. The linearized equations of motion for the cranked Hamiltonian admit solutions which describe an a-planar motion. One derives analytical expressions for the three wobbling frequencies, for both the cranked and chirally transformed Hamiltonians. Then one builds up two chiral bands whose properties are analyzed by means of the corresponding contour plots. Thu one proves that the two signatures, wobbling and chiral, may coexist.

In the last section, the odd system is treated by using a new boson expansion for the total angular momentum components while the angular momentum of the odd nucleon is frozen in a principal plane. The maximal moment of inertia corresponds to the axis 2. The classical trajectories are quantized by writing the boson expanded Hamiltonian in the Bargmann representation. The resulting Schrödinger Hamiltonian comprises a potential having a complex structure exhibiting several symmetric deepest minima and local minima. It seems that the local minimum is a chirally transformation image of the deepest minimum. The harmonic approximation is used to interpret the experimental data concerning the excitation energies and the transition probabilities in three bands of ^{135}Pr. A comment about the transition fro the traversal to the longitudinal regime, is included.

The results of this chapter recommend the semi-classical formalism as an useful tool for describing the main features of wobbling and chiral motion, in a realistic fashion.

Bibliography

1. H. Kuratsuji and T. Suzuki, J. Math. Phys. **21**, 472 (1980).
2. T. Oguchi, Progr. Th. Phys, **25**,721 (1961).
3. V. Bargmann, Rev. Mod. Phys. **34**, 829 (1962).
4. B. Jancovici and D. H. Schiff, Nucl. Phys. **58**, 678 (1964).
5. D. Janssen, F. Dönau, S. Frauendorf and R. V. Jolos, Nucl. Phys. **A 172**, 145 (1971).
6. H. Bateman and A. Erdely, *Higher Transcedental Functions*, vol. II, Chap. 15 (McGraw-Hill, NY, 1953).
7. W. E. Byerly, *An Elementary Treatise on Fourier's Series, and Spherical, Cylindrical and Ellipsoidal Harmonics, with Application to Problems in Mathematical Physics*, New York: Dover, 1959, p. 255.
8. G. A. Lalazissis, S. Raman, P. Ring, Atomic Data and Nuclear Data Tables **71**, 1 (1999).
9. P. Ring and P. Schück, *The Nuclear Many Body Problem*, Springer-Velag, Berlin, Heidelberg, New York, p.18.
10. I. Y. Lee *et al.*, Phys. Rev. Lett. **38**, 1454 (1977).
11. M. A. Riley *et al.*, Phys. Lett. **B 177**, 624 (1986).
12. E. M. Beck *et al.*, Phys. Lett. **215**, 624 (1988).
13. A. A. Raduta and R. Budaca, Phys. Rev. **C 84**, 044323 (2011).
14. Yue Shi *et al.*, Phys. Rev. Lett. **108**, 092501 (2012).
15. A. A. Raduta, R. Budaca and C. M. Raduta, Phys. Rev. **C 76**,064309 (2007).
16. N. Nica, NDS **141**, 1 (2017).
17. A. A. Raduta, R. Poenaru and L. Gr. Ixaru, Phys. Rev. **C 96**, 054320 (2017).
18. A. A. Raduta, R. Poenaru and Al. H. Raduta, J. Phys. G; Nucl. and Part. Phys. **45** 105104 (2018).
19. R. Poenaru and A. A. Raduta, Int. Jour. Mod. Phys. E **30**, 2150033 (2021).
20. A. A. Raduta, R. Poenaru and C. M. Raduta, Phys. Rev. **C 101**, 014302 (2020).
21. E. A. Lawrie, O. Shirinda and C. M. Petrache, Phys. Rev. C **101**, 034306 (2020).
22. D. R. Jensen *et al.*, Eur. Phys. J. **A 19**, 173 (2004).
23. G. B. Hagemann, Eur. Phys. J. **A 20**, 183 (2004).
24. M. Matsuzaki, Y. R. Shimizu, K. Matsuyanagi, Phys. Rev. **C 65**, 041303(R) (2002); Yoshifumi R. Shimizu, Masayuki Matsuzaki and Kenichi Matsuyanagi, arXiv:nucl-th/0404063v1, 22 Apr. 2004.
25. S. W. Ødegård, *et al.*, Phys. Rev. Lett. **86**, 5866 (2001).
26. A. Görgen et al., Phys. Rev **C 69** (2004) 031301(R).
27. N. Lo. Iudice, A. A. Raduta and D. S. Delion, Phys. Rev. C **50**, 127 (1993).
28. H. J. Jensen *et al.*, Nucl. Phys. A **695**, 3 (2001).
29. C. W. Reich and B. Singh, Nucl. Data Sheets **111**, 1211 (2010).
30. G. B. Hagemann, *et al.*, Nucl. Phys. A **424**, 365 (1984).
31. D.R.Jensen *et al.*, Nucl. Phys. **A 703**, 3 (2002).
32. G. Schönwaßer *et al.*, Phys. Lett. **B 552**, 9 (2003).
33. H. Amro *et al.*, Phys. Lett. **B 553**, 197 (2003).
34. D. R. Jensen *et al.*, Phys. Rev. Lett. **89**,142503 (2002).
35. A. Gheorghe, A. A. Raduta and V. Ceausescu, Nucl. Phys. **A 637**, 201 (1998).

36. A. C. Davydov, *Teoria atomnova yadra*, Moscva, 1958 (in russian), chapters 19, 20.

37. H. Frisk and R. Bengtsson, Phys. Lett. B **196** , 14(1987).

38. R. Budaca, Phys. Rev. C **97** 069801 (2018).

39. A. A. Raduta, R. Poenaru and C. M. Raduta, Phys. Rev. C **101**, 014302 (2020).

40. J. T. Matta *et al.*, Phys. Rev. Lett. **114**, 082501 (2015).

41. N. Senasharma *et al.*, Phys. Lett. B **792**, 170 (2019).

42. A. A. Raduta, Progress in Particle and Nuclear Physics **90**, 241 (2016).

43. S. Frauendorf and F. Dönau, Phys. Rev. C **89**, 014322, (2014).

44. Kosai Tanabe and Kazuko Sugawara-Tanabe, Phys. Rev. C **95**, 064315, (2017).

45. S. Frauendorf, Phys. Rev. C **97**, 069801 (2018).

46. Kosai Tanabe and Kazuko Sugawara-Tanabe, Phys. Rev. C **97**, 069802 (2018).

47. J. Goldstone, Nuovo Cimento **19**, 154 (1961).

9 Nuclear chiral dynamics within time dependent covariant density functional theory

Bo Li
Peking University, Beijing, China

Peng-Wei Zhao
Peking University, Beijing, China

9.1 INTRODUCTION

Chirality is an important symmetry in many physical systems. The nuclear chirality was originally suggested by Frauendorf and Meng in 1997 [1]. It represents a novel feature of triaxial nuclei rotating around an axis, which lies outside the three planes spanned by the principal axes of the triaxial ellipsoidal density distribution, i.e., aplanar rotation. The total angular momentum consists of three components along the short, intermediate, and long principal axes of a triaxial nucleus in the body-fixed frame, and they form either a left- or right-handed system. The two chiral systems are related by the chiral operator $TR(\pi)$ that combines time reversal T and spatial rotation by π. The broken chiral symmetry in the body-fixed frame should be restored in the laboratory frame. This leads to the so-called chiral doublet bands, which consist of a pair of nearly degenerate $\Delta I = 1$ rotational bands [1, 2]. If multiple pairs of chiral doublet bands exist in a single nucleus, it corresponds to the so-called MχD [3], which provides the evidence of triaxial shape coexistence.

The evidences of chiral doublet bands have been reported experimentally in many mass regions. For $A \sim 80$ mass region, possible chiral doublet bands in ^{80}Br [4] have been reported. In the subsequent experiment, two pairs of positive-and negative-parity doublet bands, which are interpreted as MχD with octupole correlations, have been identified in ^{78}Br [5]. For the first example of chiral candidate in $A \sim 100$ mass region, a pair of chiral doublet bands was identified in ^{104}Rh [6]. Subsequently, chiral candidates have also been reported in other Rh isotopes [7–11] and other nuclides [12–15]. For $A \sim 130$ mass region, the chiral candidates have been reported experimentally in many nuclei, such as ^{128}Cs [16], ^{133}Ce [17], ^{134}Pr [18–21], ^{135}Nd [22–24], and ^{136}Nd [25]. For the mass region $A \sim 190$, possible chiral doublet bands have been reported in ^{188}Ir [26] and ^{198}Tl [27]. More details can be seen from the data tables [28].

The manifestation of chirality in realistic nuclei is usually more complicated [29, 30]. In most cases, the chiral doublet bands are more separated in energy at low spins than at high spins. At low spins, the energy difference between the states with the same spin in the chiral partners is caused by rapid vibration between the left- and right-handed configurations, i.e., chiral vibration. When the

"

spin grows above a certain critical value, the energy splittings between the partner bands decrease, and the left-right mode changes from soft chiral vibration to tunneling between well-established chiral configurations, i.e., static chirality.

On the theoretical side, many phenomenological approaches have been employed to understand such an intriguing phenomenon. For example, the particle rotor model [1, 31–35] can show an evolution from chiral vibration toward tunneling between static left- and right- handed configurations with increasing angular momentum. Although the quantal nature allows the particle rotor model to account well for this aspect of symmetry breaking, they are based on several assumptions, the validity of which may not be assured. The interacting boson fermion-fermion model [19, 21, 36] describes the chiral structure of nuclei in laboratory frame and the results can be directly compared with experimental data, but it introduces a cubic interaction to simulate the triaxiality of nuclei, while the corresponding strength is unknown and needs to be fitted to data. The generalized coherent state model [37] studies chiral nuclei by describing a set of nucleons moving in a shell model mean-field and interacting among themselves with pairing, as well as with a particle-core interaction of spin-spin type. The parameters in the Hamiltonian need also to be fixed by fitting to the experimental energies.

A microscopic treatment is thus important for a better understanding of nuclear chirality. The tilted axis cranking (TAC) approach allows one to study nuclear chirality based on a microscopic mean field. Although it well describes the energies and the intraband transition rates of the lower one of the chiral doublet bands, it gives either one achiral solution or two degenerate chiral ones [38], so it cannot describe the left-right excitation mode. This is a well-known deficiency of the mean-field solution when it spontaneously breaks a symmetry [2]. Consequently, the random-phase approximation method [23, 39], the collective Hamiltonian method [40, 41], and the projected shell model [42–45] have been developed to describe the chiral excitations. However, these approaches are based on single-particle potentials combined with either the schematic single-j Hamiltonian [41] or the pairing plus quadrupole-quadrupole model [46]. Self-consistent methods based on more realistic two-body interactions are required for a more fundamental investigation, including all important effects such as core polarization and nuclear currents [47, 48].

The nuclear density functional theory (DFT) starts from a universal energy density functional and can achieve a self-consistent description for almost all nuclei. Covariant density functional theory (CDFT) further exploits the Lorentz symmetry, so it includes the complicated interplay between the large Lorentz scalar and vector self-energies [49, 50]. It also allows the self-consistent treatment of the spin degrees of freedom and the nuclear currents induced by the spatial parts of the vector self-energies, which play an essential role in rotating nuclei; see Refs. [48, 51] for details. Both relativistic and nonrelativistic DFTs have been extended with the TAC method to investigate nuclear chirality [52, 53]. In particular, the recent TAC-CDFT has been widely used to investigate the chirality in nuclei ^{106}Rh [53], ^{135}Nd [54], ^{136}Nd [25] and ^{106}Ag [55]. However, in all these works, only the lower band of the chiral doublet bands can be calculated. Note that the recently developed Relativistic Configuration-interaction Density functional (ReCD) theory [56, 57], which combines the advantages of large-scale configuration-interaction shell model and relativistic DFT, can describe both the lower and upper bands of the chiral doublet bands.

All these studies are based on static approaches. The dynamics of chiral excitations can be studied within the time-dependent DFT, which is a dynamical extension of DFT, and has been widely applied to various nuclear structure and reaction processes, see recent reviews [58–60] for details. The solution of the time-dependent covariant density functional theory (TDCDFT) in three-dimensional lattice space is feasible only recently [61–65] thanks to the progress for solving CDFT in the same space [66–71]. Recently, in Ref. [72], the dynamics of chiral nuclei has been investigated for the first time with the time-dependent and tilted axis cranking covariant density functional theories on a three-dimensional space lattice in a microscopic and self-consistent way. A novel mechanism, i.e., chiral precession, is revealed from the microscopic dynamics of the total angular momentum in the body-fixed frame, whose harmonicity is associated with a transition from the planar into aplanar rotations with the increasing spin.

9.2 THEORETICAL FRAMEWORK

9.2.1 TILTED AXIS CRANKING COVARIANT DENSITY FUNCTIONAL THEORY

The starting point of CDFT is a standard Lagrangian density which, in the point-coupling form, can be written as [73]:

$$
\begin{aligned}
\mathcal{L} &= \mathcal{L}^{\text{free}} + \mathcal{L}^{\text{4f}} + \mathcal{L}^{\text{hot}} + \mathcal{L}^{\text{der}} + \mathcal{L}^{\text{em}} \\
&= \bar{\psi}(i\gamma_\mu\partial^\mu - m_N)\psi - \frac{1}{2}\alpha_S(\bar{\psi}\psi)(\bar{\psi}\psi) - \frac{1}{2}\alpha_V(\bar{\psi}\gamma_\mu\psi)(\bar{\psi}\gamma^\mu\psi) - \frac{1}{2}\alpha_{TV}(\bar{\psi}\vec{\tau}\gamma_\mu\psi)\cdot(\bar{\psi}\vec{\tau}\gamma^\mu\psi) \\
&\quad - \frac{1}{2}\alpha_{TS}(\bar{\psi}\vec{\tau}\psi)\cdot(\bar{\psi}\vec{\tau}\psi) - \frac{1}{3}\beta_S(\bar{\psi}\psi)^3 - \frac{1}{4}\gamma_S(\bar{\psi}\psi)^4 - \frac{1}{4}\gamma_V[(\bar{\psi}\gamma_\mu\psi)(\bar{\psi}\gamma^\mu\psi)]^2 \\
&\quad - \frac{1}{2}\delta_S\partial_\nu(\bar{\psi}\psi)\partial^\nu(\bar{\psi}\psi) - \frac{1}{2}\delta_V\partial_\nu(\bar{\psi}\gamma_\mu\psi)\partial^\nu(\bar{\psi}\gamma^\mu\psi) - \frac{1}{2}\delta_{TV}\partial_\nu(\bar{\psi}\vec{\tau}\gamma_\mu\psi)\cdot\partial^\nu(\bar{\psi}\vec{\tau}\gamma^\mu\psi) \\
&\quad - \frac{1}{2}\delta_{TS}\partial_\nu(\bar{\psi}\vec{\tau}\psi)\partial^\nu(\bar{\psi}\gamma^\mu\psi) - \frac{1}{4}F_{\mu\nu}F^{\mu\nu} - e\bar{\psi}\gamma^\mu\frac{1-\tau_3}{2}A_\mu\psi.
\end{aligned}
\tag{9.1}
$$

It includes the Lagrangian density $\mathcal{L}^{\text{free}}$ for free nucleons, the four-fermion point-coupling terms $\mathcal{L}^{\text{4f}}$, the higher-order terms $\mathcal{L}^{\text{hot}}$ accounting for the medium effects, the derivative terms $\mathcal{L}^{\text{der}}$ to simulate the finite-range effects that are crucial for a quantitative description of nuclear density distributions, and the electromagnetic interaction terms $\mathcal{L}^{\text{em}}$. Thus one can build the energy density functional for a nuclear system,

$$
\begin{aligned}
E_{\text{tot}} &= \int d^3r \left\{ \sum_{k=1}^{A} \bar{\psi}_k(r)(-i\gamma\cdot\nabla + m_N)\psi_k(r) + \frac{\alpha_S}{2}\rho_S(r)^2 + \frac{\beta_S}{3}\rho_S(r)^3 + \frac{\gamma_S}{4}\rho_S(r)^4 \right. \\
&\quad + \frac{\delta_S}{2}\rho_S(r)\Delta\rho_S(r) + \frac{\alpha_V}{2}j^\mu(r)j_\mu(r) + \frac{\gamma_V}{4}(j^\mu(r)j_\mu(r))^2 + \frac{\delta_V}{2}j^\mu(r)\Delta j_\mu(r) \\
&\quad \left. + \frac{\alpha_{TV}}{2}j_{TV}^\mu(r)\cdot[j_{TV}(r)]_\mu + \frac{\delta_{TV}}{2}j_{TV}^\mu(r)\cdot\Delta[j_{TV}(r)]_\mu + \frac{e^2}{2}j_p^\mu(r)A_\mu(r) \right\}.
\end{aligned}
\tag{9.2}
$$

The local densities and currents ρ_S, j^μ, j_{TV}^μ and j_p^μ are given by

$$
\rho_S(r) = \sum_{k=1}^{A} \bar{\psi}_k(r)\psi_k(r),
\tag{9.3a}
$$

$$
j^\mu(r) = \sum_{k=1}^{A} \bar{\psi}_k(r)\gamma^\mu\psi_k(r),
\tag{9.3b}
$$

$$
j_{TV}^\mu(r) = \sum_{k=1}^{A} \bar{\psi}_k(r)\gamma^\mu\tau_3\psi_k(r),
\tag{9.3c}
$$

$$
j_p^\mu(r) = \sum_{k=1}^{A} \bar{\psi}_k(r)\gamma^\mu\frac{1-\tau_3}{2}\psi_k(r),
\tag{9.3d}
$$

where τ_3 is the isospin Pauli matrix with the eigenvalues $+1$ for neutrons and -1 for protons.

For nuclear chirality, the functional is transformed into a body-fixed frame rotating with a constant angular velocity vector ω, which is along an arbitrary direction in space, i.e, the three-dimensional TAC-CDFT [53]. The corresponding Kohn-Sham equation for nucleons is a static equation in the rotating frame

$$
[\alpha\cdot(\hat{p}-V)+V^0+\beta(m_N+S)-\omega\cdot\hat{J}]\psi_k(r) = \varepsilon_k\psi_k(r)
\tag{9.4}
$$

Here, m_N is the nucleon mass, and the scalar $S(r)$ and four-vector $V^\mu(r)$ potentials are determined by the densities and currents [see Eq. (9.3)], and the explicit expressions for the potentials read

$$S(r) = \alpha_S \rho_S + \beta_S \rho_S^2 + \gamma_S \rho_S^3 + \delta_S \Delta \rho_S, \tag{9.5a}$$

$$V^\mu(r) = \alpha_V j^\mu + \gamma_V (j^\mu j_\mu) j^\mu + \delta_V \Delta j^\mu + \tau_3 \alpha_{TV} j_{TV}^\mu + \tau_3 \delta_{TV} \Delta j_{TV}^\mu + e \frac{1 - \tau_3}{2} A^\mu, \tag{9.5b}$$

where A^μ is the electromagnetic vector potential determined by Poisson's equation. The magnitude of the angular velocity ω is connected to the angular momentum quantum number I by the semiclassical relation $\langle \hat{J} \rangle \cdot \langle \hat{J} \rangle = I(I+1)$, and its orientation is determined by minimizing the total Routhian self-consistently. By solving the Dirac equation Eq. (9.4) self-consistently, one can obtain the single-particle Dirac spinors $\psi_k(r)$.

9.2.2 TIME-DEPENDENT COVARIANT DENSITY FUNCTIONAL THEORY

The time evolution of a Slater determinant, which is the wave function of the nucleus with A nucleons

$$|\Phi(t)\rangle = \prod_{k=1}^{A} c_k^+(t) |-\rangle, \tag{9.6}$$

is modeled by TDCDFT [61, 62]. The initial Slater determinant $|\Phi(t = 0)\rangle$ is a solution of the self-consistent tilted axis cranking mean-field equations, with constraints on the orientation and magnitude of the angular momentum. The corresponding single-particle states $\psi_k(r,t)$ are solutions of the time-dependent Dirac equation

$$i \frac{\partial}{\partial t} \psi_k(r,t) = \hat{h}(r,t) \psi_k(r,t), \tag{9.7}$$

where the single-particle Hamiltonian $\hat{h}(r,t)$ is given by

$$\hat{h}(r,t) = \alpha \cdot (\hat{p} - V(r,t)) + V^0(r,t) + \beta(m_N + S(r,t)). \tag{9.8}$$

Here, the time-dependent scalar $S(r,t)$ and four-vector $V^\mu(r,t)$ potentials are determined by the time-dependent densities and currents in the isoscalar-scalar, isoscalar-vector, and isovector-vector channels. In the dynamic case, we employ the same point-coupling relativistic energy density functional [73] as in the static case, and the explicit expressions for the potentials read

$$S(r,t) = \alpha_S \rho_S + \beta_S \rho_S^2 + \gamma_S \rho_S^3 + \delta_S \Delta \rho_S, \tag{9.9a}$$

$$V^\mu(r,t) = \alpha_V j^\mu + \gamma_V (j^\mu j_\mu) j^\mu + \delta_V \Delta j^\mu + \tau_3 \alpha_{TV} j_{TV}^\mu + \tau_3 \delta_{TV} \Delta j_{TV}^\mu + e \frac{1 - \tau_3}{2} A^\mu. \tag{9.9b}$$

Here, the time-dependent densities and currents are defined in terms of time-dependent single-particle wave functions $\psi_k(r,t)$:

$$\rho_S(r,t) = \sum_{k=1}^{A} \bar{\psi}_k(r,t) \psi_k(r,t), \tag{9.10a}$$

$$j^\mu(r,t) = \sum_{k=1}^{A} \bar{\psi}_k(r,t) \gamma^\mu \psi_k(r,t), \tag{9.10b}$$

$$j_{TV}^\mu(r,t) = \sum_{k=1}^{A} \bar{\psi}_k(r,t) \gamma^\mu \tau_3 \psi_k(r,t). \tag{9.10c}$$

For the following calculated results, the adopted energy density functional is PC-PK1 [73]. The mesh spacing of the lattice is 0.8 fm for all directions, and the box size is $L_x \times L_y \times L_z = 24 \times 24 \times 24$ fm^3. The time-dependent Dirac equation (9.7) is solved with the predictor-corrector method. The step for the time evolution is 0.1 fm/c.

9.3 CHIRAL DYNAMICS OF ^{135}ND

In the following, we will mainly focus on the chirality in the odd-A nucleus ^{135}Nd, which is observed in Ref [22], with the configuration consisting of two proton particles and one neutron hole in the $h_{11/2}$ shell, i.e., $\pi h_{11/2}^2 \otimes v h_{11/2}^{-1}$. And, a new pair of chiral doublet bands was reported with the configuration $\pi[h_{11/2}^1(gd)^1] \otimes v h_{11/2}^{-1}$ [24], which makes it a new candidate for MχD. In Ref. [72], the experimental energies of these two pairs of the chiral doublet bands are well reproduced without any adjustable parameters beyond the well-defined density functional PC-PK1, while the analyses are illustrated only by taking the positive-parity band as examples. Here, we illustrate the analyses with both positive- and negative- parity configurations.

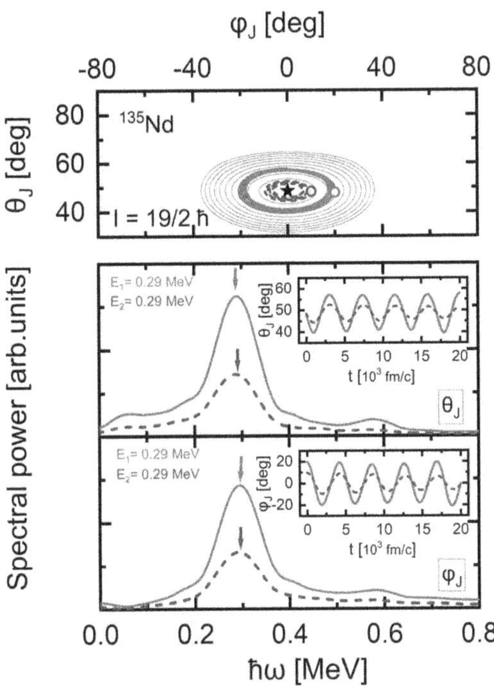

Figure 9.1 The top panel shows trajectories of the tilted angles (θ_J, ϕ_J) for the total angular momentum in the body-fixed frame by TDCDFT calculations starting from the initial states with spin $I = 19/2\hbar$ obtained from TAC-CDFT calculations with the configuration $\pi[h_{11/2}^1(gd)^1] \otimes v h_{11/2}^{-1}$. The solid and dash lines denotes two trajectories starting at two different initial points, whose (θ_J, ϕ_J) are represented by the open circles. Contours join points on the surface with the same energy, and the contour interval is 0.02 MeV. The black star denotes the location of the minimum in the energy surface. The middle (lower) panel shows spectral powers obtained by Fourier analysis on the corresponding time evolutions of tilted angle θ_J (ϕ_J). The arrows indicate the centroids energies. Insets show time evolutions of the tilted angle θ_J (ϕ_J) starting at corresponding initial states.

In the top panel of Fig. 9.1, the total energy surface is depicted in the plane of the tilted angle θ_J and ϕ_J of the total angular momentum with the configuration $\pi[h_{11/2}^1(gd)^1] \otimes v h_{11/2}^{-1}$ at $I = 19/2\hbar$. Here, θ_J is the polar angle between the total angular moment and long axis, and ϕ_J is the azimuth angle between the projection of the total angular momentum in the short-intermediate plane and the short axis. The black star denotes the location of the minima, $(\theta_J, \phi_J) = (48.5°, 0°)$, in the energy surface. This means the total angular momentum, which is mainly contributed by the proton particle and neutron hole, is in the short-intermediate plane.

Staring from the constraint TAC-CDFT solutions away from the minimum point, one can obtained the dynamical evolution of the system by means of the time dependent wave functions

$\psi_k(r,t)$. Note that the solutions of TDCDFT are in the space fixed frame, where the angular momentum is conserved during the time evolution, but the nuclear density distribution varies. Since the nuclear deformation barely changes during the time evolution, it is convenient to analyze the dynamical behavior of the nucleus in the body-fixed rotating frame, which can be easily obtained by calculating the principal axes of inertia, i.e., the short, intermediate, and long axes of the nucleus. At the initial time, the axes of the space fixed frame are set to be along with the three principal axes of inertia. The initial states corresponding to open circles are obtained by constraining the azimuth angle ϕ_J from equilibrium state to $10°$ and $20°$ respectively. The trajectories of the tilted angle θ_J and ϕ_J, represented by the dash line and solid line, are roughly ellipses centered at the location of the minima point in the energy surface. The widths of two ellipses are about $10°$ (dash) and $20°$ (solid), and the heights are about $9°$ (dash) and $18°$ (solid) in the (θ_J, ϕ_J) plane. This is a clearly precession picture, i.e., the angular momentum vector itself is around the tilted axis in the short-long plane. The oscillation amplitude of the ϕ_J is larger than the one of θ_J, because of the oblate ellipse shape of energy surface. In the ϕ_J direction, it involves both positive and negative values, which reflects the chiral nature of the precession due to the transition between the left- and right-handed sectors. Therefore, we call it "chiral precession" [72], which shows the microscopic dynamics of the chiral excitations. The spectral powers, obtained from the oscillations of θ_J with Fourier analysis [74], are shown in the middle panel of Fig. 9.1, and the time evolutions of θ_J are shown in the insert. The motions of θ_J starting at two different initial points are highly harmonic, and the corresponding spectral powers show single peak structures with same centroids. The lower panel of Fig. 9.1 shows the spectral powers, which are obtained from the oscillations of ϕ_J with Fourier analysis, and the insert shows the time evolutions of ϕ_J. One can find the motions of ϕ_J with two different initial points are also highly harmonic, and the centroids of peaks are same as the ones of θ_J. So, the motions of (θ_J, ϕ_J) are harmonic and correlated, which demonstrates the harmonic nature of chiral procession, and the centroids of the peaks correspond to the chiral excitation energy $E^* = 0.29$ MeV at $I = 19/2\hbar$.

In order to derive the chiral excitation energies between the whole chiral doublet bands, the operations at spin $I = 19/2\hbar$ are performed again for other half-integer spins. The initial states for other half-integer spins are also obtained by only constraining the azimuth angle ϕ_J from equilibrium state to $10°$ and $20°$. As shown in the top panel of Fig. 9.2, the trajectories with two different initial points at spin $I = 21/2\hbar$ are obtained by TDCDFT. The black star denotes the position of the equilibrium state with $(\theta_J, \phi_J) = (50°, 0°)$. The trajectories are also oblate ellipses, whose widths are same as the ones at $I = 19/2\hbar$, but the heights are smaller. The frequencies for the oscillations of θ_J and ϕ_J starting at two different initial points are same, and their spectral powers, shown in the middle and lower panels respectively, have single peak structures. All the peaks have same centroids and the corresponding chiral excitation energy is about 0.26 MeV, which is smaller than the one at $I = 19/2\hbar$.

As shown in the top panel of Fig. 9.3, the trajectories at $I = 23/2\hbar$ are also ellipses, whose widths are same as the ones at $I = 21/2\hbar$, but the heights are smaller. The trajectories roughly rotate around the black star $(\theta_J, \phi_J) = (51°, 0°)$, which is the position of the equilibrium state. The time evolutions of θ_J and ϕ_J starting at two different initial points (see the insets in Fig. 9.3) also have same frequencies. It is more clearly to see the spectral powers shown in the middle and lower panels of Fig. 9.3 respectively. And their corresponding chiral excitation energy is about 0.24 MeV, which is smaller than the one at $I = 21/2\hbar$.

As shown in the top panel of Fig. 9.4, the trajectories starting at two different initial states with $I = 25/2\hbar$ are also ellipses. Their widths are same as the ones at $I = 23/2\hbar$, but heights are smaller. The trajectories are along the iso-energy curve and roughly rotate around the black star $(\theta_J, \phi_J) = (52°, 0°)$, which is the position of the equilibrium state. Comparing with the energy surface at $I = 19/2\hbar$, the energy surface becomes softer in the ϕ_J direction and stiffer in the θ_J direction. It leads to constriction of the trajectories in the θ_J direction. The time evolutions of θ_J and ϕ_J, shown in

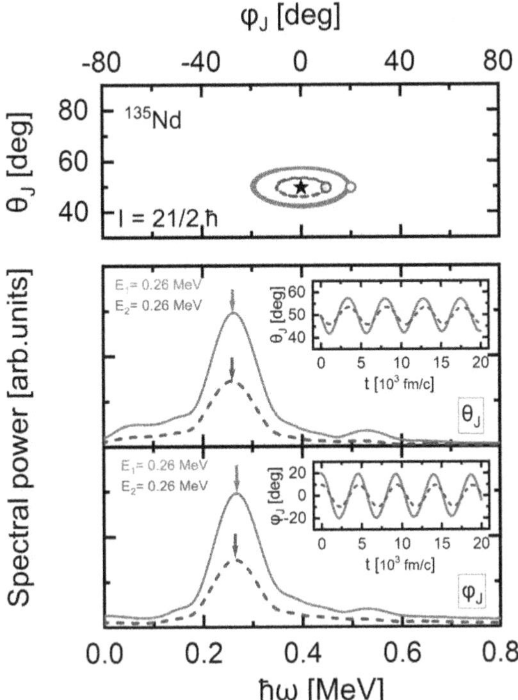

Figure 9.2 Same as Fig. 9.1, but with spin $I = 21/2\hbar$.

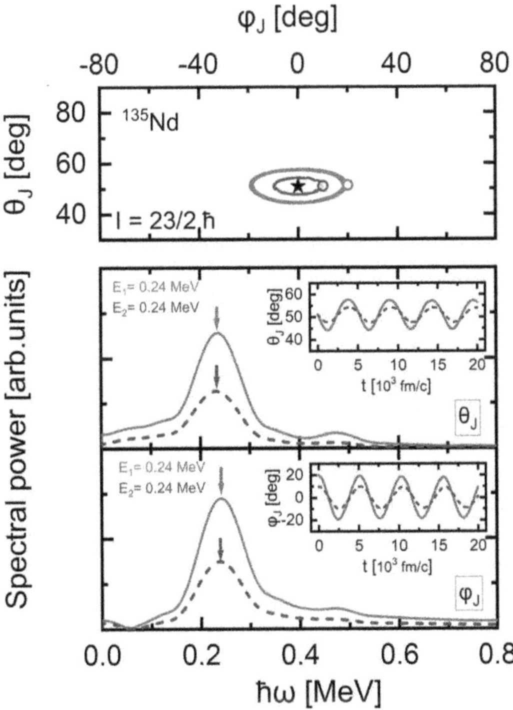

Figure 9.3 Same as Fig. 9.1, but with spin $I = 23/2\hbar$.

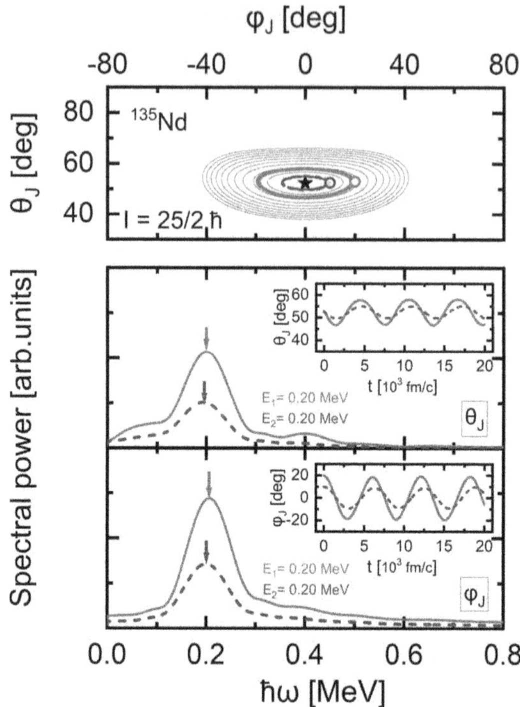

Figure 9.4   Same as Fig. 9.1, but with spin $I = 25/2\hbar$.

the insets of the middle and lower panels of Fig. 9.4 respectively, have same frequencies. It is more clearly to see their spectral powers. The peaks have the same centroids and their corresponding chiral excitation energy is about 0.20 MeV, which is smaller than the one at $I = 23/2\hbar$.

As shown in the top panel of Fig. 9.5, the trajectories at $I = 27/2\hbar$ are also ellipses, whose widths are same as the ones at $I = 25/2\hbar$, but heights are smaller. The trajectories roughly rotate around the black star $(\theta_J, \phi_J) = (54°, 0°)$, which is the position of the equilibrium state. The time evolutions of θ_J and ϕ_J starting at two different initial points (see the insets in the middle and lower panel of Fig. 9.5) have same frequencies. But, from the corresponding spectral powers, there are about 0.01 MeV differences between the centroids of peaks for trajectories starting at two different initial states. It is corresponding to the increasingly anharmonic effects during planar-aplanar transition.

Increasing spin to $29/2\hbar$, there is only one minimal point $(\theta_J, \phi_J) = (56°, 0°)$ on the energy surface (see the top panel of Fig. 9.6), which means the static chirality does not appear. But, the trajectories starting at two different initial points are quite flat, because the energy surface is extremely soft in the ϕ_J direction. The time evolutions of (θ_J, ϕ_J) starting at different initial points become anharmonic and uncorrelated. It is more clearly to see the spectral powers in the middle and low panel of Fig. 9.6. The differences between the centroids of the peaks are about 0.03 MeV, larger than the ones at lower spins. This is due to the increasing anharmonic effects.

As shown in the top panel of Fig. 9.7, there are two degenerate energy minima, separated by a barrier at $\phi_J = 0°$, are at $(59°, 25°)$ and $(59°, -25°)$ in the (θ_J, ϕ_J) plane. Since static chirality appears at $I = 31/2\hbar$, the trajectories strongly depend on the choice of the initial states. If the initial energy is lower than the barrier, the trajectory is limited in the neighborhood of the minimum with a finite ϕ_J corresponding to either the left- or the right-handed sector. However, if the initial energy is higher than the barrier, the trajectory could extend across both positive and negative ϕ_J values. As seen in the inset of the middle panel or the lower panel in the Fig 9.7, the two precessions are disparate not only in the amplitudes but also in the periods. This indicates the obvious anharmonic

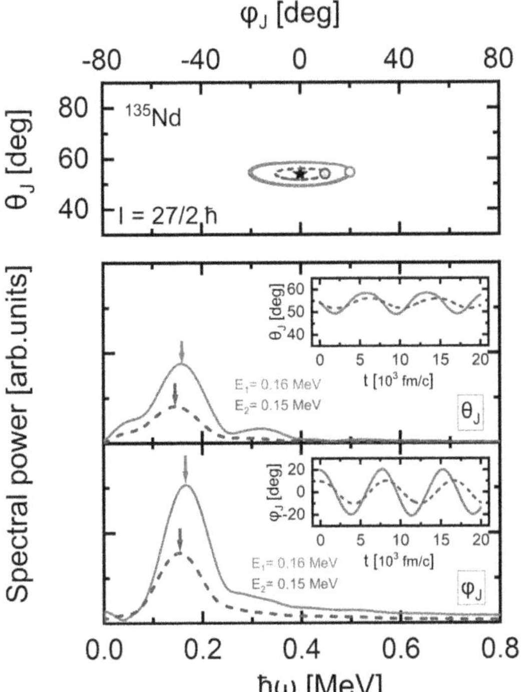

Figure 9.5   Same as Fig. 9.1, but with spin $I = 27/2\hbar$.

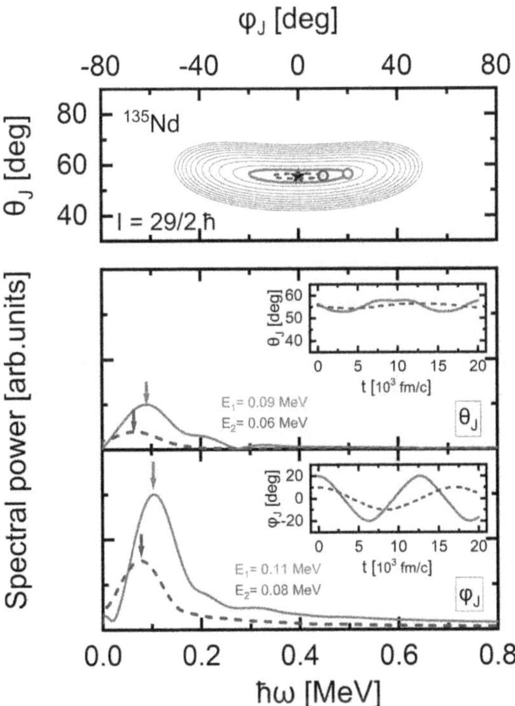

Figure 9.6   Same as Fig. 9.1, but with spin $I = 29/2\hbar$.

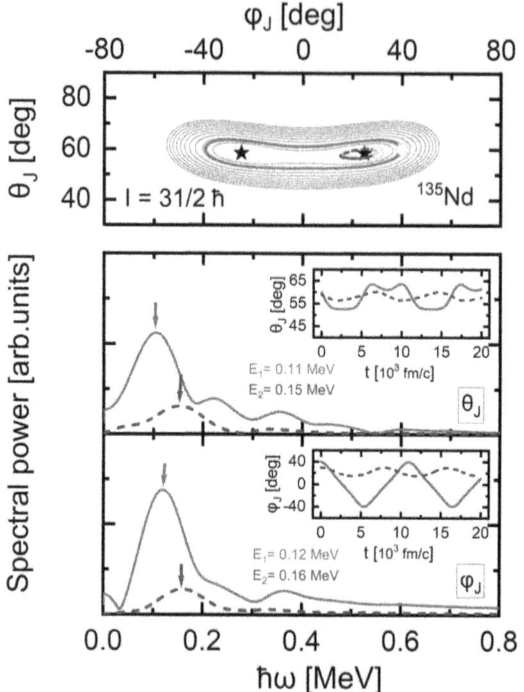

Figure 9.7   Same as Fig. 9.1, but with spin $I = 31/2\hbar$.

effects of the chiral motion, which are associated with quantum tunneling between left- and right-handed sectors. To understand this phenomenom further, to resort to the so-called requantization of the TDCDFT [75], e.g., the time-dependent generator coordinate method. However, to our knowledge, it is hard, if not impossible, to be implemented in a realistic microscopic description of nuclei due to the inaccessible computational requirements. One of the ways out is probably to derive a collective Hamiltonian with Gaussian overlap approximation (GOA) in the space of the tilted angles [40, 41]. However, there are still many open questions about the quality of the GOA, the determination of the mass inertia, the influence of the time-reversal symmetry breaking, etc. The solutions of these questions need to be further studied.

For the chiral doublet bands with configuration $\pi[h_{11/2}^2] \otimes \nu h_{11/2}^{-1}$, chiral excitation energies can also be obtained by TD-CDFT. From the $I = 27/2\hbar$ to $I = 41/2\hbar$, all energy minima are at $\theta_J = 0°$ and static chirality dose not appear. The behaviors of chiral procession in configuration $\pi[h_{11/2}^2] \otimes \nu h_{11/2}^{-1}$ are similar with the ones in configuration $\pi[h_{11/2}^1(gd)^1] \otimes \nu h_{11/2}^{-1}$ at the spins without static chirality. At $I = 27/2\hbar$, shown in the top panel of Fig. 9.8, the trajectories with two different initial states are also ellipses. The widths of two elliptical trajectories are about 10° (dash) and 20° (solid) respectively, and the heights are about 10° (dash) and 20° (solid) respectively in the (θ_J, ϕ_J) plane. The trajectories roughly rotate around the black star $(\theta_J, \phi_J) = (63°, 0°)$, which is the position of the equilibrium state. The time evolutions of θ_J and ϕ_J starting at two different initial points (see the insets in the middle and lower panel of Fig. 9.8) have same frequencies. It is more clearly to see their spectral powers shown in the middle and lower panels of Fig. 9.8. The peaks have same centroids, and the corresponding chiral excitation energy is about 0.29 MeV.

As shown in the top panel of Fig. 9.9, the trajectories starting at two different initial states are also ellipses with $I = 29/2\hbar$, whose widths are same as the ones at $I = 27/2\hbar$, but heights are smaller. The trajectories roughly rotate around the black star $(\theta_J, \phi_J) = (64°, 0°)$, which is the position of the equilibrium state. The time evolutions of θ_J and ϕ_J, shown in the insets of the middle and lower

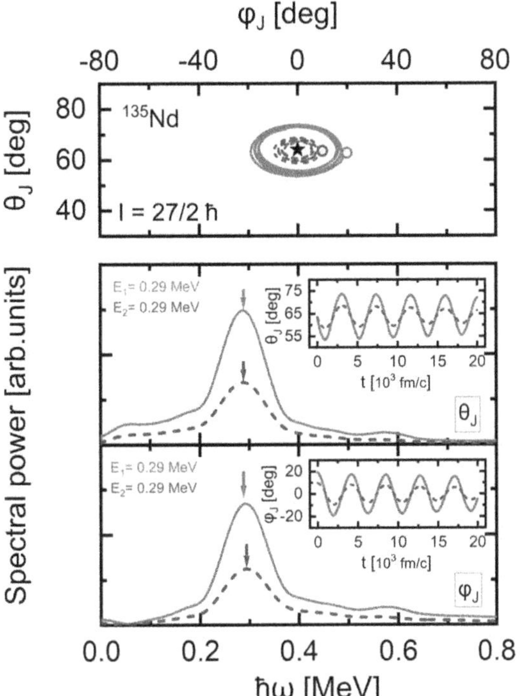

Figure 9.8   Same as Fig. 9.1, but with spin $I = 27/2\hbar$ and configuration $\pi[h_{11/2}^2] \otimes \nu h_{11/2}^{-1}$.

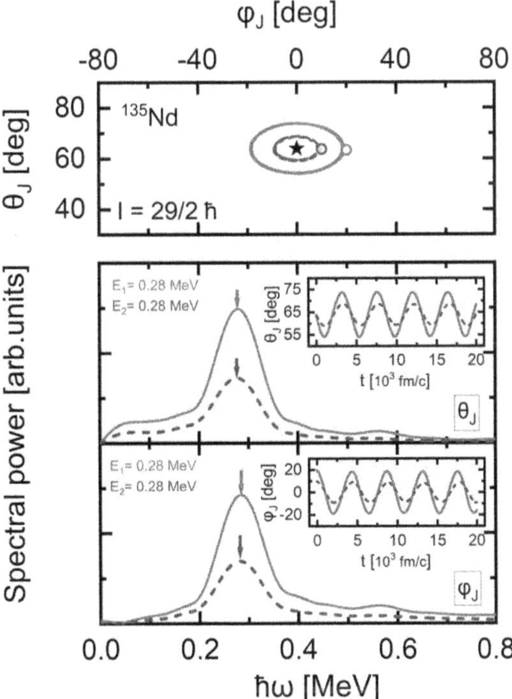

Figure 9.9   Same as Fig. 9.8, but with spin $I = 29/2\hbar$.

panels of Fig. 9.9 respectively, have the same frequencies. It is more clearly to see their spectral powers. The peaks have the same centroids and its chiral excitation energy is about 0.28 MeV, which is smaller than the one at $I = 27/2\hbar$.

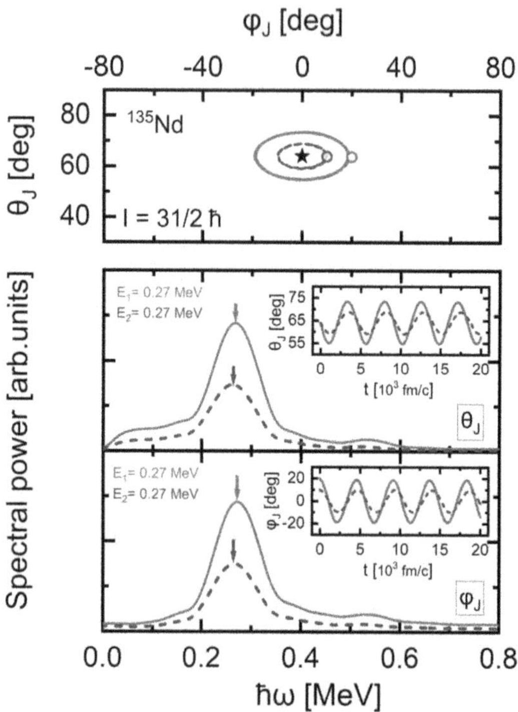

Figure 9.10   Same as Fig. 9.8, but with spin $I = 31/2\hbar$.

As shown in the top panel of Fig. 9.10, the trajectories starting at two different initial states with $I = 31/2\hbar$ are also ellipses, whose widths are same as the ones at $I = 29/2\hbar$, but heights are smaller. The trajectories roughly rotate around the black star $(\theta_J, \phi_J) = (65°, 0°)$, which is the position of the equilibrium state. The time evolutions of θ_J and ϕ_J, shown in the insets of the middle and lower panels of Fig. 9.10 respectively, have same frequencies, It is more clearly to see their spectral powers. The peaks have same centroids and its chiral excitation energy is about 0.27 MeV, which is smaller than the one at $I = 29/2\hbar$.

As shown in the top panel of Fig. 9.11, the trajectories starting at two different initial states with $I = 33/2\hbar$ are also ellipses, whose widths are same as the ones at $I = 31/2\hbar$, but heights are smaller. The trajectories roughly rotate around the black star $(\theta_J, \phi_J) = (65.5°, 0°)$, which is the position of the equilibrium state. The time evolutions of θ_J and ϕ_J, shown in the insets of the middle and lower panels of Fig. 9.11 respectively, have same frequencies. It is more clearly to see their spectral powers. The peaks have same centroids and its excitation energy is about 0.25 MeV, which is smaller than the one at $I = 31/2\hbar$.

As shown in the top panel of Fig. 9.12, the trajectories starting at two different initial states with $I = 35/2\hbar$ are also ellipses, whose widths are same as the ones at $I = 33/2\hbar$, but heights are smaller. The trajectories roughly rotate around the black star $(\theta_J, \phi_J) = (66°, 0°)$, which is the position of the equilibrium state. The time evolutions of θ_J and ϕ_J, shown in the insets of the middle and lower panels of Fig. 9.12 respectively, have same frequencies. It is more clearly to see their spectral powers. The peaks have same centroids and its excitation energy is about 0.23 MeV, which is smaller than the one at $I = 33/2\hbar$.

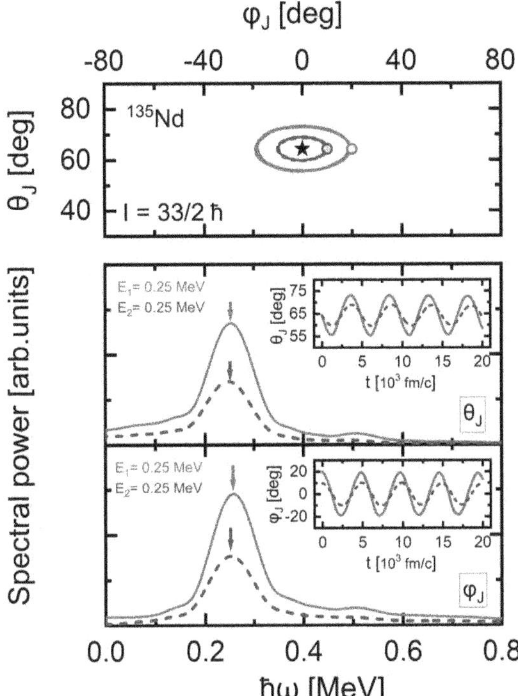

Figure 9.11 Same as Fig. 9.8, but with spin $I = 33/2\hbar$.

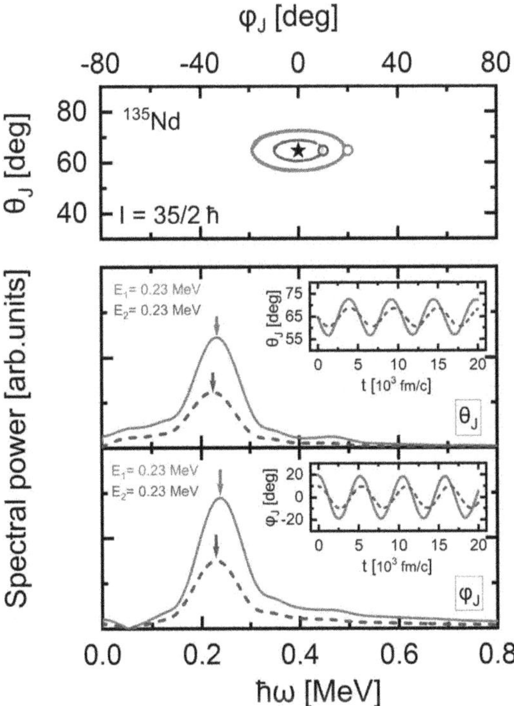

Figure 9.12 Same as Fig. 9.8, but with spin $I = 35/2\hbar$.

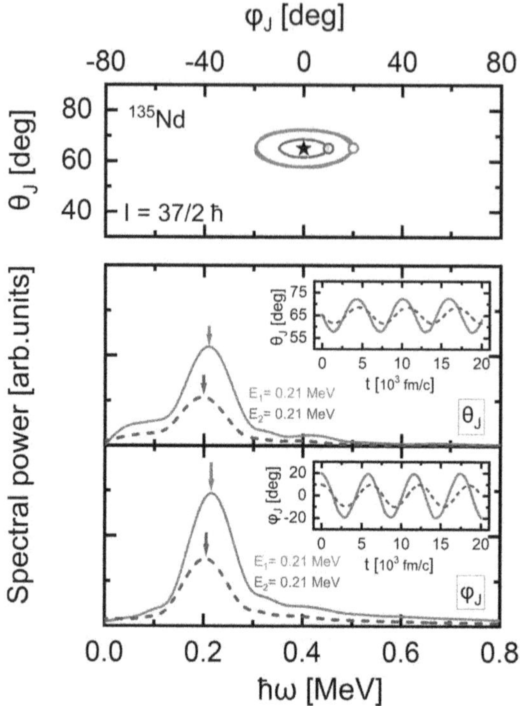

Figure 9.13 Same as Fig. 9.8, but with spin $I = 37/2\hbar$.

As shown in the top panel of Fig. 9.13, the trajectories starting at two different initial states with $I = 37/2\hbar$ are also ellipses, whose widths are same as the ones at $I = 35/2\hbar$, but heights are smaller. The trajectories roughly rotate around the black star $(\theta_J, \phi_J) = (66.5°, 0°)$, which is the position of the equilibrium state. The time evolutions of θ_J and ϕ_J, shown in the insets of the middle and lower panels of Fig. 9.13 respectively, have same frequencies. It is more clearly to see their spectral powers. The peaks have the same centroids and its excitation energy is about 0.21 MeV, which is smaller than the one at $I = 35/2\hbar$.

As shown in the top panel of Fig. 9.14, the trajectories starting at two different initial states with $I = 39/2\hbar$ are also ellipses, whose widths are same as the ones at $I = 37/2\hbar$, but heights are smaller. The trajectories roughly rotate around the black star $(\theta_J, \phi_J) = (67°, 0°)$, which is the position of the equilibrium state. The time evolutions of θ_J and ϕ_J with two different initial points (see the insets in the middle and lower panel of Fig. 9.14) have the similar frequencies. But, from the spectral powers, one can clearly see that there are about 0.01 MeV differences between the centroids of peaks for trajectories starting at two different initial states. It is corresponding to the increasingly anharmonic effects during planar-aplanar transition.

Increasing spin to $41/2\hbar$, there is only one minimal point $(\theta_J, \phi_J) = (68°, 0°)$ on the energy surface (see the top panel of Fig. 9.15), which means static chirality does not appear. The trajectories starting at two different initial points are quite flat in the ϕ_J direction. The time evolutions of (θ_J, ϕ_J) for two different initial points become anharmonic and uncorrelated. It is more clearly to see the spectral powers in the middle and lower panels of Fig. 9.15. The differences between the centroids of the peaks are about 0.03 MeV, which are larger than ones at lower spins and the anharmonic effects become larger.

The energies of chiral excitation at given spins, which are extracted by the Fourier analysis, and the TAC-CDFT solutions are shown in the Fig. 9.16. The calculated energies for two pairs of chiral doublet bands in ^{135}Nd, which are built, respectively, on the configurations $\pi[h^1_{11/2}(gd)^1] \otimes \nu h^{-1}_{11/2}$

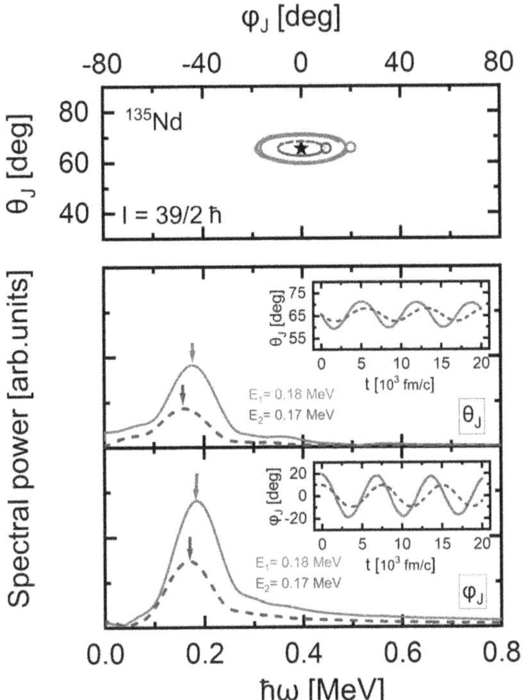

Figure 9.14 Same as Fig. 9.8, but with spin $I = 39/2\hbar$.

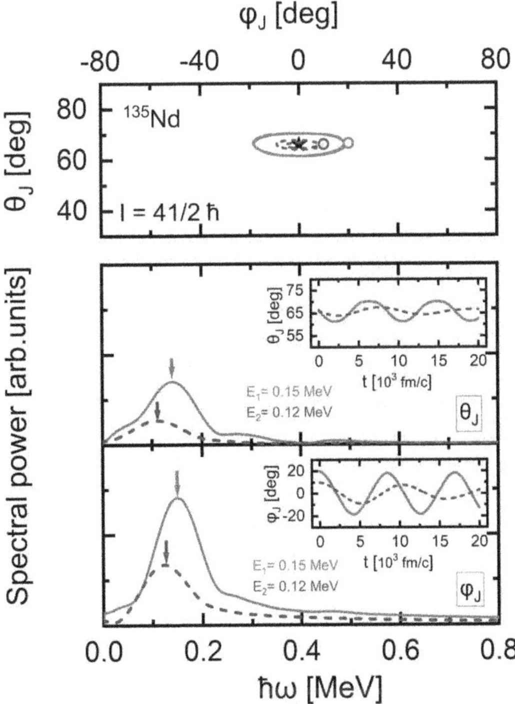

Figure 9.15 Same as Fig. 9.8, but with spin $I = 41/2\hbar$.

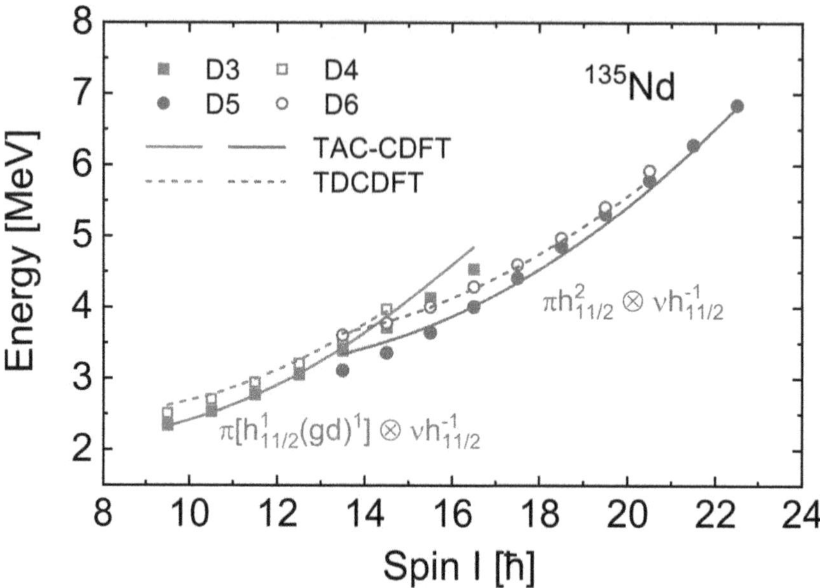

Figure 9.16 Calculated excitation energies (solid and dashed lines) for chiral doublet bands in ^{135}Nd built on the configurations $\pi[h^1_{11/2}(gd)^1] \otimes \nu h^{-1}_{11/2}$ and $\pi h^2_{11/2} \otimes \nu h^{-1}_{11/2}$ in comparison with the data (solid and open symbols) [24]. The solid lines represent the TAC-CDFT results, and the dashed ones represent the results including the chiral excitation energies obtained with the TDCDFT calculations.

and $\pi[h^2_{11/2}] \otimes \nu h^{-1}_{11/2}$, are shown in comparison with the data [24]. It can be seen that the experimental energies are well reproduced. As mentioned in Ref. [72], a fully microscopic and self-consistent description for the chiral doublet bands in the framework of DFTs is thus realized by the TDCDFT and TAC-CDFT.

9.4   SUMMARY

In summary, we illustrate the analyses of dynamic chiral picture of ^{135}Nd built on the positive- and negative- parity configurations with the microscopic time-dependent covariant density functional theory, and the initial states are obtained from the static three dimensional tilted axis cranking solutions. A novel mechanism, chiral precession, as revealed in Ref. [72], can be shown from the microscopic dynamics of the total angular momentum in the body-fixed frame. In the chiral doublet bands with the configuration $\pi[h^1_{11/2}(gd)^1] \otimes \nu h^{-1}_{11/2}$, a clear transition from the planar into aplanar rotations has been shown from the total energy surfaces with fixed spins. The obvious planar-aplanar transition corresponds to the increasingly anharmonic effects in the chiral procession. The experimental energies of the two pairs of the chiral doublet bands in ^{135}Nd, which are extracted by Fourier analysis, are well reproduced without any adjustable parameters beyond the well-defined density functional.

9.5   ACKNOWLEDGMENTS

This work was partly supported by the National Natural Science Foundation of China (Grants No. 12070131001, No. 11935003, No. 11975031, and No. 12141501), and the High-performance Computing Platform of Peking University.

Bibliography

1. S. Frauendorf and J. Meng, Nucl. Phys. A **617**, 131 (1997).
2. S. Frauendorf, Rev. Mod. Phys. **73**, 463 (2001).
3. J. Meng, J. Peng, S. Q. Zhang and S.-G. Zhou, Phys. Rev. C **73**, 037303 (2006).
4. S. Y. Wang, *et al.*, Phys. Lett. B **703**, 40 (2011).
5. C. Liu, *et al.*, Phys. Rev. Lett. **116**, 112501 (2016)
6. C. Vaman, *et al.*, Phys. Rev. Lett. **92**, 032501 (2004)
7. J. Timár, *et al.*, Phys. Lett. B **598**, 178 (2004).
8. P. Joshi, *et al.*, Phys. Lett. B **595**, 135 (2004).
9. J. Timár, *et al.*, Phys. Rev. C **73**, 011301 (2006).
10. D. Tonev, *et al.*, Phys. Rev. Lett. **112**, 052501 (2014).
11. I. Kuti, *et al.*, Phys. Rev. Lett. **113**, 032501 (2014).
12. P. Joshi, *et al.*, Phys. Rev. Lett. **98**, 102501 (2007).
13. Y. Luo, *et al.*, Phys. Lett. B **670**, 307 (2009).
14. E. O. Lieder, *et al.*, Phys. Rev. Lett. **112**, 202502 (2014).
15. N. Rather, *et al.*, Phys. Rev. Lett. **112**, 202503 (2014).
16. E. Grodner, *et al.*, Phys. Rev. Lett. **97**, 172501 (2006).
17. A. D. Ayangeakaa, *et al.*, Phys. Rev. Lett. **110**, 172504 (2013).
18. K. Starosta, *et al.*, Phys. Rev. Lett. **86**, 971 (2001).
19. D. Tonev, *et al.*, Phys. Rev. Lett. **96**, 052501 (2006).
20. C. M. Petrache, G. B. Hagemann, I. Hamamoto and K. Starosta, Phys. Rev. Lett. **96**, 112502 (2006).
21. D. Tonev, *et al.*, Phys. Rev. C **76**, 044313 (2007).
22. S. Zhu, *et al.*, Phys. Rev. Lett. **91**, 132501 (2003).
23. S. Mukhopadhyay, *et al.*, Phys. Rev. Lett. **99**, 172501 (2007).
24. B. F. Lv, *et al.*, Phys. Rev. C **100**, 024314 (2019).
25. C. M. Petrache, *et al.*, Phys. Rev. C **97**, 041304 (2018).
26. D. L. Balabanski, *et al.*, Phys. Rev. C **70**, 044305 (2004).
27. E. A. Lawrie, *et al.*, Phys. Rev. C **78**, 021305 (2008).
28. B. Xiong and Y. Wang, At. Data Nucl. Data Tables **125**, 193 (2019).
29. J. Meng and S. Q. Zhang, J. Phys. G **37**, 064025 (2010).
30. J. Meng and P. Zhao, Phys. Scr. **91**, 053008 (2016).
31. T. Koike, K. Starosta and I. Hamamoto, Phys. Rev. Lett. **93**, 172502 (2004).
32. J. Peng, J. Meng and S. Q. Zhang, Phys. Rev. C **68**, 044324 (2003).
33. S. Q. Zhang, B. Qi, S. Y. Wang and J. Meng, Phys. Rev. C **75**, 044307 (2007).
34. B. Qi, S. Q. Zhang, J. Meng, S. Y. Wang and S. Frauendorf, Phys. Lett. B **675**, 175 (2009).
35. Q. B. Chen, B. F. Lv, C. M. Petrache and J. Meng, Phys. Lett. B **782**, 744 (2018).
36. S. Brant, D. Tonev, G. de Angelis and A. Ventura, Phys. Rev. C **78**, 034301 (2008).
37. A. A. Raduta, A. H. Raduta and C. M. Petrache, J. Phys. G **43**, 095107 (2016).
38. V. I. Dimitrov, S. Frauendorf and F. Dönau, Phys. Rev. Lett. **84**, 5732 (2000).
39. D. Almehed, F. Dönau and S. Frauendorf, Phys. Rev. C **83**, 054308 (2011).
40. Q. B. Chen, S. Q. Zhang, P. W. Zhao, R. V. Jolos and J. Meng, Phys. Rev. C **87**, 024314 (2013).
41. Q. B. Chen, S. Q. Zhang, P. W. Zhao, R. V. Jolos and J. Meng, Phys. Rev. C **94**, 044301 (2016).
42. G. H. Bhat, J. A. Sheikh and R. Palit, Phys. Lett. B **707**, 250 (2012).
43. F. Q. Chen, Q. B. Chen, Y. A. Luo, J. Meng and S. Q. Zhang, Phys. Rev. C **96**, 051303(R) (2017).
44. F. Q. Chen, J. Meng and S. Q. Zhang, Phys. Lett. B **785**, 211 (2018).
45. Y. K. Wang, F. Q. Chen, P. W. Zhao, S. Q. Zhang and J. Meng, Phys. Rev. C **99**, 054303 (2019).
46. S. Frauendorf, Nucl. Phys. A **677**, 115 (2000).
47. P. W. Zhao, J. Peng, H. Z. Liang, P. Ring and J. Meng, Phys. Rev. Lett. **107**, 122501 (2011).

48. J. Meng, J. Peng, S.-Q. Zhang and P.-W. Zhao, Front. Phys. **8**, 55 (2013).

49. B. D. Serot and J. D. Walecka, Adv. Nucl. Phys. **16**, 1 (1986).

50. Z. X. Ren and P.W. Zhao, Phys. Rev. C **102**, 021301(R) (2020).

51. *Relativistic Density Functional for Nuclear Structure, International Review of Nuclear Physics*, edited by J. Meng (World Scientific, Singapore, 2016), Vol. 10.

52. P. Olbratowski, J. Dobaczewski, J. Dudek and W. Płóciennik, Phys. Rev. Lett. **93**, 052501 (2004).

53. P. W. Zhao, Phys. Lett. B **773**, 1 (2017).

54. J. Peng and Q. B. Chen, Phys. Lett. B **810**, 135795 (2020).

55. P. W. Zhao, Y. K. Wang and Q. B. Chen, Phys. Rev. C **99**, 054319 (2019).

56. P. W. Zhao, P. Ring and J. Meng, Phys. Rev. C **94**, 041301 (2016).

57. Y. K. Wang, P. W. Zhao and J. Meng, Phys. Lett. B **848**, 138346 (2024).

58. T. Nakatsukasa, K. Matsuyanagi, M. Matsuo and K. Yabana, Rev. Mod. Phys. **88**, 045004 (2016).

59. C. Simenel and A. S. Umar, Prog. Part. Nucl. Phys. **103**, 19 (2018).

60. P. D. Stevenson and M. C. Barton, Prog. Part. Nucl. Phys. **104**, 142 (2019).

61. Z. X. Ren, P. W. Zhao and J. Meng, Phys. Lett. B **801**, 135194 (2020).

62. Z. X. Ren, P. W. Zhao and J. Meng, Phys. Rev. C **102**, 044603 (2020).

63. B. Li, D. Vretenar, Z. X. Ren, T. Nikšić, J. Zhao, P. W. Zhao and J. Meng, Phys. Rev. C **107**, 014303 (2023).

64. B. Li, D. Vretenar, T. Nikšić, P. W. Zhao and J. Meng, Phys. Rev. C **108**, 014321 (2023).

65. B. Li, D. Vretenar, T. Nikšić, J. Zhao, P. W. Zhao and J. Meng, Front. Phys. **19**, 44201 (2024).

66. Y. Tanimura, K. Hagino and H. Z. Liang, Prog. Theor. Exp. Phys. **2015**, 073D01 (2015).

67. Z. X. Ren, S. Q. Zhang and J. Meng, Phys. Rev. C **95**, 024313 (2017).

68. Z. X. Ren, S. Q. Zhang, P. W. Zhao, N. Itagaki, J. A. Maruhn, and J. Meng, Sci. China: Phys., Mech. Astron. **62**, 112062 (2019).

69. Z. X. Ren, P. W. Zhao, S. Q. Zhang and J. Meng, Nucl. Phys. A **996**, 121696 (2020).

70. B. Li, Z. X. Ren and P. W. Zhao, Phys. Rev. C **102**, 044307 (2020).

71. F. F. Xu, B. Li, Z. X. Ren and P. W. Zhao, Phys. Rev. C **109**, 014311 (2024).

72. Z. X. Ren, P. W. Zhao and J. Meng, Phys. Rev. C **105**, L011301 (2022).

73. P. W. Zhao, Z. P. Li, J. M. Yao and J. Meng, Phys. Rev. C **82**, 054319 (2010).

74. P.-G. Reinhard, P. D. Stevenson, D. Almehed, J. A. Maruhn and M. R. Strayer, Phys. Rev. E **73**, 036709 (2006).

75. P. Ring and P. Schuck, *The Nuclear Many-Body Problem* (Springer Science and Business Media, New York, 2004).

10 Study of wobbling motion within CDFT+PRM approach

Qi-Bo Chen

East China Normal University, Shanghai, China

Hua-Ming Dai

East China Normal University, Shanghai, China

Ying-Zhi Ji

East China Normal University, Shanghai, China

10.1 WOBBLING MOTION

The nuclear wobbling mode, intrinsically linked to the triaxiality of nuclear shape, has obtained significant attention in recent years both experimentally and theoretically. This phenomenon was originally proposed by Bohr and Mottelson [1] for a triaxial rotor, specifically for even-even nuclei without coupling quasiparticles, where the sole angular momentum in the system is the total angular momentum. In this framework, the nucleus undergoes rotation around the principal axis with the largest moment of inertia, typically the intermediate axis, leading to harmonic oscillations about the space-fixed angular momentum vector. The anticipated energy spectra arising from this motion are typified by a sequence of rotational $E2$ bands corresponding to distinct oscillation quanta (n). Transitions between these bands predominantly involve $\Delta I = 1$ transitions exhibiting an $E2$ character, as the wobbling motion stems from the collective motion of the entire triaxial charge density relative to the angular momentum vector.

While Bohr and Mottelson originally described this motion for even-even nuclei without intrinsic angular momentum, the direct experimental verification of this simple form is still lacking. However, the inclusion of angular momentum stemming from intrinsic single-particle motion can significantly enrich and potentially facilitate the observation of the nuclear wobbling mode, particularly in odd-A mass nuclei. When a triaxial rotor is coupled with a high-j valence particle, Frauendorf and Dönau introduced the concept of two distinct wobbling modes: longitudinal wobbling (LW) and transverse wobbling (TW) [2]. In the LW, the angular momentum of the high-j valence particles aligns parallel to the principal axis with the largest moment of inertia, whereas in the TW, it aligns perpendicular to this axis. Recently, Chen and Frauendorf extended on this classification by associating the wobbling motion with the topology of classical orbits visualized through spin coherent state (SCS) maps [3]. The LW involves the total angular momentum J revolving around the axis with the largest moment of inertia, while the TW entails J revolving around an axis perpendicular to this principal axis. The distinction between TW and LW is primarily observed in the excitation energy and $E2$ transition properties. The excitation energy of LW states increases with rising angular momentum, whereas for TW, it decreases. Both TW and LW modes exhibit enhanced $I \rightarrow I - 1$ $E2$ transitions between adjacent wobbling bands.

DOI: 10.1201/9781032691602-10

10.2 EXPERIMENTAL PROGRESS

The nuclear wobbling phenomena observed to date are compiled in Fig. 10.1 together with the observed chiral nuclei, which is another fingprint of triaxial deformation. The majority of the identified wobbling nuclei are characterized by an odd number of protons. Notable examples include [161]Lu [4], [163]Lu [5, 6], [165]Lu [7], [167]Lu [8], [167]Ta [9] and [151]Eu [10] within the $A \approx 160$ mass region, as well as [135]Pr [11, 12], [133]La [13], [130]Ba [14, 15], [127]Xe [16], [133]Ba [17], [136]Nd [18] and [125]Xe [19] in the $A \approx 130$ mass region. Among these, [135]Pr represents the first example of TW observed at normal quadrupole deformation [11, 12], while [130]Ba stands as the first case of two-quasiparticle wobbling bands within an even-even nucleus [14, 15, 20]. In the heavier $A \approx 190$ mass region, the candidates include [187]Au [21] and [183]Au [22]. In the lighter $A \approx 100$ mass region, the sole candidate is [105]Pd, which possesses an odd neutron number $N = 59$ and has been proposed as a wobbling nucleus with the excited state based on a quasi-neutron configuration [23]. Very recently, the wobbling candidate was further reported in the lighter $A \approx 80$ mass region in [74]Br [24]. It is important to recognize that some of these wobblers are subject to ongoing debate [25–32].

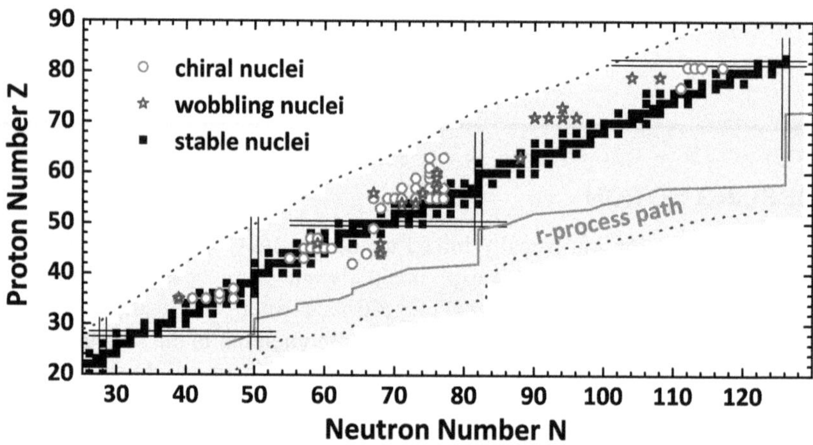

Figure 10.1   Reported chiral and wobbling candidates spreading over the $A \approx 80$, 100, 130, 160 and 190 mass regions.

10.3 THEORETICAL PROGRESS

From a theoretical standpoint, a variety of models have been developed in elucidating the dynamics of nuclear wobbling motion. The triaxial rotor model [1, 33] and the particle rotor model (PRM) [2, 3, 15, 34–38] are among the prominent approaches that have been employed. Moreover, approximation solutions derived from angular momentum coherent states [39–45] have provided valuable insights into the underlying mechanisms of wobbling. The cranking model combined with random phase approximation (RPA) [46–54] has also been pivotal in this regard. Additionally, the collective Hamiltonian framework, which is predicated on the tilted axis cranking model [55, 56] and the semi-classical approach [57], have offered novel perspectives on the wobbling phenomenon. Lastly, the projected shell model [20, 58, 59] has been used to understand the wobbling motion.

By utilizing a framework that recognizes the total angular momentum as a good quantum number, the PRM reinstates rotational symmetry within the system. This approach facilitates a comprehensive quantization of energy levels and electromagnetic transitions. Compared to alternative models, the PRM stands out for its simplicity and reduced computational demand, rendering it particularly advantageous for systematic investigations into wobbling motion. As a result, the present study will adopt the PRM for its analytical framework.

10.3.1 PARTICLE ROTOR MODEL HAMILTONIAN

In our study, we employ the PRM, in which the angular momentum is a good quantum number. Within the PRM framework, the nucleus is decomposed into two distinct components: the triaxial rotor and the valence quasiparticle. The PRM's total Hamiltonian is given by [1]

$$\hat{H}_{\text{PRM}} = \hat{H}_{\text{coll}} + \hat{H}_{\text{intr}}, \tag{10.1}$$

where $\hat{H}_{\text{coll}}$ represents the collective rotor Hamiltonian, expressed as

$$\hat{H}_{\text{coll}} = \sum_{k=1}^{3} \frac{\hat{R}_k^2}{2\mathcal{J}_k} = \sum_{k=1}^{3} \frac{(\hat{I}_k - \hat{j}_k)^2}{2\mathcal{J}_k}, \tag{10.2}$$

with the index k ranging from 1 to 3 corresponds to the three principal axes within the body-fixed reference frame. In this context, $\hat{R}_k$ and $\hat{I}_k$ denote the angular momentum operators for the collective rotor and the entire nucleus, respectively, while $\hat{j}_k$ represents the angular momentum operator for a single valence nucleon. Furthermore, the quantities $\mathcal{J}_k$ correspond to the moments of inertia for the three principal axes.

For the intrinsic Hamiltonian of the valence nucleons, we adopt a single-j shell model, which provides an effective approximation for high-j intruder orbitals and is well-suited for the model analysis in this work. The single-j shell Hamiltonian is defined as

$$\hat{h}_p = \frac{1}{2}C \left\{ \cos\gamma \left(\hat{j}_3^2 - \frac{j(j+1)}{3} \right) + \frac{\sin\gamma}{2\sqrt{3}} (\hat{j}_+^2 + \hat{j}_-^2) \right\}, \tag{10.3}$$

where the angle γ acts as the parameter for triaxial deformation, and the coefficient C is directly proportional to the quadrupole deformation parameter β. Furthermore, the effects of pairing are incorporated through the BCS-quasiparticle energy expression

$$\varepsilon_{\text{intr}}^{(v)} = \sqrt{(e_v - \lambda)^2 + \Delta^2}. \tag{10.4}$$

In this equation, the single-particle energy e_v is derived from the solution of the single-j shell Hamiltonian $\hat{h}_p$. The parameters λ and Δ represent the Fermi surface and the pairing gap, respectively. As λ increases from the bottom to the top of a j-shell, the valence quasiparticle transitions from a particle-like to a hole-like character.

We proceed to diagonalize the PRM Hamiltonian using the BCS one-quasiparticle states for the particle wave functions. This approach limits our investigation to moderate spin states; however, it is adequate for addressing the issues at hand in this paper. Consequently, the total nuclear Hamiltonian $\hat{H}_{\text{PRM}}$ is solved by diagonalization within a strong-coupling basis, given by

$$|IMKjk\rangle = \sqrt{\frac{2I+1}{16\pi^2}} D_{MK}^{I}(\omega)|jk\rangle, \tag{10.5}$$

where $D_{MK}^{I}(\omega)$ represents the standard Wigner functions, which depend on the three Euler angles $\omega = (\psi', \theta', \varphi')$. Here, I is the total angular momentum quantum number for the odd-mass nuclear system (rotor plus particle), while K (M) represents the projection onto the 3-axis of the intrinsic (laboratory) frame. Additionally, k denotes the 3-axis component of the particle's angular momentum j in the intrinsic frame.

The PRM eigenstates $|IMv\rangle$ are subsequently expanded as

$$|IMv\rangle = \sum_{K,k} C^{(v)}_{IKk}|IMKjk\rangle. \tag{10.6}$$

Considering the D_2 symmetry inherent to a triaxial nucleus, the summation over K and k, which range from $-I$ to I and from $-j$ to j, respectively, must satisfy the condition that $K-k$ is even. This symmetry imposes a relationship on half of the coefficients, such that

$$C^{(v)}_{I-K-k} = (-1)^{I-j}C^{(v)}_{IKk}. \tag{10.7}$$

It is also important to note that, in constructing the matrix elements of the PRM Hamiltonian, each single-particle matrix element must be scaled by a pairing factor $u_\mu u_v + v_\mu v_v$, where v is the conventional occupation number and $u^2 + v^2 = 1$. This adjustment accounts for the effects of pairing, which are essential for an accurate representation of the nuclear structure.

10.3.2 ELECTROMAGNETIC TRANSITION PROBABILITIES

The probabilities of electromagnetic transitions, specifically the electric quadrupole ($E2$) and magnetic dipole ($M1$) transitions, can be derived from the PRM wave function using the respective $E2$ and $M1$ operators.

For $E2$ transitions, the pertinent operator is expressed as

$$\mathcal{M}(E2,\mu) = \sqrt{\frac{5}{16\pi}}\hat{Q}_{2\mu}, \tag{10.8}$$

$$\hat{Q}_{2\mu} = D^2_{\mu 0}Q'_{20} + \left(D^2_{\mu 2} + D^2_{\mu -2}\right)Q'_{22}, \tag{10.9}$$

where the intrinsic quadrupole moments are defined as

$$Q'_{20} = Q_0\cos\gamma, \tag{10.10}$$

$$Q'_{22} = \frac{1}{\sqrt{2}}Q_0\sin\gamma, \tag{10.11}$$

and Q_0 represents the intrinsic charge quadrupole moment.

For $M1$ transitions, the corresponding operator is given by

$$\mathcal{M}(M1,\mu) = \sqrt{\frac{3}{4\pi}}\frac{e\hbar}{2Mc}(g_p - g_R)\hat{j}_\mu, \tag{10.12}$$

with $\hat{j}_\mu$ representing the spherical tensor of the quasiparticle's angular momentum in the laboratory frame, defined as

$$\hat{j}_\mu = \left(\hat{j}_0 = \hat{j}_3, \quad \hat{j}_{\pm 1} = \frac{\mp\left(\hat{j}_1 \pm i\hat{j}_2\right)}{\sqrt{2}}\right), \tag{10.13}$$

and the gyromagnetic factors for the quasiparticle g_p and the rotor g_R.

10.4 INFLUENCES OF FERMI SURFACE AND DEFORMATION PARAMETERS ON THE WOBBLING MOTION

It is well-established that two essential conditions must be met for the emergence of wobbling or chiral modes in nuclear systems. These conditions are significant triaxial deformation and the presence of high-j particles and/or holes in the nuclear configuration. The degree of triaxial deformation is crucial in determining the very existence of wobbling [1] or chiral modes [60], whereas the specific configuration of particles and holes dictates the classification of the wobbling mode [2, 3] or the chiral mode. For example, from a qualitative standpoint, a triaxially deformed configuration that features particles at the bottom of the deformed j-shell or holes at the apex of the shell is characteristic of a TW mode. Conversely, the presence of quasiparticles in the mid-shell is indicative of a LW mode. In the subsequent analysis, we will present a quantitative assessment of how the Fermi surface affects the wobbling motion.

The energy spectra for the yrast bands, designated as B1 for the signature $\alpha = -1/2$ and B2 for $\alpha = +1/2$, are calculated for various Fermi surfaces, namely $\lambda = e_1, e_2, e_3, e_4, e_5$ and e_6, for the $\pi(1h_{11/2})^1$ configuration. These spectra are plotted with respect to a rigid-rotor reference in Fig. 10.3 for two sets of triaxial deformation parameters: $\gamma = 30°$ and $\gamma = 20°$ [61].

The energy difference between bands B2 and B1 is computed using the formula

$$\Delta E = E_{B2}(I) - \frac{1}{2}[E_{B1}(I+1) + E_{B1}(I-1)]. \tag{10.14}$$

The results are displayed in Figs. 10.2(a) and 10.2(b).

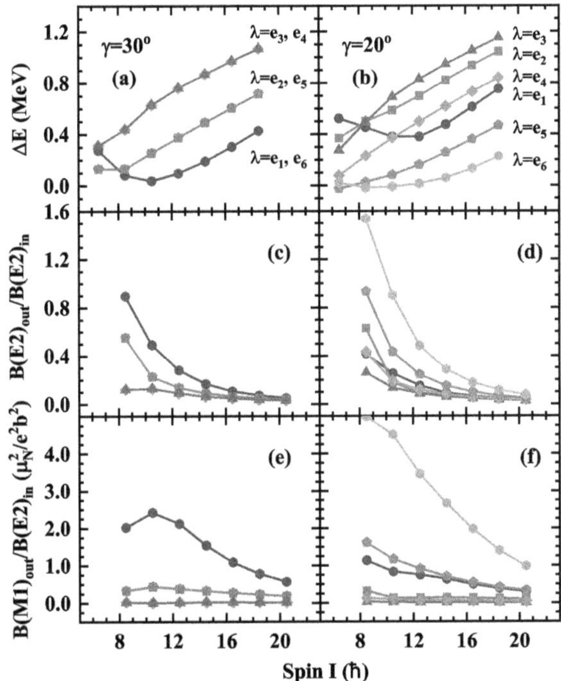

Figure 10.2   The energy difference between the bands B2 and B1 ΔE as well as the ratios of $B(E2)_{out}/B(E2)_{in}$ and $B(M1)_{out}/B(E2)_{in}$ calculated by PRM with the Fermi surface $\lambda = e_1, e_2, e_3, e_4, e_5$ and e_6 for $\gamma = 30°$ and $\gamma = 20°$. Here, the subscript "out" indicates the $\Delta I = 1$ connecting transitions from band B2 to B1, while "in" the $\Delta I = 2$ transitions that connect the band B2. Adapted from Ref. [61].

It is important to observe that the total Hamiltonian used in this study remains invariant under the transformation $\gamma \to 60° - \gamma$ and $\lambda = e_\nu \to \lambda = e_{7-\nu}$. Consequently, for $\gamma = 30°$, the energy spectra and their corresponding differences for Fermi surfaces e_ν and $e_{7-\nu}$ are found to be identical (compare 10.2(a)).

As the Fermi surface λ transitions from e_1 to e_6, the coupled valence particle evolves from the base to the apex of the $h_{11/2}$ shell. This evolution signifies a transition in the particle's characteristics from particle-like to hole-like. When $\lambda = e_1$, corresponding to a particle configuration, bands B2 and B1 exhibit a pattern of approaching and then separating with increasing spin for both $\gamma = 30°$ and $20°$. Consequently, the energy difference ΔE between these bands initially diminishes and subsequently increases. The critical spin values at which this transition occurs are $I = 21/2\hbar$ for $\gamma = 30°$ and $I = 25/2\hbar$ for $\gamma = 20°$. The higher critical spin value observed for $\gamma = 20°$ is attributed to the larger ratio of moments of inertia between the short (s) and intermediate (m) axes, $\mathcal{J}_s/\mathcal{J}_m$.

For a nucleus exhibiting wobbling behavior, the energy difference ΔE is interpreted as the wobbling energy. As highlighted in Ref. [2], a decreasing ΔE is indicative of a TW mode, whereas an increasing ΔE corresponds to a LW mode. Therefore, the calculated results suggest that the wobbling mode undergoes a transition from transverse to longitudinal as the spin increases. This type of wobbling system has been observed experimentally, as exemplified by the case of ^{135}Pr [11].

When the Fermi surface is at $\lambda = e_6$, corresponding to a hole configuration, bands B2 and B1 exhibit a similar behavior: they approach each other and then separate as the spin increases for both $\gamma = 30°$ and $20°$. As previously mentioned, for $\gamma = 30°$, the energy spectra and critical spin values are identical to those when $\lambda = e_1$, with the critical spin being $I = 21/2\hbar$. However, for $\gamma = 20°$, the critical spin is lower, reaching only $I = 17/2\hbar$. This reduction in critical spin is attributed to the decrease in the ratio $\mathcal{J}_l/\mathcal{J}_m$ from 0.25 at $\gamma = 30°$ to approximately 0.12 at $\gamma = 20°$.

It is worth noting that, as discussed in Ref. [2], the existence of TW in a hole configuration was theoretically predicted. However, to date, no corresponding experimental evidence has been documented. The present study confirms the existence of TW in a hole configuration, thereby suggesting that further experimental research in this area is desirable.

When the Fermi surface deviates from a pure particle or hole configuration, specifically within the range $\lambda = e_2$ to e_5, the energy difference ΔE increases with spin, as depicted in Figs. 10.2(a)–(b), for both $\gamma = 30°$ and $20°$ cases. Notably, when $\lambda = e_3$, ΔE reaches its maximum value. This increasing trend of ΔE is characteristic of a LW mode, as indicated in Ref. [2]. In contrast to the scenario where $\lambda = e_1$ and LW modes emerge at high spin values, the LW mode in this instance is observed at lower spin values. This particular manifestation of LW has indeed been observed experimentally, as exemplified by the case of ^{187}Au [21]. In this instance, the wobbling band is established on a $h_{9/2}$ configuration, with the Fermi surface situated near the second energy level of the $h_{9/2}$ shell.

A definitive experimental indicator of wobbling phenomena in electromagnetic transition probabilities is the observed collective enhancement of $B(E2)$ values for the $n \to n-1$ transitions between adjacent wobbling excitations. This enhancement is a result of the wobbling motion of the entire charged nucleus. In contrast, for signature partner bands, the connecting transitions are predominantly of the magnetic dipole ($M1$) variety. In Figs. 10.2(c)–(f), we display the ratios of $B(E2)_{\text{out}}/B(E2)_{\text{in}}$ and $B(M1)_{\text{out}}/B(E2)_{\text{in}}$ for the $\Delta I = 1$ transitions that connect band B2 to B1. These ratios are calculated for Fermi surfaces $\lambda = e_1$ through e_6 at triaxial deformation angles of $\gamma = 30°$ and $20°$. The subscript "out" refers to the $\Delta I = 1$ transitions from band B2 to B1, while "in" denotes the $\Delta I = 2$ transitions within band B2 itself. It is generally observed that the ratios $B(E2)_{\text{out}}/B(E2)_{\text{in}}$ and $B(M1)_{\text{out}}/B(E2)_{\text{in}}$ exhibit a decreasing trend with increasing spin. This behavior can be attributed to the amplitude of the wobbling motion, which is proportional to $1/\sqrt{I}$, where I represents the spin angular momentum.

At a triaxial deformation angle of $\gamma = 30°$, the ratio $B(E2)_{\text{out}}/B(E2)_{\text{in}}$ exhibits a collective enhancement when the Fermi surface is at $\lambda = e_1$. In this scenario, the $B(M1)_{\text{out}}/B(E2)_{\text{in}}$ values are relatively large, of the order of $1\mu_N^2/e^2b^2$, which is not characteristic of an ideal wobbling motion. This discrepancy is attributed to the coupling of the wobbling motion with the vibrations of the proton and neutron currents against each other, specifically the scissor mode. This mode can reduce the $M1$ strength [54], an effect not accounted for in the current PRM calculations. When the Fermi surface moves further away from the bottom of the shell, the value of $B(E2)_{\text{out}}/B(E2)_{\text{in}}$ diminishes. Similarly, the $B(M1)_{\text{out}}/B(E2)_{\text{in}}$ values also decrease. Notably, when the Fermi surface reaches the middle of the shell ($\lambda = e_3$ or e_4), the transition probabilities associated with the magnetic dipole moment approach zero and become independent of the total spin. Upon further increasing the Fermi surface's location, both $B(E2)_{\text{out}}/B(E2)_{\text{in}}$ and $B(M1)_{\text{out}}/B(E2)_{\text{in}}$ begin to increase. This trend is anticipated since the total PRM Hamiltonian used in these calculations is invariant under the transformation $\lambda = e_v \rightarrow e_{7-v}$ for the triaxial deformation $\gamma = 30°$ considered herein.

The trends observed for the ratios $B(E2)_{\text{out}}/B(E2)_{\text{in}}$ and $B(M1)_{\text{out}}/B(E2)_{\text{in}}$ at a triaxial deformation angle of $\gamma = 30°$ are similarly discernible at $\gamma = 20°$. Specifically, both ratios decrease as the Fermi surface λ progresses up to e_3, after which they exhibit an increase upon further increment of λ. The decrement of these two ratios with the growth of spin I is most pronounced when $\lambda = e_6$, corresponding to a hole configuration.

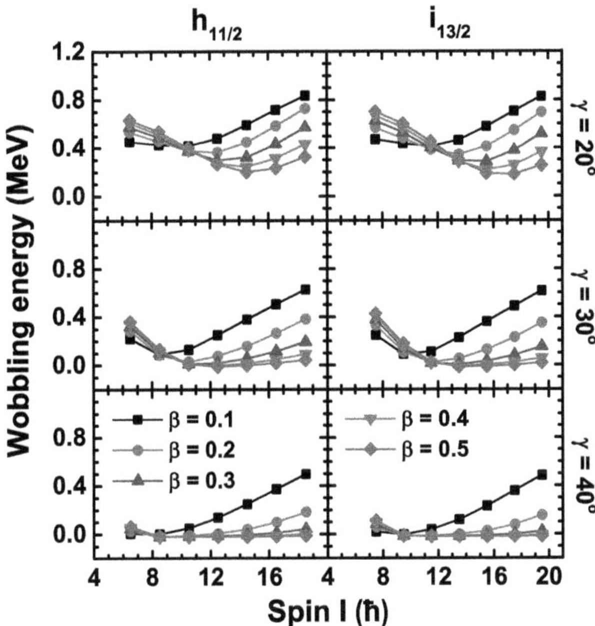

Figure 10.3 The calculated wobbling energies as functions of spin I by the PRM for the $h_{11/2}$ and $i_{13/2}$ configurations with different deformation parameters β and γ.

Furthermore, the deformation parameters exert significant influence on the wobbling bands. As illustrated in Fig. 10.3, based on the PRM, the energy differences between the wobbling bands vary with deformation. As previously discussed, the variation in ΔE can indicate the type of wobbling motion; a decreasing ΔE suggests the presence of TW, while an increasing ΔE is indicative of LW [2, 3]. For the $h_{11/2}$ configuration at $\gamma = 20°$, ΔE initially decreases and then increases, signifying TW in the low spin region and a transition to LW in the high spin region, aligning with prior findings [2, 3, 61]. The magnitude of ΔE is sensitive to the quadrupole deformation β. With a small β, for instance, $\beta = 0.1$, the change in ΔE at low spin is minimal, demonstrating subdued TW characteristics. Moreover, the minimum of ΔE emerges at $I = 10.5\hbar$, transitioning to LW at higher spins. A smaller β results in a smaller ΔE at low spins. For a fixed γ, such as $\gamma = 20°$, an increase in β makes the nucleus more prone to exhibit TW at low spins. Additionally, the inflection points of ΔE for various β values occur at different spins; for $\beta = 0.2$ and 0.3, it is at $I = 12.5\hbar$, while for $\beta = 0.4$ and 0.5, it is at $I = 14.5\hbar$. It is important to note that these TW spin regions are significantly shorter than those observed in Lu isotopes, which extend up to $40\hbar$ [62]. This discrepancy is attributed to the ratio $\mathcal{J}_s/\mathcal{J}_m$ used in the calculations, which is considerably smaller (approximately 0.12, 0.25 and 0.43) than the values observed in realistic scenarios (approximately

0.85) [2]. As the spin increases, the magnitude of ΔE for larger β diminishes, presenting distinct behaviors compared to the low spin region cases.

In contrast to the behavior of ΔE at $\gamma = 20°$, the energy differences ΔE at $\gamma = 30°$ exhibit a pronounced decrease in the low spin range for all considered values of quadrupole deformation β, including $\beta = 0.1$. Furthermore, the inflection point marking the transition from TW to LW shifts toward lower spin values. When examining wobbling bands with larger β, it is observed that ΔE increases gradually in the high spin region, and the characteristics of LW become less distinct. As the triaxial deformation angle γ increases to $40°$, TW for the $h_{11/2}$ configuration vanishes. For a nucleus with $\beta = 0.1$, LW is present throughout the entire spin range, and if β increases, LW gradually diminishes, indicating the absence of wobbling motion. Notably, instances where $\Delta E < 0$ at certain spins suggest that the wobbling excitation is unstable. Similar traits are observed in the wobbling energy of the $i_{13/2}$ configuration. In a short, within the range of deformations investigated, TW is more likely to form at larger values of β and smaller values of γ. Compared to the influence of triaxial deformation γ, LW displays a heightened sensitivity to variations in β.

10.5 CDFT+PRM APPROACH

In the quest for wobbling or chiral candidates within the nuclear chart, a robust theoretical framework is essential for elucidating the configuration and deformation characteristics of specific nuclei. The Covariant Density Functional Theory (CDFT) serves as a pivotal instrument in this regard, offering a comprehensive and microscopic description for a broad spectrum of nuclear phenomena [63–67]. Notably, the development and widespread application of the adiabatic and configuration fixed constraint triaxial CDFT [68] have been developed in investigating the potential configurations and corresponding deformations in nuclei, thereby exploring the potential manifestations of nuclear chirality or multiple chirality, which are indicative of triaxiality or the coexistence of triaxiality [60, 68].

In recent years, the synergistic application of the quantum PRM and CDFT methodologies, referred to as the CDFT+PRM approach [65, 66, 68], has been successfully employed to characterize the wobbling candidates reported in nuclei such as ^{105}Pd [23], ^{130}Ba [15], ^{187}Au [21] and ^{183}Au [22], as well as the chiral and MχD candidates [69–78].

Within this framework, initial adiabatic and configuration fixed constraint triaxial CDFT calculations are conducted to map out the potential energy surfaces in the β-γ plane, trace the potential energy curves along the β axis, and ascertain the single-particle energy levels. This analysis aids in identifying the possible configuration and deformation parameters. Subsequently, these parameters are input into the PRM to compute the energy spectra, electromagnetic transition probabilities, and the angular momentum geometries associated with the wobbling motion. Ultimately, this approach facilitates the revelation of the intrinsic physics underlying the observed rotational bands.

In this work, we demonstrate the utility of the CDFT+PRM approach in dissecting the intricacies of wobbling motion.

10.6 EXPLORATION OF WOBBLING MODES IN AN ISOTONE CHAIN: $N = 59$ CASE

As previously discussed, in the mass region approximately $A \approx 100$, the nucleus ^{105}Pd with an odd neutron number $N = 59$ stands as a singular candidate. It has been proposed as a TW candidate, marking the first instance where the wobbling excited state is founded on a quasi-neutron configuration [23]. Despite the scarcity of wobbling candidates reported within this mass region, a plethora of evidence supports the existence of nuclear chiral doublet bands or multiple chiral doublet bands (cf., e.g., Refs. [70, 71, 79–81]). This evidence suggests that the $A \approx 100$ mass region is characteristically triaxially deformed. Consequently, it is of considerable interest to explore the potential presence of additional wobbling nuclei in this mass region, particularly along the chain of $N = 59$ isotones.

In this section, we employ constrained CDFT to investigate the configurations and triaxial deformations of the $N = 59$ isotones within the $A \approx 100$ mass region. The isotones under scrutiny include ^{95}Kr $(Z = 36)$, ^{97}Sr $(Z = 38)$, ^{99}Zr $(Z = 40)$, ^{101}Mo $(Z = 42)$, ^{103}Ru $(Z = 44)$, ^{105}Pd $(Z = 46)$ and ^{107}Cd $(Z = 48)$, in an effort to identify the possible existence of the wobbling mode [82]. The objective of this study is to furnish theoretical insights that will inform forthcoming experimental endeavors focused on triaxial deformation and wobbling motion in the $N = 59$ isotone chain.

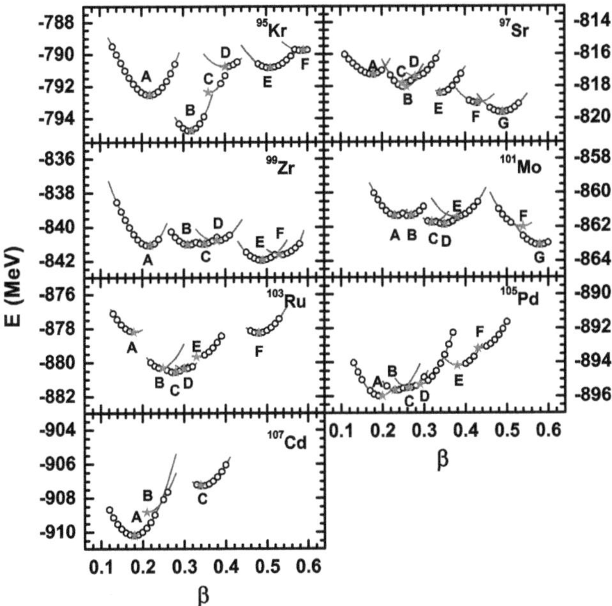

Figure 10.4   The potential energy curves as functions of deformation β in adiabatic (open circles) and configuration fixed (solid lines) constrained triaxial CDFT calculation for ^{95}Kr, ^{97}Sr, ^{99}Zr, ^{101}Mo, ^{103}Ru, ^{105}Pd and ^{107}Cd. The local minima in the energy surfaces for the fixed configuration are represented as stars and labeled as A, B, C, ... in accordance with the increasing β value. Adapted from Ref. [82].

The potential energy curves (PECs), which represent the energy as a function of the quadrupole deformation parameter β, for the isotones ^{95}Kr, ^{97}Sr, ^{99}Zr, ^{101}Mo, ^{103}Ru, ^{105}Pd and ^{107}Cd have been computed using adiabatic constrained CDFT calculations, as depicted in Fig. 10.4 [82]. In these calculations, the constraint is applied to the expectation value of $\langle \hat{Q}_{20}^2 + 2\hat{Q}_{22}^2 \rangle$, effectively allowing for the automatic determination of the triaxial deformation through energy minimization. In Fig. 10.4, the open circles indicate the outcomes of the adiabatic calculations, which produce irregular energy curves. Some local minima are indistinct, making them challenging to discern. The configuration fixed calculations effectively address this issue [68], and the corresponding results are plotted as lines in Fig. 10.4. Each curve's minimum is marked by a red star and labeled with a capital letter (A, B, C, ...) in order of increasing β value. The ground state deformation parameters derived from the configuration fixed calculation results are noted, such as local minimum B in ^{95}Kr, G in ^{97}Sr, E in ^{99}Zr, G in ^{101}Mo, C in ^{103}Ru, A in ^{105}Pd and A in ^{107}Cd.

Inspecting Fig. 10.4 for ^{95}Kr reveals the presence of several excited states featuring distinct configurations beyond the ground state. These include state A $(\beta = 0.22,\ \gamma = 35.9°)$, C $(\beta = 0.36,\ \gamma = 60°)$, D $(\beta = 0.40,\ \gamma = 0°)$, E $(\beta = 0.51,\ \gamma = 5.85°)$ and F $(\beta = 0.59,\ \gamma = 0°)$. Notably, state A exhibits a pronounced triaxial deformation, which is conducive to the formation of wobbling motion. This observation is significant as it suggests that other nuclei may also exhibit wobbling modes, given the presence of excited states with substantial triaxial deformation.

The excited states with significant triaxial deformation in various nuclei are as follows: A ($\beta = 0.22$, $\gamma = 35.9°$) in ^{95}Kr, C ($\beta = 0.25$, $\gamma = 37.9°$) and D ($\beta = 0.28$, $\gamma = 30.0°$) in ^{97}Sr, B ($\beta = 0.31$, $\gamma = 23.3°$) in ^{99}Zr, A ($\beta = 0.23$, $\gamma = 25.9°$), B ($\beta = 0.27$, $\gamma = 22.9°$) and C ($\beta = 0.32$, $\gamma = 21.6°$) in ^{101}Mo, B ($\beta = 0.25$, $\gamma = 20.4°$) in ^{103}Ru, as well as B ($\beta = 0.23$, $\gamma = 22.2°$), C ($\beta = 0.27$, $\gamma = 24.9°$) and D ($\beta = 0.29$, $\gamma = 30.7°$) in ^{105}Pd.

It is important to highlight that wobbling bands arising from state C in ^{105}Pd have been experimentally observed [23], validating the theoretical predictions and underscoring the relevance of these findings to experimental nuclear physics.

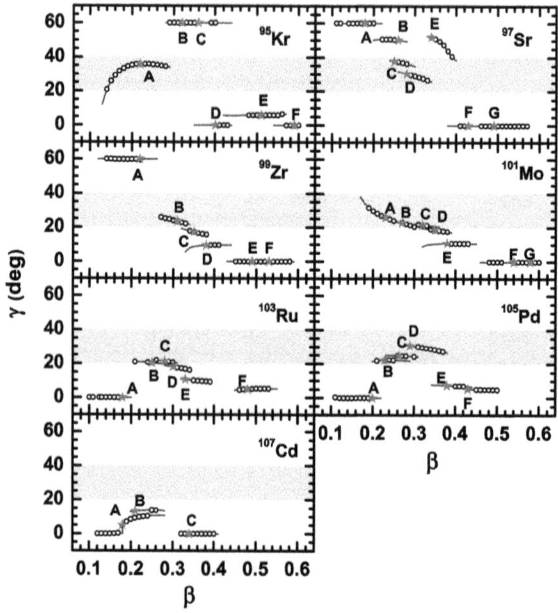

Figure 10.5 The triaxial deformations γ as functions of deformation β in adiabatic (open circles) and configuration fixed (solid lines) constrained triaxial CDFT calculation for ^{95}Kr, ^{97}Sr, ^{99}Zr, ^{101}Mo, ^{103}Ru, ^{105}Pd and ^{107}Cd. The local minima in the energy surfaces for the fixed configuration are represented as stars and labeled as A, B, C, ... in accordance with the increasing β value. The shaded areas figure represent the triaxial deformation beneficial to the wobbling mode. Adapted from Ref. [82].

To provide a clearer visualization of the nuclear triaxial deformation's evolution with the deformation parameter β, Fig. 10.5 illustrates the variations of the triaxial deformation γ with respect to β [82]. The figure presents the results obtained from adiabatic (open circles) and configuration fixed (solid lines) constrained triaxial CDFT calculations for the isotones ^{95}Kr, ^{97}Sr, ^{99}Zr, ^{101}Mo, ^{103}Ru, ^{105}Pd and ^{107}Cd. The shaded regions in each subplot of the figure denote the range of triaxial deformation that is favorable for the wobbling mode. The triaxial deformation γ exhibits a smooth variation with β for each configuration, reinforcing the notion that the deformation γ is predominantly dictated by the specific configuration in question. Furthermore, it is observable that all nuclei, with the exception of ^{107}Cd, demonstrate the potential for significant triaxial deformation, which could be instrumental in facilitating the wobbling mode.

It is important to highlight that the configuration involving the $h_{11/2}$ neutron orbital has been previously identified in the deformed negative parity bands observed in ^{97}Sr and ^{99}Zr [83]. Specifically, the $\nu h_{11/2}$ configuration was attributed to the $7/2^-$ band in ^{97}Sr and the $11/2^-$ band in ^{99}Zr. The experimental excitation energies for the bandhead states are reported as 0.771 MeV ($I = 7/2\hbar$) for ^{97}Sr and 0.821 MeV ($I = 11/2\hbar$) for ^{99}Zr. These values exhibit a slight deviation from the theoretical prediction of 2.2 MeV for ^{97}Sr, while the experimental value for ^{99}Zr aligns well with the theoretical prediction of 0.92 MeV.

To delve into the rotational characteristics of the configurations under investigation and assess the potential for wobbling modes in nuclei, the PRM [3, 35–37, 61, 84] is utilized. The PRM calculations incorporate deformation parameters (β, γ) for ^{97}Sr and ^{99}Zr, which are extracted from the CDFT calculations, with specific values of $(0.28, 30.4°)$ and $(0.31, 23.3°)$, respectively. The single-j shell Hamiltonian parameter C is determined using the formula:

$$C = \left(\frac{123}{8}\sqrt{\frac{5}{\pi}}\right)\frac{2N+3}{j(j+1)}A^{-1/3}\beta. \tag{10.15}$$

The moment of inertia for the irrotationional flow type is given by:

$$\mathcal{J}_k(\gamma) = \mathcal{J}_0 \sin^2\left(\gamma - \frac{2k\pi}{3}\right), \tag{10.16}$$

where $\mathcal{J}_0$ is set to 30 $\hbar^2/$MeV. For the electromagnetic transitions, empirical values are adopted: the intrinsic quadrupole moment is given by $Q = (3/\sqrt{5\pi})R_0^2 Z\beta$, and the gyromagnetic ratios are taken as $g_R = Z/(A-1)$ and $g_\nu(h_{11/2}) = -0.21$.

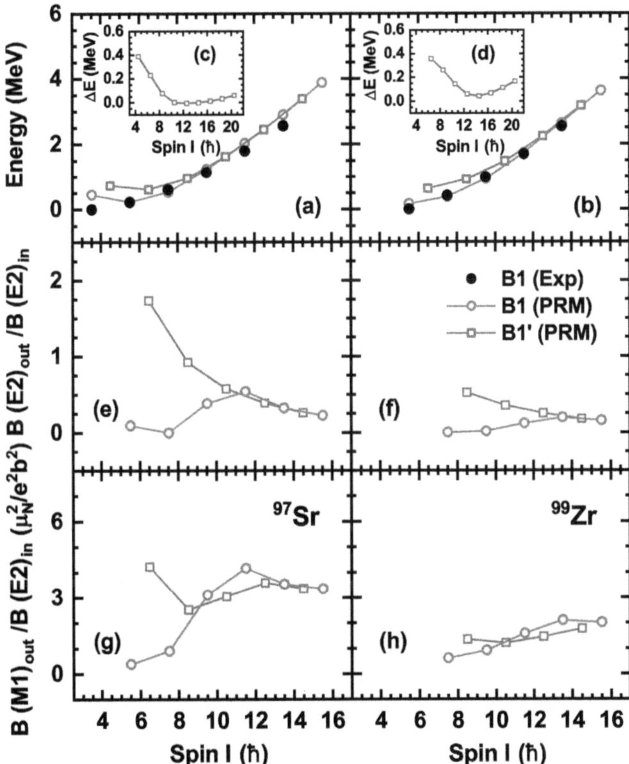

Figure 10.6 The lowest bands B1 and B1′ based on the configuration $\nu h_{11/2}^1$ in ^{97}Sr and ^{99}Zr. (a)-(b) The calculated energy spectra by PRM are compared with the experimental data in Ref. [83]. (c)-(d) Theoretical energy difference between the doublet bands B1 and B1′. (e)-(h) The B$(E2)_{\text{out}}$/B$(E2)_{\text{in}}$ and B$(M1)_{\text{out}}$/B$(E2)_{\text{in}}$ values for the interband transitions from band B1′ to B1 calculated by the PRM. Adapted from Ref. [82].

The calculated energy spectra, designated as bands B1 and B1′, along with the ratios of $B(E2)_{\text{out}}/B(E2)_{\text{in}}$ and $B(M1)_{\text{out}}/B(E2)_{\text{in}}$ for ^{97}Sr and ^{99}Zr, are presented in Fig. 10.6 alongside the corresponding experimental energy spectra [82]. The calculations are found to be in good agreement with the experimental findings, demonstrating the accuracy of the adopted theoretical approach. Furthermore, this study employs the HFA method [2] to investigate the wobbling mode. The results derived from the HFA method are depicted by blue solid lines in Fig. 10.6, providing an additional perspective on the wobbling motion and its characteristics.

The energy difference $\Delta E(I)$ serves as a critical metric for characterizing the nature of wobbling motion. In the low spin regime, $\Delta E(I)$ typically decreases with increasing spin, which is a hallmark characteristic of the TW mode [2, 3]. In contrast, for the higher spin region, the wobbling energy within the PRM framework is observed to increase, akin to the behavior seen in ^{135}Pr (refer to Fig. 16 of Ref. [2]). It is important to note that the wobbling energies calculated using the PRM are generally somewhat lower compared to those observed in ^{135}Pr (refer to Fig. 16 of Ref. [2]). This discrepancy can be rationalized by considering the close relationship between wobbling energy and the moment of inertia, specifically, the inverse proportionality between wobbling energy and the s-axis moment of inertia (refer to Eq. (24) of Ref. [2]). The s-axis moment of inertia for ^{135}Pr is $\mathcal{J}_s = 6 \ \hbar^2/$MeV (refer to Table I of Ref. [2]), which is lower than that of ^{97}Sr ($\mathcal{J}_s = 7.9 \ \hbar^2/$MeV) and ^{99}Zr ($\mathcal{J}_s = 10.7 \ \hbar^2/$MeV). Consequently, the wobbling energies for ^{97}Sr and ^{99}Zr are slightly reduced compared to those in ^{135}Pr.

The electromagnetic transition probability ratios $B(E2)_{\text{out}}/B(E2)_{\text{in}}$ and $B(M1)_{\text{out}}/B(E2)_{\text{in}}$ for bands B1$'$ and B1, as determined by the PRM, are depicted in Figures 10.6(e)-(h) [82]. The PRM-derived ratio $B(E2)_{\text{out}}/B(E2)_{\text{in}}$ exhibits a substantial value (≥ 0.5) across the entire spin region, signifying that the $E2$ transitions from band B1 to band B1$'$ are highly collective. This feature is characteristic of TW, indicative of the triaxial charge density's wobbling motion relative to the angular momentum vector. Notably, this ratio diminishes as the angular momentum I increases.

The variation of the $B(M1)_{\text{out}}/B(E2)_{\text{in}}$ ratio with the angular momentum I in ^{97}Sr and ^{99}Zr is not monotonic, exhibiting an initial decrease followed by an increase. Within the framework of the PRM, the inflection point occurs at $I = 17/2\hbar$ for ^{97}Sr and $I = 21/2\hbar$ for ^{99}Zr. This behavior suggests that the $B(M1)_{\text{out}}/B(E2)_{\text{in}}$ ratio can be utilized to pinpoint the collapse of the TW mode. It is observed that the $B(M1)_{\text{out}}/B(E2)_{\text{in}}$ values do not diminish as expected for a purely wobbling motion. This discrepancy can be ascribed to the coupling of the wobbling motion with the scissor mode vibrations of proton and neutron currents, which can enhance the $M1$ strength [54]. However, this effect is not considered in the PRM calculations. In Fig. 10.6, the $B(M1)_{\text{out}}/B(E2)_{\text{in}}$ and $B(E2)_{\text{out}}/B(E2)_{\text{in}}$ values for the transitions from band B1 to B1$'$ are further displayed. A marked difference between the transitions of B1$' \to$ B1 and B1 $\to$ B1$'$ is noted, which provides additional evidence supporting their classification as wobbling partner bands [47].

Further analysis has confirmed that the calculated mixing ratios $\delta = \langle E2\rangle_{\text{out}}/\langle M1\rangle_{\text{out}}$ are positive for the studied nuclei. A positive δ value is indicative of a TW mode established on a neutron configuration [23], as opposed to a TW mode built on a proton configuration, which would exhibit a negative δ value [11, 12]. This distinction arises from the opposing signs of the $g_n - g_R$ and $g_p - g_R$ factors within the $M1$ operator [23].

Despite these theoretical insights, additional experimental work is necessary to expand the energy level scheme for the partner band and to obtain the relevant electromagnetic transition probability data. Such data are crucial for discerning the presence of wobbling motion and ruling out other potential explanations.

10.7 EXPLORATION OF WOBBLING MODES IN EVEN-EVEN NUCLEI: ^{130}BA CASE

In the subsequent analysis, we utilize ^{130}Ba as a prime example to elucidate the computations conducted within the framework of CDFT+PRM methodology.

The ^{130}Ba stands out as the first instance showcasing two-quasiparticle wobbling bands within an even-even nucleus [14, 15, 20]. Previous studies [15] have shown the calculated energy spectra using PRM against experimental data. Here, we present the obtained configurations and corresponding deformation parameters in Table 10.1, acquired via adiabatic and configuration-fixed constraint triaxial CDFT computations. Our calculations adopt the PC-PK1 effective interaction [85], simplifying by neglecting pairing correlations. The Dirac equation is numerically resolved utilizing a spherical harmonic oscillator basis comprising 12 major shells. As illustrated in Table 10.1, there are several local minima in the potential energy curve, in which the protons and neutrons are also zero-quasiparticle configuration. The excited two-quasiparticle configuration a involving $\pi h_{11/2}^2$ has a triaxially deformed shape with $\beta = 0.24$ and $\gamma = 21.5°$, which provides the precondition for the establishment of wobbling motion. In addition, its excitation energy of 3.13 MeV with respect to the ground state A is comparable with the experimental excitation energy of 3.79 MeV of the $I = 10\hbar$ state of the yrast wobbling band [14].

Table 10.1

Energies (in MeV), deformation parameters β and γ, as well as the corresponding configurations (both valence nucleon and unpaired nucleon) of the local minima in ^{130}Ba obtained by the adiabatic and configuration-fixed constrained triaxial CDFT calculations.

State	Valence-cfg.	Unpaired-cfg.	Energy	β	γ
A	$\pi[g_{7/2}^4 2d_{5/2}^2] \otimes \nu[2d_{3/2}^2 h_{11/2}^{-4}]$	—	-1088.74	0.23	13.9°
B	$\pi[g_{7/2}^4 h_{11/2}^2] \otimes \nu[2d_{3/2}^2 h_{11/2}^{-4}]$	—	-1087.10	0.25	25.6°
C	$\pi[g_{7/2}^4 h_{11/2}^2] \otimes \nu[2d_{5/2}^2 h_{11/2}^{-4}]$	—	-1085.05	0.34	44.1°
D	$\pi[g_{9/2}^{-2} g_{7/2}^{-2} h_{11/2}^2] \otimes \nu[2d_{5/2}^2 h_{11/2}^{-4} 2f_{5/2}^2 3p_{3/2}^2]$	—	-1082.73	0.47	2.9°
a	$\pi[g_{7/2}^4 h_{11/2}^2] \otimes \nu[2d_{3/2}^2 h_{11/2}^{-4}]$	$\pi h_{11/2}^2$	-1085.61	0.24	21.5°

The obtained deformation parameters and configuration information are then input to the PRM to describe the experimental energy spectra, energy difference between the two bands, as well as the available electromagnetic transition probabilities $B(M1)_{\text{out}}/B(E2)_{\text{in}}$ and $B(E2)_{\text{out}}/B(E2)_{\text{in}}$ [84].

Fig. 10.7 shows the probability density distributions $\mathcal{P}(\theta, \varphi)$ for the orientation of the angular momenta J with respect to the body-fixed frame (spin coherent state plots [3] or azimuthal plots [35, 86, 87]) at selected spins. The figure corroborates the scenario of stable TW. Due to the D_2 symmetry, $\mathcal{P}(\theta, \varphi)$ is an even function of φ with a period of π. The distributions $\mathcal{P}(\theta, \varphi)$ are centered at $\theta = 90°$, corresponding to the very small l-axis angular momentum component. The yrast band S1 has a maximum at $\varphi = 0°$ whereas the excited S1$'$ has a minimum there. These are the expected distributions for the φ-symmetric zero-phonon and φ-antisymmetric one-phonon states. The distributions extend toward $\varphi = \pm 90°$ with increasing spin. The fact that the distributions centered at $\varphi = 0°$ and $\varphi = \pm 180°$ do not merge at $\varphi = \pm 90°$ indicates that the TW mode is stable.

Figure 10.7 additionally presents the distributions $\mathcal{P}(\theta, \varphi)$ corresponding to the subsequent higher bands. As denoted by the labeling, there is a natural inclination to interpret these bands as two and three phonon wobbling excitations. Particularly noteworthy is the nearly constant energy spacing observed for $12 \leq I \leq 18$, which lends credence to the multi-wobbling hypothesis. In line with this interpretation, the $\mathcal{P}(\theta = 90°, \varphi)$ distribution for the $n = 2$ band attains a maximum at $\varphi = 0$ flanked by symmetrically positioned minima, indicative of a φ-symmetric wave function with two nodes. Conversely, the distribution $\mathcal{P}(\theta, \varphi)$ for the $n = 3$ band displays a minimum at $\varphi = 0°$ between two symmetric maxima and two symmetric minima farther out, characteristic of a wave function with three nodes, bolstering the three-phonon wobbling excitation interpretation. However, it should be noted that interpreting these excitations solely as vibrations centered at $\varphi = 0°, \pm 180°$ overlooks the merging of the two branches at $\varphi = \pm 90°$, indicating a more intricate underlying structure.

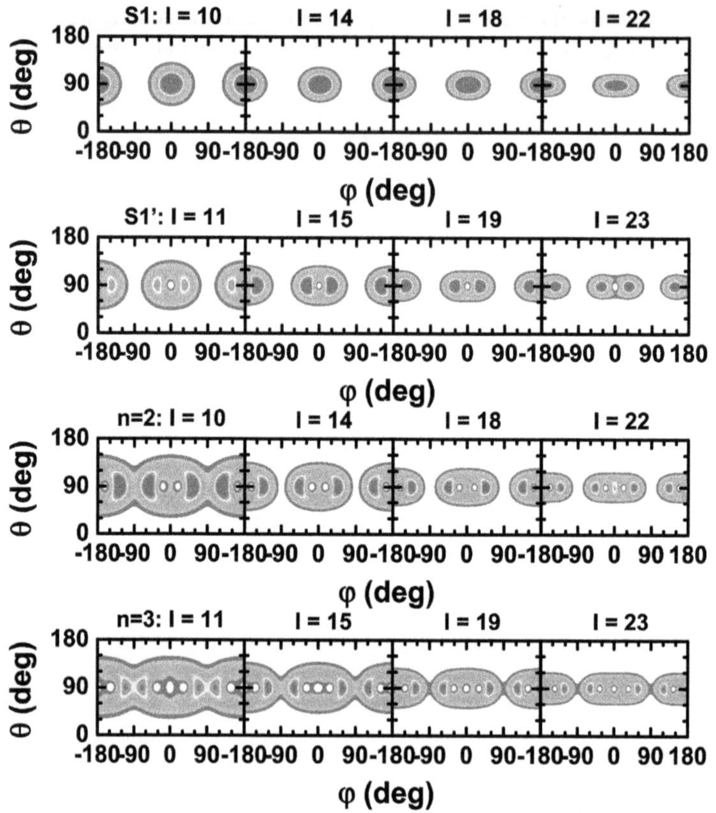

Figure 10.7 Distributions of the probability $\mathcal{P}(\theta, \varphi)$ for the orientation of the angular momentum J with respect to the body-fixed frame calculated for selected values of angular momentum. Here, θ is the angle between J and the l-axis, and φ is the angle between the projection of J onto the sm-plane and the s-axis.

It is therefore very interesting to explore more wobbling candidates in the even-even nuclei.

10.8 SUMMARY

In summary, we first examine the impact of the Fermi surface on nuclear wobbling motion. It is revealed that both the position of the Fermi surface and the characteristics of the valence quasiparticle play a decisive role in shaping the behavior of the energy difference ΔE with respect to spin and the associated electromagnetic transition probabilities. A quasiparticle in a deformed aligned (DALs) proton or deformed aligned hole (DALl) configuration exhibits a parabolic ΔE characteristic of wobbling energy. In contrast, a Fermi aligned (FAL) quasiparticle is observed to display a monotonic increase in ΔE. Notably, DALs particles or DALl holes significantly enhance the inter-band transition probabilities, while the presence of a FAL quasiparticle tends to suppress these probabilities. The trends observed in the ratios $B(E2)_{\text{out}}/B(E2)_{\text{in}}$ and $B(M1)_{\text{out}}/B(E2)_{\text{in}}$ differ from the variations in the energy difference ΔE between bands as the Fermi surface location changes. Higher ΔE values are correlated with lower ratios of $B(E2)_{\text{out}}/B(E2)_{\text{in}}$ and $B(M1)_{\text{out}}/B(E2)_{\text{in}}$, whereas lower ΔE values are associated with higher ratios.

Subsequent to this, the effects of deformation on the wobbling mode are explored through an analysis of wobbling energy. Within the range of deformations investigated, larger quadrupole deformation β and smaller triaxial deformation γ are more conducive to the formation of TW. A smaller β is linked to a smaller transverse wobbling energy, while a larger γ corresponds to a larger LW energy. The LW modes show a higher sensitivity to changes in β compared to γ.

Finally, adiabatic and configuration fixed constrained triaxial CDFT calculations are employed to probe the potential for wobbling modes along the $N = 59$ isotones ^{95}Kr, ^{97}Sr, ^{99}Zr, ^{101}Mo, ^{103}Ru, ^{105}Pd and ^{107}Cd in the mass region $A \approx 100$ as well as the wobbling candidates in the even-even nuclei. The corresponding potential energy curves, deformation parameters, and configurations for both ground and excited states are determined. From the adiabatic constraint calculations, several states meeting the criteria of high-j valence particles and appropriate triaxial deformation are identified in ^{97}Sr, ^{99}Zr, ^{101}Mo and ^{105}Pd. Intriguingly, ^{101}Mo may be a candidate for triaxial shape coexistence, with possible evidence of multiple wobbling modes. Moreover, rotational bands based on the $\nu(1h_{11/2})^1$ configuration in ^{97}Sr and ^{99}Zr are analyzed using the PRM to study the nature of the wobbling motion. The PRM successfully reproduces the experimental energy spectra. The energy differences between doublet bands, the enhanced $B(E2)_{\mathrm{out}}/B(E2)_{\mathrm{in}}$ values, and the one-phonon oscillation characteristic of the total angular momentum suggest that these two nuclei are TW candidates, indicating the potential for a broader region of wobbling modes near the $A \approx 100$ mass region. In addition, ^{130}Ba was identified as the first example wobbling candidates in the even-even nuclei with two-quasiparticle configuration. Further experimental work to extend the level schemes and extract electromagnetic transition probabilities data is essential to validate these theoretical predictions.

10.9 ACKNOWLEDGMENTS

We thank Sihua Li and Xianrong Zhou for their contributions on the results presented here. This work was supported by the National Natural Science Foundation of China under Grant No. 12205103.

Bibliography

1. A. Bohr and B. R. Mottelson, Nuclear structure, Vol. II (Benjamin, New York, 1975).
2. S. Frauendorf and F. Dönau, Phys. Rev. C **89**, 014322 (2014).
3. Q. B. Chen and S. Frauendorf, Eur. Phys. J. A **58**, 75 (2022).
4. P. Bringel *et al.*, Eur. Phys. J. A **24**, 167 (2005).
5. S. W. Ødegård *et al.*, Phys. Rev. Lett. **86**, 5866 (2001).
6. D. R. Jensen *et al.*, Phys. Rev. Lett. **89**, 142503 (2002).
7. G. Schönwaßer *et al.*, Phys. Lett. B **552**, 9 (2003).
8. H. Amro *et al.*, Phys. Lett. B **553**, 197 (2003).
9. D. J. Hartley *et al.*, Phys. Rev. C **80**, 041304(R) (2009).
10. A. Mukherjee *et al.*, Phys. Rev. C **107**, 054310 (2023).
11. J. T. Matta *et al.*, Phys. Rev. Lett. **114**, 082501 (2015).
12. N. Sensharma *et al.*, Phys. Lett. B **792**, 170 (2019).
13. S. Biswas *et al.*, Eur. Phys. J. A **55**, 159 (2019).
14. C. M. Petrache *et al.*, Phys. Lett. B **795**, 241 (2019).
15. Q. B. Chen, N. Kaiser, U.-G. Meißner, and J. Meng, Phys. Rev. C **99**, 064326 (2019).
16. S. Chakraborty *et al.*, Phys. Lett. B **811**, 135854 (2020).
17. K. Rojeeta Devi *et al.*, Phys. Lett. B **823**, 136756 (2021).
18. B. F. Lv *et al.*, Phys. Rev. C **105**, 034302 (2022).
19. M. Prajapati *et al.*, Phys. Rev. C **109**, 034301 (2024).
20. Y. K. Wang, F. Q. Chen and P. W. Zhao, Phys. Lett. B **802**, 135246 (2020).
21. N. Sensharma *et al.*, Phys. Rev. Lett. **124**, 052501 (2020).
22. S. Nandi *et al.*, Phys. Rev. Lett. **125**, 132501 (2020).
23. J. Timár *et al.*, Phys. Rev. Lett. **122**, 062501 (2019).

24. R. J. Guo *et al.*, Phys. Rev. Lett. **132**, 092501 (2024).

25. S. Frauendorf, Phys. Rev. C **97**, 069801 (2018).

26. K. Tanabe and K. Sugawara-Tanabe, Phys. Rev. C **97**, 069802 (2018).

27. E. A. Lawrie, O. Shirinda and C. M. Petrache, Phys. Rev. C **101**, 034306 (2020).

28. B. F. Lv *et al.*, Phys. Rev. C **103**, 044308 (2021).

29. K. Tanabe and K. Sugawara-Tanabe, Phys. Rev. C **95**, 064315 (2017).

30. B. F. Lv *et al.*, Phys. Lett. B **824**, 136840 (2022).

31. S. Guo *et al.*, Phys. Lett. B **828**, 137010 (2022).

32. K. Nomura and C. M. Petrache, Phys. Rev. C **105**, 024320 (2022).

33. W. X. Shi and Q. B. Chen, Chin. Phys. C **39**, 054105 (2015).

34. I. Hamamoto, Phys. Rev. C **65**, 044305 (2002).

35. E. Streck, Q. B. Chen, N. Kaiser and U.-G. Meißner, Phys. Rev. C **98**, 044314 (2018).

36. Q. B. Chen, S. Frauendorf, N. Kaiser, U.-G. Meißner and J. Meng, Phys. Lett. B **807**, 135596 (2020).

37. C. Broocks, Q. B. Chen, N. Kaiser and U.-G. Meißner, Eur. Phys. J. A **57**, 161 (2021).

38. H. Zhang, B. Qi, X. D. Wang, H. Jia and S. Y. Wang, Phys. Rev. C **105**, 034339 (2022).

39. A. A. Raduta, R. Poenaru and L. G. Ixaru, Phys. Rev. C **96**, 054320 (2017).

40. A. A. Raduta, R. Poenaru and A. H. Raduta, J. Phys. G: Nucl. Part. Phys. **45**, 105104 (2018).

41. R. Budaca, Phys. Rev. C **97**, 024302 (2018).

42. A. A. Raduta, C. M. Raduta and R. Poenaru, J. Phys. G: Nucl. Part. Phys. **48**, 015106 (2021).

43. C. M. Raduta, A. A. Raduta, R. Poenaru and A. H. Raduta, J. Phys. G: Nucl. Part. Phys. **49**, 025105 (2022).

44. R. Budaca, Phys. Rev. C **103**, 044312 (2021).

45. R. Budaca and C. M. Petrache, Phys. Rev. C **106**, 014313 (2022).

46. E. R. Marshalek, Nucl. Phys. A **331**, 429 (1979).

47. Y. R. Shimizu and M. Matsuzaki, Nucl. Phys. A **588**, 559 (1995).

48. M. Matsuzaki, Y. R. Shimizu and K. Matsuyanagi, Phys. Rev. C **65**, 041303 (2002).

49. M. Matsuzaki, Y. R. Shimizu and K. Matsuyanagi, Phys. Rev. C **69**, 034325 (2004).

50. Y. R. Shimizu, M. Matsuzaki and K. Matsuyanagi, Phys. Rev. C **72**, 014306 (2005).

51. D. Almehed, R. G. Nazmitdinov and F. Dönau, Phys. Scr. **T125**, 139 (2006).

52. Y. R. Shimizu, T. Shoji and M. Matsuzaki, Phys. Rev. C **77**, 024319 (2008).

53. T. Shoji and Y. R. Shimizu, Progr. Theor. Phys. **121**, 319 (2009).

54. S. Frauendorf and F. Dönau, Phys. Rev. C **92**, 064306 (2015).

55. Q. B. Chen, S. Q. Zhang, P. W. Zhao and J. Meng, Phys. Rev. C **90**, 044306 (2014).

56. Q. B. Chen, S. Q. Zhang and J. Meng, Phys. Rev. C **94**, 054308 (2016).

57. Q. B. Chen and S. Frauendorf, Phys. Rev. C **109**, 044304 (2024).

58. M. Shimada, Y. Fujioka, S. Tagami and Y. R. Shimizu, Phys. Rev. C **97**, 024318 (2018).

59. F.-Q. Chen and C. M. Petrache, Phys. Rev. C **103**, 064319 (2021).

60. S. Frauendorf and J. Meng, Nucl. Phys. A **617**, 131 (1997).

61. S. H. Li, H. M. Dai, Q. B. Chen and X.-R. Zhou, Chin. Phys. C **48**, 034102 (2024).

62. D. J. Hartley *et al.*, Phys. Rev. C **83**, 064307 (2011).

63. P. Ring, Prog. Part. Nucl. Phys. **37**, 193 (1996).

64. D. Vretenar, A. Afanasjev, G. Lalazissis and P. Ring, Phys. Rep. **409**, 101 (2005).

65. J. Meng, H. Toki, S. Zhou, S. Zhang, W. Long and L. Geng, Prog. Part. Nucl. Phys. **57**, 470 (2006).

66. J. Meng, ed., *Relativistic Density Functional for Nuclear Structure*, vol. 10 of International Review of Nuclear Physics (World Scientific, Singapore, 2016).

67. J. Meng and P. W. Zhao, *Relativistic Density Functional Theories* (Springer Nature Singapore, Singapore, 2023), pp. 2111.

68. J. Meng, J. Peng, S. Q. Zhang and S.-G. Zhou, Phys. Rev. C **73**, 037303 (2006).

69. A. D. Ayangeakaa *et al.*, Phys. Rev. Lett. **110**, 172504 (2013).
70. E. O. Lieder *et al.*, Phys. Rev. Lett. **112**, 202502 (2014).
71. I. Kuti *et al.*, Phys. Rev. Lett. **113**, 032501 (2014).
72. C. Liu *et al.*, Phys. Rev. Lett. **116**, 112501 (2016).
73. C. M. Petrache *et al.*, Phys. Rev. C **94**, 064309 (2016).
74. E. Grodner *et al.*, Phys. Rev. Lett. **120**, 022502 (2018).
75. Q. B. Chen, B. F. Lv, C. M. Petrache and J. Meng, Phys. Lett. B **782**, 744 (2018).
76. J. Peng and Q. B. Chen, Phys. Lett. B **793**, 303 (2019).
77. B. F. Lv *et al.*, Phys. Rev. C **100**, 024314 (2019).
78. J. Peng and Q. B. Chen, Phys. Lett. B **806**, 135489 (2020).
79. C. Vaman *et al.*, Phys. Rev. Lett. **92**, 032501 (2004).
80. N. Rather *et al.*, Phys. Rev. Lett. **112**, 202503 (2014).
81. D. Tonev *et al.*, Phys. Rev. Lett. **112**, 052501 (2014).
82. H. M. Dai, Q. B. Chen and X.-R. Zhou, Phys. Rev. C **108**, 054306 (2023).
83. W. Urban *et al.*, Nucl. Phys. A **689**, 605 (2001).
84. Q. B. Chen, S. Frauendorf and C. M. Petrache, Phys. Rev. C **100**, 061301(R) (2019).
85. P. W. Zhao, Z. P. Li, J. M. Yao and J. Meng, Phys. Rev. C **82**, 054319 (2010).
86. F. Q. Chen, Q. B. Chen, Y. A. Luo, J. Meng and S. Q. Zhang, Phys. Rev. C **96**, 051303(R) (2017).
87. Q. B. Chen, K. Starosta and T. Koike, Phys. Rev. C **97**, 041303(R) (2018).

11 Recent developments and novel applications of the interacting boson-fermion model

Kosuke Nomura
Hokkaido University, Sapporo, Japan

11.1 INTRODUCTION

The low-energy collective excitations of medium-heavy and heavy atomic nuclei are mostly of quadrupole type. The quadrupole collectivity is characterized by the anharmonic vibrations of the nuclear surface, and by the rotational motions when the nucleus is strongly deformed [1]. While the quadrupole deformation is considered axially symmetric prolate or oblate, i.e., invariant under the rotation around the symmetry axis of the intrinsic frame of reference, in many cases this symmetry is broken. The nuclei with nonaxial, or triaxially degrees of freedom display many intriguing nuclear structure aspects that include the quantum shape phase transitions [2], and shape coexistence [3]. The nonaxial, γ-soft, shape has been often described by the two different collective models: the one in terms of the rigid triaxial rotor picture, in which the collective potential has a definite minimum at γ different from $\gamma = 0°$ and $60°$ [4], while the other is based on the γ-unstable rotor picture, in which the potential is completely flat along the γ deformation [5].

While the vast majority of the experimental and theoretical investigations of the nuclear deformations is concentrated on the even-even nuclei, there is also wealth of spectroscopic data on those nuclei having odd number(s) of neutrons or/and protons. Several exotic collective dynamics, such as wobbling motions [1] and chirality [6], have been indeed recognized in the odd-mass and odd-odd nuclei with nonaxial quadrupole deformations. A microscopic description of the structure of odd-nucleon systems has been a challenging task for nuclear structure theory, since for these nuclear systems both the collective and single-particle motions should be treated on the same footing. The difficulty of computing spectroscopic properties of the odd nuclei accurately and systematically becomes even more significant for heavy mass regions.

The interacting boson model (IBM) [7], a model in which correlated nucleon pairs are represented by effective boson degrees of freedom, has been remarkably successful in the phenomenological description of the low-energy quadrupole collective states of medium-heavy and heavy even-even nuclei [8]. Extension of the IBM to deal with odd-mass nuclear systems has been made [9] in a similar spirit to the particle-core coupling scheme. Namely, an even-even core nucleus is described in terms of the bosonic degrees of freedom in the IBM, which is then coupled to the (unpaired) single-particle degrees of freedom through appropriate boson-fermion interactions. The extended model is referred to as the interacting boson-fermion model (IBFM), and has been employed in a number of phenomenological studies [10]. In the following, recent developments and implementations of the IBFM in the studies of the odd-mass and odd-odd heavy nuclei are illustrated, with a

DOI: 10.1201/9781032691633-11

"

special focus on its novel applications to wobbling bands in odd-mass nuclei [11], to chiral doublet bands in γ-soft $A \approx 130$ nuclei [12], and to double-β decay properties [13].

11.2   THE INTERACTING BOSON-FERMION MODEL

11.2.1   BASIC IDEAS

The Hamiltonian of the IBFM system can be expressed as

$$\hat{H} = \hat{H}_{\mathrm{B}} + \hat{H}_{\mathrm{F}} + \hat{V}_{\mathrm{BF}} , \tag{11.1}$$

where $\hat{H}_{\mathrm{B}}$ and $\hat{H}_{\mathrm{F}}$ stand for the Hamiltonians for the IBM and single fermion, respectively. $\hat{V}_{\mathrm{BF}}$ represents the boson-fermion interaction. In what follows a version of the IBM that distinguishes between the neutron and proton degrees of freedom (IBM-2) [8, 14] is considered as it is more realistic than its simpler version (IBM-1), without the distinction between neutron and proton bosons. The building blocks of the IBM-2 are the neutron s_v (with spin and parity $J = 0^+$) and d_v ($J = 2^+$) bosons, and the proton s_π and d_π bosons. Here s_v (s_π) and d_v (d_π) bosons represent the collective monopole and quadrupole pairs of valence neutrons (protons), respectively. The following form of the IBM-2 Hamiltonian is commonly used.

$$\hat{H}_{\mathrm{B}} = \varepsilon_d (\hat{n}_{d_v} + \hat{n}_{d_\pi}) + \kappa \hat{Q}_v \cdot \hat{Q}_\pi . \tag{11.2}$$

Here $n_{d_\rho} = d_\rho^\dagger \cdot \tilde{d}_\rho$ stands for the number operator for neutron ($\rho = v$) or proton ($\rho = \pi$) bosons, with ε_d being single-d-boson energy. The second term on the right-hand side of Eq. (11.2) represents the quadrupole-quadrupole interaction between neutron and proton bosons, with the quadrupole operator for neutrons or protons, $\hat{Q}_\rho = s_\rho^\dagger \tilde{d}_\rho + d_\rho^\dagger \tilde{s}_\rho + \chi_\rho (d_\rho^\dagger \times \tilde{d}_\rho)^{(2)}$. κ is the strength parameter, and χ_v and χ_π are dimensionless parameters. The Hamiltonian of Eq. (11.2) represents the most essential features of quadrupole collective states. It also has a link to the underlying nuclear structure, so that the first term corresponds to the pairing-like interactions between identical particles, keeping the nucleus spherical, while the second term represents the quadrupole-quadrupole interaction between nonidentical nucleons and is responsible for deformations. The competition between these terms determines the degrees and types of quadrupole collectivity for a given nucleus.

The single-nucleon Hamiltonian, $\hat{H}_{\mathrm{F}}$ is identified as

$$\hat{H}_{\mathrm{F}} = -\sum_{j_\rho} \varepsilon_{j_\rho} \sqrt{2 j_\rho + 1} (a_{j_\rho}^\dagger \times \tilde{a}_{j_\rho})^{(0)} \equiv \sum_{j_\rho} \varepsilon_{j_\rho} \hat{n}_{j_\rho} , \tag{11.3}$$

where ε_{j_ρ} stands for the single-particle energy of the odd neutron or proton orbital, j_ρ. $a_{j_\rho}^{(\dagger)}$ represents particle annihilation (creation) operator, and $\tilde{a}_{j_\rho}$ is defined by $\tilde{a}_{j_\rho m_\rho} = (-1)^{j_\rho - m_\rho} a_{j_\rho - m_\rho}$. On the right-hand side of Eq. (11.3), $\hat{n}_{j_\rho}$ stands for the number operator for the unpaired nucleon.

The boson-fermion interaction term, $\hat{V}_{\mathrm{BF}}$, should in principle take a highly complicated form. However, the following simplified form has been shown to be sufficient to describe low-energy properties of odd-mass nuclei [10].

$$\hat{V}_{\mathrm{BF}} = \Gamma_\rho \hat{V}_{\mathrm{dyn}}^\rho + \Lambda_\rho \hat{V}_{\mathrm{exc}}^\rho + A_\rho \hat{V}_{\mathrm{mon}}^\rho , \tag{11.4}$$

where the first, second, and third terms represent the quadrupole dynamical, exchange, and monopole interactions, with the strength parameters Γ_ρ, Λ_ρ, and A_ρ, respectively. The dynamical and exchange interactions play a dominant role in determining the low-lying states of odd-mass nuclei. Especially, the latter term reflects that bosons are made of nucleon pairs. On the other hand, the monopole term has only a minor effect of either compressing or stretching the entire energy spectrum. In the generalized seniority considerations [10, 15], each term in the above expression has been shown to have the j dependent forms:

$$\hat{V}^{\rho}_{\text{dyn}} = \sum_{j_\rho j'_\rho} \gamma_{j_\rho j'_\rho} (a^{\dagger}_{j_\rho} \times \tilde{a}_{j'_\rho})^{(2)} \cdot \hat{Q}_{\rho'}, \tag{11.5}$$

$$\hat{V}^{\rho}_{\text{exc}} = -\left(s^{\dagger}_{\rho'} \times \tilde{d}_{\rho'}\right)^{(2)} \cdot \sum_{j_\rho j'_\rho j''_\rho} \sqrt{\frac{10}{N_\rho (2j_\rho + 1)}} \beta_{j_\rho j'_\rho} \beta_{j''_\rho j_\rho}$$

$$: \left[(d^{\dagger}_{\rho} \times \tilde{a}_{j''_\rho})^{(j_\rho)} \times (a^{\dagger}_{j'_\rho} \times \tilde{s}_{\rho})^{(j'_\rho)}\right]^{(2)} : + (\text{H.c.}), \tag{11.6}$$

$$\hat{V}^{\rho}_{\text{mon}} = \hat{n}_{d_\rho} \hat{n}_{j_\rho}, \tag{11.7}$$

where the factors $\gamma_{j_\rho j'_\rho} = (u_{j_\rho} u_{j'_\rho} - v_{j_\rho} v_{j'_\rho}) q_{j_\rho j'_\rho}$ and $\beta_{j_\rho j'_\rho} = (u_{j_\rho} v_{j'_\rho} + v_{j_\rho} u_{j'_\rho}) q_{j_\rho j'_\rho}$, with $q_{j_\rho j'_\rho} = \langle \ell_\rho \frac{1}{2} j_\rho \| Y^{(2)} \| \ell'_\rho \frac{1}{2} j'_\rho \rangle$ being matrix element of the fermion quadrupole operator in the single-particle basis. $\hat{Q}_{\rho'}$ in Eq. (11.5) denotes the quadrupole operator in the boson system (11.2). The notation $: (\cdots) :$ in Eq. (11.6) stands for normal ordering. It is also mentioned that among the boson-fermion interactions those between unlike bosons, i.e., between neutron (proton) bosons and the odd proton (neutron), dominate the dynamical and exchange terms, while the interactions between like bosons, i.e., between neutron (proton) bosons and the odd neutron (proton), are relevant to the monopole type interaction [10]. The form of the IBFM-2 Hamiltonian, with the boson-fermion interaction being given by the expressions in Eqs (11.5), (11.6), and (11.7), has been frequently employed in realistic calculations for odd-mass nuclei [10].

11.2.2 MICROSCOPIC FORMULATION

The IBM should have certain microscopic foundations on the underlying multi-nucleon dynamics [14, 16, 17]. Microscopic derivation of the IBM-2 Hamiltonian has been extensively studied until 1990's in terms of a schematic shell model, that is, matrix elements of physical observables in the low-seniority states of the nuclear shell model consisting, e.g., of the monopole (S) and quadrupole (D) pairs, are mapped onto the corresponding quantities in the sd-boson system. This approach has been tested and shown to be valid for nearly spherical and weakly deformed regions [14, 16]. Since the late 2000's, a method of deriving the IBM-2 parameters using the results of the self-consistent mean-field (SCMF) calculations has been developed [17, 18]. In this approach, the triaxial quadrupole potential energy surface (PES), $E_{\text{SCMF}}(\beta, \gamma)$, which is obtained from the constrained SCMF calculations within the framework of nuclear density functional theory (DFT), is mapped onto the equivalent PES in the boson system, $E_{\text{IBM}}(\beta, \gamma)$. The Hamiltonian of the IBM-2 is completely determined so that the relation,

$$E_{\text{SCMF}}(\beta, \gamma) \approx E_{\text{IBM}}(\beta, \gamma), \tag{11.8}$$

is satisfied, i.e., the IBM-2 PES becomes similar to the SCMF one. Here, $E_{\text{IBM}}(\beta, \gamma)$ is obtained as the expectation value of the IBM-2 Hamiltonian in state $|\Phi\rangle$, i.e., $E_{\text{IBM}}(\beta, \gamma) = \langle \Phi | \hat{H}_{\text{B}} | \Phi \rangle / \langle \Phi | \Phi \rangle$. $|\Phi\rangle$ denotes the coherent state [19] defined by

$$|\Phi\rangle = \Pi_{\rho = v, \pi} \left[s^{\dagger}_\rho + \beta_\rho \cos \gamma_\rho d^{\dagger}_0 + \frac{1}{\sqrt{2}} \beta_\rho \sin \gamma_\rho (d^{\dagger}_{+2} + d^{\dagger}_{-2}) \right] |0\rangle, \tag{11.9}$$

where β_ρ and γ_ρ are boson analogs of the quadrupole deformations β and γ. They can be related to each other through the formulas [8, 17, 19], $\beta_v = \beta_\pi \propto \beta$ and $\gamma_v = \gamma_\pi = \gamma$. The ket $|0\rangle$ denotes the inert core. The IBM-2 with the parameters determined by the mapping procedure of Eq. (11.8) has been shown to be valid for describing low-energy quadrupole collective states corresponding to the nearly spherical [17, 18], γ-soft [18, 20], and strongly deformed [21] shapes, and further applied to study octupole deformations and collectivity [22], and the shape coexistence [23].

For odd-mass nuclei, the spherical single(quasi)-particle energies, ε_{j_ρ} in the Hamiltonian $\hat{H}_F$, and the occupation probabilities, $v_{j_\rho}^2$, which appear in the expressions for the boson-fermion interactions $\hat{V}_{BF}$, are computed by the same constrained SCMF calculation as that used to determine the IBM-2 Hamiltonian for the even-even nucleus [24]. The three strength parameters of $\hat{V}_{BF}$ (11.4), Γ_ρ, Λ_ρ and A_ρ, are treated as free parameters, and are fit to reproduce a few of the observed lowest-energy levels for each odd-mass nucleus. This semi-microscopic version of the IBFM(-2) framework, with most of the ingredients being specified by the nuclear EDF-SCMF calculations, has been applied in several nuclear structure studies, including the quantum phase transitions in the axially-symmetric [25] and γ-soft [26] odd-mass nuclei, octupole correlations in neutron-rich Ba isotopes [27], and the calculations of single-β [28–31] and double-β decay nuclear matrix elements [13].

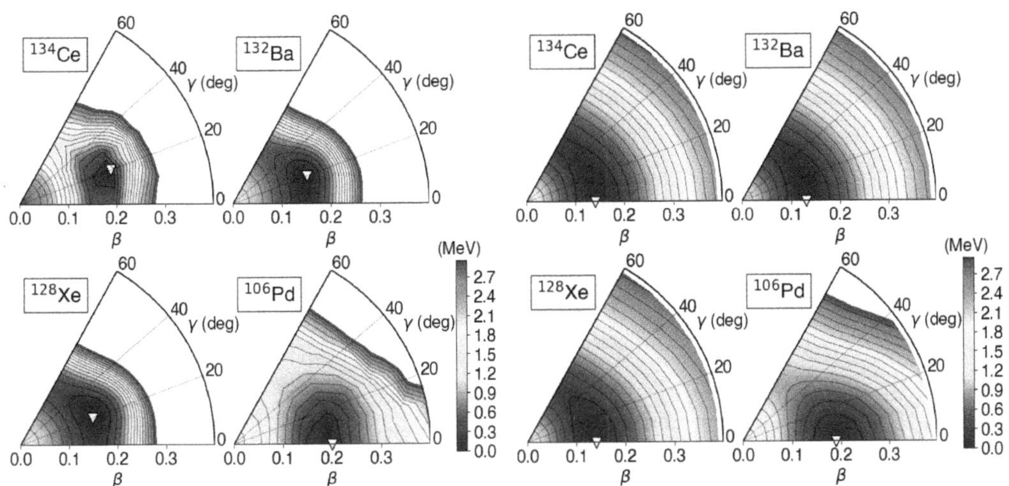

Figure 11.1 Potential energy surfaces for the even-even core nuclei obtained from the SCMF calculations with constraints on triaxial quadrupole deformations β and γ. The Gogny D1M [40] (for ^{132}Ba, and ^{128}Xe), and the DD-PC1 [42] (for ^{134}Ce and ^{106}Pd) EDFs are employed. The total mean-field energies are plotted up to 3 MeV, normalized with respect to the global minimum, represented by an open triangle. The energy difference between the neighboring contours is 0.2 MeV.

11.3 ALTERNATIVE INTERPRETATION OF WOBBLING BANDS

The wobbling motion of nuclei has traditionally been recognized in high spin bands of odd-mass Lu and Ta nuclei with mass $A \approx 160$. Whether or not new regions of the wobbling motions are found is a hot topic that has been under active investigations for the past several years. More recent experiments have actually suggested that wobbling bands occur at relatively low-spin regimes of odd-mass nuclei, e.g., ^{135}Pr [32, 33], ^{133}La [34], ^{105}Pd [35], ^{127}Xe [36], ^{187}Au [37], and ^{183}Au [38]. A criterion of interpreting the non-yrast bands of these nuclei to be wobbling bands has been based on the $E2$ to $M1$ mixing ratio δ. The wobbling bands should exhibit a dominance of the connected electric quadrupole transitions to the yrast bands over the magnetic dipole ones, corresponding to the ratio, δ larger than 1. New measurements on excited bands of ^{187}Au and ^{135}Pr, however, have found $M1$ dominance of the interband transitions. Theoretical descriptions of the recently found wobbling bands have been mostly based on the particle-rotor picture [32–35]. It would be thus interesting to attempt alternative theoretical interpretations to shed lights upon the nonyrast bands that are proposed to be wobbling bands. In Ref. [11], an IBFM-2 interpretation of the newly assigned wobbling bands of the odd-mass nuclei of interest, ^{135}Pr, ^{133}La, ^{127}Xe, and ^{105}Pd, has been presented.

11.3.1 POTENTIAL ENERGY SURFACES

The first step of the analysis is the (β, γ)-deformation constrained SCMF calculations of the PESs for the even-even nuclei ^{134}Ce, ^{132}Ba, ^{128}Xe, and ^{106}Pd, which were taken [11] as the boson cores for ^{135}Pr, ^{133}La, ^{127}Xe, and ^{105}Pd, respectively. The SCMF calculations have been carried out for ^{132}Ba and ^{128}Xe within the Hartree-Fock-Bogoliubov (HFB) model [39] employing the Gogny force with the D1M parametrization [40], and for ^{134}Ce and ^{106}Pd within the relativistic Hartree-Bogoliubov (RHB) method [41] employing the density-dependent point-coupling (DD-PC1) EDF [42] and a separable pairing force of finite range [43]. The calculated PESs, shown on the left-hand side of Fig. 11.1, suggest that most of these even-even nuclei are soft along the γ deformation. For each even-even nucleus, the SCMF PES is mapped onto the IBM-2 one to obtain the corresponding boson Hamiltonian. In the mapped IBM-2 PESs, depicted on the right-hand side of Fig. 11.1, one could see that basic characteristics of the SCMF PESs, including the softness along the γ deformation, are reproduced.

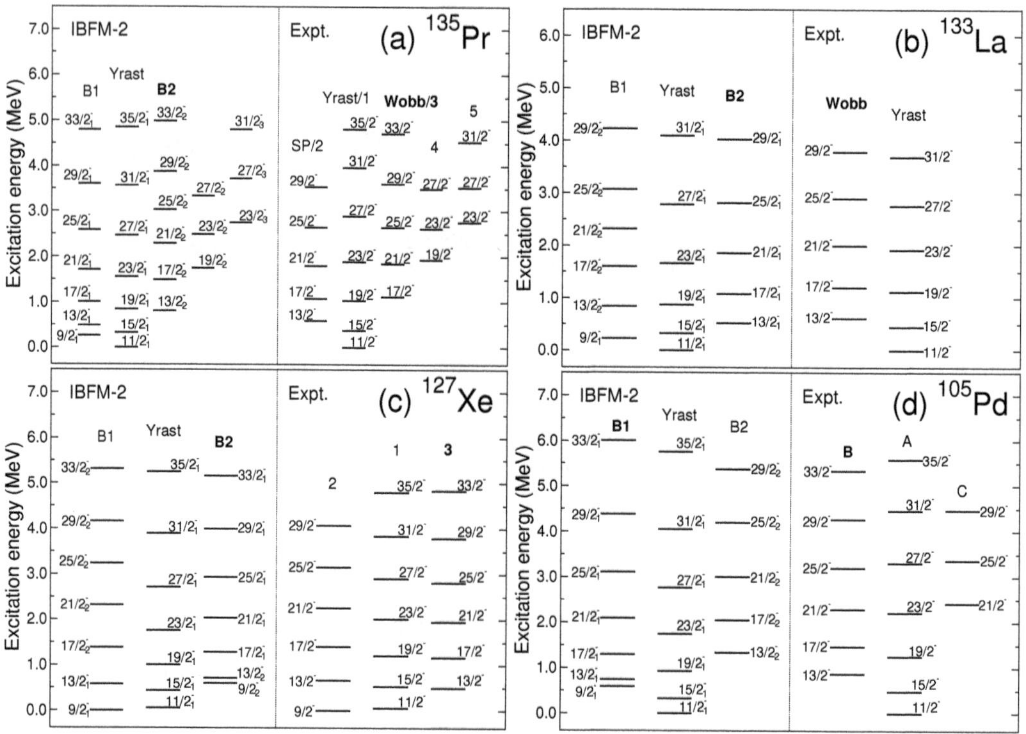

Figure 11.2 Comparisons between the IBFM-2 and experimental low-spin bands for the ^{135}Pr, ^{133}La, ^{127}Xe, and ^{105}Pd nuclei. The predicted lowest, second-lowest, and third lowest bands are labeled as "Yrast", "B1", and "B2", respectively. Each experimental band is labeled a name that is defined in the corresponding reference. That is, for ^{135}Pr, "SP1" and "SP2" are adopted from Ref. [32], and the notations "1" to "5" from [44]. The notations of the experimental bands for ^{133}La, ^{127}Xe, and ^{105}Pd are according to Refs. [34–36], respectively. Those experimental, and theoretical bands with labels in thick letters correspond to the proposed wobbling bands.

11.3.2   ENERGY LEVELS

Having the IBM-2 Hamiltonian fixed, the single-fermion Hamiltonian (11.3), and the boson-fermion interaction (11.4) in the IBFM-2 Hamiltonian are determined by the procedure described earlier. Figure 11.2 shows the low-lying negative-parity bands obtained from the IBFM-2 in comparison with the experimental energy spectra [32, 34–36, 44]. The single-particle spaces for the considered odd-mass nuclei consist only of $1h_{11/2}$ for the odd proton (for ^{135}Pr, ^{133}La and ^{105}Pd) and odd neutron (^{127}Xe). All the negative-parity bands are therefore described by the single orbital $h_{11/2}$, which is coupled to the even-even boson core nuclei. The theoretical bands have been organized so that levels are connected by $\Delta I = 2$ $E2$ transitions. In general, the IBFM-2 reproduces nicely the experimental energy levels.

In the experiment of Ref. [32], the third lowest band of ^{135}Pr, with the bandhead being the $17/2_2^-$ state, has been identified as an one-phonon wobbling band [indicated as "Wobb" in Fig. 11.2(a)]. Besides this wobbling band, the first excited band built on the $13/2_1^-$ state has been considered unfavored signature partner (SP2) band. The theoretical $\Delta I = 2$ band, built on the second $13/2^-$ state (band "B2"), corresponds to the proposed wobbling band. The theoretical counterpart of the SP band is the first excited ("B1") band. One sees that the IBFM-2 calculation yields two additional bands, built on the $19/2_2^-$ and $23/2_3^-$ states, in agreement with the experimental bands 4 and 5 [44]. The predicted interband $E2$ transitions between the $19/2_2^-$ and $23/2_3^-$ bandheads are much weaker than the inband $E2$ transition, $23/2_2^- \to 19/2_2^-$, within the $19/2_2^-$ band. This finding contradicts the earlier experiment of Ref. [33], which interpreted the $19/2_2^-$, $23/2_3^-$ and $27/2_3^-$ states to form a single band of two-phonon wobbling nature, but agrees with the more recent measurement of Ref. [44]. The two additional bands studied in Ref. [44] do not follow the rotor formula, i.e., the $I(I+1)$ sequence of energy levels, but appear to be more vibrational, thus exhibiting a fingerprint of the γ softness in this mass region.

For ^{133}La [Fig. 11.2(b)], the second lowest band, built upon the $13/2^-$ level and denoted as "Wobb", has been interpreted as wobbling band by the experiment of Ref. [34]. The theoretical counterpart is the band "B2" shown on the left-hand side of Fig. 11.2(b). The assignment of the theoretical "B2" band to be the observed wobbling band is mainly based on the fact that the calculated $E2$ to $M1$ ratios agree with the data.

^{127}Xe is here the only nucleus in which the odd particle is neutron. Experimentally, the second excited negative-parity band ["band 3" in Fig. 11.2(c)] has been assigned to be wobbling band. The IBFM-2 band "B2" is the theoretical counter part of band 3 for ^{127}Xe.

In Fig. 11.2(d), the first excited band for ^{105}Pd, denoted as "B", has been interpreted as wobbling band [35], and it is the first evidence for a low-spin wobbling band in the mass $A \approx 100$ region. The corresponding IBFM-2 band is the one built on the $9/2_1^-$ state.

11.3.3   MIXING RATIOS

The $E2$ to $M1$ mixing ratio, δ, is computed by the formula,

$$\delta = 0.835 \times E_\gamma \frac{\langle I_f \| \hat{T}^{(E2)} \| I_i \rangle}{\langle I_f \| \hat{T}^{(M1)} \| I_i \rangle}, \tag{11.10}$$

where $E_\gamma = E_{I_i} - E_{I_f}$ is the difference between the excitation energies of the initial I_i and final I_f states, and $\langle I_f \| \hat{T}^{(E2)} \| I_i \rangle$ and $\langle I_f \| \hat{T}^{(M1)} \| I_i \rangle$ are the reduced $E2$ and $M1$ transition matrices, respectively. The corresponding transition operators for the IBFM-2 are defined as

$$\hat{T}^{(E2)} = e_\nu^B \hat{Q}_\nu + e_\pi^B \hat{Q}_\pi$$

$$- \frac{1}{\sqrt{5}} \sum_{j_\rho j_\rho'} (u_{j_\rho} u_{j_\rho'} - v_{j_\rho} v_{j_\rho'}) \left\langle \ell_\rho \frac{1}{2} j_\rho \left\| e_\rho^F r^2 Y^{(2)} \right\| \ell_\rho' \frac{1}{2} j_\rho' \right\rangle (a_{j_\rho}^\dagger \times \tilde{a}_{j_\rho'})^{(2)}, \tag{11.11}$$

where the terms in the first and second lines are the bosonic and fermionic $E2$ operators, respectively. $\hat{Q}_\rho$ is the quadrupole operator for neutron or proton bosons, defined in Eq. (11.2). We assume that the effective $E2$ charges for proton e_π^B and neutron e_ν^B bosons are equal to each other, $e_\pi^B = e_\nu^B \equiv e^B$, and fix e^B so that the experimental $B(E2; 2_1^+ \to 0_1^+)$ rate of each even-even core nucleus is reproduced. For the fermion part, standard effective charge $e^F = 1.5\,(0.5)\,eb$ is adopted for the odd proton (neutron). The $M1$ transition operator reads:

$$\hat{T}^{(M1)} = \sqrt{\frac{3}{4\pi}}\left(g_\pi^B \hat{L}_\pi + g_\nu^B \hat{L}_\nu\right)$$

$$- \sqrt{\frac{1}{4\pi}} \sum_{j_\rho j_\rho'} (u_{j_\rho} u_{j_\rho'} + v_{j_\rho} v_{j_\rho'}) \langle j_\rho \| g_l^\rho \ell + g_s^\rho \mathbf{s} \| j_\rho' \rangle (a_{j_\rho}^\dagger \times \tilde{a}_{j_\rho'})^{(1)}, \qquad (11.12)$$

where the terms in the first and second lines are bosonic and fermionic contributions, respectively. The effective neutron boson g_ν^B and proton boson g_π^B factors are chosen to be $g_\pi^B \approx 1.0\,\mu_N$, and $g_\nu^B \approx 0\,\mu_N$ (see Ref. [11]). For the odd neutron (neutron) g-factors, the Schmidt values $g_l = 0\,\mu_N$ and $g_s = -3.82\,\mu_N$ ($g_l = 1.0\,\mu_N$ and $g_s = 5.58\,\mu_N$) are used, with g_s quenched by 30% with respect to the free value.

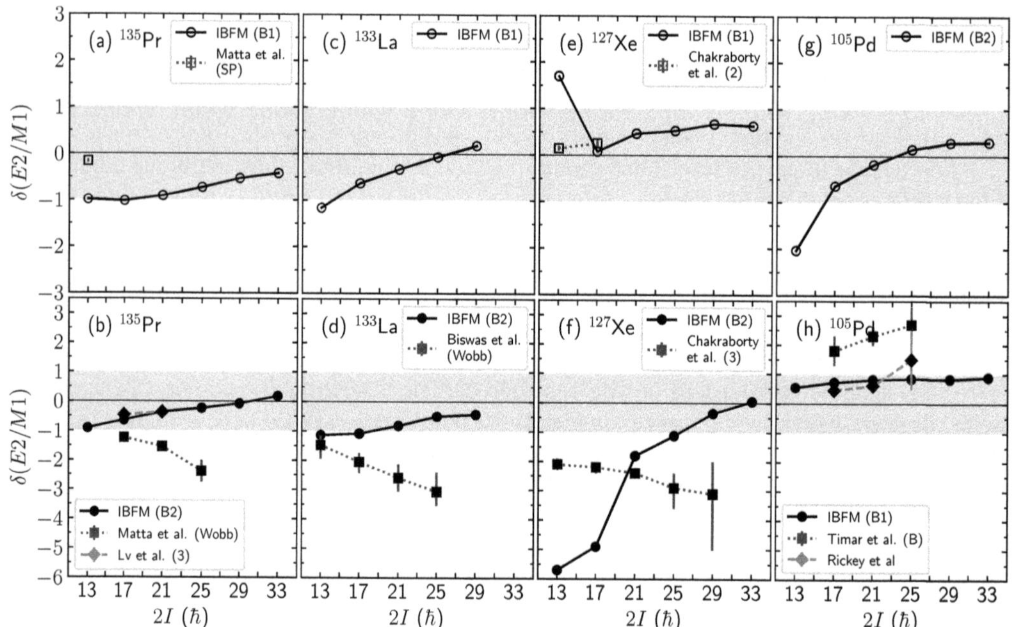

Figure 11.3 The mixing ratios δ of the $E2$ to $M1$ $\Delta I = 1$ transitions from yrare to yrast bands for the odd-mass nuclei ^{135}Pr, ^{133}La, ^{127}Xe, and ^{105}Pd. The shaded area in each panel indicates $|\delta| < 1$. The notations of the experimental bands follow those used in Refs. [32, 34–36, 44, 45]. The calculated and experimental δ's represented by solid symbols correspond to those for the proposed wobbling bands.

Figure 11.3 compares between the calculated and experimental δ for the interband $\Delta I = 1$ transitions from the yrare to yrast bands. Firstly one notices that the predicted δ's for both the B1 $\to$ Yrast and B2 $\to$ Yrast transitions are, in most cases, small in magnitude, i.e., within the range $|\delta| < 1$.

For ^{135}Pr, the calculated δ's are close to zero for both the B1 $\to$ Yrast and B2 $\to$ Yrast band transitions. In Fig. 11.3(b) the measured δ values for the Wobb $\to$ Yrast bands in Ref. [32] increase in magnitude to be larger than $|\delta| = 1$. The recent experiment of Ref. [44], on the other hand, gives much smaller mixing ratios for the same band (band 3), and agrees nicely with the IBFM-2 prediction.

Similarly, the calculated δ values for ^{133}La are generally within the range $|\delta| < 1$. Especially the δ ratios for the B2 $\rightarrow$ Yrast transitions from the IBFM-2 are much smaller in magnitude than those experimental values of Ref. [34] for the band "Wobb" assigned to be wobbling band, but are rather close to those predicted by the quasiparticle-plus-rotor model calculations performed also in Ref. [34].

The absolute values of the δ mixing ratios for the B1 $\rightarrow$ Yrast transitions for the ^{127}Xe nucleus are small. As seen in Fig. 11.3(f), absolute values of the corresponding δ ratios for the states with $I \geqslant 21/2$ for the B2 $\rightarrow$ Yrast interband transitions are also less than 1. However, too large $|\delta|$ absolute values, $|\delta| \approx 5$, are obtained at lower spins for the B2 $\rightarrow$ Yrast transition for ^{127}Xe. This is due to the fact that the corresponding $M1$ matrix elements have been calculated to be too small. The Gogny-HFB calculation here performed for ^{127}Xe suggests that the $\nu h_{11/2}$ orbital is half filled, i.e., $v^2_{h_{11/2}} \approx 0.5$, in which case the dynamical and exchange terms in the boson-fermion interaction becomes rather sensitive to the choice of the corresponding strength parameters. The adopted strength parameters for these terms might have yielded considerable amounts of configuration mixing in low-spin states, thus leading to the too small $M1$ rates.

For ^{105}Pd, the calculated δ's with the IBFM-2 for the interband transitions of both the B1 and B2 bands are all less than 1, except for the B2 $\rightarrow$ Yrast transition at $I = 13/2^-$. The experimentally proposed wobbling band [35] is band B, the δ's of which measured for the $I = 17/2$, $21/2$, and $25/2$ states are much larger than the IBFM-2 counterparts. The δ values obtained with the IBFM-2 for the B1 $\rightarrow$ Yrast transitions are rather close to the earlier experiment reported by Rickey *et al.* [45].

11.4 CHIRAL BANDS IN ODD-ODD NUCLEI

Another characteristic of collective motions in γ-soft (triaxial) nuclei is the appearance of chiral doublet bands and various selection rules of the $E2$ and $M1$ transitions of high-spin states. Candidate nuclei for the chiral bands have been suggested in several mass regions, e.g., odd-odd Rh and Ag nuclei with $A \approx 100$, and odd-odd Cs and La nuclei with $A \approx 130$. See Ref. [46], and a compilation of experimental data on the chiral doublet candidate nuclei is found. In particular, the chiral bands in the odd-odd nuclei around the $A \approx 130$ Cs regions are empirically expected to be formed by the two-quasiparticle pair of the odd neutron hole and odd proton, i.e., $[(\nu h_{11/2})^{-1} \otimes (\pi h_{11/2})^1]^{(J)}$, coupled to even-even Xe nuclei.

11.4.1 INTERACTING BOSON FERMION-FERMION MODEL

The interacting boson-fermion-fermion model (IBFFM-2) [10, 47] is an extension of the IBFM-2 to study odd-odd nuclei. In the IBFFM-2, both an unpaired neutron and an unpaired proton degrees of freedom, and neutron-proton interactions are considered. The Hamiltonian of the IBFFM-2 is then expressed as

$$\hat{H} = \hat{H}_B + \hat{H}_F^{\nu} + \hat{H}_F^{\pi} + \hat{V}_{BF}^{\nu} + \hat{V}_{BF}^{\pi} + \hat{V}_{\nu\pi} \ . \tag{11.13}$$

Here $\hat{H}_B$, $\hat{H}_F^{\rho}$ and $\hat{V}_{BF}^{\rho}$ are the IBM-2 Hamiltonian (11.2), single-neutron (or proton) Hamiltonian (11.3), and the terms representing the interaction between the single-neutron (or proton) and the IBM-2 core (11.4), respectively. The last term, $\hat{V}_{\nu\pi}$, is the residual interaction between the odd neutron and proton, which is given as

$$\hat{V}_{\nu\pi} = 4\pi(\nu_d + \nu_{ssd}\sigma_\nu \cdot \sigma_\pi)\delta(\mathbf{r})\delta(|\mathbf{r}_\nu| - \mathbf{R_0})\delta(|\mathbf{r}_\pi| - \mathbf{R_0})$$

$$- \frac{1}{\sqrt{3}}\nu_{ss}\sigma_\nu \cdot \sigma_\pi + \nu_T\left[\frac{3(\sigma_\nu \cdot \mathbf{r})(\sigma_\pi \cdot \mathbf{r})}{r^2} - \sigma_\nu \cdot \sigma_\pi\right] \ . \tag{11.14}$$

In the above equation, the first term consists of the δ, and spin-spin δ terms, while the second and third terms represent, respectively, the spin-spin, and tensor interactions. v_d, v_{ssd}, v_{ss}, and v_T are strength parameters. $\mathbf{r} = \mathbf{r}_\nu - \mathbf{r}_\pi$ is the relative coordinate of the neutron and proton, and $R_0 = 1.2A^{1/3}$ fm.

The IBFFM-2 has previously been employed for realistic nuclear structure studies such as the chirality [48], single-β [49], and double-β decays [50]. These studies have been made in purely phenomenological ways, so that all the model parameters including those for the even-even nuclei are obtained from experiment. Another remarkable consequence of the IBFFM-2 is that it gives rise to dynamical supersymmetries that simultaneously describe an even-even, and odd-mass and odd-odd neighboring nuclei based on group theory [51–53].

As in the case of the IBFM-2 for the odd-mass systems, most of the building blocks of the IBFFM-2 have been shown [54] to be obtained from the SCMF calculations employing the same EDF as the one used for the neighboring even-even and odd-mass nuclei. The coupling constants of the interaction $\hat{V}^\nu_{\mathrm{BF}}$ ($\hat{V}^\pi_{\mathrm{BF}}$) can be determined by fitting to low-energy spectra of the neighboring odd-N (odd-Z) nucleus [24]. The parameters in (11.14), v_d, v_{ssd}, v_{ss}, and v_T, have also been determined to reproduce a few low-lying states of each odd-odd nucleus. In a number of studies using the IBFFM-2, however, it has been found that only one or two terms in $\hat{V}_{\nu\pi}$ are enough to reproduce energy spectra of odd-odd nuclei.

The Gogny-HFB SCMF calculations performed for the Ba and Xe isotopes in the mass $A \approx 130$ region have suggested [12, 26] many instances exhibiting notable γ softness in their triaxial quadrupole PESs. In Ref. [12], the odd-odd nuclei $^{124-132}$Cs have been considered, with the even-even core nuclei being taken to be $^{124-132}$Xe. The single-particle energies and occupation probabilities for the neighboring odd-mass Cs and Xe and odd-odd Cs nuclei, which are needed to build the IBFFM-2 Hamiltonian, have been provided by the same Gogny-HFB calculations as for the even-even core nuclei. In this section, the odd-odd Cs are mainly considered, while the results for the even-even Xe, and odd-mass Xe and Cs nuclei are found in Refs. [12, 26].

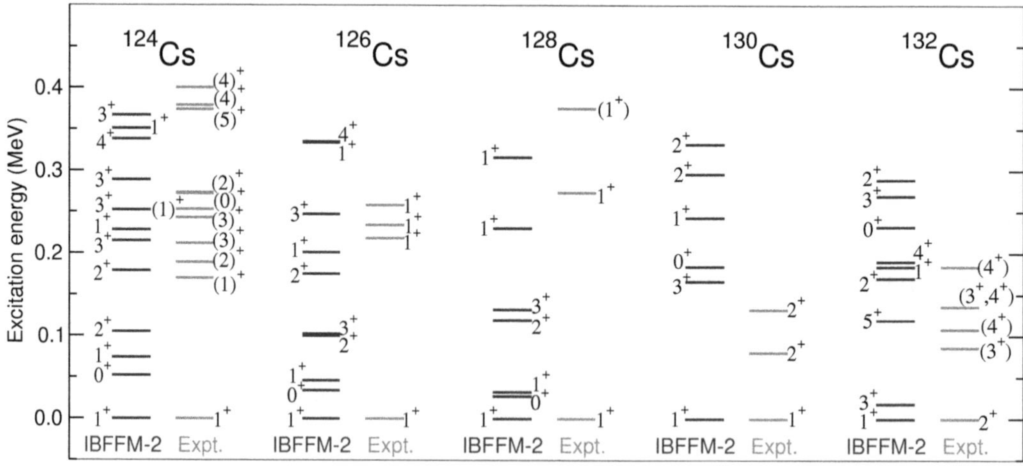

Figure 11.4 Calculated and experimental [55] energy spectra for low-spin low-energy positive-parity states of odd-odd Cs nuclei.

11.4.2 ENERGY SPECTRA FOR ODD-ODD CS

Figure 11.4 depicts the calculated energy spectra of low-spin low-energy positive-parity states of the odd-odd nuclei $^{124-132}$Cs obtained from the IBFFM-2, as compared with the experimental data.

Most of those low-lying positive-parity states below the excitation energy of $E_x \lesssim 0.4$ MeV are made of the pair configurations of the type $[(\nu sdg)^{-1} \otimes (\pi sdg)^1]^{(J)}$ coupled to the even-even boson core nuclei $^{124-132}$Xe. For most of these Cs nuclei, the ground-state spin 1^+ is correctly reproduced by the IBFFM-2.

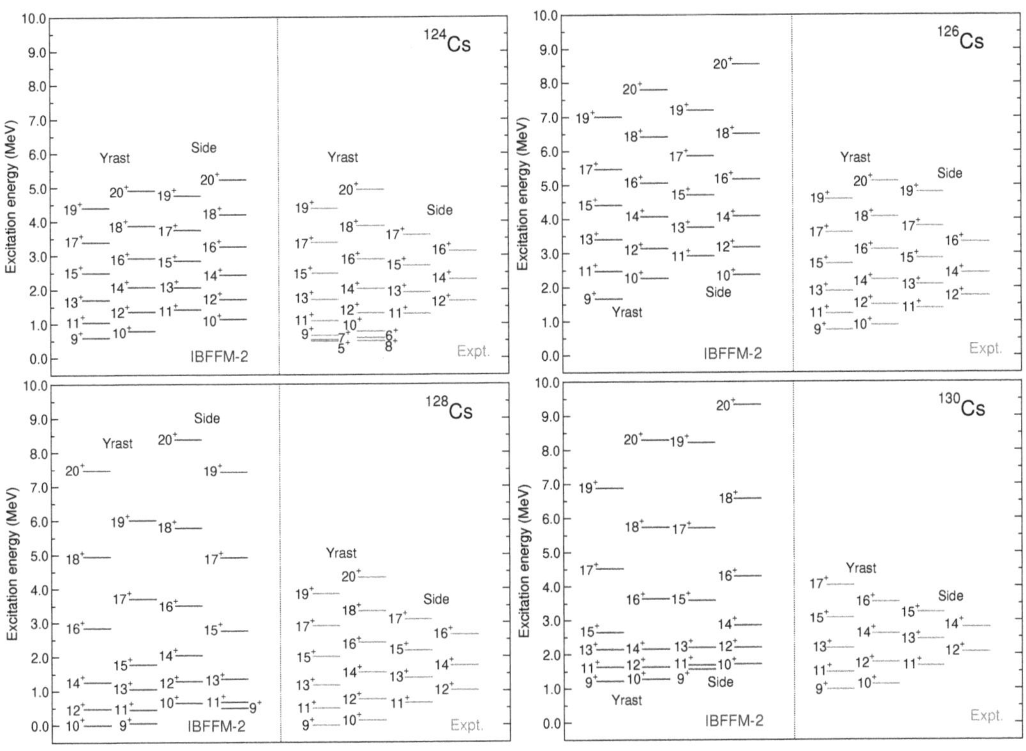

Figure 11.5   Calculated and experimental [55] higher-spin bands of the odd-odd nuclei ^{124}Cs, ^{126}Cs, ^{128}Cs and ^{130}Cs.

Higher-spin band structures of the $^{124-132}$Cs nuclei are shown in Fig. 11.5. Here, for both the theoretical and experimental bands of each nucleus, two pairs of bands are associated with the yrast and side bands, respectively. The theoretical bands are here organized so that the lowest energy states with spin I_1, and the second lowest states with spin I_2 belong to the yrast and side bands, respectively. One sees from Fig. 11.5 that the IBFFM-2 produces doublet-like bands, where close-lying states with the same spin I appear.

The IBFFM-2 reproduces reasonably the experimental band structure of ^{124}Cs. For ^{126}Cs, however, the excitation energies of the bandhead states are overestimated by 1-2 MeV, while the observed band structure is described rather well.

As for ^{128}Cs, absolute energies of the high spin positive-parity bands have not been established, hence the corresponding experimental bands in Fig. 11.5 are shown with respect to the energy of the 9_1^+ yrast state. Accordingly, the theoretical bands for ^{128}Cs are plotted with respect to the excitation energy of the 10_1^+ state, which is calculated to be 543 keV.

As compared to the observed bands, which look harmonic, each predicted band for ^{128}Cs seems rather rotational and is stretched in energy. This is also the case with ^{130}Cs, whereas the bandhead energies of each band is reproduced well. The discrepancy reflects the fact that the configuration space for the IBFFM-2 becomes more restricted as one approaches the neutron major shell closure $N = 82$. Similar energy spectra as ^{130}Cs are obtained for ^{132}Cs.

11.4.3 ELECTROMAGNETIC PROPERTIES

In addition to the energy levels and band structure of higher-spin states, the electromagnetic transition properties serve as an indicator for the chirality. One of the best candidates for the chirality is the nucleus ^{128}Cs, where the observed $B(E2; I \rightarrow I-2)$ and $B(M1; I \rightarrow I-1)$ intraband and interband transitions of the yrast and side bands exhibit odd-even-spin staggering patterns as functions of spin [56]. Using a schematic particle-rotor model [57], selection rules for the intraband and interband $E2$ and $M1$ transition rates have been considered. These are used as a benchmark for more realistic calculations.

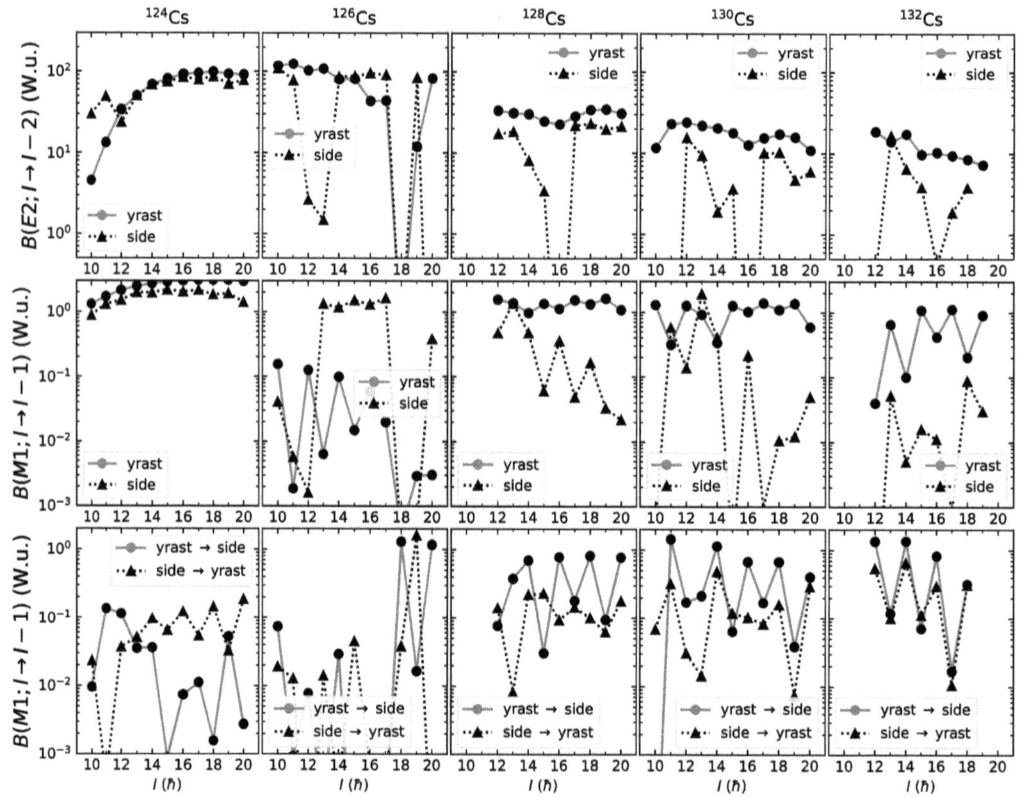

Figure 11.6 Calculated $B(E2; I \rightarrow I-2)$ and $B(M1; I \rightarrow I-1)$ transition strengths in Weisskopf units (W.u.) for the positive-parity bands of $^{124-132}$Cs, corresponding to (upper row) the $\Delta I = 2$ intraband $E2$ transitions for the yrast and side bands, (middle row) the $\Delta I = 1$ intraband $M1$ transitions for the yrast and side bands, and (lower row) the interband $\Delta I = 1$ $M1$ transitions from the yrast to side bands, and vice versa.

In the upper row of Fig. 11.6, calculated intraband $B(E2; I \rightarrow I-2)$ transition rates for the yrast and side bands of the considered odd-odd Cs nuclei are displayed as functions of spin. In contrast to the simplified particle-rotor model prediction of Ref. [57], the IBFFM-2 results for the $B(E2)$ rates hardly show a definite staggering pattern, but rather exhibit irregular behaviors, as they vanish at some particular spins. On the other hand, one could identify, in the middle and lower rows of Fig. 11.6, certain odd-even-spin staggering in the behaviors of the intraband and interband $B(M1; I \rightarrow I-1)$ transitions. For the chiral doublet candidate nucleus, ^{128}Cs, in particular, the predicted $B(M1)$ values agree reasonably well with the measured ones of Ref. [58] both qualitatively and quantitatively. There are, however, also some irregularities in the predicted $B(M1)$ rates. For instance, the calculated $B(M1; I \to I-1)$ intraband transitions for the yrast and side bands of ^{124}Cs, and the intraband transitions for the side bands of ^{126}Cs only gradually change or stay constant for spins within the range $13^+ \leqslant I \leqslant 17^+$.

The major reason why the irregularities occur in the predicted $B(E2)$ and $B(M1)$ systematic is that, for the higher-spin states, considerable mixing among different pair configurations are present in the IBFFM-2 calculations, i.e., in addition to the configuration $[(vh_{11/2})^{-1} \otimes (\pi h_{11/2})^1]^{(J)}$, there may arise contributions from other pair configurations that contain $s_{1/2}$, $d_{3/2}$, $d_{5/2}$ and $g_{7/2}$ single-particle orbitals. In fact, for ^{126}Cs main components of the IBFFM-2 wave functions for the yrast states with spin up to $I \leqslant 17^+$ are based on the pair configurations constructed by the neutron and proton normal parity sdg orbitals. The effect of the configuration mixing could be so significant that in some cases it may not be adequate to simply assign the lowest energy states with spin I_1 and the second lowest energy states with I_2 into the yrast and side bands, respectively.

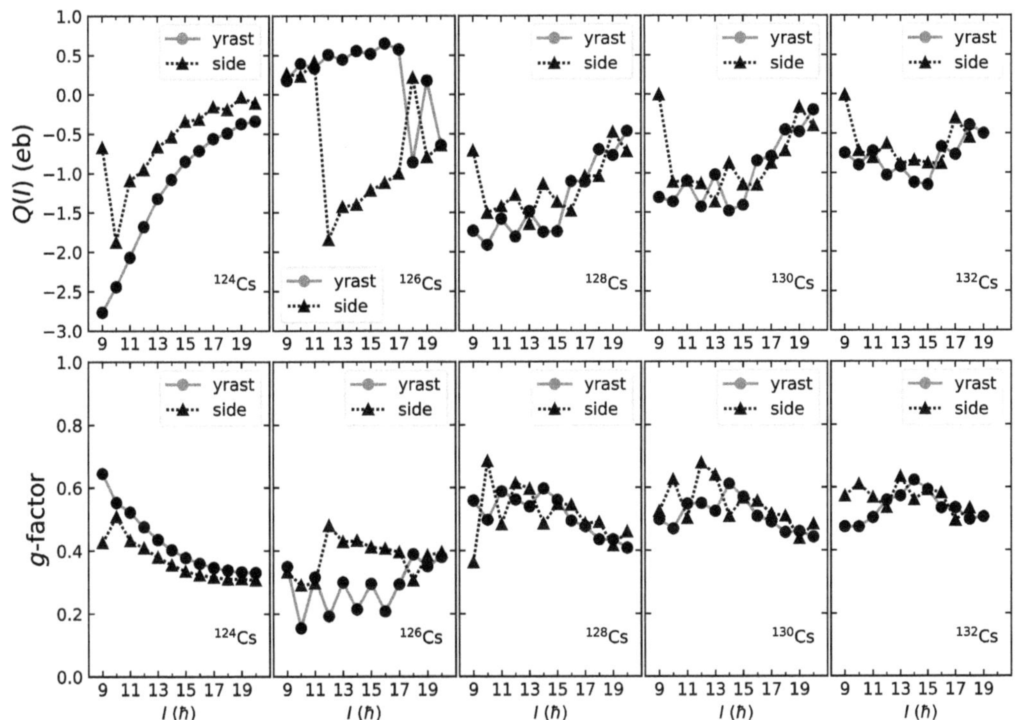

Figure 11.7 The calculated electric quadrupole $Q(I)$ and g factors for the higher-spin positive-parity yrast and side bands for $^{124-132}$Cs. Experimental g factor of $+0.59 \pm 0.01$ for the 9_1^+ yrast state of ^{128}Cs [58] is indicated by an open circle.

Furthermore, in Fig. 11.7 shows calculated electric quadrupole moment $Q(I)$ and g factors of the yrast and side bands of the odd-odd Cs. Of particular interest is the fact that both $Q(I)$ moments and g factors are similar between the yrast and side bands. This feature could be another signature that these bands comprise partner bands of the chiral doublets. An exception is ^{126}Cs, for which the calculated $Q(I)$ moments for the yrast and side bands are substantially different for the spins with $12^+ \leqslant I \leqslant 17^+$. This deviation correlates with the fact that, for this particular nucleus, the nature of the IBFFM-2 wave functions happens to be different between the two bands. That is, the side band is determined mainly by the $[(vh_{11/2})^{-1} \otimes (\pi h_{11/2})^1]^{(J)}$ particle-hole pair configuration, whereas the yrast band is dominated by the pair components composed of the normal-parity orbitals. It is also worth to note that the calculated g factor for the 9^+ yrast state of ^{128}Cs is in a very good agreement with the experimental value of $+0.59 \pm 0.01$ [58].

11.5 CALCULATION OF DOUBLE-β DECAY MATRIX ELEMENTS

The quantitative descriptions of low-lying states of odd-mass and odd-odd nuclei are also essential to compute nuclear decay processes. In particular, accurate theoretical calculations, as well as precise measurement by experiments, on the β decay rates are vital to model astrophysical processes producing heavy chemical elements. Methods of studying the single-β decay could be further utilized to investigate double-β decay, a rare process in which two successive β decays occur between neighboring even-even heavy nuclei. Particularly, the neutrino-less double-β decay ($0\nu\beta\beta$), a type of the double-β decay that does not emit neutrinos, is fundamental importance, since observation of this process would imply violations of various symmetry requirements for the electroweak interaction, and help deepening our current understanding of the nature of neutrinos [59]. Both the single- and double-β decay processes should be sensitive to the wave functions for the initial and final nuclei computed with a given nuclear structure model, and hence serve as a testing ground for nuclear structure models.

By using as microscopic input the EDF-SCMF calculations, the IBFM-2 and IBFFM-2 have been used to analyze the β properties for the even- [29] and odd-mass [28] γ-soft nuclei in the mass $A \approx 130$ region, for the transitional As and Ge nuclei in the mass $A \approx 70$ region [30], and for the neutron-rich Rh and Pd nuclei with the mass $A \approx 100 - 120$ [31]. This section highlights an application of the mapped IBM-2 and IBFFM-2 [13] to calculate the two-neutrino double-β decay ($2\nu\beta\beta$) properties of a large number of the candidate nuclei. In that study, in addition to the initial and final even-even nuclei the intermediate states of the neighboring odd-odd nuclei have been calculated using the IBFFM-2, without assuming closure approximation.

As the starting point, the constrained RHB calculations using the relativistic functional DD-PC1 [42] and a separable pairing force of finite range [43] have been performed to provide inputs to determine the IBM-2 Hamiltonian for initial and final even-even nuclei, and quasiparticle energies and occupation probabilities for the intermediate odd-odd nuclei. The mapped IBM-2 and IBFFM-2 have been shown to provide the wave functions of the even-even and odd-odd nuclei involved in the $2\nu\beta\beta$ decay processes, which give reasonable description of low-lying excitation spectra and electromagnetic transition properties of each nucleus (see, Ref. [13] for further details).

The $2\nu\beta\beta$ decay nuclear matrix element (NME), $M_{2\nu}$, is given as

$$M_{2\nu} = g_A^2 \cdot m_e c^2 \left[M_{2\nu}^{\mathrm{GT}} - \left(\frac{g_V}{g_A} \right)^2 M_{2\nu}^{\mathrm{F}} \right], \tag{11.15}$$

with $g_V = 1$ and $g_A = 1.269$ the vector and axial vector coupling constants, respectively. $M_{2\nu}^{\mathrm{GT}}$ and $M_{2\nu}^{\mathrm{F}}$ are Gamow-Teller (GT) and Fermi (F) matrix elements, respectively, and are given by

$$M_{2\nu}^{\mathrm{GT}} = \sum_N \frac{\langle 0_F^+ \| \hat{t}^+ \hat{\sigma} \| 1_N^+ \rangle \langle 1_N^+ \| \hat{t}^+ \hat{\sigma} \| 0_{1,I}^+ \rangle}{E_N - E_I + \frac{1}{2}(Q_{\beta\beta} + 2m_e c^2)}, \tag{11.16}$$

$$M_{2\nu}^{\mathrm{F}} = \sum_N \frac{\langle 0_F^+ \| \hat{t}^+ \| 0_N^+ \rangle \langle 0_N^+ \| \hat{t}^+ \| 0_{1,I}^+ \rangle}{E_N - E_I + \frac{1}{2}(Q_{\beta\beta} + 2m_e c^2)}, \tag{11.17}$$

where $\hat{t}^{\pm}$ represents the isospin raising or lowering operator, $\hat{\sigma}$ is the spin operator, $Q_{\beta\beta}$ is the double-β decay Q value, and E_I (E_N) stands for the energy of the initial (intermediate) state. The sums in Eqs. (11.16) and (11.17) have been taken over all the intermediate states 1_N^+ and 0_N^+ with the excitation energies E_x below 10 MeV. The operators for the fermionic systems, $\hat{t}^{\pm}$ and $\hat{t}^{\pm}\hat{\sigma}$, are mapped onto the IBFFM-2 counterparts:

$$\hat{t}^{\pm}\sigma \mapsto \hat{T}^{GT} = \sum_{j_v j_\pi} \eta^{GT}_{j_v j_\pi} \left(\hat{P}_{j_v} \times \hat{P}_{j_\pi}\right)^{(1)} , \quad \hat{t}^{\pm} \mapsto \hat{T}^{F} = \sum_{j_v j_\pi} \eta^{F}_{j_v j_\pi} \left(\hat{P}_{j_v} \times \hat{P}_{j_\pi}\right)^{(0)} , \qquad (11.18)$$

where the coefficients η's read

$$\eta^{GT}_{j_v j_\pi} = -\frac{1}{\sqrt{3}} \left\langle \ell_v \frac{1}{2} j_v \left\| \sigma \right\| \ell_\pi \frac{1}{2} j_\pi \right\rangle \delta_{\ell_v \ell_\pi} , \quad \eta^{F}_{j_v j_\pi} = -\sqrt{2 j_v + 1}\, \delta_{j_v j_\pi} , \qquad (11.19)$$

and $\hat{P}_{j_\rho}$'s are particle transfer operators with coefficients that were shown [60] to be functions of the occupation probabilities $v^2_{j_\rho}$ for the intermediate odd-odd nuclei, which are provided also by the RHB-SCMF calculations. No adjustable parameter is introduced in this formalism.

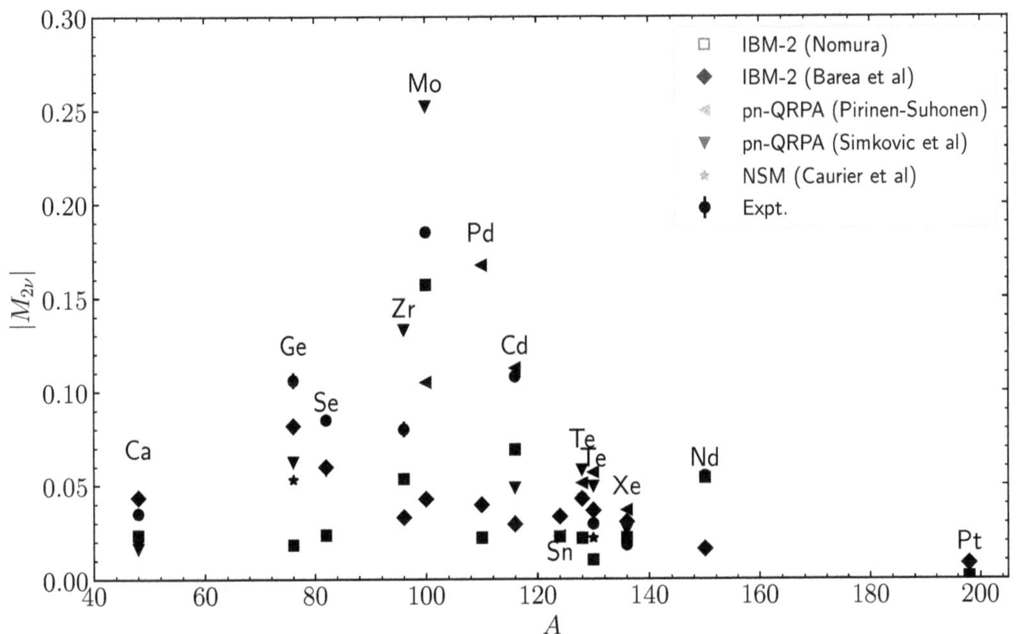

Figure 11.8   Calculated $2\nu\beta\beta$-decay NMEs (11.15) in Ref. [13] ["IBM-2 (Nomura)"]. The experimental NMEs and those obtained from the pn-QRPA [62, 63], from the IBM-2 [64], and from the NSM [65] are also shown.

Figure 11.8 depicts the NMEs obtained from the mapped IBFFM-2 ["IBM-2 (Nomura)"] for the $0^+_1 \rightarrow 0^+_1$ $2\nu\beta\beta$ decays of the 26 even-even nuclei ^{48}Ca, ^{76}Ge, ^{82}Se, ^{96}Zr, ^{100}Mo, ^{110}Pd, ^{116}Cd, ^{124}Sn, ^{128}Te, ^{130}Te, ^{136}Xe, ^{150}Nd, and ^{198}Pt. Here the $Q_{\beta\beta}$ values in Eqs. (11.16) and (11.17) have been taken from experiment, and the g_A factor in Eq. (11.15) has been quenched so that it depends on the mass A exponentially: $g_{A,\text{eff}} = g_A \cdot e^{-0.008A}$. The calculated values are compared to the earlier theoretical predictions within the proton-neutron quasiparticle random phase approximation (pn-QRPA) [62, 63], phenomenological IBM-2 with closure approximation [64], and nuclear shell model (NSM) [65]. For all the theoretical NMEs shown in Fig. 11.8, both the GT and Fermi matrix elements have been quenched with the effective g_A factors defined in those references. The experimental data correspond to the effective NMEs, extracted from the half-lives $\tau^{(2\nu)}_{1/2}$ data of Ref. [66].

A well-known fact is that different theoretical approaches predict double-β NMEs that differ from each other by a factor of 2-3. This also seems to be the case with the $2\nu\beta\beta$ NME predictions indicated in Fig. 11.8. One sees in the figure that the NMEs obtained from the mapped IBM-2 are not very far from, and show similar mass dependence to, the earlier theoretical values. Exceptions are found for the ^{76}Ge$\to^{76}$Se and ^{82}Se$\to^{82}$Kr decay NMEs, for which the mapped IBM-2 gives considerably smaller values than the other theoretical methods, as well as the experimental data. For these decays, a much larger amount of quenching of the g_A factor seems to be necessary within the mapped IBM-2 framework. The necessity of the quenching, in general, implies certain deficiencies of a given nuclear model, which could arise from various approximations and simplifications considered in the model. In the present case, the too small $M_{2\nu}$ values could be attributed to the restricted forms and chosen parameters for the IBM-2 Hamiltonian and boson-fermion interactions, to the limited model spaces for both the IBM-2 and IBFFM-2, to the particular choice of the EDF used for the SCMF calculations, or combinations of them [13].

11.6 SUMMARY

To summarize, characteristic features of γ-soft nuclei – wobbling and chirality, have been studied within the revised version of the IBFM. The model has not often been used to study these collective properties. The low-lying structure of even-even nuclei is here described in terms of bosons that reflect collective nucleon pairs. For the odd-mass and odd-odd nuclei, unpaired nucleon degrees of freedom are explicitly taken into account, and the interactions between the even-even boson core and odd nucleons are modeled in a similar fashion to the particle-core coupling. The IBM-2 Hamiltonian for the even-even core nuclei, single(quasi)-particle energies and occupation probabilities of the odd particle in the odd-mass and odd-odd systems are determined by using the results of the SCMF calculations with a choice of the EDF.

The IBFM-2 calculations have been performed for the low-spin negative-parity bands of the odd-mass nuclei ^{135}Pr, ^{133}La, ^{127}Xe, and ^{105}Pd, which were identified as wobbling bands by earlier measurements. The predicted $E2/M1$ mixing ratios for the yrare to yrast bands for these nuclei are small, $\delta \approx 1$, suggesting the $M1$ dominance of the $\Delta I = 1$ transitions of the yrare bands in question to the yrast bands. This result contradicts the wobbling interpretations of these bands, but are rather in good agreement with the more recent measurements for ^{135}Pr and the much earlier data for ^{105}Pd. This application sheds light upon the nature of the low-spin bands of γ-soft nuclei, while questioning their wobbling interpretation.

Furthermore, the IBFFM-2 has been employed to analyze high-spin positive-parity bands of odd-odd Cs nuclei, that are considered candidates for the chiral doublet bands. The predicted $B(M1;I \to I-1)$ intraband and interband transitions of the yrast and side bands in many cases exhibit odd-even-spin staggering patterns with increasing spin. Such a systematic behavior agrees with the selection rule obtained from a simple particle-rotor model. The results illustrates that the IBFFM-2 is capable of studying such collective dynamics as chirality.

The IBFM-2 and IBFFM-2 have been further applied in a systematic analysis of the $2\nu\beta\beta$ decay rates [13]. The calculated $2\nu\beta\beta$-decay NMEs for the candidate nuclei have been shown to be similar to, or fallen into the range of, those values reported in previous theoretical calculations using different nuclear structure models. This result has paved a way of calculating consistently the low-lying states and decay processes based largely on the nuclear EDF.

11.7 ACKNOWLEDGMENT

The results reported in this chapter are based on the collaborations with C.M. Petrache, R. Rodríguez-Guzmán, L.M. Robledo, D. Vretenar, and T. Nikšić.

Bibliography

1. A. Bohr and B.R. Mottelson, *Nuclear Structure* (Benjamin, New York, 1975), Vols. I and II.
2. P. Cejnar, J. Jolie and R.F. Casten, Rev. Mod. Phys. **82**, 2155 (2010).

3. K. Heyde and J.L. Wood, Rev. Mod. Phys. **83**, 1467 (2011).
4. A.S. Davydov and G.F. Filippov, Nucl. Phys. **8**, 237 (1958).
5. L. Wilets and M. Jean, Phys. Rev. **102**, 788 (1956).
6. S. Frauendorf and J. Meng, Nucl. Phys. A **617**, 131 (1997).
7. A. Arima and F. Iachello, Phys. Rev. Lett. **35**, 1069 (1975).
8. F. Iachello and A. Arima, *The Interacting Boson Model* (Cambridge University Press, Cambridge, 1987).
9. F. Iachello and O. Scholten, Phys. Rev. Lett. **43**, 679 (1979).
10. F. Iachello and P. Van Isacker, *The Interacting Boson-Fermion Model* (Cambridge University Press, Cambridge, 1991).
11. K. Nomura and C.M. Petrache, Phys. Rev. C **105**, 024320 (2022).
12. K. Nomura, R. Rodríguez-Guzmán, and L.M. Robledo, Phys. Rev. C **101**, 014306 (2020).
13. K. Nomura, Phys. Rev. C **105**, 044301 (2022).
14. T. Otsuka, A. Arima and F. Iachello, Nucl. Phys. A **309**, 1 (1978).
15. O. Scholten, Prog. Part. Nucl. Phys. **14**, 189 (1985).
16. T. Mizusaki and T. Otsuka, Prog. Theor. Phys. Suppl. **125**, 97 (1996).
17. K. Nomura, N. Shimizu and T. Otsuka, Phys. Rev. Lett. **101**, 142501 (2008).
18. K. Nomura, N. Shimizu, T. Otsuka, Phys. Rev. C **81**, 044307 (2010).
19. J.N. Ginocchio and M.W. Kirson, Nucl. Phys. A **350**, 31 (1980)
20. K. Nomura, N. Shimizu, D. Vretenar, T. Nikšić and T. Otsuka, Phys. Rev. Lett. **108**, 132501 (2012)
21. K. Nomura, T. Otsuka, N. Shimizu, and L. Guo, Phys. Rev. C **83**, 041302(R) (2011).
22. K. Nomura, D. Vretenar, T. Nikšić and B-N. Lu, Phys. Rev. C **89**, 024312 (2014).
23. K. Nomura, T. Otsuka and P. Van Isacker, J. Phys. G **43**, 024008 (2016).
24. K. Nomura, T. Nikšić and D. Vretenar, Phys. Rev. C **93**, 054305 (2016).
25. K. Nomura, T. Nikšić and D. Vretenar, Phys. Rev. C **94**, 064310 (2016).
26. K. Nomura, R. Rodríguez-Guzmán, and L. M. Robledo, Phys. Rev. C **96**, 064316 (2017).
27. K. Nomura, T. Nikšić and D. Vretenar, Phys. Rev. C **97**, 024317 (2018).
28. K. Nomura, R. Rodríguez-Guzmán, and L. M. Robledo, Phys. Rev. C **101**, 024311 (2020).
29. K. Nomura, R. Rodríguez-Guzmán, and L.M. Robledo, Phys. Rev. C **101**, 044318 (2020).
30. K. Nomura, Phys. Rev. C **105**, 044306 (2022).
31. K. Nomura, L. Lotina, R. Rodríguez-Guzmán, and L.M. Robledo, Phys. Rev. C **106**, 064304 (2022).
32. J.T. Matta *et al.*, Phys. Rev. Lett. **114**, 082501 (2015).
33. N. Sensharma *et al.*, Phys. Lett. B **792**, 170 (2019).
34. S. Biswas *et al.*, Eur. Phys. J. A **55**, 159 (2019).
35. J. Timár *et al.*, Phys. Rev. Lett. **122**, 062501 (2019).
36. S. Chakraborty *et al.*, Phys. Lett. B **811**, 135854 (2020).
37. N. Sensharma *et al.*, Phys. Rev. Lett. **124**, 052501 (2020).
38. S. Nandi *et al.*, Phys. Rev. Lett. **125**, 132501 (2020).
39. L.M. Robledo, T.R. Rodríguez, and R.R. Rodríguez-Guzmán, J. Phys. G: Nucl. Part. Phys. **46**, 013001 (2019).
40. S. Goriely, S. Hilaire, M. Girod, and S. Péru, Phys. Rev. Lett. **102**, 242501 (2009).
41. D. Vretenar, A.V. Afanasjev, G.A. Lalazissis and P. Ring, Phys. Rep. **409**, 101 (2005).
42. T. Nikšić, D. Vretenar and P. Ring, Phys. Rev. C **78**, 034318 (2008).
43. Y. Tian, Z.Y. Ma and P. Ring, Phys. Lett. B **676**, 44 (2009).
44. B.F. Lv *et al.*, Phys. Lett. B **824**, 136840 (2022).
45. F.A. Rickey *et al.*, Phys. Rev. C **15**, 1530 (1977).
46. B. Xiong and Y. Wang, At. Data Nucl. Data Tables **125**, 193 (2019).
47. S. Brant, V. Paar and D. Vretenar, Z. Phys. A **319**, 355 (1984).

48. S. Brant, D. Tonev, G. de Angelis and A. Ventura Phys. Rev. C **78**, 034301 (2008).
49. S. Brant, N. Yoshida and L. Zuffi, Phys. Rev. C **74**, 024303 (2006).
50. N. Yoshida and F. Iachello, Prog. Theor. Exp. Phys. **2013**, 043D01 (2013).
51. F. Iachello, Phys. Rev. Lett. **44**, 772 (1980).
52. A.B. Balantekin, I. Bars and F. Iachello, Phys. Rev. Lett. **47**, 19 (1981).
53. P. Van Isacker, J. Jolie, K. Heyde and A. Frank, Phys. Rev. Lett. **54**, 653 (1985).
54. K. Nomura, R. Rodríguez-Guzmán, and L. M. Robledo, Phys. Rev. C **99**, 034308 (2019).
55. Brookhaven National Nuclear Data Center, http://www.nndc.bnl.gov.
56. E. Grodner *et al.*, Phys. Rev. Lett. **97**, 172501 (2006).
57. T. Koike, K. Starosta and I. Hamamoto, Phys. Rev. Lett. **93**, 172502 (2004).
58. E. Grodner *et al.*, Phys. Rev. Lett. **120**, 022502 (2018).
59. M. Agostini, G. Benato, J. Menendéz, and F. Vissani, Rev. Mod. Phys. **95**, 025002 (2023).
60. F. Dellagiacoma and F. Iachello, Phys. Lett. B **218**, 399 (1989).
61. J. Barea and F. Iachello, Phys. Rev. C **79**, 044301 (2009).
62. P. Pirinen and J. Suhonen, Phys. Rev. C **91**, 054309 (2015).
63. F. Šimkovic, A. Smetana and P. Vogel, Phys. Rev. C **98**, 064325 (2018).
64. J. Barea, J. Kotila and F. Iachello, Phys. Rev. C **91**, 034304 (2015).
65. E. Caurier, F. Nowacki and A. Poves, Int. J. Mod. Phys. E **16**, 552 (2007).
66. A. Barabash, Universe **6**, 159 (2020).

12 Microscopic investigation of wobbling motion in atomic nuclei using TPSM

Javid Ahmad Sheikh
University of Kashmir and Islamic University of Science and Technology, Kashmir, India

Sheikh Jehangir
Islamic University of Science and Technology, Kashmir, India

Gow-Har Bhat
Government Degree College Shopian, Kashmir, India

12.1 INTRODUCTION

Atomic nuclei depict a rich diversity of shapes and structures, primarily driven by the quantum mechanical shell effects and to elucidate these features microscopically is one of main research themes in nuclear physics [1]. The information regarding the nuclear shape is inferred through a comparison of the measured quantities with the predictions of a theoretical model that assumes a geometrical shape for the atomic nucleus in the intrinsic frame of reference. The unified model of Bohr and Mottelson has played a key role in our understanding of nuclear shapes [2]. In this model, the shape is assumed to have dominant quadrupole degree of freedom and is parametrized in terms of axial and non-axial deformation parameters of β and γ. Most of the deformed nuclei have prolate shapes with $\gamma = 0°$, and there are also some nuclei with oblate spheroidal shape having $\gamma = 60°$. These nuclei are referred to as axial and have the projection of the angular momentum along the symmetry axis, designated by "K", a conserved quantum number. There are selection rules based on this conserved quantity, for instance, Alaga rules and nuclei in many regions of the periodic table are known to follow the selection rules based on the K-quantum number [2].

Nevertheless, there are also a few regions in nuclear chart for which axial symmetry is predicted to be broken with $0 < \gamma° < 60$. The regions include $A \sim 160$ [3–10] and $A \sim 130$ [11–15]. These nuclei, referred to as triaxial systems, have unequal moments of inertia along the three principle axes (short axis s, medium axis m and long axis l) and the rotational motion can lead to very special band structures corresponding to chiral and wobbling phenomena. The chiral symmetry is expected for nuclei where total angular momentum have components along all the three mutually perpendicular axes with the collective rotor angular momentum directed along the medium axis, which has the largest moment of inertia within the irrotational model and $0 \leq \gamma \leq 60$. The angular momentum vectors of the valence particles align their angular momentum vectors along the short and long axis for particle- and hole-orbits, respectively. This arrangement of the three mutually perpendicular angular momentum vectors form two coordinate systems with opposite chirality, namely, with left and right handedness, corresponding to the left and right hand orientations of the three vectors. The operator transforming these two systems known as chiral operator ($\hat{\chi} = \hat{T}\hat{R}(\pi)$), first performs a rotation through $180°$ and then the time reversal operation, changing the directions of the angular momenta. The spontaneous breaking of chiral symmetry occurs in the intrinsic or body-fixed reference frame and in the laboratory frame with the restoration of this broken symmetry, it manifests in the appearance of chiral doublet bands, i.e., a pair $\Delta I = 1\hbar$ bands with the same parity.

Another observation that is only possible for triaxial systems is the wobbling motion and occurs when the rotation about the intermediate or medium axis having largest moment of inertia precesses about the fixed angular momentum vector because of the perturbations from the rotation of short and long axis. The wobbling motion, the classical analog of which is the spinning motion of an asymmetric top [16], is an excitation mode unique to a triaxial body. The rotation about the medium axis with the largest moment of inertia has the lowest energy and gives rise to the yrast states. The excited states near the yrast line are generated by transferring some angular momentum from the m-axis to s- and l-axes. It was shown by Bohr and Mottelson that for large angular momentum $I >> 1$, a family of rotational bands are obtained which are characterized by the wobbling quantum number (n_ω). Considering the D_2 symmetry, one obtains the selection rule : $(-1)^{n_\omega} = (-1)^I$ [2] for even-even systems. Therefore, the band structures with even n_ω ($n_\omega = 0, 2, 4..$) contain only even-spin states, and with odd n_ω ($n_\omega = 1, 3, ...$) have only odd-spin members. The key characteristic feature of the wobbling bands is that the interband I->(I-1) transitions are predominantly E2 as compared with the transitions between the two signatures in normal rotational bands which are predominantly M1.

The wobbling excitation mode was first identified in ^{163}Lu [3] and subsequently was observed in ^{161,165,167}Lu [4, 5, 17] and ^{165}Tm isotopes [18]. Recently, the wobbling mode has also been observed in other mass regions [7, 11–15]. The purpose of the present work is to perform a systematic investigation of the wobbling band structures observed in odd-mass isotopes using the microscopic approach of the triaxial projected shell model (TPSM) [19]. This model is now well established to provide a unified description of the properties of deformed and transitional nuclei [20–27]. The TPSM approach has been employed to perform a systematic investigation of the chiral band structures observed in various mass regions of the periodic table [28]. In a more recent work, a systematic analysis has been carried out for the γ bands observed in deformed nuclei and it has been demonstrated that staggering of the energies and $B(E2)$ transition probabilities of the γ bands can provide important information on the nature of the collective motion in atomic nuclei [29].

The present chapter is organized in the following manner. In the next section, we provide some basic elements of the wobbling motion as discussed in the original work of Bohr and Mottelson for even-even systems, and generalized to odd-mass systems by Hamamoto [16] and Frauendorf and Dönau [30]. In Section 12.3, a survey of experimental results on the wobbling motion is provided. In Section 12.4, the microscopic TPSM approach used in the present work is briefly discussed. In Section 12.5, the results of the TPSM are presented and discussed in light of the available experimental measurements. In Section 12.6, the present investigation on wobbling mode is summarized with some future perspectives.

12.2 BASIC CONCEPTS OF WOBBLING MODE

Bohr and Mottelson in 1975 showed that wobbling mode in atomic nuclei for large angular-momentum, $I >> 1$, will give rise to a family of band structures characterized by the phonon quantum number [2]. The properties of the wobbling bands were obtained using the triaxial particle-rotor model (TPRM) and in the present section, we shall follow the original work of Bohr and Mottelson [2].

Considering that the triaxial deformed field is time-reversal invariant, the leading order rotor Hamiltonian for an even-even system can be written as

$$\widehat{H}_{rot} = \sum_{\kappa=1}^{3} A_\kappa \hat{I}_\kappa^2 \quad , \tag{12.1}$$

where the coefficients A_κ are inverse of the moments of inertia ($\mathcal{I}_\kappa$) and are given by

$$A_\kappa = \frac{\hbar^2}{2\mathcal{I}_\kappa} \; . \tag{12.2}$$

The axes are labeled such that $A_1 < A_2 < A_3$ or accordingly $\mathcal{I}_1 > \mathcal{I}_2 > \mathcal{I}_3$ and the yrast states have $|I_1| \sim I$.

The wobbling excitations are small amplitude oscillations of the angular momentum vector about the axis having the largest moment of inertia. The energy values are given by a harmonic spectrum of wobbling quanta [2]

$$E(n_\omega, I) = A_1 I(I+1) + (n_\omega + \frac{1}{2})\, \hbar\omega, \tag{12.3}$$

where the quantum number n_ω gives the number of wobbling quanta and $\hbar\omega$ is the wobbling energy defined as the energy associated with wobbling excitation and is given by

$$\hbar\omega = 2I((A_2 - A_1)(A_3 - A_1))^{1/2} \tag{12.4}$$

The quadrupole transition probabilities are given by [2]

$$B(E2; n_\omega I \rightarrow n_\omega, I \pm 2) \approx \tfrac{5}{16\pi} e^2 Q_2^2 \tag{12.5}$$

$$B(E2; n_\omega I \rightarrow n_\omega - 1, I - 1) = \tfrac{5}{16\pi} e^2 \tfrac{n_\omega}{I} (\sqrt{3} Q_0 x - \sqrt{2} Q_2 y)^2 \tag{12.6}$$

$$B(E2; n_\omega I \rightarrow n_\omega + 1, I - 1) = \tfrac{5}{16\pi} e^2 (\tfrac{n_\omega + 1}{I})(\sqrt{3} Q_0 y - \sqrt{2} Q_2 x)^2 \tag{12.7}$$

where, Q_0 and Q_2 are intrinsic quadrupole moments.

The excitations in wobling mode are based on the quantum phonon model and have very special characteristics. In particular, the transitions from the $n_\omega = 2$ to the $n_\omega = 1$ should be a factor of two stronger as compared to the transitions from $n_\omega = 1$ to $n_\omega = 0$. On the other hand, the direct transition from $n_\omega = 2$ to $n_\omega = 0$ band are forbidden.

It was shown in the Refs. [16] and [30] that the presence of an odd-particle will modify the properties of wobbling motion in odd-mass systems. In particular, if the odd-particle occupies a low-K orbital of a high-j intruder orbital, the wobbling frequency will decrease [30] with angular-momentum, I and is referred to as the transverse wobbling (TW). This is in comparison to the longitudnal wobbling (LW), originally proposed by Bohr and Mottelson, for which the wobbling frequency increases with I. For odd-mass systems, the LW mode occurs if the odd-particle occupies a high-K orbital.

12.3   SURVEY OF EXPERIMENTAL DATA ON WOBBLING MODE

Although wobbling motion was originally predicted [2] for an even-even nucleus, but it was first experimentally established in the odd-proton nucleus ^{163}Lu [3]. For this system, several triaxial strongly deformed (TSD) band structures were known to exist [31], however, the structure of these bands remained obscure as the transitions between the bands were not measured. In the seminal work [3], nine interconnecting transitions were measured and it was possible to establish the structure of the two TSD bands, referred to as TSD1 and TSD2. It was demonstrated that these two bands have similar intrinsic structure with alignments (i_x) and moments of inertia ($J^{(2)}$) being almost identical. In the normal cranking picture [32], these bands are signature partner bands with TSD1 being the favored partner having $\alpha = 1/2$ and TSD2 having $\alpha = -1/2$ with i_x of the later being different from the former as in the unfavored branch, the aligned angular-momentum of the particle is tilted with respect to the rotational axis. In the case of the favored signature band, the angular-momentum vector of the particle is aligned along the rotational axis. Further, in the cranking model the unfavored signature band is expected at a higher excitation energy greater than 1 MeV and in the experimental data, its excitation energy relative to TSD1 band is less than 0.3 MeV. It has been shown using the particle-rotor model description that the observed features in ^{163}Lu could be explained by using three moments of inertia that correspond to the wobbling motion of a triaxial nucleus [33, 34].

Electromagnetic properties provide a stringent test on the nature of the excitation mechanism observed in atomic nuclei [2]. In the cranking picture, $B(M1)$ transitions between $\alpha = -1/2$ and $+1/2$ signature partner bands are enhanced, whereas for the wobbling motion these $I \to (I-1)$ transitions are dominated by $B(E2)$. The angular-correlation, angular-distribution and polarization data [3] were found to be consistent with $M1/E2$ multipolarity for the connecting transitions with the mixing ratio (δ) of $(90.6 \pm 1.3)\%$ for the $E2$ and $(9.4 \pm 1.3)\%$ for the $M1$ transition. It was also noted that the phase of the zigzag pattern observed for $B(E2)_{out}/B(E2)_{in}$ and $B(M1)_{out}/B(E2)_{in}$ was opposite to that calculated in the cranking model and was consistent with the phase obtained for the wobbling mode in particle-rotor model calculations [3]. Moreover, in the quantal phonon model of the wobbling mode, which is valid for $I >> 1\hbar$, the inter-band transitions follow special features with $B(E2)$ of the inter-band $I \to (I-1)$ transitions competing with those of the in-band $I \to (I-2)$ transitions. Further, the inter-band $B(E2)$ strength from the $n_\omega = 2$ to the $n_\omega = 1$ band should be a factor of two larger than the $B(E2)$ strength from the $n_\omega = 1$ to the $n_\omega = 0$ band [2, 33, 34]. These expected properties of the phonon wobbling mode have been experimentally verified for ^{163}Lu [35].

Candidate wobbling bands have also been reported in other neighboring odd-A lutetium isotopes of ^{161,165,167}Lu [4–7], although the angular correlation and linear polarization measurements were not possible to confirm the $B(E2)$ character of the inter-band transitions as has been done for the ^{163}Lu isotope. Nevertheless, the similarity of the decay pattern of the TSD band structures of the three Lu isotopes with ^{163}Lu and large values obtained for the $B(E2)_{out}/B(E2)_{in}$ transitions, the structures have been designated as wobbling bands. The search for wobbling in other nuclides in this mass region failed to establish any more candidates, except for ^{167}Ta [18]. It was demonstrated using the tilted axis cranking approach that the density of states for Lu isotopes is quite low for the TSD minimum and the wobbling bands could be identified. In other neighboring nuclides, the density of sates is quite high and particle-hole excitation modes are more favorable [36]. For ^{167}Ta, the similarity of the linking transitions between the TSD2 and TSD1 bands with the corresponding transitions in ^{163,165,167}Lu isotopes, the two TSD bands observed have been grouped as wobbling structures [18]. We would like to mention that TSD1 band in ^{167}Ta depicts a bandcrossing feature with $J^{(2)}$ showing a peak at about $\hbar\omega = 0.35$ MeV. This bandcrossing feature is also observed in ^{167}Lu, but not in other Lu isotopes.

In all the above systems, the wobbling bands have been established for the TSD minimum configuration with axial deformation value of ~ 0.40. The wobbling mode for a normal deformed nuclide was reported for the first time in ^{135}Pr with $\varepsilon \sim 0.16$ [12]. Wobbling and signature partner (SP) band structures were delineated from the analysis of the angular distribution and polarization data. The mixing ratios determined from these measurements show large $B(E2)$ mixing for the transitions from $n_\omega = 1$ wobbling band to the yrast structure and for the SP band it is dominated by $B(M1)$. The evidence for the existence of a two-phonon wobbling band was provided in a subsequent work [13]. It needs to be added that new measurements have been performed and from the analysis of this new data, it has been shown that mixing ratio, $|\delta| < 1$, contradicting the wobbling interpretation of the observed bands in ^{135}Pr [37].

It was predicted [30] using the particle-rotor model that for the TW wobbling mode, the wobbling frequency initially will increase up to some critical angular-momentum and then will begin decreasing. However, in all the TW cases identified before 2020, wobbling frequency decreased with spin. It was shown in 2020 [38] that for one of the two TW wobbling bands identified in ^{183}Au, the frequency decreases initially and then after some spin values it increases. For the heavier ^{187}Au, a pair of longitudinal wobbling structures have been observed with the SP band [39].

The appearance of wobbling bands have also been reported in ^{127}Xe and ^{133}La through angular distributions (and/or correlation) and linear polarization measurements. The bands observed have been classified as $n_\omega = 0, 1, 2$ and SP ($n_\omega = 2$, SP bands have not been observed in ^{133}La) bands [11, 40]. In both these nuclei, the wobbling frequency increases with spin and, therefore, has longitudinal character with the angular-momentum of the particle along the m-axis. In ^{133}Ba, transverse wobbling mode has been proposed as for ^{135}Pr with wobbling frequency decreasing as a function of spin [41].

12.4   TRIAXIAL PROJECTED SHELL MODEL APPROACH

Triaxial projected shell model approach is conceptually similar to the spherical shell model (SSM) and differing only in the way the basis space is chosen. TPSM employs the deformed intrinsic states of the triaxial Nilsson potential, which incorporate essential long-range correlations, as the basis configuration [42]. The truncation of the many-body basis in TPSM is very efficient as not only the numerical effort is drastically reduced, but also makes physical interpretation more transparent. The recently generalized TPSM approach with the inclusion of higher-order quasiparticle states has emerged as a powerful tool for exploring the triaxial characteristics of atomic nuclei as computational resources involved are quite modest and it is feasible to systematically investigate a broad range of atomic nuclei. As a matter of fact, several systematic investigations have been performed for chiral and γ vibrational band structures observed in triaxial nuclei [20–27]. The model space in the TPSM approach is spanned by multiquasiparticle basis states from different oscillator shells [43, 44]. This allows to investigate high-spin band structures in well-deformed and transitional nuclei all across the nuclear chart.

The flow chart of the TPSM calculations can primarily be divided into three stages. In the first stage, the deformed basis are constructed by solving the triaxial Nilsson potential with a realistic set of quadrupole deformation parameters of ε and ε' for a given system under consideration. These deformation values lead to an accurate Fermi surface and it allows one to choose an optimum set of the basis states around the Fermi surface for a realistic description of the system. The standard Bardee-Cooper-Schriefer (BCS) procedure is then carried out to include the pairing correlations with the parameters given in Table 12.1.

The rotational symmetry is not conserved by the intrinsic states generated from the deformed Nilsson calculations and in the second stage this symmetry is restored through three – dimensional angular-momentum projection technique [45, 46]. In this technique, good angular-momentum basis states are projected out from the Nilsson + BCS states using the explicit three- dimensional angular-momentum projection operator "P^I_{MK}" given by [47] :

$$\hat{P}^I_{MK} = \frac{2I+1}{8\pi^2} \int d\Omega\, D^I_{MK}(\Omega)\, \hat{R}(\Omega), \tag{12.8}$$

with the rotation operator

$$\hat{R}(\Omega) = e^{-i\alpha\hat{J}_z} e^{-i\beta\hat{J}_y} e^{-i\gamma\hat{J}_z}. \tag{12.9}$$

Here, "Ω" represents the set of Euler angles ($\alpha, \gamma = [0, 2\pi], \beta = [0, \pi]$) and $\hat{J}$'s are the angular-momentum operators. The angular-momentum projection operator in Eq. (12.8) apart from projecting the good angular-momentum, also projects states with good K-values. The projected multi-quasiparticle basis states for different systems are given for odd-proton nuclei as :

$$\{\hat{P}^I_{MK}\, a^\dagger_{\pi_1}|\Phi\rangle; \hat{P}^I_{MK}\, a^\dagger_{\pi_1} a^\dagger_{\nu_1} a^\dagger_{\nu_2}|\Phi\rangle; \hat{P}^I_{MK}\, a^\dagger_{\pi_1} a^\dagger_{\pi_2} a^\dagger_{\pi_3}|\Phi\rangle;$$

$$\hat{P}^I_{MK}\, a^\dagger_{\pi_1} a^\dagger_{\pi_2} a^\dagger_{\pi_3} a^\dagger_{\nu_1} a^\dagger_{\nu_2}|\Phi\rangle\} \tag{12.10}$$

and for odd-neutron nuclei :

$$\{\hat{P}^I_{MK}\, a^\dagger_{\nu_1}|\Phi\rangle; \hat{P}^I_{MK}\, a^\dagger_{\nu_1} a^\dagger_{\pi_1} a^\dagger_{\pi_2}|\Phi\rangle; \hat{P}^I_{MK}\, a^\dagger_{\nu_1} a^\dagger_{\nu_2} a^\dagger_{\nu_3}|\Phi\rangle;$$

$$\hat{P}^I_{MK}\, a^\dagger_{\nu_1} a^\dagger_{\nu_2} a^\dagger_{\nu_3} a^\dagger_{\pi_1} a^\dagger_{\pi_2}|\Phi\rangle\} \tag{12.11}$$

where $|\Phi\rangle$ in above equations represent the triaxial quasiparticle vacuum state and $a^{\dagger}_{\nu_i}, a^{\dagger}_{\pi_i}$ are quasi-particle creation operators, with the index $\nu_i(\pi_i)$, denoting the neutron (proton) quantum numbers and running over the selected single-quasiparticle states.

In the third and final stage, projected basis states given by Eqns. (12.10) and (12.11) are then used to diagonalize the shell model Hamiltonian. A two-body Hamiltonian in terms of separable forces is adopted which consists of the modified harmonic oscillator single-particle Hamiltonian and a residual two-body interaction comprising of quadrupole-quadrupole, monopole pairing and quadrupole pairing terms. These terms represent specific correlations which are considered to be essential to describe the low-energy nuclear phenomena [48]. The Hamiltonian has the following form :

$$\hat{H} = \hat{H}_0 - \frac{1}{2}\chi\sum_{\mu}\hat{Q}^{\dagger}_{\mu}\hat{Q}_{\mu} - G_M\hat{P}^{\dagger}\hat{P} - G_Q\sum_{\mu}\hat{P}^{\dagger}_{\mu}\hat{P}_{\mu}. \tag{12.12}$$

In the above equation, $\hat{H}_0$ is the spherical single – particle Hamiltonian containing the proper spin–orbit force for correct shell closures [49]. The QQ-force strength, χ, in Eq. (12.12) is related to the quadrupole deformation ε as a result of the self-consistent Hartree-Fock-Bogoliubov condition and the relation is given by [42]:

$$\chi_{\tau\tau'} = \frac{\frac{2}{3}\varepsilon\hbar\omega_{\tau}\hbar\omega_{\tau'}}{\hbar\omega_n\langle\hat{Q}_0\rangle_n + \hbar\omega_p\langle\hat{Q}_0\rangle_p}, \tag{12.13}$$

where $\omega_{\tau} = \omega_0 a_{\tau}$, with $\hbar\omega_0 = 41.4678A^{-\frac{1}{3}}$ MeV, and the isospin-dependence factor a_{τ} is defined as

$$a_{\tau} = \left[1 \pm \frac{N-Z}{A}\right]^{\frac{1}{3}},$$

with $+ (-)$ for τ = neutron (proton). The harmonic oscillation parameter is given by $b^2_{\tau} = b^2_0/a_{\tau}$ with $b^2_0 = \hbar/(m\omega_0) = A^{\frac{1}{3}}$ fm^2. The monopole pairing strength G_M (in MeV) is of the standard form

$$G_M = \frac{G_1 \mp G_2\frac{N-Z}{A}}{A}, \tag{12.14}$$

where the minus (plus) sign applies to neutrons (protons). The values of G_1 and G_2 are chosen such that the calculated gap parameters reproduce the experimental mass differences. The quadrupole pairing strength G_Q is assumed to be proportional to G_M, the proportionality constant being fixed as usual to be 0.16. The single-particle space for the TPSM usually includes three major harmonic oscillator shells (N) each for neutrons and protons in a calculation for deformed heavy nuclei. This large size of single-particle space accommodates sufficiently large number of active nucleons as well as all the important orbits and ensures the microscopic description of the collective motion. For different mass regions the oscillator shells employed are: N = 4, 5, 6 for neutrons and N = 3, 4, 5, for protons for A =160 and 180 region; N = 3, 4, 5 for neutrons and N = 3, 4, 5, for protons for A =130 region; N = 3, 4, 5 for neutrons and N = 2, 3, 4 for protons for A =100 and 110 regions. The shell model Hamiltonian, Eq. (12.12) is diagonalized in the angular-momentum projected basis states, Eqns. (12.10) and (12.11), by following the Hill-Wheeler prescription [42]. The generalized eigen-value equation is given by

$$\sum_{\kappa'K'}\{\mathcal{H}^I_{\kappa K\kappa'K'} - E\mathcal{N}^I_{\kappa K\kappa'K'}\}f^{\sigma I}_{\kappa'K'} = 0, \tag{12.15}$$

where the Hamiltonian and norm kernels are given by

$$\mathcal{H}^I_{\kappa K\kappa'K'} = \langle\Phi_{\kappa}|\hat{H}\hat{P}^I_{KK'}|\Phi_{\kappa'}\rangle,$$
$$\mathcal{N}^I_{\kappa K\kappa'K'} = \langle\Phi_{\kappa}|\hat{P}^I_{KK'}|\Phi_{\kappa'}\rangle.$$

The Hill-Wheeler wave function is given by

$$\psi^{\sigma}_{IM} = \sum_{\kappa,K} f^{\sigma I}_{\kappa K}\, \hat{P}^{I}_{MK}|\,\Phi_{\kappa}\rangle. \tag{12.16}$$

where $f^{\sigma I}_{\kappa K}$ are the variational coefficients and index "κ" designates the basis states of Eqns. (12.10) and (12.11). The wave-function is then used to evaluate the electromagnetic transition probabilities. The reduced electric transition probabilities $B(EL)$ from an initial state (σ_i, I_i) to a final state (σ_f, I_f) are given by [50]

$$B(EL, I_i \to I_f) = \frac{1}{2I_i + 1}|\langle \psi^{\sigma_f I_f}||\hat{Q}_L||\psi^{\sigma_i I_i}\rangle|^2, \tag{12.17}$$

and the reduced matrix element can be expressed as

$$\langle\,\psi^{\sigma_f I_f}||\quad \hat{Q}_L \quad ||\psi^{\sigma_i I_i}\rangle$$

$$= \sum_{\kappa_i,\kappa_f,K_i,K_f} f^{\sigma_i I_i}_{\kappa_i K_i} f^{\sigma_f I_f}_{\kappa_f K_f} \sum_{M_i,M_f,M} (-)^{I_f - M_f}$$

$$\times \begin{pmatrix} I_f & L & I_i \\ -M_f & M & M_i \end{pmatrix}$$

$$\times \langle\Phi|\hat{P}^{I_f}_{K_f M_f}\hat{Q}_{LM}\hat{P}^{I_i}_{K_i M_i}|\Phi\rangle$$

$$= 2 \sum_{\kappa_i,\kappa_f,K_i,K_f} f^{\sigma_i I_i}_{\kappa_i K_i} f^{\sigma_f I_f}_{\kappa_f K_f}$$

$$\times \sum_{M',M''} (-)^{I_f - K_f}(2I_f + 1)^{-1} \begin{pmatrix} I_f & L & I_i \\ -K_f & M' & M'' \end{pmatrix}$$

$$\times \int d\Omega\, D^{I_i}_{M'' K_i}(\Omega)\langle\Phi_{\kappa_f}|\hat{O}_{LM'}\hat{R}(\Omega)|\Phi_{\kappa_i}\rangle$$

Table 12.1

The axial deformation parameter (ε) and triaxial deformation parameter ε' employed in the calculation for odd-A mass nuclei. The axial deformation ε is taken from Refs. [4, 5, 16, 18, 51, 52]. The asterisk ($*$) on ε is the deformation value for the positive parity bands in the ^{183}Au nucleus.

	^{161}Lu	^{163}Lu	^{165}Lu	^{167}Lu	^{167}Ta	^{105}Pd	^{127}Xe	^{131}Cs	^{133}Ba	^{133}La	^{135}Pr	^{151}Eu	^{183}Au	^{185}Au	^{187}Au
ε	0.400	0.400	0.380	0.430	0.370	0.257	0.150	0.140	0.150	0.150	0.160	0.200	0.280	0.270*	0.220
ε'	0.110	0.100	0.110	0.110	0.100	0.110	0.100	0.100	0.100	0.110	0.100	0.100	0.110	0.100	0.110
γ	15^0	14^0	16^0	14^0	15^0	23^0	34^0	36^0	34^0	36^0	32^0	27^0	21^0	20^0	27^0

12.5   RESULTS AND DISCUSSION

The TPSM calculations for ^{161,163,165,167}Lu and ^{167}Ta have been performed with the axial and non-axial deformation parameters given in Table 12.1. These quantities have been adopted from the earlier studies, in particular, from the ultimate cranking calculations [4, 5, 16, 18, 51] which predicted TSD shapes for these nuclei.

The TPSM energies obtained after diagonalization of the shell model Hamiltonian are compared with the available experimental data in Figs. 12.1–12.5. The calculated bands are depicted for the lowest two favored ($\alpha = +1/2$) and two unfavored ($\alpha = -1/2$) signature band structures. For ^{161}Lu, two TSD bands have been identified and are compared with the TPSM calculated band structures in Fig. 12.1. The experimental bands, labeled as TSD1 and TSD2, have been categorized as $n_\omega = 0$ and 1 wobbling bands, based on the similarity of their properties with the corresponding band structures

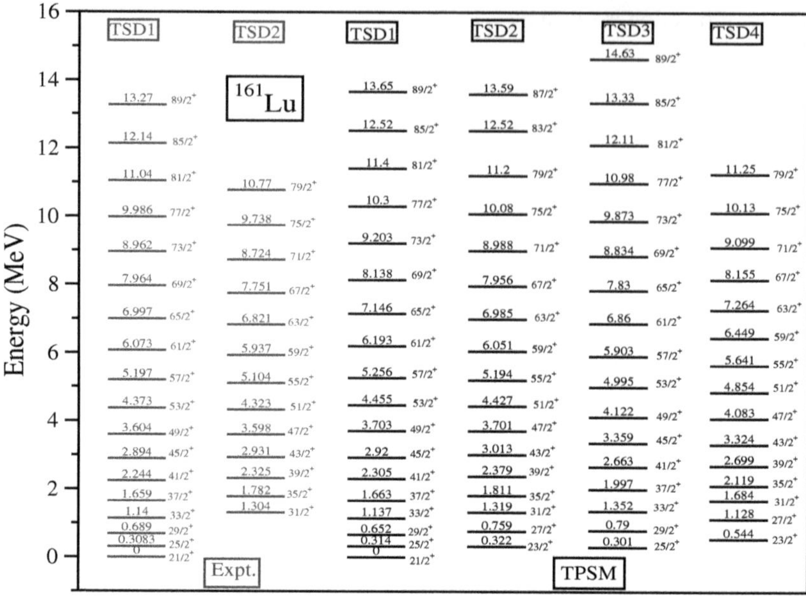

Figure 12.1   TPSM energies for the lowest four bands after configuration mixing are plotted along with the available experimental data for the ^{161}Lu isotope Data is taken from [6].

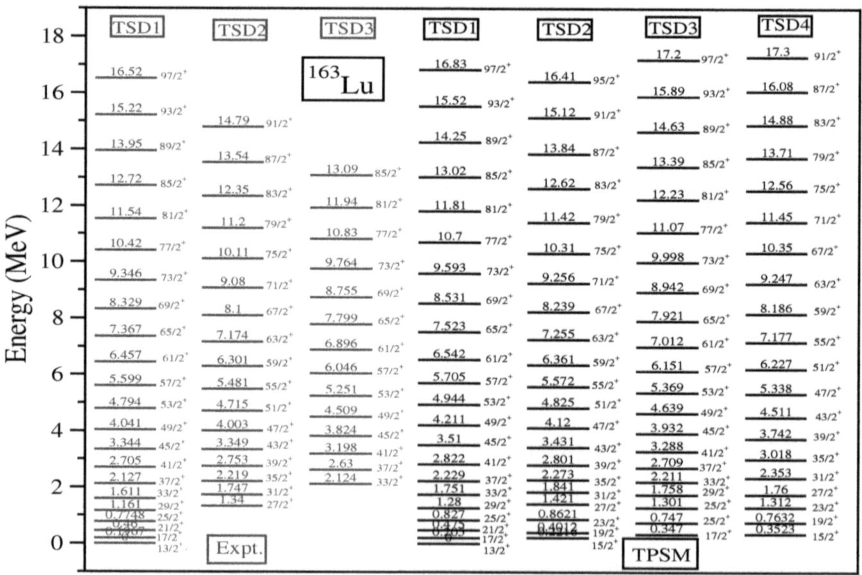

Figure 12.2   TPSM energies for the lowest four bands after configuration mixing are plotted along with the available experimental data for the ^{163}Lu isotope. Data is taken from [3].

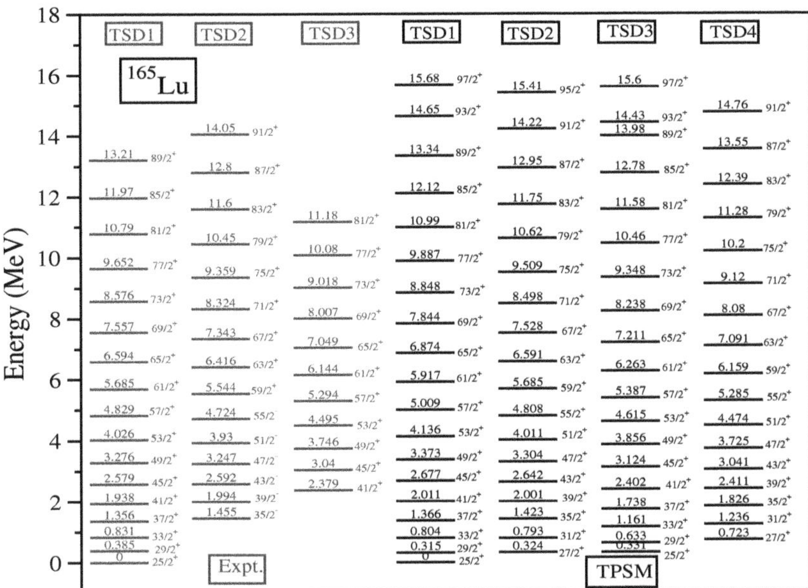

Figure 12.3 TPSM energies for the lowest four bands after configuration mixing are plotted along with the available experimental data for the ^{165}Lu isotope. Data is taken from [4].

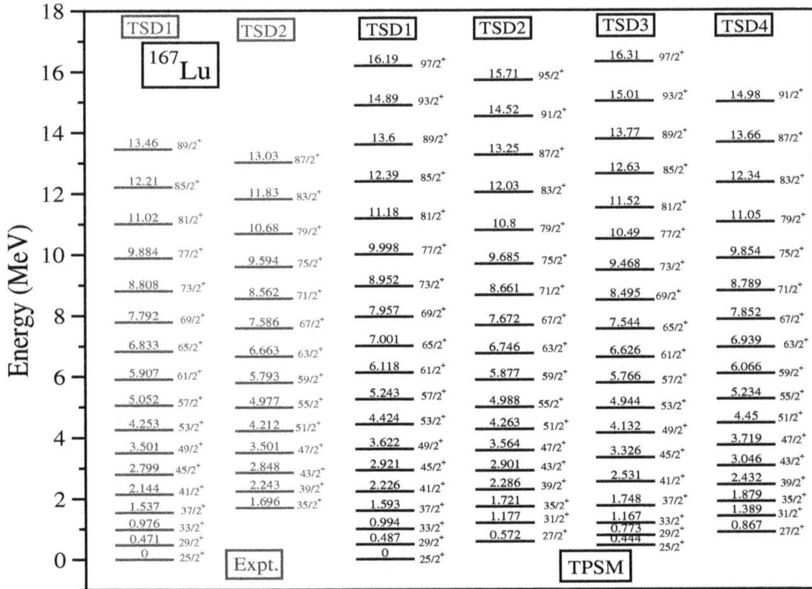

Figure 12.4 TPSM energies for the lowest four bands after configuration mixing are plotted along with the available experimental data for the ^{167}Lu isotope. Data is taken from [7].

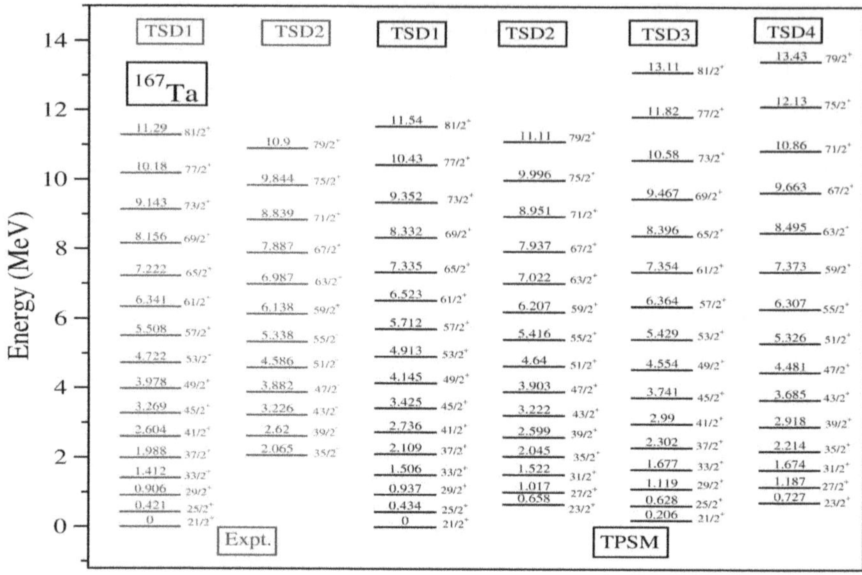

Figure 12.5 (Color online) TPSM energies for the lowest four bands after configuration mixing are plotted along with the available experimental data for the ^{167}Ta isotope. Data is taken from [18].

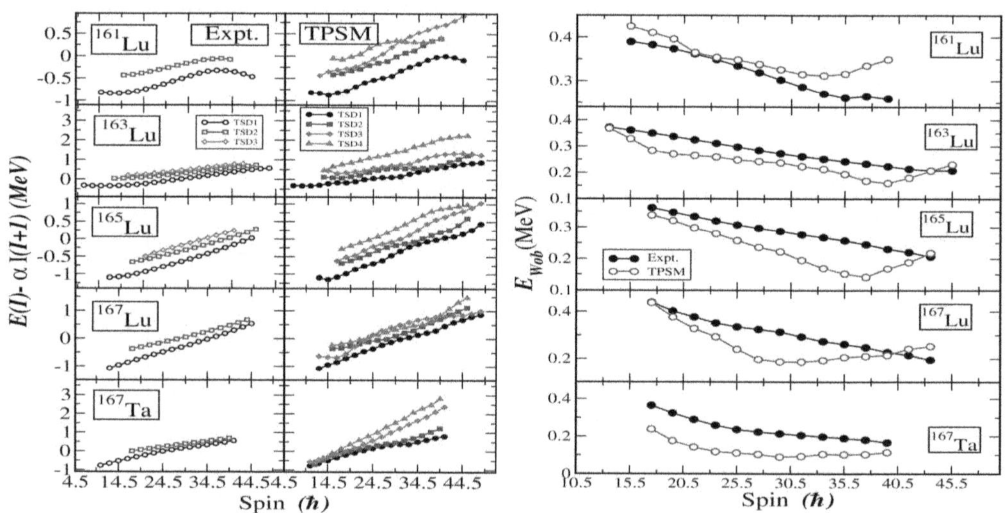

Figure 12.6 (Color online) (Left panel) TPSM energies for the lowest four bands after configuration mixing are plotted along with the available experimental data for ^{161,163,165,167}Lu and ^{167}Ta isotopes. The scaling factor $\alpha = 32.322A^{-5/3}$. (Right panel) TPSM wobbling energies are compared with the experimental values obtained from the bands TSD1 and TSD2 for $^{161-167}$Lu and ^{167}Ta

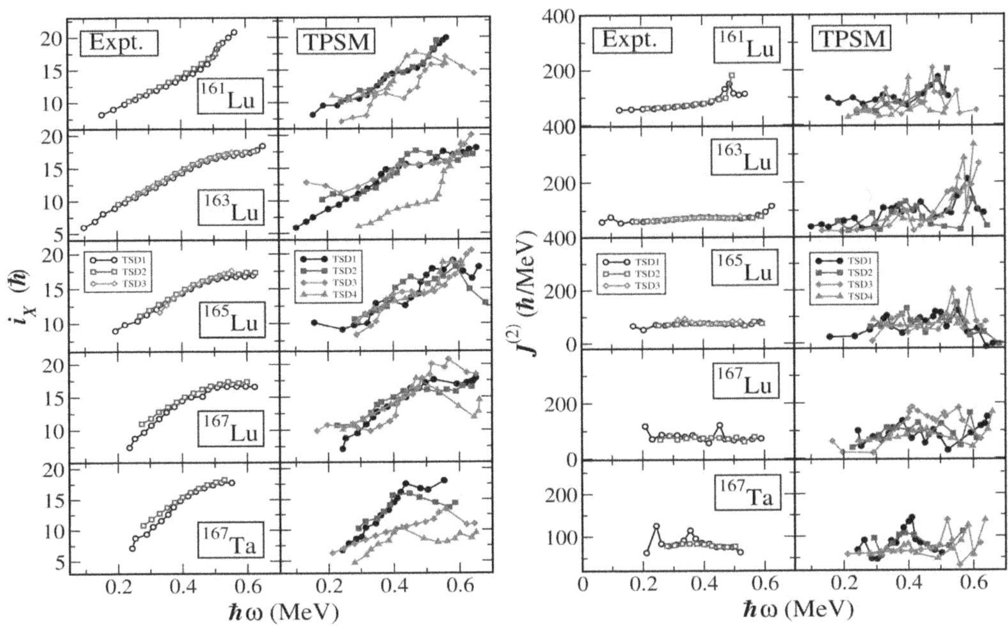

Figure 12.7 (Color online) (Left panel) Comparison of the aligned angular momentum, $i_x = I_x(\omega) - I_{x,ref}(\omega)$, where $\hbar\omega = \frac{E_\gamma}{I_x^i(\omega)-I_x^f(\omega)}$, $I_x(\omega) = \sqrt{I(I+1)-K^2}$ and $I_{x,ref}(\omega) = \omega(J_0 + \omega^2 J_1)$, obtained from the measured energy levels as well as those calculated from the TPSM results, for $^{161-167}$Lu and ^{167}Ta nuclei. (Right Panel) Comparison between experimental and calculated dynamic moment of inertia, $\mathcal{J}^{(2)} = \frac{4}{E_\gamma(I)-E_\gamma(I-2)}$, for the TSD1, TSD2, TSD3 and TSD4 for $^{161-167}$Lu and ^{167}Ta nuclei. The reference band Harris parameters used are J_0=30 and J_1=40 [4], obtained from the measured energy levels as well as those calculated from the TPSM results, for $^{161-167}$Lu and ^{167}Ta nuclei.

in ^{163}Lu. It is noted from Fig. 12.1 that TPSM energies for TSD1 and TSD2 bands are in good agreement with the experimental energies. The deviation of the TPSM energies is small at low-spin, but with increasing spin the difference is about 500 keV for the higher observed spin states. The TPSM energies for ^{163}Lu are compared with the experimental quantities in the Fig. 12.2 and it is again evident that TPSM approach reproduce the known energies quite well with a maximum deviation of about 550 keV at the highest spin, $I = 85/2$. For ^{163}Lu, three TSD bands have been identified and have been assigned as n_ω=0,1 and 2. This is the first system where the occurrence of wobbling excitation mode was confirmed for the first time with the measurement of transition probabilities that will be discussed in detail later.

The calculated TPSM energies for ^{165}Lu and ^{167}Lu are compared with the experimental energies in Figs. 12.3 and 12.4. Three and two TSD bands have been identified for ^{165}Lu and ^{167}Lu, respectively and it is noted again that TPSM approach reproduces the data quite well. There are differences of about 300 keV for the highest angular momentum states in both the nuclei. The energies for ^{167}Ta are compared in Fig.12.5 for the two known TSD bands and it is noted that TPSM values are in reasonable agreement with the known energies.

To analyze the relative excitation energies of the wobbling bands, for the energies presented in Figs. 12.1–12.5 a core contribution have been subtracted, and the resulting energies are presented in Fig. 12.6 (left panel). In the wobbling description, $n_\omega = 0$ TSD1 band is the result of the rotation of the system about the axis that has the largest moment of inertia, which is generally the m-axis. The excited $n_\omega = 1,2$ and 3 wobbling bands are obtained when the angular-momentum from m-axis is transferred to l- and s-axis. The excitation energy of the first wobbling band is about 500 keV and the $n_\omega = 2$ band is less than 500 keV from the $n_\omega = 1$ band.

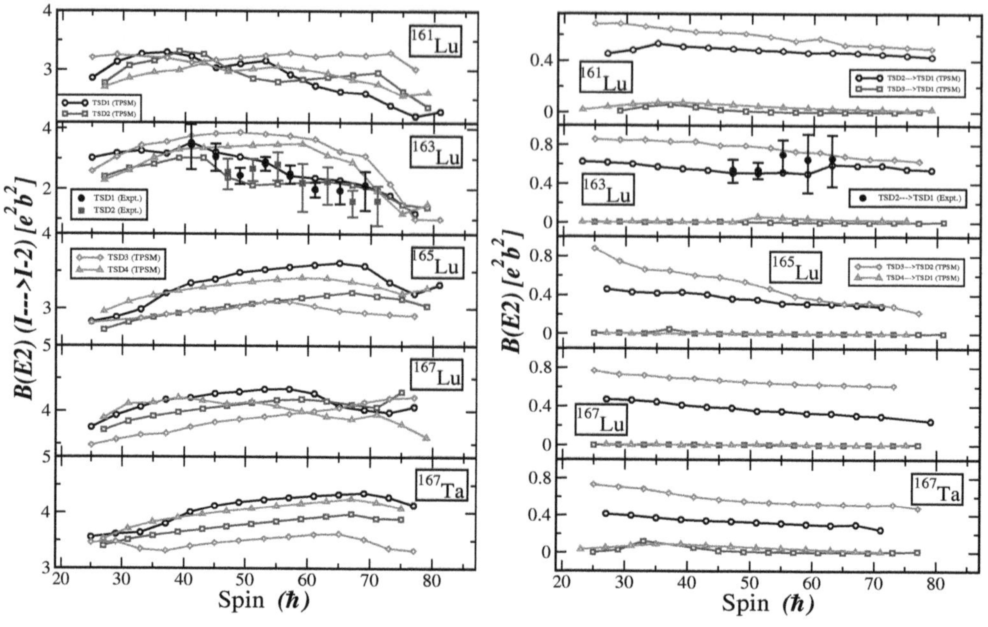

Figure 12.8 (Color online) Comparison between experimental and calculated $B(E2)_{in}$ (left panel) and $B(E2)_{out}$ (right panel) vs. spin for the TSD1, TSD2, TSD3 and TSD4 wobbling bands for $^{161-167}$Lu and ^{167}Ta nuclei.

For the longitudinal motion, the rotational axis is along the long-axis and the wobbling frequency, defined as,

$$E_{wob}(I) = E_{n_\omega=1}(I) - [E_{n_\omega=0}(I+1) + E_{n_\omega=0}(I-1)]/2, \tag{12.18}$$

increases with spin [30]. For higher values of n_ω, the rotational axis moves away from the m-axis and will have smaller K-values along the m-axis. For the transverse wobbling motion, m-axis is replaced by s-axis and the wobbling frequency decreases with spin [30]. The wobbling frequency has been calculated from the excitation energies spectra and are displayed in Fig. 12.6 (right panel) for the studied Lu- and Ta nuclides. The calculated results are in good agreement with the experimental values and, in particular, the decreasing tendency of the wobbling energy as a function of angular momentum is reproduced in all the cases.

As the wobbling bands are based on the same intrinsic configuration, the aligned angular-momentum i_x and moments of inertia $J^{(2)}$ are expected to be identical. These quantities are depicted in Fig. 12.7 for the five nuclei. It is noted from the two figures that i_x and $J^{(2)}$ obtained from the measured quantities are almost the same for all the wobbling bands. These quantities calculated using the TPSM values are also similar for the lowest two bands TSD1 and TSD2. However, the calculated quantities for the excited bands TSD3 and TSD4 are somewhat different in comparison to the lowest two bands.

The transition probabilities provide the important information on the nature of the excitation mechanism [2]. For the wobbling motion, the probabilities have the characteristic property that transitions from $n_\omega=1$ to $n_\omega=0$ are dominated by $B(E2)$ as compared to $B(M1)$ in the normal cranking picture for the transitions from the SP to the yrast band. Further, in the harmonic wobbling limit, the transitions from $n_\omega=2$ to $n_\omega=1$ are predicted to be a factor of two larger as compared to the transitions from $n_\omega=1$ to $n_\omega=0$, and direct transition from $n_\omega=2$ to $n_\omega=0$ are forbidden.

Before discussing the inter-band transitions, we first demonstrate the predictive power of the in-band $B(E2)$ transitions as these are measured to a good accuracy in some cases. These transitions are shown in Fig. 12.8 (left panel) for the lowest four bands of the five nuclei studied in the present work.

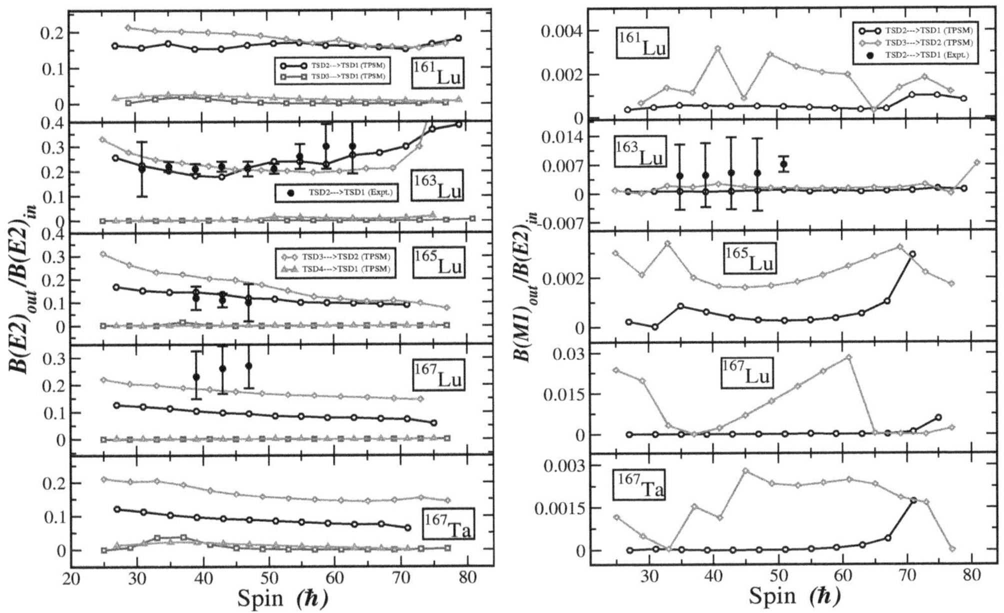

Figure 12.9   (Color online) Comparison between experimental and calculated $B(E2)_{out}/B(E2)_{in}$ (left panel) and $B(M1)_{out}/B(E2)_{in}$ (right panel) vs. spin for the TSD1, TSD2, TSD3 and TSD4 wobbling bands for $^{161-167}$Lu and ^{167}Ta nuclei.

The measured values are known for TSD1 and TSD2 bands of ^{163}Lu and the TPSM predicted values are noted to be in reasonable agreement with the known transitions. It is evident from the figure that for ^{161}Lu and ^{163}Lu, TSD1 and TSD2 in-band transitions are similar, however, for other bands these transitions vary for the four bands. It would be quite interesting to perform the experimental measurements of the the transition probabilities for other isotopes in order to confirm the varying nature of the predicted values.

The inter-band transitions among the four bands for the five isotopes are depicted in Fig. 12.8 (right panel). The $B(E2)$ transitions from TSD3 to TSD2 are clearly enhanced as compared to the transitions from TSD2 to TSD1 and direct transitions from TSD4 to TSD1, and TSD3 to TSD1 are retarded as expected for the harmonic wobbling mode. For ^{163}Lu, it is noted that TPSM predicted transitions for TSD2 to TSD1 are in good agreement with the measured transitions.

The TPSM calculated $B(E2)_{out}/B(E2)_{in}$ transition ratios are displayed in Fig. 12.9 (left panel) and are quite large as expected for the wobbling bands. The experimental ratios have been deduced from the DCO and polarization measurements for ^{163}Lu ^{165}Lu and ^{167}Lu and are also displayed in Fig. 12.9. It is noted from the figure that for ^{163}Lu and ^{165}Lu, the TPSM values are in good agreement with the measured values, however, for ^{167}Lu the TPSM predicted values deviate for the three known data points. The $B(M1)_{out}/B(E2)_{in}$ ratios are plotted in Fig. 12.9 (right panel) and are seen to be quite small as compared to the $B(E2)_{out}/B(E2)_{in}$ ratios. These ratios have been measured for ^{163}Lu and TPSM calculated ratios are in good agreement.

To explore further that the above discussed wobbling bands are transverse in character, Fig. 12.10 depicts the energies of the bands projected from the quasiparticle configurations before the TPSM Hamiltonian is diagonalized for ^{163}Lu as an illustrative example. In the calculations, the short-axis is chosen as the quantization axis and "K" denotes the angular momentum projection on this axis. This is in contrast to the TPSM calculations published so far, where the long-axis is considered as the quantization axis. This choice simplifies the interpretation since in the TW regime, the odd quasiparticle tends to align its angular momentum along the short axis. Obviously, such a change of

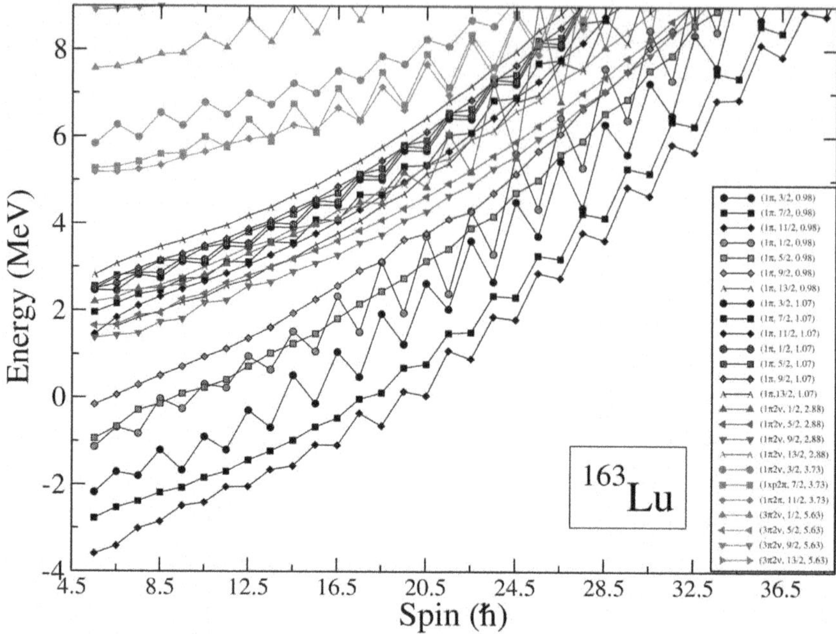

Figure 12.10   (Color online) Energies of the projected K-configurations with the short axis chosen as the quantization axis to which "K" quantum number refers. The curves are labeled by three quantities: quasiparticle character, K quantum number and energy of the quasiparticle state. For instance, $(1p, 3/2, 0.98)$ designates a one quasiproton state with $K = 3/2$ having intrinsic energy of 0.98 MeV.

the quantization axis leaves the observables unchanged. Numerically, it is achieved by changing the triaxiality parameter γ to another equivalent sector that interchanges s- and l-axis. For this sector, the γ value is changed to the new value of $(-120° - \gamma)$. It is noted from the figure that the yrast configuration corresponding to TSD1 ($\alpha = +1/2$) and TSD2 ($\alpha = -1/2$) band structures have $K=11/2$, which signifies that odd-proton is almost aligned toward the s-axis. The first excited band structures corresponding to TSD3 and TSD4 bands have $K=7/2$, indicating that for these band structures the rotational axis has moved away from the principle s-axis.

In the present work, we have also performed the TPSM calculations for ten wobbling bands observed in normal deformed nuclei with the deformation parameters employed listed in Table 12.1. On the left panel of Fig. 12.11, the excitation energies for the lowest bands for all the normal deformed wobbling cases are depicted against the known data and the behavior of the wobbling frequency as a function of spin is also presented in Fig. 12.11 on the right panel. The frequency decreases for the cases of ^{131}Cs, ^{135}Pr, ^{151}Eu, ^{183}Au (negative parity), ^{133}Ba and ^{105}Pd, signifying that these nuclei have transverse wobbling mode. For ^{133}La, ^{187}Au (negative parity) and ^{127}Xe, the wobbling frequency increases with spin, which indicates that the collective motion has longitudinal character. For the positive parity wobbling band in ^{183}Au, the frequency first increases and then it decreases. It has been predicted for the TW motion that there is a critical spin up to which the frequency will increase and then after this spin value, the frequency will start decreasing [30]. It is noted that for the negative parity band in ^{183}Au, the TPSM predicts that the frequency will increase after I=33/2, and it would be interesting to verify this prediction in future experimental investigations. The derived rotational frequencies $\hbar\omega$ vs spin (I) are compared with the TPSM values in Fig. 12.12. The calculations reproduce the experimental data for all the ten wobbling bands observed in normal deformed nuclei.

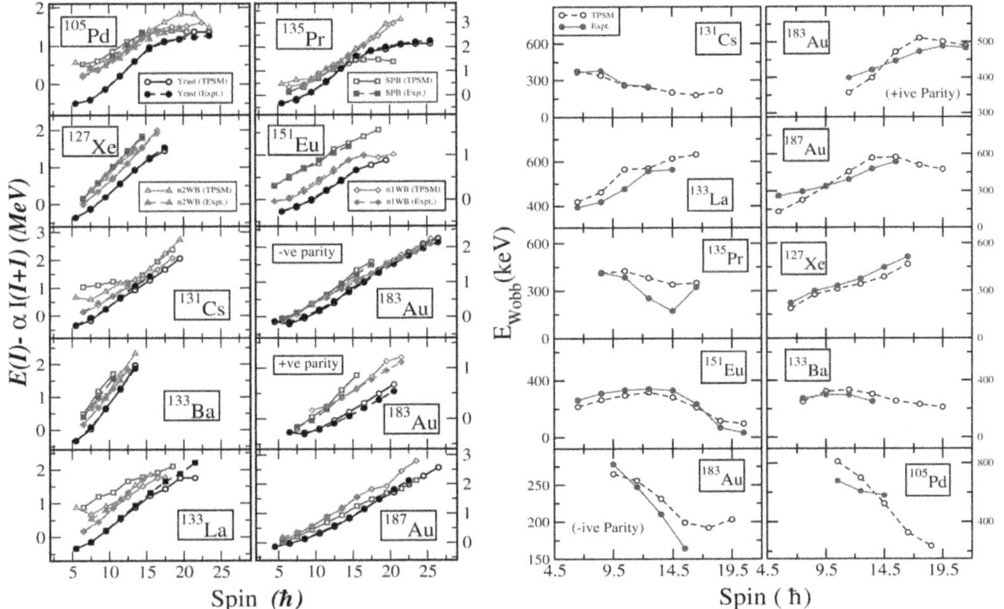

Figure 12.11 (Color online) (Left panel) TPSM energies for the lowest bands after configuration mixing are plotted along with the available experimental data for ^{105}Pd, ^{127}Xe, ^{131}Cs, ^{133}Ba, ^{133}La, ^{135}Pr, ^{151}Eu, ^{183}Au (-ve and +ve parity) and ^{187}Au nuclei. The scaling factor $\alpha = 32.322A^{-5/3}$. (Right panel) TPSM wobbling energies are compared with the experimental values obtained from the bands TW/LW for ^{105}Pd, ^{127}Xe, ^{131}Cs, ^{133}La, ^{133}Ba, ^{135}Pr, ^{151}Eu, ^{183}Au (negative parity), and ^{183}Au (positive parity). Data is taken from [11–15].

To probe further the quasiparticle structures of the observed band structures, we have analyzed the alignments of the bands as a function of the spin and the results are presented in Fig. 12.13. The observed values are in the reasonable agreement with data. $B(E2)_{out}/B(E2)_{in}$ transitions for all the 10 nuclides are depicted in Fig. 12.14 (left panel). As already discussed, these transitions are crucial to establish the wobbling nature of the band structures. It is evident from the ratios that $B(E2)_{out}$ and $B(E2)_{in}$ have similar values which establishes the wobbling nature of these bands. On the other hand, the ratios $B(M1)_{out}/B(E2)_{in}$ displayed in Fig. 12.14 (right panel) are quite small as expected.

12.6   SUMMARY AND FUTURE PERSPECTIVES

In the present work, we have performed a systematic investigation of the wobbling band structures observed in odd-mass nuclei. The analysis has been performed for fourteen nuclides using the tri-axial projected shell model approach. This model is now well established as a method of choice to study the high-spin band structures in deformed and transitional nuclei. The TPSM approach employs the triaxial basis configurations and is well suited to investigate the properties of triaxial nuclei. It has been already used to perform a systematic study of the chiral band structures in atomic nuclei, which is a fingerprint of the triaxial deformation. The wobbling motion is another excitation mode which is only possible for triaxial shapes. This mode was originally predicted by Bohr and Mottelson [2] for an even-even system and it was shown that wobbling motion in the large I limit will give rise to a family of band structures, designated by the harmonic oscillator quantum number, n_ω. The characteristic feature of the wobbling mode is that transitions $I \to (I-1)$ from $n_\omega = 1$ to $n_\omega = 0$ bands are dominated by $B(E2)$ rather than $B(M1)$ as in the standard cranking model.

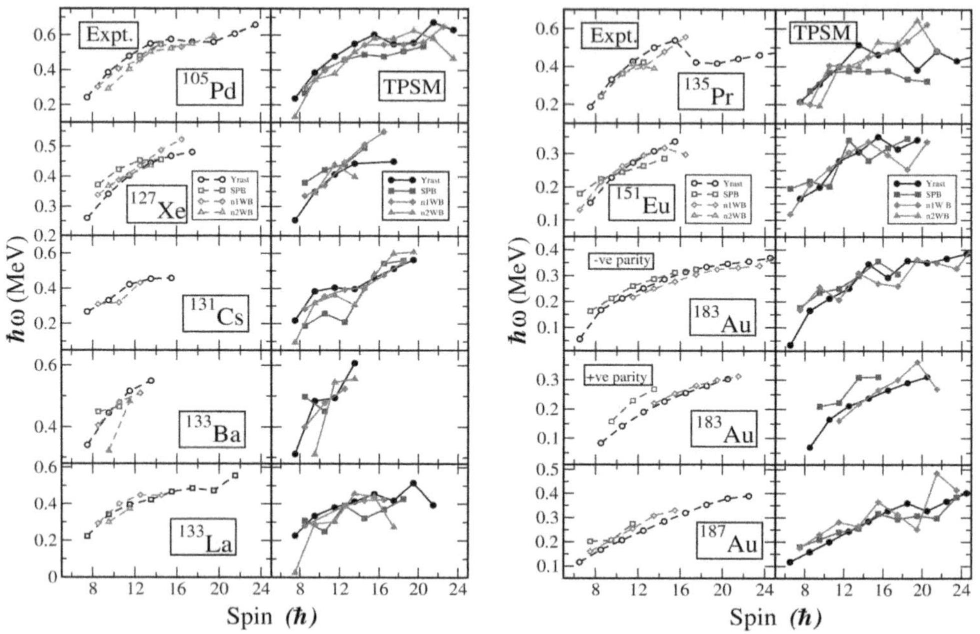

Figure 12.12   Experimental and calculated values from TPSM for the rotational frequency as functions of the spin I for the Bands $n_\omega = 0, 1$ and 2.

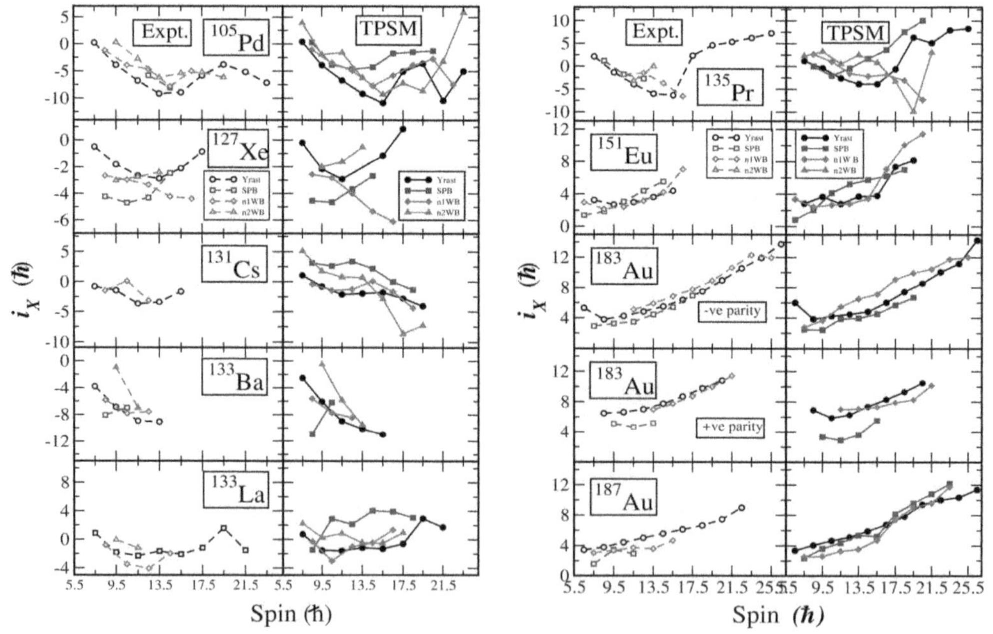

Figure 12.13   Comparison of the aligned angular momentum, $i_x = I_x(\omega) - I_{x,ref}(\omega)$, where $\hbar\omega = \dfrac{E_\gamma}{I_x^i(\omega) - I_x^f(\omega)}$, $I_x(\omega) = \sqrt{I(I+1) - K^2}$ and $I_{x,ref}(\omega) = \omega(J_0 + \omega^2 J_1)$, obtained from the measured energy levels as well as those calculated from the TPSM results, for ^{105}Pd, ^{127}Xe, ^{131}Cs, ^{133}La, ^{133}Ba, ^{135}Pr, ^{151}Eu, ^{183}Au (negative parity), and ^{183}Au (positive parity) nuclei . The reference band Harris parameters used are J_0=16 and J_1=15 [54].

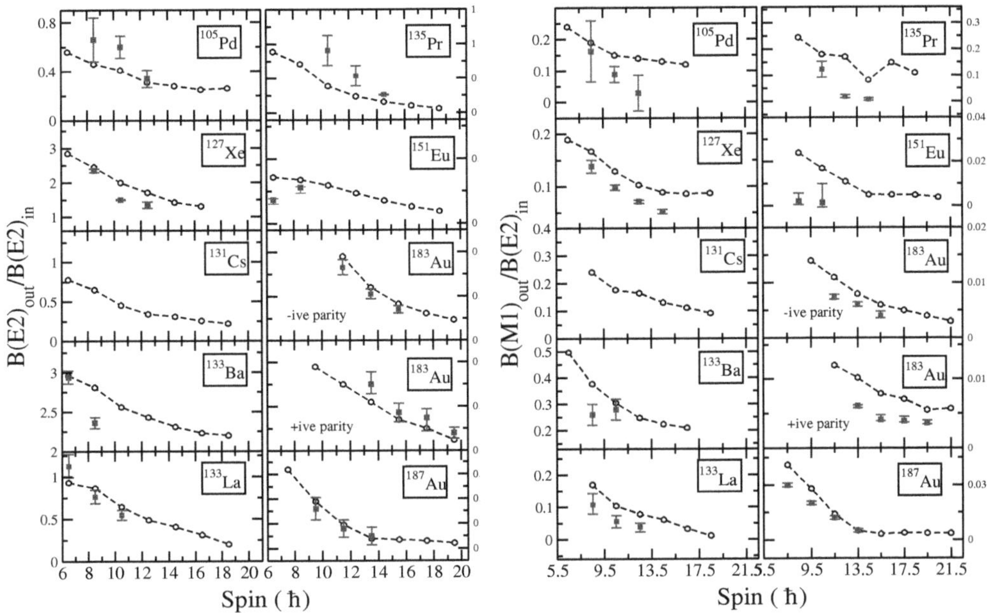

Figure 12.14 Comparison between experimental and calculated $B(E2)_{out}/B(E2)_{in}$ (left panel) and $B(M1)_{out}/B(E2)_{in}$ (right panel) vs. spin for the bands TW/LW for ^{131}Cs, ^{133}La, ^{135}Pr, ^{151}Eu, ^{183}Au (*negative* parity), ^{183}Au (*positive* parity), ^{127}Xe, ^{133}Ba and ^{105}Pd nuclei.

The wobbling mode was first identified in the odd-proton ^{163}Lu nucleus [3] and it was shown that the observed band structures obey all the characteristic features expected for the wobbling motion. Subsequently, the wobbling bands have also been observed in other nuclides and in the present work we have studied all these cases using the TPSM approach. It has been demonstrated that TPSM provides a reasonable description of the observed properties. In particular, the behavior of the wobbling frequency with spin is very well reproduced in all the studied cases. The in-band and inter-band transition probabilities have also been evaluated using the TPSM wavefunctions, and these have been shown to be consistent with the measured data.

In the normal TPSM analysis, long-axis is employed as the quantization axis. In order to analyze the results, we have used the short-axis as the quantization axis and is achieved by choosing the proper deformation values in another equivalent sector. This change of the axis simplifies the interpretation for cases where transverse wobbling is expected. For the TW motion, s-axis is the rotational axis and the resulting wavefunction will have large components along the s-axis. This has been shown for ^{163}Lu as an illustrative example of TW motion.

Further, the advantage of the TPSM approach is that it can describe the three dimensional wobbling mode and the standard cranking motion in a unified manner. In the normal cranking mode, SP bands are expected which have dominant $B(M1)$ transition probabilities, and we have demonstrated that the existence of wobbling and SP band structures identified in some nuclei can be simultaneously described using the TPSM approach. In the particle-rotor model picture, the wobbling and SP bands structures have very different geometry. For the case of excited wobbling bands, the rotational angular-momentum vector is tilted with respect to the principle axis and in the case of SP bands, it is the angular-momentum of the valence particle that is tilted and not the rotational axis. It is evident that the TPSM approach assimilates both these scenarios in a microscopic manner. In a more recent work [53], the TPSM approach has been used to substantiate the existence of a doublet wobbling excitation mode in ^{105}Pd.

In future, we intend to investigate the existence of wobbling motion in even-even systems. Several observed band structures in even-even nuclei have been proposed as candidate wobbling bands. These systems include, ^{130}Ba, ^{112}Ru, ^{134}Ce, ^{104}Pd, ^{114}Pd and ^{136,138}Nd. In a few even-even systems, the odd-spin branch of the γ band is lower than the even-spin branch [30, 55–57] in comparison to a large class of systems where it is opposite. These few nuclei have been categorized as γ rigid and it will be interesting to explore whether these systems have wobbling characteristics. The common feature is that the first excited band in these few nuclei is the odd-spin branch of the γ band and for the wobbling motion, it is also the odd-spin band having the harmonic oscillator quantum number, $n_\omega=1$.

The authors would like to acknowledge Nazira Nazir and S.P. Rouoof for their help in the preparation of some of the figures presented in the manuscript.

Bibliography

1. S. Frauendorf, Int. J. Mod. Phys. E **24**, 1541001 (2015).
2. A. Bohr and B. R. Mottelson, *Nuclear Structure*, Vol. II (Benjamin Inc., New York, 1975).
3. S. W. Ødegård, *et al.*, Phys. Rev. Lett. **86**, 5866 (2001).
4. G. Schönwaßer, *et al.*, Phys. Lett. B **552**, 9 (2003).
5. H. Amro, *et al.*, Phys. Lett. B **553**, 197 (2003).
6. P. Bringel, *et al.*, Eur. Phys. J. A **24**, 167 (2005).
7. H. Amro, *et al.*, Phys. Rev. C **71**, 011302 (2005).
8. S. Aberg, Nucl. Phys. A **520**, 35c (1990) and references therein.
9. I. Ragnarsson, Phys. Rev. Lett. **62**, 2084 (1989).
10. T. Bengtsson, Nucl. Phys. A **496**, 56 (1989); 512, 124 (1990).
11. S. Biswas, *et al.*, Eur. Phys. J. A **55**, 159 (2019).
12. J. T. Matta, *et al.*, Phys. Rev. Lett. **114**, 082501 (2015).
13. N. Sensharma, *et al.*, Phys. Lett. B **792**, 170 (2019).
14. A. Mukherjee, *et al.*, Phys. Rev. C **107**, 054310 (2023).
15. J. Timár, *et al.*, Phys. Rev. Lett. **122**, 062501 (2019).
16. I. Hamamoto, *et al.*, Acta Phys. Pol. **B 32**, 2545 (2001).
17. P. Bringel, *et al.*, Phys. Rev. C **73**, 054314 (2006).
18. D. J. Hartley, *et al.*, Phys. Rev. C **80**, 041304(R) (2009).
19. J. A. Sheikh and K. Hara, Phys. Rev. Lett. **82**, 3968 (1999).
20. G. H. Bhat, J. A. Sheikh and R. Palit, Phys. Lett. B **707**, 250 (2012).
21. G. H. Bhat, W. A. Dar, J. A. Sheikh and Y. Sun, Phys. Rev. C **89**, 014328 (2014).
22. S. Jehangir, G. H. Bhat, J. A. Sheikh, S. Frauendorf, S. N. T. Majola, P. A. Ganai and J. F. Sharpey-Schafer, Phys. Rev. C **97**, 014310 (2018).
23. S. Jehangir, I. Maqbool, G. H. Bhat, J. A. Sheikh, R. Palit and N. Rather, Eur. Phys. J. A **56**, 197 (2020).
24. S. Jehangir, G. H. Bhat, N. Rather, J. A. Sheikh and R. Palit, Phys. Rev. C **104**, 044322 (2021).
25. S. Jehangir, G. H. Bhat, J. A. Sheikh, S. Frauendorf, W. Li, R. Palit and N. Rather, Eur. Phys. J. A **57**, 308 (2021).
26. S. Jehangir, N. Nazir, G. H. Bhat, J. A. Sheikh, N. Rather, S. Chakraborty and R. Palit, Phys. Rev. C **105**, 054310 (2022).
27. N. Nazir, S. Jehangir, S. P. Rouoof, G. H. Bhat, J. A. Sheikh, N. Rather and S. Frauendorf, Phys. Rev. C **107**, L021303 (2023).
28. J. A. Sheikh, G. H. Bhat, W. A. Dar, S. Jehangir and P. A. Ganai, Phys. Scr. **94**, 063015 (2016).
29. S. P. Rouoof, Nazira Nazir, S. Jehangir, G. H. Bhat, J. A. Sheikh, N. Rather and S. Frauendorf, Eur. Phys. J. A **60**, 40 (2024).
30. S. Frauendorf and F. Dönau, Phys. Rev. C **89**, 014322 (2014).
31. H. Schnack-Petersen, *et al.*, Nucl. Phys. A **594**, 175 (1995).

32. S. Frauendorf, Rev. Mod. Phys. **73**, 463 (2001).
33. I. Hamamoto, Phys. Rev. C **65**, 044305 (2002).
34. I. Hamamoto and G. B. Hagemann, Phys. Rev. C **67**, 014319 (2003).
35. D. R. Jensen, *et al.*, Phys. Rev. Lett. **89**, 142503 (2002).
36. N. S. Pattabiraman, *et al.*, Phys. Lett. B **647**, 243 (2007).
37. B. Lv and C. M. Petrache, Symmetry **15**, 1075 (2023).
38. S. Nandi, *et al.*, Phys. Rev. Lett. **125**, 132501 (2020).
39. N. Sensharma, *et al.*, Phys. Rev. Lett. **124**, 052501 (2020).
40. S. Chakraborty, *et al.*, Phys. Lett. B **811**, 135854 (2020).
41. K. Rojeeta Devi, *et al.*, Phys. Lett. B **823**, 136756 (2021).
42. K. Hara and Y. Sun, Int. J. Mod. Phys. E **04**, 637 (1995).
43. Nazira Nazir, S. Jehangir , S. P. Rouoof , G. H. Bhat , J. A. Sheikh , N. Rather and Manzoor A. Malik, Phys. Rev. C **108**, 044308 (2023).
44. B. M. Musangu, *et al.*, Phys. Rev. C **104**, 064318 (2021).
45. K. Hara and S. Iwasaki, Nucl. Phys. A **332**, 61 (1979)
46. K. Hara and S. Iwasaki, Nucl. Phys. A **348**, 200 (1980)
47. P. Ring and P. Schuck, *The Nuclear Many-Body Problem* (Springer-Verlag, New York), (1980).
48. K. Kumar and M. Baranger, Nucl. Phys. A**122**, 273 (1968).
49. S. G. Nilsson, C. F. Tsang, A. Sobiczewski, Z. Szymanski, S. Wycech, C. Gustafson, I. Lamm, P. Moller and B. Nilsson, Nucl. Phys. **A 131**, 1 (1969).
50. Y. Sun and J. L. Egido, Nucl. Phys. A **580**, 1 (1994).
51. R. Bengtsson and H. Ryde, Eur. Phys. J **A 22**, 355 (2004).
52. P. Moller, J. R. Nix, W. D. Myers and W. J. Swiatecki, At. Data Nucl. Data Tables **59**, 185 (1995).
53. A. Karmakar, *et al.*, https://doi.org/10.48550/arXiv.2403.08235 (2024).
54. C. M. Petrache, *et al.*, Nucl. Phys. **A 597**, 106-126 (1996).
55. Y. K. Wang, *et al.*, Nucl. Phys. A **834**, 28c (2010).
56. C. M. Petrache, *et al.*, Phys. Lett. **B 795**, 241 (2019).
57. Q. B. Chen, S. Frauendorf and C. M. Petrache, Phys. Rev. C **100**, 061301(R) (2019).

13 Spin dynamics of triaxial nuclei with quasiparticle alignments

Radu Budaca
National Institute for Physics and Nuclear Engineering, Magurele, Romania

13.1 INTRODUCTION

Atomic nuclei exhibit predominantly spherical or axially symmetric shapes in their ground state. Deviations from axial symmetry are rare, and mostly of the dynamical nature. Therefore, the search of nuclear shapes with rigid triaxiality is a continuous quest of nuclear structure studies. Although it cannot be measured directly, triaxiality substantially affects the nucleon separation energies, fragmentation of the large amplitude collective resonances, fission barriers, disintegration probabilities, and other nuclear properties. Triaxial deformation of nuclei can be ascertained by spectral characteristics such as the signature inversion [1] or γ band staggering [2], which are restricted to some particular nuclear systems. The identification of stable triaxiality in nuclear systems with interacting collective and single particle degrees of freedom is quite difficult, but it is indisputable when chiral symmetry breaking [3] or wobbling motion [4, 5] is involved. The obvious tool for the study and interpretation of these phenomena is the particle-rotor model (PRM) [6], whose quantum structure treats the single-particle and collective degrees of freedom on the same footing. The dynamics of the system, usually described in classical terms, is extracted from quantum averages or semiclassical mappings. As the final observables are indexed by the total angular momentum, it is then desirable to have a description of the system as a whole dynamical object. This can be realized by considering the single-particle degrees of freedom as perturbing effects. In this way the PRM Hamiltonian is reduced to a cranked triaxial rotor, whose cranking terms are just phenomenologically interpreted as being generated by quasiparticle alignments. Such general quadratic Hamiltonians occur naturally in classical mechanics of gyroscopes, motion of atoms in external fields, molecular physics, nonlinear optics, celestial and galactic dynamics [7]. An equivalent classical picture can be obtained from the quantum problem through a time-dependent variational principle. One can then extract valuable information about the system's dynamics from the well established classical treatment of quadratic Hamiltonians on the unit sphere. This procedure is presented in the next section. Section 13.3 discusses concrete numerical applications on wobbling and chiral bands observed experimentally in few nuclei. Last section is devoted to conclusions.

13.2 THEORETICAL APPROACH

The interplay of collective and single-particle degrees of freedom associated with a system composed of a triaxial core and a set of nucleons is described by means of a general PRM Hamiltonian [6]. In order to treat in a unified manner the emergence of wobbling and chiral bands in triaxial nuclei, one will consider two quasiparticles with spins $\vec{j}$ and $\vec{j}'$ contributing to a total angular mo-

mentum $\hat{I}$. The PRM Hamiltonian for such a case can be written as:

$$H = H_R + H_{qp} + H'_{qp},\tag{13.1}$$

which is a sum of contributions coming from separate degrees of freedom. As the total angular momentum is the final observable, the triaxial rotor part of the total Hamiltonian is expressed as

$$H_R = \sum_{k=1,2,3} A_k(\hat{I}_k - \hat{j}_k - \hat{j}'_k)^2\tag{13.2}$$

in terms of its angular momentum $\vec{R} = \vec{I} - \vec{j} - \vec{j}'$. Its inertial parameters $A_k = 1/(2\mathcal{J}_k)$ are usually defined by moments of inertia (MOI) in the hydrodynamic nuclear model [6]:

$$\mathcal{J}_k = \frac{4}{3}\mathcal{J}_0 \sin^2\left(\gamma - \frac{2}{3}k\pi\right).\tag{13.3}$$

This definition corresponds to the semi-axis lengths of the core

$$R_k = R_0\left[1 + \sqrt{\frac{5}{4\pi}}\beta\cos\left(\gamma - \frac{2\pi}{3}k\right)\right],\tag{13.4}$$

where β is the axial deformation. Within this correspondence, the direction of the total angular momentum vector relative to the ellipsoidal density distribution of the core, can be tracked by referring to the principal body-fixed axes as the long (l), short (s) and medium (m). For $\gamma \in (60°, 120°)$, one has $R_2 < R_3 < R_1$, such that axes 1, 2, and 3 become long, short and respectively medium. This interval of γ deformation will be predominantly used in what follows, because it assures that the maximal MOI is always along the third axis. As a consequence, the triaxial core will favor rotations around the third (medium) intrinsic principal axis. In what concerns the alignments of the single particle spins, these are dictated by the their associated quasiparticle nature [5]. More precisely, a hole will align its spin along the l axis, a particle along the s axis, while a nucleon in the vicinity of the Fermi surface aligns its spin along the m axis.

The spins of the quasiparticles are considered rigidly aligned along the principal planes 1-3 and 2-3. This alignment geometry together with the adopted interval of γ deformation establish the conditions for the emergence of transverse wobbling and chiral configurations. Other geometrical arrangements can be accommodated by a suitable change in the interval of the γ deformation [8, 9]. The rigid or frozen alignment (FA) approximation amounts to the replacement of the quasiparticle spin operators with their real expectation values:

$$\begin{aligned}
\hat{j}_1 &= j_1\cos\alpha,\ \hat{j}'_1 = 0,\\
\hat{j}_2 &= 0,\ \hat{j}'_2 = j'\cos\alpha',\\
\hat{j}_3 &= j\sin\alpha,\ \hat{j}'_3 = j'\sin\alpha'.
\end{aligned}\tag{13.5}$$

The α and α' angles account for the deviation of the quasiparticle spin alignments in respect to the l and s axes. The FA approximation renders the quasiparticle spin contribution as constant, reducing the relevant part of the PRM Hamiltonian to [10]:

$$\begin{aligned}
H_{align} &= A_1\hat{I}_1^2 + A_2\hat{I}_2^2 + A_3\hat{I}_3^2 - 2A_1 j\hat{I}_1\cos\alpha - 2A_2 j'\hat{I}_2\cos\alpha'\\
&\quad - 2A_3(j\sin\alpha + j'\sin\alpha')\hat{I}_3.
\end{aligned}\tag{13.6}$$

The above operator represents a rotor Hamiltonian for a total spin $\hat{I}$ which is cranked by quasiparticle alignments.

13.2.1 SEMICLASSICAL FORMALISM

The semiclassical description of the rotational motion described by the quantum Hamiltonian (13.6), starts by applying a time-dependent variational principle [11]:

$$\delta \int_0^t \left\langle \psi \left| H_{align} - \frac{\partial}{\partial t'} \right| \psi \right\rangle dt' = 0. \tag{13.7}$$

The variational state ψ must be constructed according to the problem under consideration and parametrized by a restricted set of complex variables [12]. Integrating the equations of motion for the complex variables provided by the variational principle is equivalent to solving the eigenvalue problem for the quantum Hamiltonian, if the variational state spans the whole Hilbert space of the quantum system. This condition is naturally satisfied by the coherent states, due to their over-completeness property. Moreover their continuous character induces a natural transition between quantum and classical pictures [13]. Therefore, one will use here a coherent state for the $SU(2)$ algebra of the angular momentum operators [14, 15]

$$|\psi_{IM}(x,\varphi)\rangle = \sum_{K=-I}^{I} \frac{1}{(2I)^I} \sqrt{\frac{(2I)!}{(I-K)!(I+K)!}} (I+x)^{\frac{I-K}{2}} (I-x)^{\frac{I+K}{2}} e^{i\varphi(I+K)} |IMK\rangle, \tag{13.8}$$

stereographically parametrized by the azimuth angle φ and a projection variable $x = I\cos\theta$ related to the polar angle θ of the total angular momentum vector direction in the intrinsic frame of reference in respect to the quantization axis. The third axis is chosen as a quantization axis, and therefore the $|IMK\rangle$ functions are the eigenstates of the intrinsic angular momentum operators $\hat{I}^2$ and $\hat{I}_3$ and their counterparts from the laboratory frame of reference, while $\hat{I}_-$ is the usual lowering operator.

Developing the variational principle for the Hamiltonian (13.6) using the variational state (13.8), one arrives at a classical energy function ($\hbar = 1$)

$$\begin{aligned}
\mathcal{H}(x,\varphi) &= \langle \Psi_{IM}(x,\varphi)|H_{align}|\Psi_{IM}(x,\varphi)\rangle \\
&= \frac{(2I-1)(I^2-x^2)}{2I}(A_1\cos^2\varphi + A_2\sin^2\varphi - A_3) \\
&\quad -2\sqrt{I^2-x^2}(A_1 j\cos\alpha\cos\varphi + A_2 j'\cos\alpha'\sin\varphi) \\
&\quad -2A_3 x(j\sin\alpha + j'\sin\alpha') + \frac{I}{2}(A_1+A_2) + A_3 I^2,
\end{aligned} \tag{13.9}$$

and a set of equations of motion expressed with the help of the Poisson bracket as:

$$\{\mathcal{H},x\} = \dot{x}, \quad \{\mathcal{H},\varphi\} = \dot{\varphi}. \tag{13.10}$$

The canonical form of the equations of motion assures that the full structure of the original quantum Hamiltonian system is reproduced and allows the identification of x with the generalized momentum and φ with the generalized coordinate, having the relationship $\{\varphi,x\} = 1$. The classical energy function is obtained with the use of the following coherent state averages

$$\langle \hat{I}_k^2 \rangle = \frac{I}{2} + \frac{2I-1}{2I}\langle \hat{I}_k \rangle^2, \ k = 1,2,3, \tag{13.11}$$

where

$$\begin{aligned}
\mathcal{I}_1 = \langle \hat{I}_1 \rangle &= \sqrt{I^2-x^2}\cos\varphi, \\
\mathcal{I}_2 = \langle \hat{I}_2 \rangle &= \sqrt{I^2-x^2}\sin\varphi, \\
\mathcal{I}_3 = \langle \hat{I}_3 \rangle &= x,
\end{aligned} \tag{13.12}$$

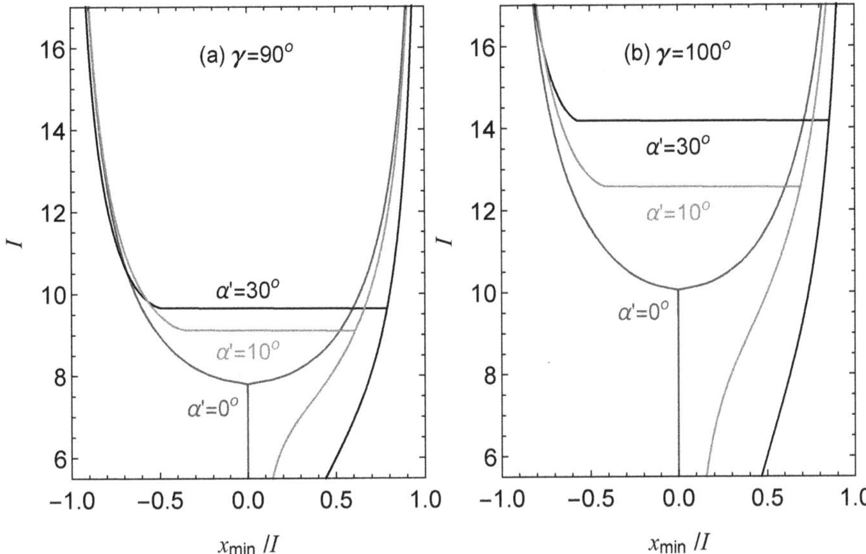

Figure 13.1 The dependence on the total angular momentum I of the projection coordinate x_m of the classical energy minima for $\gamma = 90°$ (a) and $\gamma = 100°$ (b) with few values of titling angle $\alpha' = 0°, 10°, 30°$ associated with quasiparticle spins $j = 0$ and $j' = 11/2$.

are the classical projection components of the angular momentum vector in polar coordinates satisfying $\sum_k \langle \hat{I}_k \rangle^2 = I^2$. Note that the non-vanishing deviation $(\Delta \hat{I}_k)^2 = \langle \hat{I}_k^2 \rangle - \langle \hat{I}_k \rangle^2$ leads to a maximal uncertainty principle $(\Delta \hat{I}_i)^2 (\Delta \hat{I}_j)^2 = |\langle \hat{I}_k \rangle|^2/4$, which explains the emergence of classical energy terms linear in angular momentum I. On the other hand, $\sum \langle \hat{I}_k^2 \rangle = I(I+1)$ shows that the variational state is an eigenfunction for the $\hat{I}^2$ operator [11].

13.2.2 CLASSICAL DYNAMICS

The classical orbits of the total angular momentum vector are constructed around stationary points which are minima of the classical energy function. The orbits are usually obtained by integrating the equations of motion. A more intuitive visualization of the classical trajectories can be realized as a function of the classical components $\mathcal{I}_k$ of the total angular momentum (13.12). Within this space, the classical energy function reads:

$$\mathcal{H}(\mathcal{I}_1, \mathcal{I}_2, \mathcal{I}_3) = \frac{I}{2}(A_1 + A_2 + A_3) + \frac{(2I-1)}{2I}(A_1 \mathcal{I}_1^2 + A_2 \mathcal{I}_2^2 + A_3 \mathcal{I}_3^2)$$
$$- 2A_1 \mathcal{I}_1 j \cos \alpha - 2A_2 \mathcal{I}_2 j' \cos \alpha' \qquad (13.13)$$
$$- 2A_3 \mathcal{I}_3 (j \sin \alpha + j' \sin \alpha').$$

The intersection of the surfaces corresponding to the two constants of motion, total angular momentum squared $\sum_k \mathcal{I}_k^2 = I^2$ and the energy $\mathcal{H}(\mathcal{I}_k) = E$, determines then the classical trajectories. The two surfaces are a sphere and respectively a displaced ellipsoid, whose relative position depends on the deformation and alignment configuration [9, 16].

The coordinates of the stationary points designate the average direction of the total angular momentum vector. When dealing with a single quasiparticle spin, the problem is essentially planar and the azimuth angle of the minima is constant, $\varphi = 0$ for hole and $\varphi = \pi/2$ for particle alignments. The projection coordinate of the minima is, however, spin-dependent. More precisely, as can be

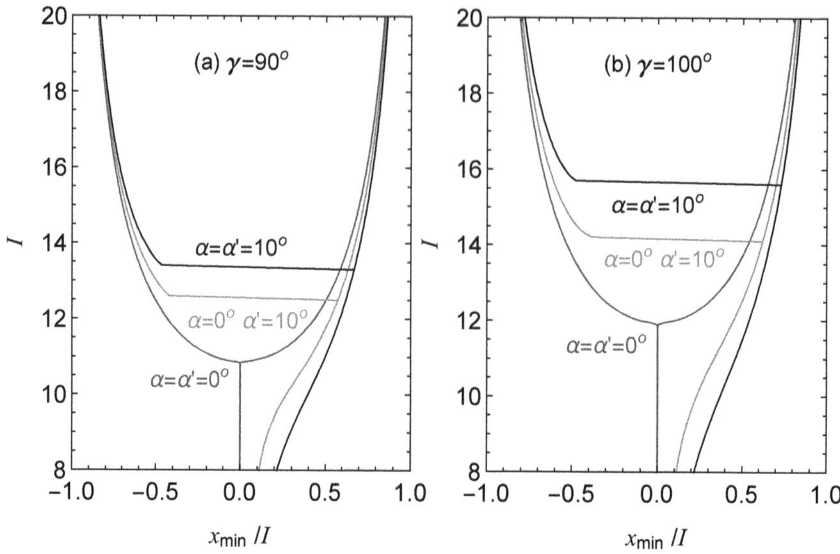

Figure 13.2 The dependence on the total angular momentum I of the projection coordinate x_m of the classical energy minima for $\gamma = 90°$ (a) and $\gamma = 100°$ (b) with few combinations of titling angles $\alpha = \alpha' = 0°, 10°$ and $\alpha = 0°$, $\alpha' = 10°$ associated with quasiparticle spins $j = j' = 11/2$.

seen from Fig.13.1, the single minimum splits in two at a certain critical value of angular momentum. The two branches increase their separation with spin. The discontinuity of both solutions at the critical point from the case of axial symmetry is retained only for one branch in the general case of non-zero tilting. The triaxiality deformation defines the position of the critical point, raising it for higher γ (prolate) deformation when the alignment is considered for a particle (j'), and for lower γ values (oblate) in case of a hole alignment (j) [8].

Fig.13.2 shows that the evolution of the projection coordinate of the classical energy minima is similar in the case of two mutually perpendicular quasiparticle spin alignments. There are, however, few particularities worth to mention. The critical angular momentum will be minimal for the triaxiality $\gamma = 90°$ and will increase for both oblate and prolate directions. The two deformation phases will have different behaviors when the two alignments are not symmetric and $\gamma \neq 90°$. Another distinguishing aspect of the geometry with three vectors is that the associated azimuth angle of the stationary points become spin-dependent when $\gamma \neq 90°$. Indeed, as one can see from Fig.13.3, $\varphi_{min} = 45° = const.$ when $\gamma = 90°$, lower for a larger particle tilting and higher for a predominant tilting of the hole spin. In the $\gamma \neq 90°$ case, the evolution of the azimuth angle φ_{min} for axial alignments, that is a perfect chiral geometry, is well separated into two distinct phases, planar and aplanar. In the first phase, the total angular momentum vector moves in the plane of the two quasiparticle spins toward the short axis for prolate deformation and respectively to the long axis for the more oblate deformation. After the critical point delimiting the two phases, the azimuth angle corresponding now to two minima remains constant. The introduction of tilting in this picture smoothes this transition into a single spin-dependence for the azimuth coordinate of the lowest energy minimum. The second minimum of higher energy starts at the critical point with a larger (smaller) value for prolate (oblate) deformations, and decreases (increases) asymptotically with spin to the same value of the lowest energy minimum.

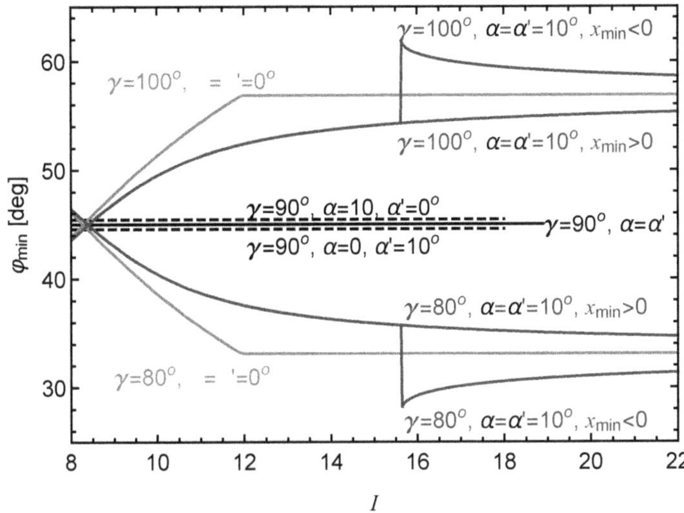

Figure 13.3 The dependence on the total angular momentum I of the azimuth coordinate φ_m of the classical energy minima for $\gamma = 80°, 90°, 100°$ with few combinations of titling angles $\alpha(\alpha') = 0°, 10°$ associated with quasiparticle spins $j = j' = 11/2$.

13.2.3 QUANTIZED SYSTEM FOR A PROJECTION VARIABLE

As the observables associated to nuclei are of quantum nature, one must quantize the new phase space in order to keep the connection between the conclusions drown from the classical picture and the quantum spectroscopy. The classical energy function always has a single minimum in one of the generalized coordinates. In the presently adopted alignment conditions, this coordinate is φ. The classical energy function to be quantized can then be approximated by its harmonic expansion around the corresponding minimum points in $\varphi_0(x)$ for fixed values of x:

$$\tilde{\mathcal{H}}(x,\varphi) \approx \mathcal{H}(x,\varphi_0(x)) + \frac{1}{2}\left(\frac{\partial^2 \mathcal{H}}{\partial \varphi^2}\right)_{\varphi_0(x)} [\varphi - \varphi_0(x)]^2. \tag{13.14}$$

The function $\varphi_0(x)$ is determined as the minimum energy path along the projection variable x and is obtained from the condition $\partial \mathcal{H}(x,\varphi)/\partial \varphi = 0$. It reduces to a quartic equation for $\cos \varphi_0(x)$:

$$(2I-1)\sqrt{I^2-x^2}(A_2-A_1)\cos \varphi_0(x) \sin \varphi_0(x)$$
$$= 2I\left[A_2 j' \cos \varphi_0(x)\cos \alpha' - A_1 j \sin \varphi_0(x)\cos \alpha\right]. \tag{13.15}$$

There are some trivial solutions of the above equation for specific alignment and deformation conditions. For $\gamma = 90°$ and two quasiparticle spins $j, j' \neq 0$, one has $\tan \phi_0 = \frac{j' \cos \alpha'}{j \cos \alpha} = const$. Whereas for a single quasiparticle spin $j(j')$, the solution is constant $\phi_0 = 0(\pi/2)$ regardless of the triaxiality and the tilting. In the most general case, the above equation has a single physical solution [15].

The harmonic approximation is meant to achieve a second order function in variable φ, which can be quantized to become a Schrödinger equation in the momentum space. This is achieved through the correspondence principle $\varphi = i\frac{d}{dx}$ with a prior symmetrization of the mixed products with co-ordinate and momentum functions $u(x)w(\phi) = [u(x)w(\phi) + u(x)w(\phi)]/2$. The resulting differential equation reads:

$$-\frac{1}{2B(x)}\frac{d^2 F(x)}{dx^2}+\left(\frac{B'(x)}{2\,[B(x)]^2}-\frac{i\varphi_0(x)}{B(x)}\right)\frac{dF(x)}{dx}+$$

$$\left[\mathcal{H}(x,\varphi_0(x))+\frac{[\varphi_0(x)]^2-i\varphi_0(x)}{2B(x)}+\frac{i\varphi_0(x)B'(x)}{2\,[B(x)]^2}+\right.$$

$$\left.\frac{B''(x)}{4\,[B(x)]^2}-\frac{[B'(x)]^2}{2\,[B(x)]^3}\right]F(x)=EF(x),\tag{13.16}$$

where

$$\begin{aligned}B(x)&=\left[\frac{\partial^2\mathcal{H}(x,\varphi)}{\partial\varphi^2}\right]_{\varphi_0(x)}^{-1}\tag{13.17}\\[4pt]&=\left\{\frac{2I-1}{I}(I^2-x^2)(A_2-A_1)\cos 2\varphi_0(x)+2\sqrt{I^2-x^2}\right.\\[4pt]&\quad\left.\times\left[A_1 j\cos\varphi_0(x)\cos\alpha+A_2 j'\sin\varphi_0(x)\cos\alpha'\right]\right\}^{-1},\end{aligned}$$

with prime and double prime denoting x derivatives. Making the change of function

$$f(x)=[B(x)]^{-\frac{1}{4}}F(x)e^{i\int\varphi_0(x)dx},\tag{13.18}$$

the above differential equation acquires a Schrödinger form

$$\left[-\frac{1}{2}\frac{1}{\sqrt{B(x)}}\frac{d}{dx}\frac{1}{\sqrt{B(x)}}\frac{d}{dx}+V(x)\right]f(x)=Ef(x),\tag{13.19}$$

with a particular definition of the kinetic term involving coordinate-dependent effective mass. Here, the function $B(x)$ plays the role of the coordinate dependent effective mass. While the potential is defined as

$$V(x)=\mathcal{H}(x,\varphi_0(x))+\frac{B''(x)}{8\,[B(x)]^2}-\frac{9\,[B'(x)]^2}{32\,[B(x)]^3},\tag{13.20}$$

and follows closely the minimum valley of the classical energy function. Similar collective Hamiltonians were employed for the description of both wobbling [17, 18] and chiral modes [19–21], with mass functions and potentials obtained from a tilted axis cranking formalism. Here, the whole model is naturally arising solely from the deformation and the alignment geometry of the system.

Given the fact that the associated eigenvalue problem is by default bounded by $|x|\leq I$, the quantum states are obtained within a diagonalization procedure using a basis of particle in the box wavefunctions:

$$\begin{aligned}f_I^s(x)&=F_I^s(x)\,[B(x)]^{-\frac{1}{4}}\tag{13.21}\\[4pt]&=\frac{[B(x)]^{-\frac{1}{4}}}{\sqrt{I}}\left\{\sum_{n=1}^{n_{Max}}A_n^s\cos\left[\frac{(2n-1)\pi x}{2I}\right]+\sum_{n=1}^{n_{Max}}B_n^s\sin\left[\frac{2n\pi x}{2I}\right]\right\},\end{aligned}$$

with Dirichlet boundary condition $F_I^s(I)=F_I^s(-I)=0$. The diagonalization solutions indexed by the order s are further used to define the density probability for the active projection variable

$$\rho_I^s(x)=|F_I^s(x)|^2=|f_I^s(x)|^2\sqrt{B(x)},\tag{13.22}$$

from which one can ascertain the system's quantum dynamics and can determine the coefficients

$$\Lambda_{IKs} = \left[\sum_{K=-I}^{I} \rho_I^s(K) \right]^{-\frac{1}{2}} F_I^s(K), \tag{13.23}$$

of the total wave function expansion in rotation states

$$|\Psi_{IMs}\rangle = \sum_{K=-I}^{I} \Lambda_{IKs}|IKM\rangle. \tag{13.24}$$

This representation is useful for the calculation of $E2$ and $M1$ electromagnetic transitions calculated with the operators

$$T_{1\mu}(M1) = \sqrt{\frac{3}{4\pi}} \mu_N \sum_{\nu=0,\pm1} \left(g_{eff} j_\nu + g'_{eff} j'_\nu \right) D^1_{\mu\nu}, \tag{13.25}$$

$$T_{2\mu}(E2) = t_1 q_{2\mu} + t_2 \left[q \times q \right]_{2\mu}, \tag{13.26}$$

where $g_{eff} = g_j - g_R$ and $g'_{eff} = g_{j'} - g_R$ are effective gyromagnetic factors gathering the corresponding values of the core and of the single-particle orbitals, while t_1 and t_2 are the coefficients of the second order expansion in quadrupole moments $q_{2\mu}$. The electric transition probability can then be explicitly given by the formula:

$$B(E2; Is \to I's') = \frac{5}{16\pi} \left| \sum_{K,K'} \Lambda^*_{IKs} \Lambda_{I'K's'} \left[\tilde{Q}_0 C^{I\,2\,I'}_{K0K'} + \frac{\tilde{Q}_2}{\sqrt{2}} \left(C^{I\,2\,I'}_{K2K'} + C^{I\,2\,I'}_{K-2K'} \right) \right] \right|^2, \tag{13.27}$$

where

$$\tilde{Q}_0 = Q \left(\cos\gamma - \chi\beta \sqrt{\frac{2}{7}} \cos 2\gamma \right), \quad \tilde{Q}_2 = Q \left(\sin\gamma + \chi\beta \sqrt{\frac{2}{7}} \sin 2\gamma \right), \tag{13.28}$$

are redefined quadrupole components, with Q being an empirical quadrupole moment, β is the quadrupole deformation, while $\chi = t_2/t_1$ is an adjustable parameter of the relative contribution of the two terms defining the $E2$ transition operator.

For the calculation of the $M1$ matrix elements, the eigenvalues of the spherical components of the quasiparticle spin operators are approximated with values consistent with FA [16]:

$$j_\pm = \mp\frac{j\cos\alpha}{\sqrt{2}}, \quad j'_\pm = -\frac{ij'\cos\alpha'}{\sqrt{2}}, \quad j_0 = j\sin\alpha, \quad j'_0 = j'\sin\alpha'. \tag{13.29}$$

This leads to the following expression for the magnetic transition probability:

$$B(M1; Is \to I's') = \frac{3}{4\pi} \mu_N^2 \left| \sum_{K,K'} \Lambda^*_{IKs} \Lambda_{I'K's'} \left[\frac{g_{eff} j\cos\alpha}{\sqrt{2}} \left(C^{I\,1\,I'}_{K1K'} - C^{I\,1\,I'}_{K-1K'} \right) + \right. \right.$$

$$\left. \left. \frac{ig'_{eff} j'\cos\alpha'}{\sqrt{2}} \left(C^{I\,1\,I'}_{K1K'} + C^{I\,1\,I'}_{K-1K'} \right) - \left(g_{eff} j\sin\alpha + g'_{eff} j'\sin\alpha' \right) C^{I\,1\,I'}_{K0K'} \right] \right|^2. \tag{13.30}$$

The evaluation of effective gyromagnetic factors usually involves empirical quenching for the single-particle part and the rather crude rigid rotor estimate $g_R = Z_R/A_R$ of the core contribution. For the case with a single quasiparticle spin, it can be fixed from data on the mixing ratio $\delta_{I \to I-1}$.

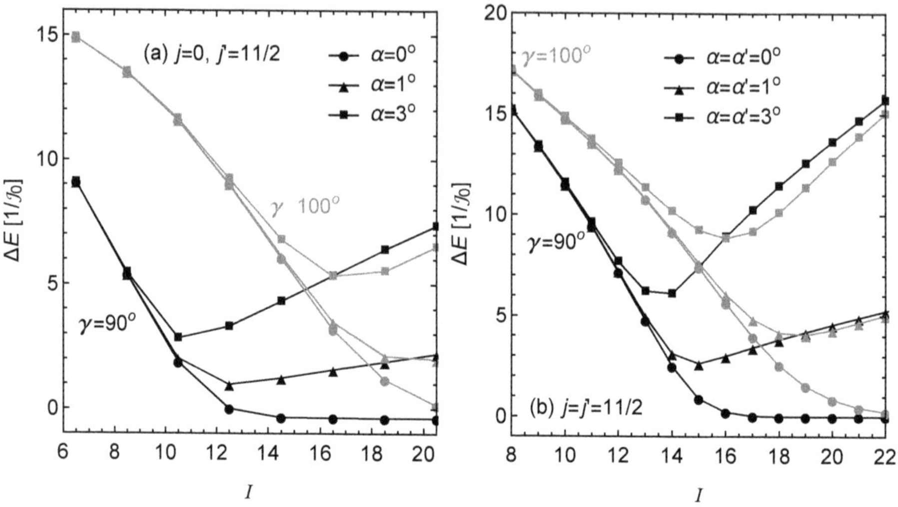

Figure 13.4 Wobbling energy (a) for the case $j = 0$ and $j' = 11/2$ and the energy splitting of the chiral bands (b) with $j = j' = 11/2$ as a function of angular momentum for $\gamma = 90°, 100°$ and a tilting $\alpha(\alpha') = 0°, 1°, 3°$.

13.3 APPLICATIONS TO WOBBLING AND CHIRAL BANDS

The total energy formula applied for a quantitative description of wobbling and chiral partner bands is:

$$E_{I,s} = E_{Is}^{diag}(\mathcal{J}_0, \gamma, \alpha, \alpha') + CI(I+1) + E_0. \tag{13.31}$$

The first term denotes the eigenvalue of order s determined from the diagonalization of Eq.(13.19), which is parametrized by the triaxiality γ, tilting angles α and α', and the scaling MOI $\mathcal{J}_0$. This energy is amended with a rotational contribution which is meant to reproduce the experimental rotation spectrum without compromising the structure of total quantum system because the total wave function is an eigenfunction of the $\hat{J}^2$ operator. Despite the relatively large number of parameters, the aspects of wobbling excitation and chirality, as for example the energy difference

$$\Delta E(I) = \begin{cases} E_{I,2} - E_{I,1}, & \text{chiral bands}, \\ E_{I,2} - \frac{1}{2}\left(E_{I+1,1} + E_{I-1,1}\right), & \text{wobbling bands}, \end{cases} \tag{13.32}$$

depends up to a scaling factor only on the triaxiality and the tilting angles.

The evolution with spin of these quantities is presented in Fig.13.4 for few demonstrative values of γ and tilting angles $\alpha(\alpha')$. In the absence of tilting, the energy splitting $\Delta E(I)$ for single and two distinct quasiparticle alignments decreases with spin to the zero value which marks the establishment of two equivalent static rotational configurations constructed as symmetric and antisymmetric superpositions of positive and negative K solutions. In the case of axial alignments, the two emerging rotational configurations are related through a simple rotation transformation in the case of a single quasiparticle spin. While in the case of the chiral arrangement, an additional time inversion is required for the same transformation. Due to the restoration of the rotational symmetry in the laboratory frame of reference, one then expects to observe in this high spin regime, a termination of the wobbling excitations and respectively two degenerate bands for the static chirality [22]. The decreasing part of $\Delta E(I)$ is more important for configurations with single quasiparticle alignments, being associated with the transverse wobbling motion. The same regime for two perpendicular quasiparticle spins is understood as chiral vibration [23], which transforms into a static chiral configuration when the two solutions become degenerate. The tilting of quasiparticle spins shortens the interval of the decreasing energy difference. Moreover, the presence of tilting breaks the degeneracy between the yrast and yrare states, changing the well known invariance with spin of the corresponding energy difference into an increasing dependence on spin. This increasing part of the energy difference has a higher slope for larger tilting and is mostly unaffected by the change in triaxiality. The triaxiality, however, dictates the angular momentum of the energy difference minimum which marks the transition between the two distinct dynamical modes. All these correlations of the triaxiality deformation and the tilting of the quasiparticle alignments with specific spectral characteristics were used to describe the dynamical evolution of states belonging to wobbling and chiral bands. In what follows, one will present the main findings of these studies separately for the phenomena related to wobbling motion and the chiral geometry.

13.3.1 ANHARMONIC WOBBLING MOTION IN EVEN-EVEN AND ODD MASS NUCLEI

Rotational bands based on quasiparticle alignments which exhibit characteristics of wobbling excitations are presently found in a dozen of odd mass nuclei [24–36] and even in a handful of even-even isotopes [37–42]. Moreover, there are expectations of wobbling excitations also in lighter nuclei [43, 44], enforcing thus the idea of a more widespread phenomenon. The extension of experimental realization for the wobbling motion engages questions regarding the specific systematic or sudden changes of the underlying phenomenology in distinct regions of the nuclide chart. The experimental observations are predominantly associated with the transverse wobbling mechanism [4, 5], where the nucleus precesses transversally to the body fixed axis with the maximal MOI. This is possible due to the quasiparticle alignments which enhance transversal effective MOI. Although the reported transverse wobbling bands are build on particle alignments, there are possible candidates of wobbling excitations based also on hole alignments [8, 35, 41]. The relatively small spins of the observed states imply the presence of important anharmonicities in their transverse wobbling motion, which hampers their description within the simplified harmonic wobbling model [5, 45, 46]. These are, however, negligible in the case of the longitudinal wobbling proposed for example in the ^{133}La [31], ^{187}Au [33], and ^{127}Xe [34] nuclei. The transverse or longitudinal identification of the wobbling motion is strictly related to the axial quasiparticle alignments [5, 45] and can be misleading when dealing with realistic quasipraticle configurations. The model presented in the previous section is especially suited for the study of the wobbling dynamics in the presence of non-axial alignments and to make a natural connection between the two extreme wobbling modes through the consistent treatment of the anharmonicities.

The first incursion of the model beyond the harmonic approximation was aimed for the study of wobbling excitation in even-even nuclei [8], where the single particle alignment was due to a strongly bound pair of proton quasiparticles $h^2_{11/2}$ contributing to a $j = 10\hbar$ spin and assumed axial. Thus, the concept of wobbling motion was extended toward the critical angular momentum marking the emergence of tilted axis minima in the potential energy. The experimental continuation of the wobbling bands into this high spin regime is, however, fraught with additional quasiparticle alignments. One of the best examples of the model's performance is given in this case by the ^{130}Ba nucleus. The evolution with spin of its measured wobbling energy and of the $B(M1)_{out}/B(E2)_{in}$ ratio, depicted in Fig.13.5(a,d), is very well reproduced. The theory suggests a parabolic evolution of the ratio between $M1$ and $E2$ transition probabilities, which is supported by the experimental data. Note that this trend is solely due to the structure of the wave function. The higher order term used for the $E2$ transition operator do not affect this aspect. The analysis of the critical angular momentum governing the existence condition for the distinct rotational phases, suggested that wobbling bands built on particles can coexist with bands constructed on holes. This is possible in a very short range of angular momentum states, if the core is close to maximal triaxiality. Obviously, the two alignments must be for different nucleon species. A possible experimental realization for such a case was forwarded in Ref.[8, 41] for the even-even ^{138}Nd nucleus.

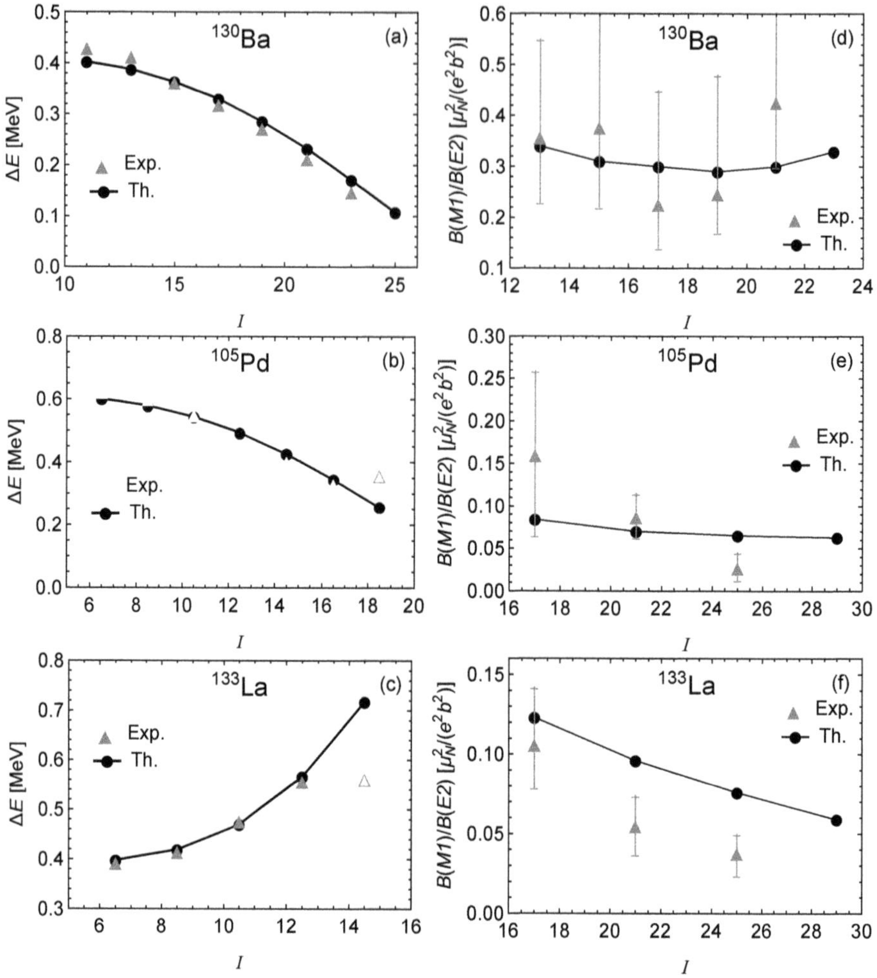

Figure 13.5 Comparison between measured data and theoretical calculations of the wobbling energy and of the transition probability ratio $B(M1)_{out}/B(E2)_{in}$ for ^{130}Ba (a,d), ^{105}Pd (b,e), and ^{133}La (c,f). Open symbols denote data points with different quasiparticle structure. Details on the theoretical calculations are given in Refs.[8, 9].

Introducing the tilting of the quasiparticle spin alignment as an adjustable parameter, offers the possibility to check whether transverse wobbling in its original definition is actually found in real nuclei. Numerical applications on the proposed wobbling bands in ^{105}Pd and ^{135}Pr confirms the axial alignment of the corresponding neutron and proton $h_{11/2}$ quasiparticle [9]. The agreement with experimental data is exemplified for ^{105}Pd nucleus in Fig.13.5(b,e). Once again, the parabolic decreasing of the wobbling energy associated with the transverse alignment, is well reproduced. The experimental $B(M1)_{out}/B(E2)_{in}$ ratio decreases with spin, a trend which is reproduced by the theory but with a very small slope.

Finally, the advanced version of the model with adjustable triaxiality and tilting of the quasiparticle alignment, provided an interesting dynamical picture for the similar bands observed in ^{133}La [9]. In previous studies [31], the measured $h_{11/2}$ quasiparticle bands were interpreted as due to longitudinal wobbling. This assignment was logically based on the increasing wobbling energy with spin. It was, however, conspicuous that the neighboring nucleus ^{135}Pr, with a very similar microscopic structure, exhibits transverse wobbling excitations. The contradiction was in a manner resolved by considering the dynamical case between the extremes of the longitudinal and transverse wobbling, called tilted axis wobbling. Indeed, model fits on the experimental data regarding the considered quasiparticle bands in ^{133}La, provided a sizeable tilting of the quasiparticle alignment. However, it is still closer to the short axis, suggesting a particle character of the aligned proton, in accordance to the similar ^{135}Pr nucleus. As can be seen in Fig.13.4(a), the tilting imply a smaller transverse contribution to the total angular momentum, and therefore the minimum in the corresponding wobbling energy lowers in spin. This behavior explains the increasing spin evolution of the wobbling energy visualized in Fig.13.5(e) for this nucleus. Fig.13.5(f) also shows that in comparison to ^{105}Pd, for this nucleus, the decreasing slope of the experimental $B(M1)_{out}/B(E2)_{in}$ ratio is well reproduced, but the actual values are overestimated. The specific dynamical effects of the non-axial alignment include the change of the $K = 0$ energy minimum of the transverse wobbling motion to $K \neq 0$ configurations consistent with tilted axis rotation. Note that, prior to the probabilistic symmetrization of the problem, the system have a single classical energy minimum in the direction of the quasiparticle alignment, whose position is moving toward higher $K(x)$ values with the raising of the total angular momentum. In the same manner, there is a critical spin where a second minimum appears, but which is asymmetric in energy and position $K(x)$. The critical angular momentum is increased considerably by the tilting. Moreover, the second minimum is sufficiently high in energy, in order for the system to favor a wobbling excitation within the lowest minimum to the detriment of the simple rotation in the higher energy minimum. In this sense, the tilted axis wobbling mode is more stable than the transverse regime, and have all features of a transitional phase toward the traditional longitudinal wobbling regime.

13.3.2 EXPERIMENTAL REALIZATION OF APPROXIMATE CHIRAL GEOMETRY IN ODD-ODD NUCLEI

It can be said that the theoretical prediction [3] and the subsequent experimental confirmation [23] of the nuclear chirality, initiated an important topic for the nuclear physics of the last decades. As a result, its manifestation is reported now in more than fifty nuclei [47, 48], based on the original one proton and one neutron perpendicular spins or generalizations to many quasiparticle alignments. Moreover, the study of this phenomenon is pushed even further by considering its effect in combination with other various nuclear structure notions, such as shape coexistence [49], γ softness [50], octupole correlations [51] or shape phase transition [52].

Ideally, the chiral geometry is identified by the presence of a degenerated doublet of bands with $\Delta I = 1$ and the same parity [22]. Nevertheless, most of the observed chiral bands are only approximately degenerated and in a restricted high spin interval, often violating the electromagnetic selection rules assumed from perfect trihedral geometry [53]. This situation is partially explained by the chiral vibration [23] commencing at low spins, which is similar to the transverse wobbling mechanism but extended to an aplanar image. On the other hand, the loss of degeneracy at higher spins is ascribed to the deviations of the all three involved angular momentum vectors from the perfect trihedral alignment. The factors inhibiting the attainment of degenerated bands stem from the fractional quasiparticle character of the aligned nucleons and from non-uniform components of the core's rotation due to the triaxial deformation and its possible softness. The present formalism can in principle account for both single-particle and collective effects which influence the emergence of the chiral geometry [10, 15]. However, fitting both triaxiality and tilting is computationally demanding. One way to overcome this obstacle is to use the eigenvalues of the cranked Hamiltonian for any values of γ, α and α', obtained easily in the angular momentum basis. As the difference between the energies computed in this way and those determined from the collective Schrödinger equation is expected to be small, the former can be used in a first fitting stage, and then improved by fine fits around the obtained parameter set with the later model. Nevertheless, considering that the phenomenon of chirality is directly related to the triaxial deformation, the present formalism with

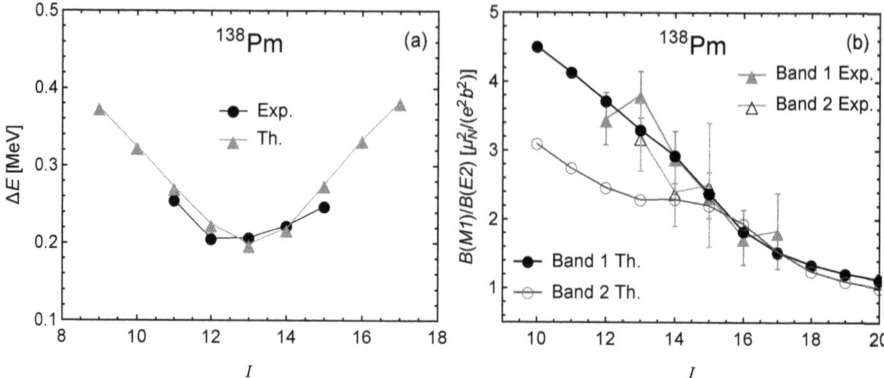

Figure 13.6 Comparison between measured data and theoretical calculations [16] of the energy difference between the chiral bands of ^{138}Pm (a) and of the transition probability ratio $B(M1)_{in}/B(E2)_{in}$ within the corresponding bands (b).

a triaxiality a priori fixed as maximal ($\gamma = 90°$) can be used for drawing meaningful conclusions regarding the system's dynamics. This is in general a sensible assumption, given the fact that the dynamical aspects do not change significantly in the vicinity of the maximal triaxiality.

From the beginning of experimental studies on nuclear chirality, the ^{134}Pr nucleus emerged as the best example, exhibiting approximately degenerated bands at a relatively low angular momentum [54]. This fact did not change even after the reconsideration of its chiral partner bands [55]. The energy splitting of the reassigned $\pi h_{11/2} \nu (h_{11/2})^{-1}$ bands is less than 280 keV for all states, supporting thus their chiral phenomenology. This is also confirmed within the present formalism, where fits on the corresponding data [16], offered very small and equal tilting of the two quasiparticle spins, $\alpha = \alpha' = 3°$. Although small, the non-vanishing tilting has important effects on the spectroscopy, successfully reproducing the small experimental energy difference between bands. Moreover, as can be seen from Fig.13.4(b), the theory also provides the correct position of the maximal closeness between bands, experimentally registered at $I = 13$. At this critical angular momentum the approximately symmetrical chiral vibration transforms into a static chirality mode, with the two bands having well defined opposite chirality. This is different from the perfect chiral configuration, where the two bands correspond to symmetric and anti-symmetric superpositions of the two distinct chiral configurations. This is due to the tilting, which makes the two static configurations different in energy. Moreover, the energy splitting between the two non-equivalent chiral configurations increases with spin.

With the introduction of tilting, the proposed model extends the concept of chirality to a more general non-trihedral arrangement of the three involved spins, facilitating the description of data with a more diverse range of spectroscopic properties. This particular ability of the model is showcased when applied to the ^{138}Pm nucleus [10, 16], with observed partner bands based on the same $\pi h_{11/2} \nu (h_{11/2})^{-1}$ configuration [56]. It was found that the energy levels of the two bands are best described when an asymmetrical tilting is considered, that is with one quasiparticle spin axially aligned and another with a non-negligible $8°$ tilting. This outcome is in total agreement for example with the PRM calculations of Ref.[57], where it was concluded that the dominant proton quasiparticle configuration has a smaller projection on the s-axis. Therefore, the tilting obtained in the present model is associated with the particle spin j'. Otherwise, one could not distinguish between results with interchanged α and α', due to the fact that for this quasiparticle configuration $j = j'$ and $A_1 = A_2$ for $\gamma = 90°$. The predicted moderate tilting implies that this nucleus undergoes a transition from an asymmetric chiral vibration with an aplanar equilibrium point, to a regime with two rotational configurations of opposite chirality and largely different orientation of the rotation axis. The two static configurations at high spins are well separated in energy. This can be observed in Fig.13.6(a), where the theoretical and experimental energy difference between partner bands is shown as a function of the total angular momentum. The evolution with spin of the experimental data is perfectly reproduced by theory, which accurately describes the attraction and then the repulsion of the two bands, as well as the angular momentum of the smallest energy difference between them. The model's performance in describing such a dynamical geometry can be further judged by comparing the experimental electromagnetic properties available for ^{138}Pm with the predictions generated by the wave-function determined before hand from fits to energy levels alone. This is done in Fig.13.6(b), using the $B(M1; I \rightarrow I-1)/B(E2; I \rightarrow I-2)$ ratios between in band magnetic and electric transition probabilities. There is a general decreasing evolution with spin of experimental data. This trend is reproduced theoretically by choosing an appropriate strength χ for the second order term in the $E2$ transition operator. Regardless of this choice, the theory predicts higher values for the first band at low spins. The ratios become approximately similar starting with the $I = 15$ states. Fig.13.6(b) demonstrates that these aspects are in a very good agreement with the measured data.

13.4 FINAL REMARKS

The semiclassical description of systems composed of a triaxial core and rigid alignments of quasiparticle spins presented here, is meant to aid the understanding of the dynamics associated with such complex quantum ensembles in terms of relatable classical concepts such as rotations, oscillations, trajectories, constants of motion, generalized coordinates and momenta. The quantum problem is framed as a triaxial rotor with cranking terms defined by the adjustable quasiparticle spin alignments. The corresponding classical picture is achieved in the standard way by means of a time-dependent variational principle. Given the particularity of the treated problems, the variational function is chosen to be a coherent state for the algebra of the total angular momentum operators. The scope of the variational principle lies in the extraction of the time-dependence of some set of variables which parametrizes the variational state and is usually reduced in comparison to the full number of degrees of freedom. The stereographic parametrization of the variational states introduces the azimuth angle and a projection variable associated with the direction of the total angular momentum vector as classical variables. As a result, a classical energy and a couple of equations of motions are obtained from solving the variational principle for the quantum Hamiltonian. The choice of variables assures that the equations of motion are canonical. This fact implies that the two classical directional variables have a canonical conjugate relationship. The dynamical features of the system is given by the classical trajectories which are solutions of the equations of motion. As a consequence, the orbits performed by the total angular momentum vector surround the stationary points of the classical energy function. Information regarding the evolution of the classical dynamics can be extracted without integrating the equations of motions. This can be realized simply by tracking the-spin dependence of the coordinates defining the stationary points. Alternatively, the classical trajectories can be obtained by the intersection of the two constants of motion. Thus it is found that the departure of the quasiparticle alignments from the principal axes as well as the change in the triaxiality of the core, greatly affects the dynamical properties of the system, leading to distinct rotational regimes with specific existence conditions.

In order to account for the quantum effects and to extract the discrete energy spectrum of the described systems, one constructed a Schrödinger equation from the classical energy function which has the total angular momentum projection as a continuous variable. This is achieved by speculating the canonical relationship between the classical coordinates. Thus, their differential representation is obtained from the correspondence principle. However, the quantization procedure is not trivial, due to the mixing of momentum and coordinate terms as well as the infinite expansion of the square root and trigonometric functions of them. The first issue can be resolved by a proper symmetrization of the mixed products, while a compact energy function is obtained by applying a harmonic approximation on the classical energy along one of the classical variables. The resulted differential equation can be brought through a change of function to a form resembling the Schrödinger equation with a kinetic term depending on a coordinate-dependent effective mass. The potential is completely defined by the geometry of the system, that is by the MOI and the assumed quasiparticle alignments. It follows closely the minimal energy path from the classical energy function and is therefore evolving from a single to double minima shapes as the angular momentum increases. Consequently, the quantum observables will exhibit specific signatures associated to the distinct dynamical phases identified in the classical analysis. Therefore, one can distinguish for a single quasiparticle spin alignment, the well known longitudinal and transverse wobbling modes and a transitional regime associated with tilted-axis wobbling. The later is conditioned by a sizable tilting of the quasiparticle spin, otherwise the transverse wobbling will degenerate into regular rotation around two distinct directions. This situation is actually what is expected from the static chirality, with the distinction that the splitting of the two rotation directions is aplanar. Similarly, it can be conjured that the chiral vibration is the generalization of the transverse wobbling mechanism to two perpendicular quasiparticle spin alignments. A selection of numerical applications was chosen here to exemplify the performance of the present formalism to reproduce experimental data on energy levels and electromagnetic properties and describe the associated phenomenology. In the case of a single quasiparticle spin, the focus is directed to the investigation of the transverse and tilted-axis wobbling motion. The consideration of two mutually perpendicular quasiparticle spins allowed the assessment of the experimental realization of chiral geometry in real nuclei and its evolution from vibrational to a static character. Following further the correspondence between wobbling and chirality, it would be interesting to investigate the possibility of a three-dimensional (aplanar) wobbling based on two quasiparticle spins with sizeable tilting. This would be the chiral equivalent of the tilted-axis wobbling.

The presented formalism is quiet involved when compared with a straightforward diagonalization of the original Hamiltonian in some angular momentum basis. It has, however, the important advantage of a Schrödinger form with a separated potential carrying the dynamical information correlated with the classical motion. Moreover, the active variable is continuous and corresponds to the total angular momentum projection, which is otherwise discrete in the phase space of angular momentum operators. This aspect provides a unique phenomenological interpretation of the quantum states in terms of quantized anharmonic oscillations of the total angular momentum vector with or without tunneling. The semiclassical origin of the model has another important consequence, which is the lack of the purely probabilistic superposition of quantum states in presence of tilting. More precisely, the model describes for example the effect of a positive quasiparticle projection along the quantization axis on both positive and negative projections of the total angular momentum without its symmetrical counterpart with negative quasiparticle projection. The two configurations have the same energy and are mutually exclusive from the physical point of view but their wave functions are automatically combined due to symmetries demanded in the quantum PRM. It is therefore imperative to take care when drawing conclusions on the system's motion from the density probability distributions.

13.5 ACKNOWLEDGMENTS

This work was supported by grants of the Ministry of Research, Innovation and Digitalization, CNCS - UEFISCDI, project number PN-III-P1-1.1-TE-2021-0109, within PNCDI III, and PN-23-21-01-01/2023.

Bibliography

1. R. Bengtsson, H. Frisk, F. May and J. Pinston, Nucl. Phys. A **415**, 189 (1984).
2. N. Zamfir and R. Casten, Phys. Lett. B **260**, 265 (1991).

3. S. Frauendorf and J. Meng, Nucl. Phys. A **617**, 131 (1997).

4. Y. R. Shimizu, M. Matsuzaki and K. Matsuyanagi, *Microscopic Study of Wobbling Motions in Hf and Lu Nuclei*, Proc. 5-th Japan-China Joint Nuclear Physics Symp. (Fukuoka, Japan) pp. 317-326, arXiv:nucl-th/0404063 (2004).

5. S. Frauendorf and F. Dönau, Phys. Rev. C **89**, 014322 (2014).

6. A. Bohr and B. R. Mottelson, *Nuclear Structure*, Vol. 2 (Benjamin, Reading, Massachusetts, 1975).

7. V. Lanchares, M. Inarrea, J. P. Salas, J. D. Sierra and A. Elipe, Phys. Rev. E **52**, 5540 (1995).

8. R. Budaca and C. M. Petrache, Phys. Rev. C **106**, 014313 (2022).

9. R. Budaca and A. I. Budaca, J. Phys. G: Nucl. Part. Phys. **50**, 125101 (2023).

10. R. Budaca, Phys. Lett. B **817**, 136308 (2021).

11. A. A. Raduta, R. Budaca and C. M. Raduta, Phys. Rev. C **76**, 064309 (2007).

12. P. Kramer and M. Saraceno, *Geometry of the Time-Dependent Variational Principle in Quantum Mechanics*, Lecture Notes in Physics **140** (Springer-Verlag, Berlin, 1981).

13. J. Klauder and B. Skagerstam, *Coherent States. Applications in Physics and Mathematical Physics* (World Scientific, Singapore, 1985).

14. R. Budaca, Phys. Rev. C **98**, 014303 (2018).

15. R. Budaca, Phys. Lett. B **797**, 134853 (2019).

16. R. Budaca, Front. Phys. **19**, 24301 (2024).

17. Q. B. Chen, S. Q. Zhang, P. W. Zhao and J. Meng, Phys. Rev. C **90**, 044306 (2014).

18. Q. B. Chen, S. Q. Zhang and J. Meng, Phys. Rev. C **94**, 054308 (2016).

19. Q. B. Chen, S. Q. Zhang, P. W. Zhao, R. V. Jolos and J. Meng, Phys. Rev. C **87**, 024314 (2013).

20. Q. B. Chen, S. Q. Zhang, P. W. Zhao, R. V. Jolos and J. Meng, Phys. Rev. C **94**, 044301 (2016).

21. X. H. Wu, Q. B. Chen, P. W. Zhao, S. Q. Zhang and J. Meng, Phys. Rev. C **98**, 064302 (2018).

22. S. Frauendorf, Rev. Mod. Phys. **73**, 463 (2001).

23. K. Starosta *et al.*, Phys. Rev. Lett. **86**, 971 (2001).

24. S. W. Ødegård *et al.*, Phys. Rev. Lett. **86**, 5866 (2001).

25. G. Schönwasser *et al.*, Phys. Lett. B **552**, 9 (2003).

26. H. Amro *et al.*, Phys. Lett. B **553**, 197 (2003).

27. P. Bringel *et al.*, Eur. Phys. J. A **24**, 167 (2005).

28. D. J. Hartley *et al.*, Phys. Rev. C **80**, 041304(R) (2009).

29. J. T. Matta *et al.*, Phys. Rev. Lett. **114**, 082501 (2015).

30. J. Timár *et al.*, Phys. Rev. Lett. **122**, 062501 (2019).

31. S. Biswas *et al.*, Eur. Phys. J. A **55**, 159 (2019).

32. S. Nandi *et al.*, Phys. Rev. Lett. **125**, 132501 (2020).

33. N. Sensharma *et al.*, Phys. Rev. Lett. **124**, 052501 (2020).

34. S. Chakraborty *et al.*, Phys. Lett. B **811**, 135854 (2020).

35. K. Rojeeta Devi *et al.*, Phys. Lett. B **823**, 136756 (2021).

36. A. Mukherjee *et al.*, Phys. Rev. C **107**, 054310 (2023).

37. C. M. Petrache *et al.*, Phys. Lett. B **795**, 241 (2019).

38. Q. B. Chen, S. Frauendorf and C. M. Petrache, Phys. Rev. C **100**, 061301(R) (2019).

39. C. M. Petrache *et al.*, Phys. Rev. C **93**, 064305 (2016).

40. F. Q. Chen and C. M. Petrache, Phys. Rev. C **103**, 064319 (2021).

41. C. M. Petrache *et al.*, Phys. Rev. C **86**, 044321 (2012).

42. B. F. Lv *et al.*, Phys. Rev. C **105**, 034302 (2022).

43. L. Hu, J. Peng and Q. B. Chen, Phys. Rev. C **104**, 064325 (2021).

44. H. M. Dai, Q. B. Chen and Xian-Rong Zhou, Phys. Rev. C **108**, 054306 (2023).

45. R. Budaca, Phys. Rev. C **97**, 024302 (2018).

46. R. Budaca, Phys. Rev. C **103**, 044312 (2021).

47. B. W. Xiong and Y. Y. Wang, At. Data Nucl. Data Tables **125**, 193 (2019).

48. Shou-Yu Wang, Chinese Phys. C **44**, 112001 (2020).
49. J. Meng, J. Peng, S. Q. Zhang and S. G. Zhou, Phys. Rev. C **73**, 37303 (2006).
50. B. F. Lv *et al.*, Phys. Rev. C **98**, 044304 (2018).
51. S. Guo *et al.*, Phys. Lett. B **807**, 135572 (2020).
52. Y. Zhang, B. Qi and S. Q. Zhang, Sci. China Phys. Mech. Astron. **64**, 122011 (2021).
53. T. Koike, K. Starosta and I. Hamamoto, Phys. Rev. Lett. **93**, 172502 (2004).
54. C. M. Petrache *et al.*, Nucl. Phys. A **597**, 106 (1996).
55. J. Timár *et al.*, Phys. Rev. C **84**, 044302 (2011).
56. K. Y. Ma *et al.*, Phys. Rev. C **97**, 014305 (2018).
57. P. Siwach, P. Arumugam, L. S. Ferreira and E. Maglione, Phys. Lett. B **811**, 135937 (2020).

14 Recent progress on nuclear chiral and wobbling motions based on the triaxial projected shell model

Ya-Kun Wang

Peking University, Beijing, China

14.1 INTRODUCTION

The spontaneous symmetry breaking in atomic nuclei leads to the occurrence of nuclear shapes with different spatial symmetries. Modern experimental measurements and theoretical calculations lead to the conclusion that most of the nuclei around the nuclear chart are deformed. Among various nuclear shapes, the axially deformed kinds characterized by deformation parameter β have been known for a long time, whose typical consequences are the observation of nuclear rotational excitations [1]. In recent decades, many efforts have been paid to study nuclei with triaxial shape that is described by the triaxial deformation parameter γ in addition to the axial one β. The existence of nuclear triaxiality can be manifested by the observation of nuclear chiral and wobbling rotations.

Nuclear chiral rotation was first predicted by Frauendorf and Meng in 1997 [2]. In such a novel rotational mode, a triaxial nucleus rotates around an axis out of the three principal planes of the ellipsoidal nuclear shape. The three components of total angular momenta induced by chiral rotation along the short, intermediate, and long principal axes of a triaxial nucleus in the intrinsic frame form either a left- or right-handed system, resulting in the spontaneous chiral symmetry breaking. The restoration of chiral symmetry in the laboratory frame leads to the observation of chiral doublet bands, i.e., a pair of nearly degenerate $\Delta I = 1$ rotational bands with the same parity. So far, around 60 chiral doublet bands have been experimentally reported; see review [3] and also the data compilation [4] for more details.

Nuclear wobbling motion is another novel rotational mode of triaxial nuclei and was first proposed by Bohr and Mottelson [1]. It represents a harmonic excitation of a triaxial nucleus on top of the uniform rotation around the intrinsic axis with the largest moment of inertia. The harmonicity results in a series of $\Delta I = 2$ rotational bands that are equidistant and characterized by the phonon number n. Moreover, as demonstrated in Ref. [1], the $\Delta I = 1$ transitions between the bands with n and $n + 1$ phonons are expected to be collectively enhanced.

The wobbling motion analyzed by Bohr and Mottelson is for a single triaxial rotor, namely, a nucleus without any valence quasiparticles, which is referred to as "simple wobbling". In cases involving quasiparticles, Frauendorf and Dönau proposed the so-called longitudinal and transverse wobblings depending on whether the quasiparticle angular momentum is oriented parallel or perpendicular to the rotor axis with the largest moment of inertia [5]. No clear evidence for the "simple wobbling" has been reported so far. Nevertheless, the experimental evidences of longitudinal wobbling have been reported in ^{133}La [6], ^{187}Au [7], ^{183}Au [8], and ^{127}Xe [9], and those of transverse

DOI: 10.1201/9781032691633-14

"

wobbling have been reported in ^{161}Lu [10], ^{163}Lu [11, 12], ^{165}Lu [13], ^{167}Lu [14], ^{167}Ta [15], ^{130}Ba [16], and ^{136}Nd [17].

The encouraging experimental observation on nuclear chiral and wobbling motions have promoted continuously theoretical studies. Various theoretical models, including the particle rotor model [2, 18–20], the generalized coherent state model [21], the interacting boson fermion-fermion model [22–24], tilted axis cranking model (TAC) [25–29], TAC with random phase approximation [5, 30, 31] and collective Hamiltonian [32, 33], and the triaxial projected shell model [34–39], have been employed to study nuclear chiral and/or wobbling rotations.

Among these models, the triaxial projected shell model (TPSM) has received a wide attention. The TPSM carries out the shell-model configuration mixing based on the deformed Nilsson mean field with the angular momentum projection technique [40]. Compared to the traditional shell model calculations, the configuration space of the TPSM is much smaller, which is realized by using the deformed Nilsson basis. Therefore, the TPSM can be applied to study the spectroscopic properties of deformed heavy nuclei, in which the chiral and wobbling motions are commonly observed. Moreover, the TPSM describes nuclear properties in the laboratory frame and, thus, the tunneling between the left- and right-handed configurations can be properly considered. Recently, the so-called *K plot* and *azimuthal plot* have been proposed by Chen *et al.* [35], from which the angular momentum geometry of nuclear chiral and wobbling motions can be clearly illustrated. Due to the aforementioned advantages and achievements, the TPSM has now become a powerful tool to study the chiral and wobbling motions.

In this contribution, I would briefly introduce the recent progress on nuclear chiral and wobbling motions based on the TPSM. The theoretical framework of the TPSM is introduced in Sec. 14.2. The theoretical descriptions of the chiral rotation in ^{136}Nd and nuclear wobbling motion in ^{130}Ba based on the TPSM are illustrated in Sec. 14.3 and Sec. 14.4, respectively. Finally, a short summary is given in Sec. 14.5.

14.2 THEORETICAL FRAMEWORK

The starting point of the present TPSM calculations is the following pairing plus quadrupole Hamiltonian [41],

$$\hat{H} = \hat{H}_0 - \frac{\chi}{2} \sum_\mu \hat{Q}_\mu^\dagger \hat{Q}_\mu - G_M \hat{P}^\dagger \hat{P} - G_Q \sum_\mu \hat{P}^\dagger \hat{P}. \tag{14.1}$$

Here, $\hat{H}_0$ represents the spherical single-particle shell-model Hamiltonian. The second term denotes the quadrupole-quadrupole interaction, whose strength χ is usually determined by the self-consistent relation and, thus, be determined by the quadrupole deformation [40]. The third term denotes the monopole paring interaction and its strength G_M is usually adjusted to reproduce the proton and neutron pairing gaps. The last term denotes the quadrupole pairing interaction, whose strength G_Q is typically assumed to be approximately 20% of G_M.

Based on the pairing plus quadrupole Hamiltonian, the intrinsic ground state of the nuclear system $|\Phi_0\rangle$ is obtained by solving the following variational equation,

$$\delta\langle\Phi_0|\hat{H} - \lambda_n\hat{N} - \lambda_p\hat{Z}|\Phi_0\rangle = 0. \tag{14.2}$$

Here, the Lagrangian multipliers λ_n and λ_p are determined respectively by applying the constraining conditions $\langle\Phi_0|\hat{N}|\Phi_0\rangle = N$ and $\langle\Phi_0|\hat{Z}|\Phi_0\rangle = Z$.

To consider the beyond mean-field correlations induced by the configuration mixing, one needs to construct a intrinsic configuration space, which consists of different kinds of multi-quasiparticle (qp) states $|\Phi_\kappa\rangle$. The configuration space including multi-qp configuration up to six-qp states for even-even nuclei can be constructed as,

$$
\begin{aligned}
\{ &|\Phi_0\rangle, \hat{\beta}^{\dagger}_{\nu_i}\hat{\beta}^{\dagger}_{\nu_j}|\Phi_0\rangle, \hat{\beta}^{\dagger}_{\pi_i}\hat{\beta}^{\dagger}_{\pi_j}|\Phi_0\rangle, \hat{\beta}^{\dagger}_{\pi_i}\hat{\beta}^{\dagger}_{\pi_j}\hat{\beta}^{\dagger}_{\nu_k}\hat{\beta}^{\dagger}_{\nu_l}|\Phi_0\rangle, \\
&\hat{\beta}^{\dagger}_{\nu_i}\hat{\beta}^{\dagger}_{\nu_j}\hat{\beta}^{\dagger}_{\nu_k}\hat{\beta}^{\dagger}_{\nu_l}|\Phi_0\rangle, \hat{\beta}^{\dagger}_{\pi_i}\hat{\beta}^{\dagger}_{\pi_j}\hat{\beta}^{\dagger}_{\pi_k}\hat{\beta}^{\dagger}_{\pi_l}|\Phi_0\rangle, \\
&\hat{\beta}^{\dagger}_{\nu_i}\hat{\beta}^{\dagger}_{\nu_j}\hat{\beta}^{\dagger}_{\nu_k}\hat{\beta}^{\dagger}_{\nu_l}\hat{\beta}^{\dagger}_{\nu_m}\hat{\beta}^{\dagger}_{\nu_n}|\Phi_0\rangle, \hat{\beta}^{\dagger}_{\pi_i}\hat{\beta}^{\dagger}_{\pi_j}\hat{\beta}^{\dagger}_{\pi_k}\hat{\beta}^{\dagger}_{\pi_l}\hat{\beta}^{\dagger}_{\pi_m}\hat{\beta}^{\dagger}_{\pi_n}|\Phi_0\rangle, \\
&\hat{\beta}^{\dagger}_{\pi_i}\hat{\beta}^{\dagger}_{\pi_j}\hat{\beta}^{\dagger}_{\nu_k}\hat{\beta}^{\dagger}_{\nu_l}\hat{\beta}^{\dagger}_{\nu_m}\hat{\beta}^{\dagger}_{\nu_n}|\Phi_0\rangle, \hat{\beta}^{\dagger}_{\nu_i}\hat{\beta}^{\dagger}_{\nu_j}\hat{\beta}^{\dagger}_{\pi_k}\hat{\beta}^{\dagger}_{\pi_l}\hat{\beta}^{\dagger}_{\pi_m}\hat{\beta}^{\dagger}_{\pi_n}|\Phi_0\rangle \},
\end{aligned}
\tag{14.3}
$$

in which $\hat{\beta}^{\dagger}_{\pi}$ and $\hat{\beta}^{\dagger}_{\nu}$ represent the qp creation operators of proton and neutron, respectively.

Different from the traditional configuration-interaction shell model, the TPSM calculations are performed based on the deformed basis, which means the rotational symmetry of the intrinsic qp states in the configuration space is broken. To restore the rotational symmetry, the projection technique is adopted. The obtained projected basis with good angular momentum I and M can be expressed as,

$$
\{\hat{P}^{I}_{MK}|\Phi_K\rangle\},
\tag{14.4}
$$

where $\hat{P}^{I}_{MK}$ is the three-dimensional angular momentum projection operator,

$$
\hat{P}^{I}_{MK} = \frac{2I+1}{8\pi^2} \int d\Omega D^{I*}_{MK}(\Omega)\hat{R}(\Omega).
\tag{14.5}
$$

The final projected wavefunctions in the TPSM are written as the linear combination of the projected basis,

$$
|\Psi^{\sigma}_{IM}\rangle = \sum_{K\kappa} f^{I\sigma}_{K\kappa}\hat{P}^{I}_{MK}|\Phi_\kappa\rangle,
\tag{14.6}
$$

and the expansion coefficients $f^{I\sigma}_{K\kappa}$ are determined by solving the following Hill-Wheeler equation,

$$
\sum_{\kappa'K'}\{\langle\Phi_\kappa|\hat{H}\hat{P}^{I}_{KK'}|\Phi_{\kappa'}\rangle - E^{I\sigma}\langle\Phi_\kappa|\hat{P}^{I}_{KK'}|\Phi_{\kappa'}\rangle\}f^{I\sigma}_{K'\kappa'} = 0,
\tag{14.7}
$$

in which σ specifies different eigenstates of the same spin I. The norm matrix $\mathcal{N}_I(K,\kappa;K',\kappa') = \langle\Phi_\kappa|\hat{P}^{I}_{KK'}|\Phi_{\kappa'}\rangle$ and the energy kernel $\langle\Phi_\kappa|\hat{H}\hat{P}^{I}_{KK'}|\Phi_{\kappa'}\rangle$ are calculated with the efficient Pfaffian algorithm proposed in Refs. [42, 43]. Based on the projected wavefunctions in (14.6), one can calculate the physical observables, such as the $M1$ and $E2$ transition probabilities.

It should be noted that the states of the projected basis in (14.4) are not orthogonal and the expansion coefficients $f^{I\sigma}_{K\kappa}$ should not be interpreted as the probability amplitudes for the state $|\Phi_\kappa\rangle$. To obtain the probability amplitudes, one needs to construct the following collective wavefunctions,

$$
g^{I\sigma}(K,\kappa) = \sum_{K'\kappa'} \mathcal{N}^{1/2}_{I}(K,\kappa;K',\kappa')f^{I\sigma}_{K'\kappa'}.
\tag{14.8}
$$

The collective wavefunctions are important for constructing the K plot, i.e., the probability distribution of the total angular momentum components in the intrinsic frame,

$$
p^{I\sigma}(|K|) = \sum_{\kappa} |g^{I\sigma}(K,\kappa)|^2 + |g^{I\sigma}(-K,\kappa)|^2.
\tag{14.9}
$$

The *azimuthal plot* depicting the probability distribution profile for the orientation of the angular momentum on the intrinsic (θ,ϕ) plane is defined as,

$$
\mathcal{P}(\theta,\phi) = \sum_{\kappa} \int_0^{2\pi} d\psi'|G^{II}(\psi',\theta,\pi-\phi,\kappa)|^2,
\tag{14.10}
$$

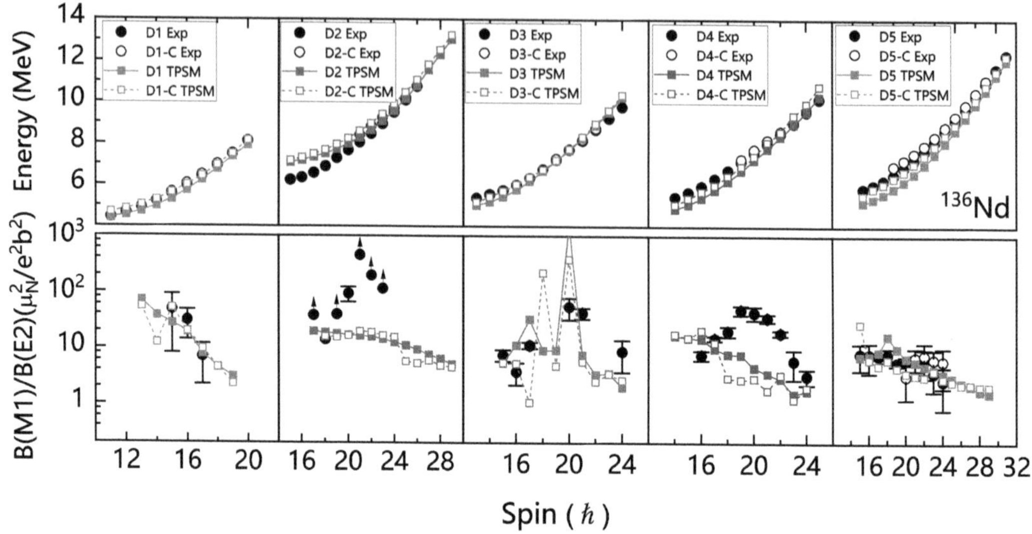

Figure 14.1 (Color online) Energy spectra and $B(M1)/B(E2)$ ratios calculated by the TPSM as functions of angular momenta, in comparison with the experimental data [44]. Taken from Ref. [37].

where (θ, ϕ) are the tilted angles of the angular momentum with respect to the intrinsic frame. The collective wavefunction $G^{II}(\psi', \theta, \pi - \phi, \kappa)$ is as follows [35],

$$G^{II}(\psi', \theta, \pi - \phi, \kappa) = \sqrt{\frac{2I+1}{8\pi^2}} \sum_K g^I(K, \kappa) D^{I*}_{IK}(\psi', \theta, \pi - \phi). \tag{14.11}$$

With the *K plot* and the *azimuthal plot*, the angular momentum geometry of nuclear chiral and wobbling motions can be clearly illustrated.

14.3 CHIRAL ROTATION IN EVEN-EVEN NUCLEUS ^{136}ND

In 2018, five pairs of nearly degenerate bands were observed in the even-even nucleus ^{136}Nd [44]. By comparing the data with the results calculated by the titled axis cranking model based on the nuclear density functional theory, these bands are suggested to be multiple chiral doublet bands.

The five possible chiral doublet bands have been then studied in Ref. [37] based on the TPSM. In this section, the obtained results are presented, and the calculated energy spectra, the transition probabilities, and also the chiral geometry will be discussed in detail.

In Fig. 14.1, the calculated energy spectra and the $B(M1)/B(E2)$ values for the chiral doublet bands and their comparison with the data are depicted. The bandhead of band D1 is taken as reference and no artificial renormalization has been made for the other bands. Generally, the calculated results reproduce satisfactorily the data. For each pair of doublets, the calculated energy spectra are nearly degenerate and the corresponding $B(M1)/B(E2)$ values are similar with each other. Therefore, the expected features for the chiral doublet bands are clearly demonstrated.

In more detail, the energy differences between bands D1 and D1-C are nearly unchanged as functions of spin, and this is in accordance with the data. The calculated $B(M1)/B(E2)$ values for both bands D1 and D1-C decrease with spin and this is also evidenced by the data.

For bands D2 and D2-C, the calculated energy spectra at the lower spin èrt' overestimate the data, and the bump of $B(M1)/B(E2)$ values around $I = 21\hbar$ cannot be reproduced by the TPSM calculations. Note that only six-qp configurations are included in the configuration space when describing the properties of D2 and D2-C. Through the analysis of qp alignments, as shown in

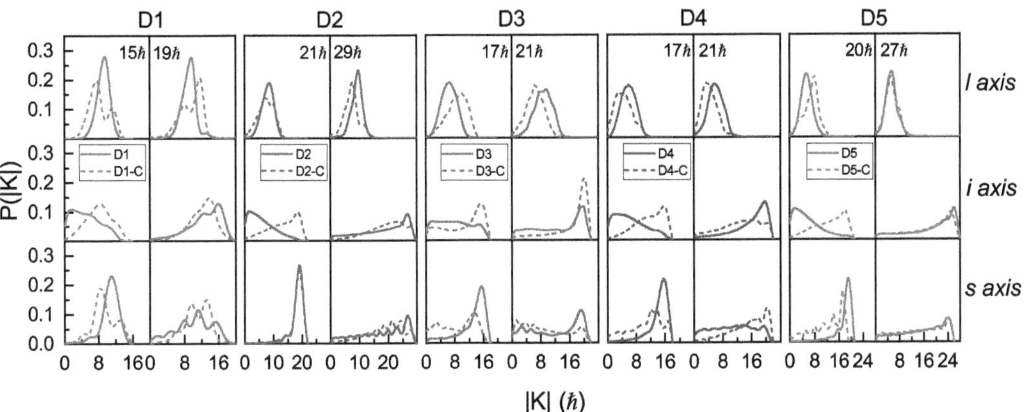

Figure 14.2   (Color online) The *K plot*, i.e., the *K* distributions for angular momenta on the three principle axes, for bands D1-D5 and their partners for selected angular momenta. The *K* distributions on the *l*, *i* and *s* axes are shown in the first, second and third rows, respectively. Taken from Ref. [37].

Ref. [44], it seems that four-qp configurations also play a role in the lower spin region of bands D2 and D2-C. Therefore, it is expected that the discrepancies between theoretical results and data would be reduced after including the four-qp configurations in the configuration space of bands D2 and D2-C.

For bands D3 and D4 together with their partners, the calculated energy spectra agree well with the data. The strong staggering behavior in $B(M1)/B(E2)$ values for bands D3 and D3-C is well reproduced by the TPSM calculations. However, the TPSM calculations somewhat underestimate the $B(M1)/B(E2)$ values for bands D4 and D4-C.

For bands D5 and D5-C, the difference between their energy spectra decreases with spin, and this behavior is well reproduced by the TPSM calculations. However, the experimental energies are slightly underestimated at lower spins, and this might be caused by the fact that the adopted deformation parameter $\beta = 0.26$ is a little bit too large. With a relatively smaller β value of 0.193, as used in Ref. [45], the data at lower spins will be overestimated. Therefore, a deformation parameter β in between is expected to better reproduce the experimental energies at lower spins for D5 and D5-C. For the $B(M1)/B(E2)$ values, the calculated results are in good agreement with the data, which have been measured for both yrast and yare bands.

As mentioned before, in the TPSM framework, the chiral geometry can be illustrated by the so-called *K plot* and *azimuthal plot*. In Fig. 14.2, the *K plots*, i.e., the *K* distributions of the total angular momenta along the three principal axes, are given for the five candidate chiral partners in ^{136}Nd.

For D1 and D1-C at $I = 15\hbar$ and $19\hbar$, the *K* distributions along *s* and *l* axes are quite similar, as expected for chiral doublet bands. However, at $I = 15\hbar$, the *K* distribution along *i* axis for D1 has a peak at $K = 0\hbar$, in contrast to the vanishing *K* distribution for D1-C. This is the typical feature of zero- and one-phonon states. Therefore, at $I = 15\hbar$, the state in D1-C is interpreted as a chiral vibration state of D1. At $I = 19\hbar$, the *K* distributions for D1 and D1-C show peaks at $K = 15\hbar$, and they are generally similar with each other. This suggests that the total angular momenta deviate from the *s-l* plane and static chirality occurs for D1 and D1-C. Similarly, the evolutions from chiral vibration to static chirality can also be well illustrated for bands D2 and D2-C, D4 and D4-C, and D5-D5-C.

For bands D3 and D3-C, however, the pattern is slightly different. As shown in Fig. 14.2, the *K* distributions at $I = 17\hbar$ cannot be interpreted by the zero- and one-phonon picture. At $I = 21\hbar$, the *K* distributions seem to be compatible with the appearance of static chirality. Nevertheless, D3 and D3-C cannot be well explained by the chiral picture and the physics behind D3 and D3-C deserves to be further explored in the future.

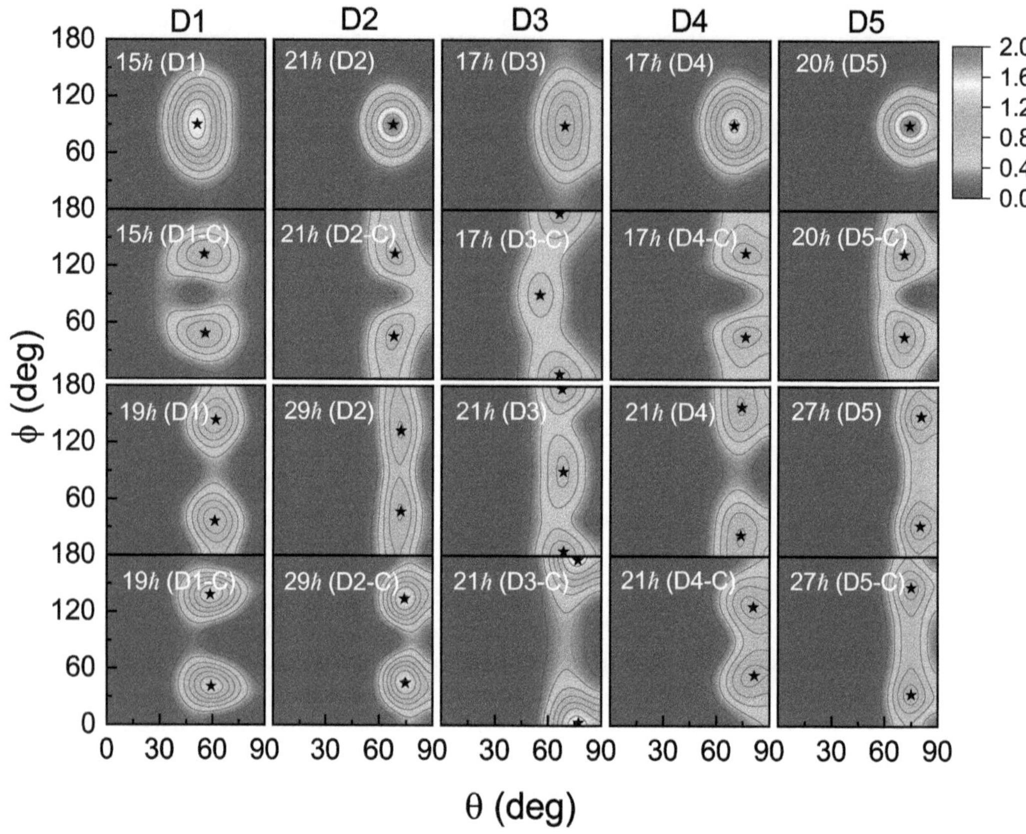

Figure 14.3 (Color online) The *azimuthal plot*, i.e., probability distribution profiles for the orientation of the angular momentum on the (θ, ϕ) plane for bands D1-D5 and their partners for selected angular momenta. The black star represents the position of a local maximum. Taken form Ref. [37].

The chiral geometry can be also examined by the *azimuthal plot*, i.e., the probability distribution profiles for the orientation of the angular momenta on the (θ, ϕ) plane, as shown in Fig. 14.3.

For bands D1 at $I = 15\hbar$, there is a single peak at $(\theta = 51°, \phi = 90°)$, indicating that the orientation of the angular momentum locates mainly in the $s - l$ plane and corresponds to a planar rotation. For bands D1-C at $I = 15\hbar$, the *azimuthal plot* show two peaks at $(\theta = 55°, \phi = 48°)$ and $(\theta = 55°, \phi = 132°)$. Therefore, the states at $I = 15\hbar$ for D1 and D1-C can be interpreted as zero- and one-phonon states, as already revealed in the *K plot*.

At $I = 19\hbar$, two peaks corresponding to the aplanar rotation occur in the *azimuthal plot* of band D1. Similar *azimuthal plots* for D1 and D1-C suggest the realization of the static chirality. The evolution from chiral vibration to static chirality can also be found through the *azimuthal plots* for bands D2 and D2-C, D4 and D4-C, and D5 and D5-C.

In comparison, the *azimuthal plots* for D3 and D3-C cannot be interpreted within in the chiral picture. As discussed in Ref. [37], the configurations of D3 and D3-C are similar to the ones of D4 and D4-C, but D3 and D3-C have higher excitation energies. Therefore, bands D3 and D3-C are actually excited bands on top of D4 and D4-C and much strong configuration mixing appears in D3 and D3-C. The stronger configuration mixing in D3 and D3-C results in more complicated structures and contaminates the chiral features of D3 and D3-C.

14.4 WOBBLING MOTION IN EVEN-EVEN NUCLEUS ^{130}BA

Most of the wobbling bands have been observed in odd-A nuclei. In 2019, the first experimental evidence for the wobbling motion in the even-even nucleus ^{130}Ba was reported. In this section, the observed two-qp bands S1 and S1' in Ref. [16], which is the first example of wobbling motion based on a two-qp configuration, are discussed within the framework of TPSM.

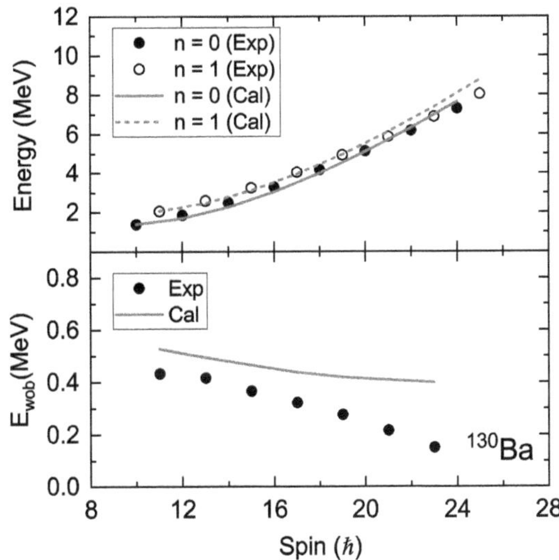

Figure 14.4   (Color online) Calculated energy spectra (top) for the zero- ($n = 0$) and one-phonon ($n = 1$) bands as well as the wobbling energy (bottom), in comparison with experimental data [16]. Taken from Ref. [38].

The calculated excitation energies for the zero- ($n = 0$) and the one-phonon ($n = 1$) states are shown in the upper panel of Fig. 14.4, in comparison with the data. The calculated excitation energy at $I = 10\hbar$ is renormalized to the experimental data. It is seen that the observed energy levels are well reproduced by the TPSM calculations.

In the lower panel of Fig. 14.4, the wobbling energies E_{wob}, defined as

$$E_{\mathrm{wob}}(I) = E_{n=1}(I) - [E_{n=0}(I+1) + E_{n=0}(I-1)]/2, \tag{14.12}$$

are depicted. The calculated E_{wob} are in good agreement with the data. Moreover, E_{wob} decrease as a function of spin, and this is in accordance with the expectation of transverse wobbling, as suggested in Ref. [5].

In Fig. 14.5, the calculated $B(M1)_{\mathrm{out}}/B(E2)_{\mathrm{in}}$ and $B(E2)_{\mathrm{out}}/B(E2)_{\mathrm{in}}$ values for transitions from the one-phonon band to the zero-phonon band are shown, together with their comparison with data. The calculated transition probabilities are in good agreement with data. However, the obtained $B(M1)_{\mathrm{out}}/B(E2)_{\mathrm{in}}$ values are comparable with the $B(E2)_{\mathrm{out}}/B(E2)_{\mathrm{in}}$ values, which is clearly different from the expectation of the ideal wobbler. In fact, the suggested configuration for the wobbling bands in ^{130}Ba is $\pi(h_{11/2})^2$. Such a two-qp configuration naturally enhances the $M1$ matrix elements. Therefore, the $\Delta I = 1$ $E2$ transitions are not anymore dominant like in the case of odd-even wobblers.

To examine the wobbling geometry, the *azimuthal plots* for the zero-(even spin) and one- (odd spin) phonon states at selected spins are shown in Fig. 14.6. For the zero-phonon states at $I = 10, 16, 22\hbar$, the *azimuthal plots* show single peaks at $\theta = 90°$ and $\phi = 90°$, indicating that the angular momentum orients mainly along the short axis. For the one-phonon states at $I = 11, 17, 23\hbar$, the angular momenta orient equally at two directions with $\phi \approx 120°$ and $\phi \approx 60°$. Moreover, the *azimuthal plots* exhibit a minimum at $\phi \approx 90°$, and this is different from the result of zero-phonon

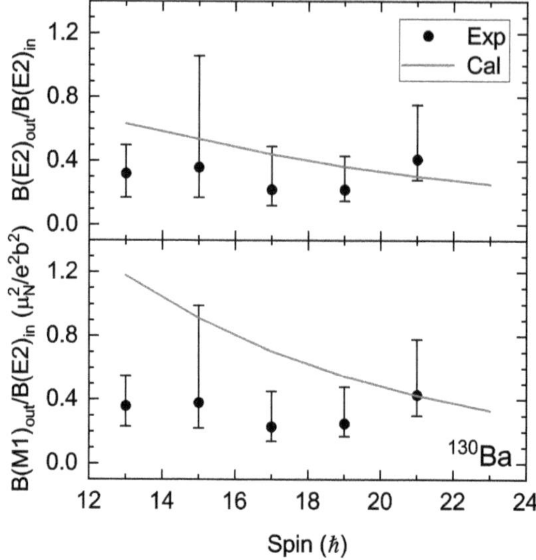

Figure **14.5** (Color online) Calculated transition probability ratios $B(E2)_{\text{out}}/B(E2)_{\text{in}}$ (top) and $B(M1)_{\text{out}}/B(E2)_{\text{in}}$ (bottom) for the transitions from the one-phonon ($n = 1$) band to the zero-phonon ($n = 0$) band in comparison with data available [16]. Taken from Ref. [38].

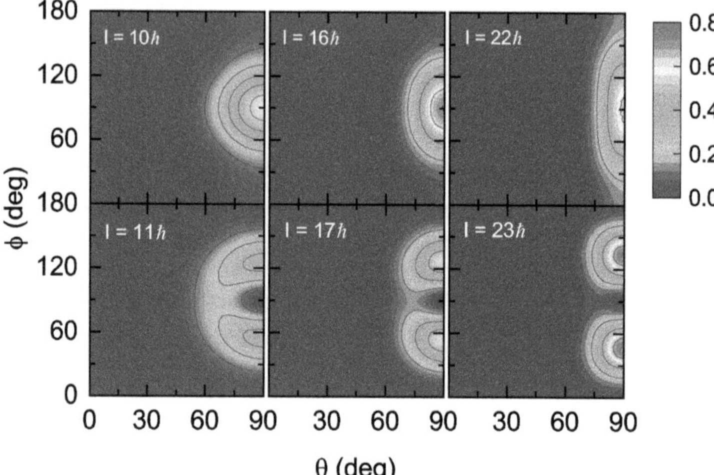

Figure **14.6** (Color online) The *azimuthal plots*, i.e., probability distribution profiles for the orientation of the angular momentum on the intrinsic (θ, ϕ) plane for the zero- (even spin) and the one- (odd spin) phonon band at several selected angular momenta. Taken from Ref. [38].

states. Therefore, the *azimuthal plots* along ϕ direction are symmetric for zero-phonon states and are asymmetric for one-phonon states. This pattern is consistent with the expectation of the wobbling motion, i.e., the precession of the total angular momentum around the short axis.

The wobbling geometry can also be illustrated by the *K plot*, as shown in Fig. 14.7. For the zero-phonon state at $I = 10\hbar$, the *K plot* along the *long* and *intermediate* axes peak mainly at $K \approx 0\hbar$, while it peaks at $K \approx 10\hbar$ on the *short* axis. This suggests that the angular momentum orients mainly along the *short* axis. For the one-phonon state at $I = 11\hbar$, the *K plot* vanishes at $K \approx 0\hbar$ while it peaks at $K \approx 6\hbar$, indicating that the orientation of the angular momentum deviates from the *short* axis.

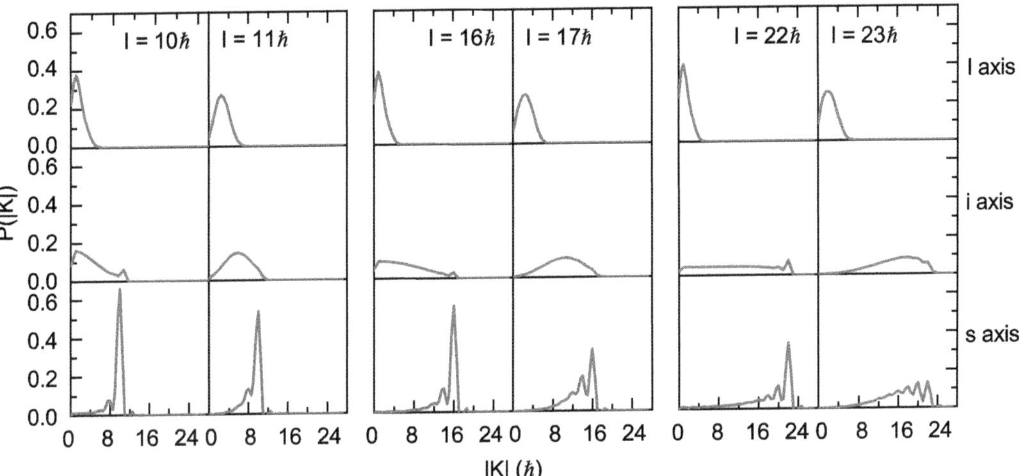

Figure 14.7   (Color online) The *K plot*, i.e., the *K* distributions of angular momenta on the three principle axes for the zero- (even spin) and one- (odd spin) phonon band at several selected angular momenta. The *K* distributions on the *long*, *intermediate*, and *short* axes are shown in the first, second, and third rows, respectively. Taken from Ref. [38].

Increasing the spin to $I = 16\hbar$ and $22\hbar$, the *K* distributions along the *long* axis are nearly unchanged, indicating that the angular momentum grows very little in the *long-axis* direction. The *K* distributions on the *intermediate* axis become very broad for states with $I = 16\hbar$ and $22\hbar$, while significant probabilities at $K_i \approx 0\hbar$ remain. This feature, together with the vanishing probabilities at $K_i \approx 0$ for states with $I = 17\hbar$ and $23\hbar$, reflects again the wobbling character of the odd-spin states.

14.5   SUMMARY

In summary, the recent progress on nuclear chiral and wobbling motions achieved by the triaxial projected shell model are illustrated. The experimental data associated with the chiral and wobbling bands are reproduced satisfactorily. By using the so-called *K plot* and *azimuthal plot*, the evolutions of chiral and wobbling geometries as functions of spins are discussed in detail. It should be emphasized that the Hamiltonian used in the triaxial projected shell model is a phenomenological pairing plus quadrupole Hamiltonian, and there are some free parameters that need to be fitted to the experimental data. Moreover, the triaxial projected shell model is performed within a limited model space and, thus, effective charges are required when calculating the transition probabilities. Recently, the triaxial projected shell model based on the universal relativistic density functional theory has been established [46, 47], and it has been already applied to study nuclear chiral rotation in a fully microscopic and quantal way [48]. By applying such method to study the wobbling motion is an interesting topic in the near future.

Bibliography

1. A. N. Bohr and B. R. Mottelson, *Nuclear Structure* (Benjamin, New York, 1975), Vol. II.
2. S. Frauendorf and J. Meng, Nucl. Phys. A **617**, 131 (1997).
3. S. Frauendorf, Rev. Mod. Phys. **73**, 463 (2001).
4. B. W. Xiong and Y. Y. Wang, At. Data Nucl. Data Tables **125**, 193 (2019).
5. S. Frauendorf and F. Dönau, Phys. Rev. C **89**, 014322 (2014).
6. S. Biswas *et al.*, Eur. Phys. Jour. A **55**, 159 (2019).
7. N. Sensharma *et al.*, Phys. Rev. Lett. **124**, 052501 (2020).

8. S. Nandi *et al.*, Phys. Rev. Lett. **125**, 132501 (2020).

9. S. Chakraborty *et al.*, Phys. Lett. B **811**, 135854 (2020).

10. P. Bringel *et al.*, Eur. Phys. Jour. A **24**, 167 (2005).

11. S. W. Ødegård *et al.*, Phys. Rev. Lett. **86**, 5866 (2001).

12. D. R. Jensen *et al.*, Phys. Rev. Lett. **89**, 142503 (2002).

13. G. Schönwaßer *et al.*, Phys. Lett. B **552**, 9 (2003).

14. H. Amro *et al.*, Phys. Lett. B **553**, 197 (2003).

15. D. J. Hartley *et al.*, Phys. Rev. C **80**, 041304 (2009).

16. C. M. Petrache *et al.*, Phys. Lett. B **795**, 241 (2019).

17. B. F. Lv *et al.*, Phys. Rev. C **105**, 034302 (2022).

18. Q. B. Chen, S. Frauendorf and C. M. Petrache, Phys. Rev. C **100**, 061301 (2019).

19. S. Q. Zhang, B. Qi, S. Y. Wang and J. Meng, Phys. Rev. C **75**, 044307 (2007).

20. B. Qi, S.Q. Zhang, J. Meng, S.Y. Wang and S. Frauendorf, Phys. Lett. B **675**, 175 (2009).

21. A. A. Raduta, A. H. Raduta and C. M. Petrache, Jour. Phys. G: Nucl. Part. Phys. **43**, 095107 (2016).

22. K. Nomura and C. M. Petrache, Phys. Rev. C **105**, 024320 (2022).

23. S. Brant, D. Tonev, G. De Angelis and A. Ventura, Phys. Rev. C **78**, 034301 (2008).

24. S. Brant and C. M. Petrache, Phys. Rev. C **79**, 054326 (2009).

25. V. I. Dimitrov, S. Frauendorf and F. Dönau, Phys. Rev. Lett. **84**, 5732 (2000).

26. P. Olbratowski, J. Dobaczewski, J. Dudek and W. Płóciennik, Phys. Rev. Lett. **93**, 052501 (2004).

27. P. W. Zhao, Phys. Lett. B **773**, 1 (2017).

28. P. W. Zhao, Y. K. Wang and Q. B. Chen, Phys. Rev. C **99**, 054319 (2019).

29. Y. P. Wang and J. Meng, Phys. Lett. B **841**, 137923 (2023).

30. D. Almehed, F. Dönau and S. Frauendorf, Phys. Rev. C **83**, 054308 (2011).

31. M. Matsuzaki, Y. R. Shimizu and K. Matsuyanagi, Phys. Rev. C **65**, 041303 (2002).

32. Q. B. Chen, S. Q. Zhang, P. W. Zhao, R. V. Jolos and J. Meng, Phys. Rev. C **87**, 024314 (2013).

33. Q. B. Chen, S. Q. Zhang, P. W. Zhao, R. V. Jolos and J. Meng, Phys. Rev. C **94**, 044301 (2016).

34. G. H. Bhat, J. A. Sheikh and R. Palit, Phys. Lett. B **707**, 250 (2012).

35. F. Q. Chen, Q. B. Chen, Y. A. Luo, J. Meng and S. Q. Zhang, Phys. Rev. C **96**, 051303 (2017).

36. F. Q. Chen, J. Meng and S. Q. Zhang, Phys. Lett. B **785**, 211 (2018).

37. Y. K. Wang, F. Q. Chen, P. W. Zhao, S. Q. Zhang and J. Meng, Phys. Rev. C **99**, 054303 (2019).

38. Y. K. Wang, F. Q. Chen and P. W. Zhao, Phys. Lett. B **802**, 135246 (2020).

39. F. Q. Chen and C. M. Petrache, Phys. Rev. C **103**, 064319 (2021).

40. K. Hara and Y. Sun, Int. Jour. Mod. Phys. E **04**, 637 (1995).

41. P. Ring and P. Schuck, *The Nuclear Many-Body Problem*, Springer Science & Business Media (2004).

42. Q. L. Hu, Z. C. Gao and Y. S. Chen, Phys. Lett. B **734**, 162 (2014).

43. G. F. Bertsch and L. M. Robledo, Phys. Rev. Lett. **108**, 042505 (2012).

44. C. M. Petrache *et al.*, Phys. Rev. C **97**, 041304 (2018).

45. J. A. Sheikh, G. H. Bhat, R. Palit, Z. Naik and Y. Sun, Nucl. Phys. A **824**, 58 (2009).

46. P. W. Zhao, P. Ring and J. Meng, Phys. Rev. C **94**, 041301 (2016).

47. Y. K. Wang, P. W. Zhao and J. Meng, Phys. Rev. C **105**, 054311 (2022).

48. Y. K. Wang, P. W. Zhao and J. Meng, Phys. Lett. B **848**, 138346 (2024).

15 Studies of nuclear Chirality in A≈80 mass region

Bin Qi

Shandong University, Weihai, China

15.1 INTRODUCTION

The occurrence of chirality in nuclear physics was originally suggested in 1997 by Frauendorf and Meng for triaxially deformed nuclei [1]. They pointed out that, in the intrinsic frame of the rotating triaxial nucleus with a few high-j valence particles and a few high-j valence holes, the total angular momentum vector may lie outside the three principal planes, referred to chiral symmetry breaking. In the laboratory reference frame, with the restoration of chiral symmetry due to quantum tunneling, the so-called chiral doublet bands, i.e., a pair of nearly degenerate $\Delta I = 1$ bands with the same parity, are expected to be observed [1, 2]. Such chiral doublet bands were first observed in 2001 in the $N = 75$ isotones [3]. In 2006, it has been suggested that multiple chiral doublet (MχD) bands can exist in a single nucleus [4]. The first experimental evidence for MχD bands was reported in ^{133}Ce in 2013, which confirmed the manifestation of triaxial shape coexistence in this nucleus [5]. So far, more than 60 experimental candidate chiral doublet bands (including several MχD) have been reported in the A $\approx$ 80, 100, 130, and 190 mass regions [6].

The present review focuses on the studies of chirality in the 80 mass region performed by the research group in Shandong University. In Section 2, we review the candidate chiral nuclei suggested based on the relativistic mean-field calculations [7, 8]. In Section 3, we review the experimental explorations of chiral doublet bands [9–16]. Finally, a summary is given in Section 4.

15.2 CANDIDATE CHIRAL NUCLEI

With an aim of searching the candidate chiral nuclei in the chain of bromine and rubidium isotopes, the triaxial deformations with corresponding configurations had been investigated using the triaxial relativistic-mean-field calculations [7, 8].

FORMALISM AND NUMERICAL DETAIL

The relativistic-mean-field (RMF) theory has obtained the great success in predicting and describing interesting physics [17, 18]. The starting point of the RMF theory is the standard effective Lagrangian density constructed with the degrees of freedom associated with the nucleon field, σ, ω, ρ meson fields, and the photon field [19]. Under "mean-field" and "no-sea" approximations, one can derive the corresponding energy density functional, from which one finds immediately the equation of motion for a single-nucleon orbit $\psi_i(r)$ with the help of the variational principle

$$\{\alpha \cdot [p - V(r)] + \beta m^*(r) + V_0(r)\}\psi_i(r) = \varepsilon_i \psi_i(r), \tag{15.1}$$

where $m^*(r)$ is defined as $m^*(r) \equiv m + g_\sigma \sigma(r)$, with m referring to the mass of the bare nucleon. The repulsive vector potential $V_0(r)$ is

$$V_0(r) = g_\omega \omega_0(r) + g_\rho \tau_3 \rho_0(r) + e\frac{1-\tau_3}{2} A_0(r), \tag{15.2}$$

where g_σ, g_ω, g_ρ are the coupling strengths of the nucleon with mesons. The time-odd fields $V(r)$ are given by the space-like components of vector fields,

$$V(r) = g_\omega \omega(r) + g_\rho \tau_3 \rho(r) + e\frac{1-\tau_3}{2} A(r). \tag{15.3}$$

The nonvanishing time-odd fields give rise to splitting between pairwise time-reversal states $\psi_{\bar{i}}$ and $\psi_{\underline{i}}(\equiv \hat{T}\psi_{\bar{i}})$, where $\hat{T}$ is the time-reversal operator. Each Dirac spinor $\psi_i(r)$ is expanded in terms of a set of three-dimensional harmonic oscillator (HO) bases in Cartesian coordinates with 12 major shells. The meson fields that provide the nuclear mean-field potentials are expanded in terms of the same HO basis as those of Dirac spinor but with 10 major shells. Due to the Pauli block effect, pairing correlations in the odd-odd nuclei are neglected.

The calculated results with parameter set PK1 [20] are shown. The constrained calculations with $\langle \hat{Q}_{20}^2 + 2\hat{Q}_{22}^2 \rangle$, i.e., β^2, are carried out to search for the ground state for a triaxially deformed nucleus here. During the β-constrained calculations, triaxial deformation is automatically obtained by minimizing the energy. First, the adiabatic constrained calculation is used to obtain the states with different configurations, and then the configuration-fixed constrained calculation is performed. Here the adiabatic constrained calculation means that the nucleons always occupy the lowest single particle levels, while the configuration-fixed constrained calculation means that the nucleons must occupy the same combination of the single particle levels during the constraint process [4, 21].

CANDIDATE CHIRAL NUCLEI IN BROMINE ISOTOPES

The calculated energy surfaces in ^{74,76,78,80,82,84}Br, based on adiabatic (open circles) and configuration-fixed (solid lines) constrained triaxial RMF theory, are presented in Fig. 15.1. The lighter isotopes are not calculated, where the Fermi surface of neutron would be close to the bottom of $g_{9/2}$ sub-shell, the chirality is difficult to occur. The minima in the energy surfaces for fixed configuration are represented as stars and labeled respectively as A, B, C, D, E, F, G, H and I.

Apart from the triaxial deformation, the proper particle and hole configuration is also necessary for the appearance of the chiral doublet bands. In Table 15.1, the calculated total energies E_{tot}, triaxial deformation parameters β and γ, and the corresponding unpaired nucleon configuration of minima are given. The present deformation parameters calculated with the triaxial RMF theory are relatively consistent with those in Ref. [22] for $^{76-84}$Br, while they are underestimated for ^{74}Br. Combining the calculated deformations and configurations, we select the suitable states for the appearance of the chirality, which are marked by the blue color in the figures and table.

For ^{74}Br, as shown in Tab. 15.1, the states D($\beta = 0.43, \gamma = 23.2°$) with $\pi g_{9/2}^1 \otimes \nu f_{5/2}^{-1}$ configuration and H(0.45, 27.5°) with $\pi g_{9/2}^1 \otimes \nu g_{9/2}^{-1}$ configuration have suitable triaxial deformation and proper configurations to form chirality. By comparing the obseved excited energies in Ref. [23], it is found that the calculated excited energies (D: 0.44MeV, F: 0.47MeV) are in very good agrement with the experimental ones (D: 0.239MeV, F: 0.307MeV).

For ^{76}Br, the states B(0.41, 20.8°) with $\pi g_{9/2}^1 \otimes \nu g_{9/2}^{-1}$ configuration and E(0.36, 32.0°) with $\pi g_{9/2}^1 \otimes \nu p_{3/2}$ have suitable triaxial deformations to form the chirality.

For ^{78}Br, the states F(0.21, 30.2°) and H(0.35, 21.4°) have the suitable triaxial deformations to form chirality. Their configurations are $\pi f_{5/2} \otimes \nu g_{9/2}^{-1}$ and $\pi g_{9/2} \otimes \nu g_{9/2}^{-1}$, which are the exact configuration suggested in the observed negative-parity chiral doublet bands and positive-parity chiral doublet bands in experiment [9].

Table 15.1

The total energies E_{tot}, triaxial deformation parameters β and γ, and their corresponding unpaired nucleon configuration of minima for states A-I in the configuration-fixed constrained triaxial RMF calculations for Br isotopes. The suitable states for the appearance of the chirality are marked by blue color. The more detailed informations can be found in Ref. [7].

Nuclei	State	Configuration (Unpaired nucleon)	E_{tot} (MeV)	(β, γ)
^{74}Br	A	$\pi g_{9/2}^{1} \otimes \nu f_{5/2}^{-1}$	-633.18	(0.32,56.9)
	B	$\pi g_{9/2}^{1} \otimes \nu g_{9/2}^{1}$	-633.08	(0.30,53.0)
	C	$\pi p_{3/2}^{1} \otimes \nu f_{5/2}^{-1}$	-632.94	(0.35,42.5)
	D	$\pi g_{9/2}^{1} \otimes \nu f_{5/2}^{-1}$	-632.74	(0.43,23.2)
	E	$\pi g_{9/2}^{1} \otimes \nu p_{3/2}^{-1}$	-632.68	(0.27,57.2)
	F	$\pi f_{5/2}^{-1} \otimes \nu p_{3/2}^{-1}$	-632.28	(0.23,51.9)
	G	$\pi p_{3/2}^{1} \otimes \nu g_{9/2}^{-1}$	-632.03	(0.19,56.2)
	H	$\pi g_{9/2}^{1} \otimes \nu g_{9/2}^{-1}$	-632.71	(0.45,27.5)
^{76}Br	A	$\pi g_{9/2}^{1} \otimes \nu g_{9/2}^{1}$	-654.68	(0.25,59.9)
	B	$\pi g_{9/2}^{1} \otimes \nu g_{9/2}^{-1}$	-654.60	(0.41,20.8)
	C	$\pi g_{9/2}^{1} \otimes \nu p_{3/2}^{-1}$	-654.53	(0.28,56.6)
	D	$\pi f_{5/2}^{-1} \otimes \nu g_{9/2}^{1}$	-654.32	(0.22,56.5)
	E	$\pi g_{9/2}^{1} \otimes \nu p_{3/2}^{-1}$	-654.26	(0.36,32.0)
	F	$\pi p_{3/2}^{1} \otimes \nu p_{1/2}^{1}$	-652.49	(0.17,56.0)
	G	$\pi p_{3/2}^{1} \otimes \nu g_{9/2}^{1}$	-651.11	(0.09,50.9)
	H	$\pi f_{5/2}^{-1} \otimes \nu g_{9/2}^{1}$	-652.09	(0.28,40.3)
^{78}Br	A	$\pi g_{9/2}^{1} \otimes \nu g_{9/2}^{-1}$	-673.50	(0.25,59.9)
	B	$\pi g_{9/2}^{1} \otimes \nu p_{3/2}^{1}$	-673.35	(0.39,21.4)
	C	$\pi p_{1/2}^{-1} \otimes \nu p_{3/2}^{1}$	-673.03	(0.34,31.9)
	D	$\pi p_{3/2}^{1} \otimes \nu g_{9/2}^{-1}$	-673.01	(0.22,52.4)
	E	$\pi p_{3/2}^{1} \otimes \nu p_{1/2}^{1}$	-672.12	(0.17,55.9)
	F	$\pi f_{5/2}^{1} \otimes \nu g_{9/2}^{-1}$	-671.85	(0.21,30.2)
	G	$\pi p_{3/2}^{1} \otimes \nu g_{9/2}^{1}$	-671.62	(0.12,55.1)
	H	$\pi g_{9/2}^{1} \otimes \nu g_{9/2}^{-1}$	-672.53	(0.35,21.4)
	I	$\pi g_{9/2}^{1} \otimes \nu g_{9/2}^{-1}$	-672.46	(0.41,24.8)
^{80}Br	A	$\pi f_{5/2}^{-1} \otimes \nu g_{9/2}^{1}$	-692.64	(0.17,0.1)
	B	$\pi g_{9/2}^{1} \otimes \nu p_{1/2}^{1}$	-692.42	(0.25,7.5)
	C	$\pi f_{5/2}^{-1} \otimes \nu p_{1/2}^{1}$	-692.41	(0.19,10.4)
	D	$\pi f_{5/2}^{1} \otimes \nu g_{9/2}^{-1}$	-691.68	(0.31,23.7)
	E	$\pi g_{9/2}^{1} \otimes \nu g_{9/2}^{-1}$	-691.10	(0.34,25.2)
^{82}Br	A	$\pi f_{5/2}^{-1} \otimes \nu g_{9/2}^{-1}$	-710.92	(0.15,0.1)
	B	$\pi p_{3/2}^{1} \otimes \nu g_{9/2}^{-1}$	-710.49	(0.10,0.2)
	C	$\pi g_{9/2}^{1} \otimes \nu g_{9/2}^{-1}$	-709.48	(0.20,0.1)
	D	$\pi g_{9/2}^{1} \otimes \nu g_{9/2}^{-1}$	-708.41	(0.25,55.4)
	E	$\pi p_{1/2}^{1} \otimes \nu g_{7/2}^{1}$	-705.62	(0.33,33.9)
	F	$\pi g_{9/2}^{1} \otimes \nu g_{7/2}^{1}$	-705.01	(0.35,24.6)
	G	$\pi g_{9/2}^{1} \otimes \nu g_{9/2}^{-1}$	-704.31	(0.41,17.5)
	H	$\pi g_{9/2}^{1} \otimes \nu g_{9/2}^{-1}$	-708.15	(0.15,33.6)
	I	$\pi g_{9/2}^{1} \otimes \nu g_{9/2}^{-1}$	-707.14	(0.27,10.2)
^{84}Br	A	$\pi p_{3/2}^{1} \otimes \nu g_{9/2}^{-1}$	-728.71	(0.06,0.3)
	B	$\pi g_{9/2}^{1} \otimes \nu g_{9/2}^{-1}$	-726.22	(0.08,34.9)
	C	$\pi g_{9/2}^{1} \otimes \nu p_{1/2}^{1}$	-725.09	(0.17,59.0)
	D	$\pi g_{9/2}^{1} \otimes \nu g_{7/2}^{1}$	-722.56	(0.27,59.9)
	E	$\pi p_{1/2}^{1} \otimes \nu g_{7/2}^{1}$	-721.25	(0.30,59.7)

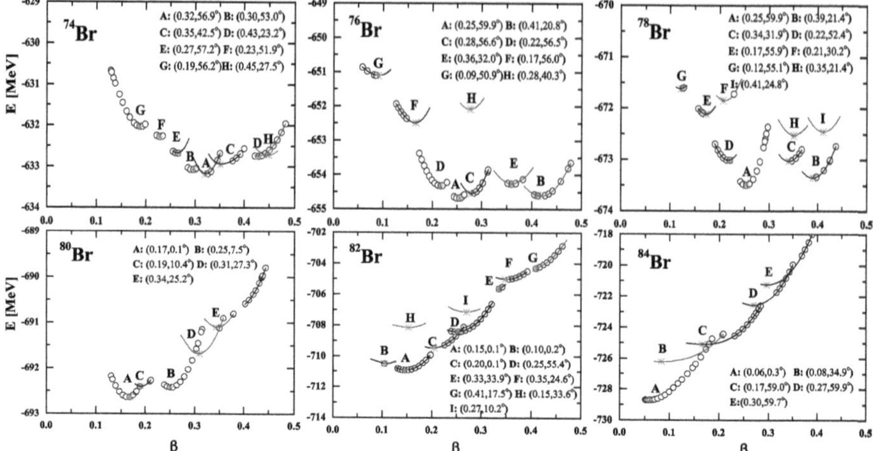

Figure 15.1   The energy surfaces in adiabatic (open circles) and configuration-fixed (solid lines) constrained triaxial RMF calculations with PK1 for ^{74,76,78,80,82,84}Br. The minima in the energy surfaces for fixed configurations are represented as stars and labeled respectively as A, B, C, D, E, F, G, H and I. Their corresponding triaxial deformation parameters β and γ are also given. The suitable states for the appearance of chirality are marked by blue color. The results and figures are taken from Ref. [7].

For ^{80}Br, the states D(0.31, 23.7°) and E(0.34, 25.2°) have the suitable triaxial deformations to form the chirality. The configuration of E is $\pi g_{9/2}^{1} \otimes \nu g_{9/2}^{-1}$, which is the exact configuration suggested in the observed positive-parity chiral doublet bands in Ref. [10].

For ^{82}Br, the calculated results infer that the excited states of H (0.15, 33.6°) and I (0.27, 10.2°) with $\pi g_{9/2}^{1} \otimes \nu g_{9/2}^{-1}$ configurations are possible candidate states to form two sets of the chiral doublet bands. State G has the triaxial deformation, however, it should unlikely be observed in experiment due to the high excited energy.

In conclusion, the boundary of chirality in bromine isotopes is investigated by using adiabatic and configuration-fixed constrained triaxial RMF theory. Several minima with the triaxial deformation and proper particle-hole configurations are obtained in ^{74,76,78,80,82}Br, where the chiral doublet bands is possible to appear. From the calculation, the possible existence of chiral doublet bands and the MχD bands are demonstrated in ^{74,76,78,80,82}Br.

The above calculations were performed in 2019 [7], when only the chiral doublet bands in ^{78}Br and ^{80}Br had been observed. It is interesting to noticed that the other predicted candidate chiral doublet bands of ^{74}Br [11], ^{76}Br [12], ^{82}Br [13] also have been observed recently.

CANDIDATE CHIRAL NUCLEI IN RUBIDIUM ISOTOPES

We also performed the adiabatic and configuration-fixed constrained triaxial RMF calculations for rubidium isotopes. The calculated energy surfaces in ^{76,78,80,82,84,86}Rb are presented in Fig. 15.2. The minima in the energy surfaces for fixed configuration are represented as stars and labeled respectively as A, B, C, D, E, F, G, and H. The present calculated deformations using the triaxial RMF theory are consistent with those values in Ref. [22] for ^{80,82,84,86}Rb, while they are overestimated by about 15% for ^{76,78}Rb. Combining the calculated deformations and configurations, we find the suitable states for the appearance of the chirality, which are marked by the blue color and asterisks in the figures. The possible candidate nuclei of chirality and MχD in odd-odd Rb isotopes are discussed in the following.

For ^{78}Rb, the states B($\beta = 0.37, \gamma = 34.0°$) and I(0.40, 44.5°) have the same unpaired nucleon configuration $\pi g_{9/2}^{1} \otimes \nu g_{9/2}^{-1}$, but have some differences for the valence nucleon configuration. In the

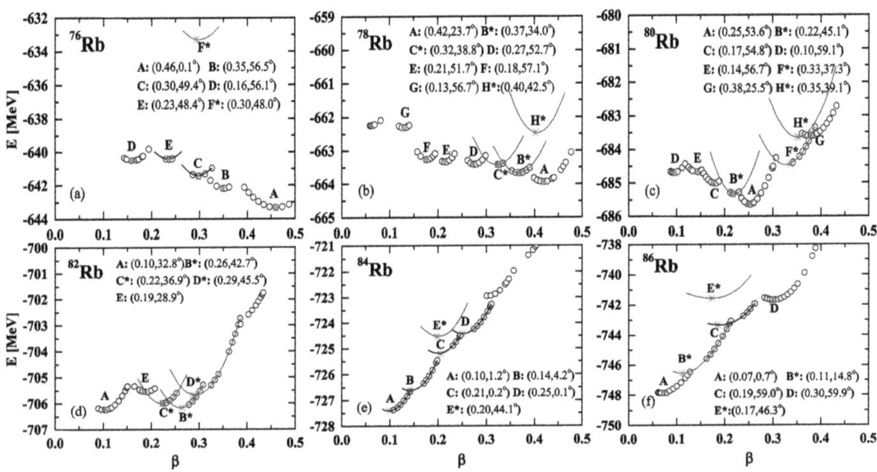

Figure 15.2 The energy surfaces in adiabatic (open circles) and configuration-fixed (solid lines) constrained triaxial RMF calculations with PK1 for ^{76,78,80,82,84,86}Rb. The minima in the energy surfaces for fixed configuration are represented as stars and labeled respectively as A, B, C, D, E, F, G, and H. Their corresponding triaxial deformation parameters β and γ are also given. The suitable states for the appearance of the chirality are marked by blue color and asterisks. The results and figures are taken from Ref. [8].

previous reported MχD in ^{103}Rh, four $\Delta I = 1$ bands (labeled as 3-6 in Ref. [24]) are based on the identical configuration. Here, it is also expected that four $\Delta I = 1$ bands based on the $\pi g^1_{9/2} \otimes \nu g^{-1}_{9/2}$ configuration (states B and I) as the MχD in ^{78}Rb, whose mechanism is different from those four bands in ^{103}Rh. Such a new type of MχD is highly expected to explore in experiment. Besides, the triaxial local minimum C(0.32, 38.8°) associates with the $\pi f^1_{5/2} \otimes \nu g^{-1}_{9/2}$ configuration, which is the similar configuration with the negative-parity chiral doublet bands in ^{78}Br [9]. Thus, MχDs combined by two sets of positive-parity chiral doublet bands and one set of negative-parity chiral doublet bands are expected to be observed in ^{78}Rb.

The triaxial states B(0.23, 45.1°) and H(0.35, 39.1°) in ^{80}Rb related to the configuration $\pi g^1_{9/2} \otimes \nu g^{-1}_{9/2}$, while the state F(0.33, 37.3°) related to the $\pi g^1_{9/2} \otimes \nu p^{-1}_{3/2}$ configuration are obtained in the calculated results. Similar with the discussion for ^{78}Rb, the present calculations support the existence of MχD bands combined by two sets of positive-parity chiral doublet bands and one set of negative-parity chiral doublet bands in ^{80}Rb.

For ^{82}Rb, the corresponding deformation parameters for the proper proton particle and neutron hole configurations of minima are B(0.26, 42.7°), C(0.22, 36.9°), and D(0.29, 45.5°), respectively. State B corresponds to the $\pi p^1_{3/2} \otimes \nu g^{-1}_{9/2}$ configuration, while states C and D associate with the $\pi g^1_{9/2} \otimes \nu g^{-1}_{9/2}$ configurations. Compared with ^{78}Rb and ^{80}Rb, the neutron Fermi surface in ^{82}Rb is closer to the top of the $g_{9/2}$ subshell, which offers the more appropriate particle-hole configuration for chirality.

Considering the above theoretical calculations for the deformations and configurations, ^{78}Rb, ^{80}Rb and ^{82}Rb should be good candidate chiral nuclei. The previously reported data of the ^{78}Rb and ^{80}Rb had shown the hints of the chiral doublet bands. The predicted best candidate chiral nucleus seems to be ^{82}Rb, in which the particle-hole configurations are more appropriate and the excited energies of triaxial deformation minima are much lower. However, no doublet bands in these nuclei was reported so far. It is highly interesting to search for the chiral doublet bands in rubidium isotopes experimentally.

15.3   EXPERIMENTAL EXPLORATION

The research group in Shandong University and the collaborators have performed the experimental exploration of the chiral nuclei in this mass region, and 8 candidate chiral nuclei (^{74}As [14], ^{74}Br [11], ^{76}Br [12], ^{78}Br [9], ^{80}Br [10], ^{82}Br [13], ^{81}Kr [15] and ^{84}Rb [16]) have been reported in this mass region so far [6]. The excitation energies relative to a rigid-rotor reference $E(I) - J*I(I+1)$, the energy staggering parameter $S(I)$, the kinematic moments of inertia MOI, alignments i_x and $B(M1)/B(E2)$ ratios for candidate chiral doublet bands in $A \sim 80$ mass region are shown in Fig. 15.3.

Experimental studies on the nuclear chirality in ^{76}Br [12], ^{84}Rb [16] were performed at the China Institute of Atomic Energy (CIAE). The detector array in CIAE consisted of several Compton-suppressed high-purity germanium (HPGe) detectors, two low-energy photon spectrometer (LEPS) detectors, and one clover detector. Experimental studies on the nuclear chirality in ^{74}As [14], ^{74,78,80,82}Br [9–11, 13] and ^{81}Kr [15], were carried out by employing the AFRODITE array in iThemba LABS. The AFRODITE array composed of eight Compton-suppressed Clover detectors. The Clover detector can measure the linear polarization of the γ-rays and identify whether the observed transitions are magnetic or electrical. It is helpful to study the simultaneous breaking of the chiral and reflection symmetries. To reduce the contamination from the side products by selecting the specific charge particle reaction channels, the CsI particle detector arrays were also used with the AFRODITE array.

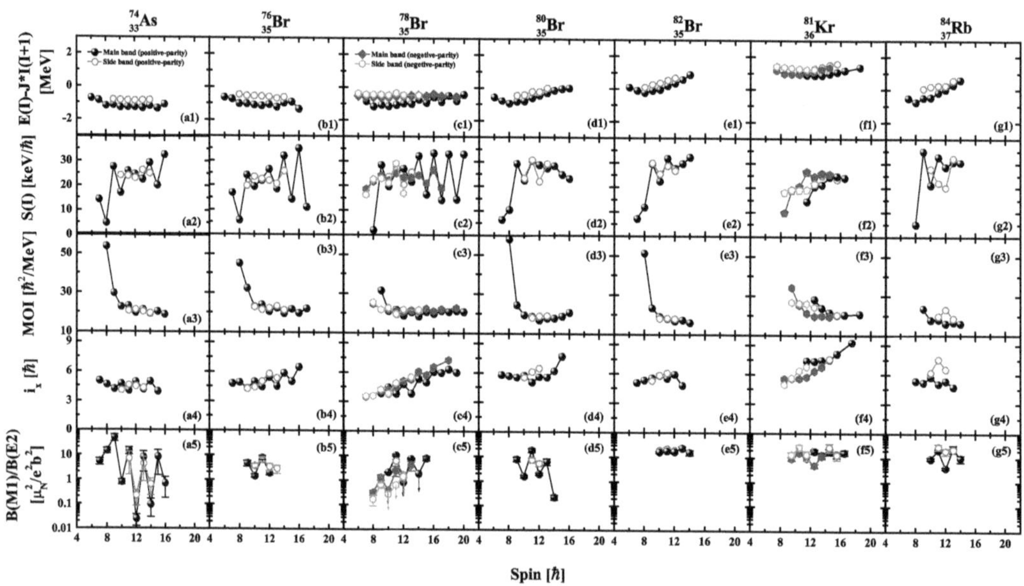

Figure 15.3   The excitation energies relative to a rigid-rotor reference $E(I) - J*I(I+1)$, the energy staggering parameter $S(I)$, the kinematic moments of inertia MOI, experimental alignments i_x and experimental $B(M1)/B(E2)$ for chiral doublet bands in ^{74}As, ^{76}Br, ^{78}Br, ^{80}Br, ^{82}Br, ^{81}Kr and ^{84}Rb. The J parameters are evaluated from the relation $J = 0.007 \times \left(\frac{158}{A}\right)^{5/3}$ MeV. The figure is taken from Ref. [6].

In 2011, a pair of candidate chiral doublet bands was observed in ^{80}Br by Wang *et al.* [10], which provided the first example for chirality in the $A \sim 80$ mass region, and gave a new chiral configuration $\pi g_{9/2} \otimes \nu g_{9/2}$. Up to now, in Br isotopes, similar positive-parity chiral doublet bands were found in ^{74}Br [11], ^{76}Br [12], ^{78}Br [9] and ^{82}Br [13] based on the $\pi g_{9/2} \otimes \nu g_{9/2}$ configurations, and a pair of negative-parity chiral doublet bands were found in ^{78}Br based on the $\pi(p_{3/2}, f_{5/2}) \otimes \nu g_{9/2}$ configuration [9]. As shown in Fig. 15.3, the positive-parity candidate chiral doublet bands with the

$\pi g_{9/2} \otimes \nu g_{9/2}$ configuration in Br isotopes maintain an energy difference ΔE of ≈ 0.4–0.6 MeV and the negative-parity candidate chiral doublet bands with the $\pi(p_{3/2}, f_{5/2}) \otimes \nu g_{9/2}$ configuration in ^{78}Br maintain an energy difference of ≈ 0.15 MeV over the observed spin range. In addition, the $S(I)$ show an almost constant value of ≈ 25 keV/$\hbar$ at $8 \leq I \leq 14$ $\hbar$ for these doublet bands. The MOI, i_x and $B(M1)/B(E2)$ values of these doublet bands are very similar. Meanwhile, the $B(M1)/B(E2)$ values show odd-even staggering as a function of spin. It should be noted that the ΔE for the candidate chiral doublet bands with the $\pi g_{9/2} \otimes \nu g_{9/2}$ configuration show a decreasing trend as N increases in Fig. 15.3. The ΔE of candidate chiral doublet bands may reflect the chiral geometry, i.e., more stable chiral geometry corresponds to the smaller energy difference. With an increase of the neutron number N in these odd-odd Br isotopes, the neutron Fermi level approaches the top of the $\nu g_{9/2}$ subshell, and the growing occupancy of neutrons in the $g_{9/2}$ orbital (gradually approaching the ideal hole) will result in more stable chiral geometry. Therefore, the smaller ΔE in odd$-$odd Br isotopes with larger N can be understood as these Br isotopes with larger N are more suitable for constructing chiral geometry than those with smaller N.

In ^{74}As, the positive-parity doublet bands were identified based on the $\pi g_{9/2} \otimes \nu g_{9/2}$ configurations [14]. As shown in Fig. 15.3, the energy difference of the two bands is approximate to 0.4 MeV. The two bands have similar $S(I)$, MOI and i_x values, and the $S(I)$ exhibits a smooth variation versus spin. The $B(M1)/B(E2)$ values of these two bands are close and show odd-even staggering. Thus, the doublet bands in ^{74}As were suggested as chiral doublet bands. This work extended the border of the chiral nuclei in the $A \approx 80$ mass region to $Z = 33$. In ^{84}Rb, the positive-parity doublet bands were also observed based on the $\pi g_{9/2} \otimes \nu g_{9/2}$ configurations [16]. In Fig. 15.3, the doublet bands of ^{84}Rb have a small energy difference and a relatively smooth variation of $S(I)$ values. The $B(M1)/B(E2)$ ratios for the doublet bands are close to each other and show odd-even staggering with the same phase as a function of spin. These experimental properties are consistent with the fingerprints of chiral doublet bands. The observation of chirality in ^{84}Rb extended the border of the chiral island in the $A \approx 80$ mass region to $Z = 37$.

Besides odd-odd nuclei, it is interesting to search for chiral doublet bands (or MχD) in odd-A or even-even nuclei in the $A \approx 80$ mass region. In 2022, candidate chiral doublet bands in odd-A nuclei had been found by investigating the medium-and high-spin states in ^{81}Kr [15]. The positive-parity doublet bands with the $\pi g_{9/2}^{2} \otimes \nu g_{9/2}^{-1}$ configuration and negative-parity doublet bands with the $\pi g_{9/2}(p_{3/2}, f_{5/2}) \otimes \nu g_{9/2}^{-1}$ configuration were identified in odd-A ^{81}Kr [15]. As shown in Fig. 15.3, the excitation energies, $S(I)$, MOI and i_x of these two bands are close, and these two bands exhibit smooth variation of S(I) as a function of spin. The $B(M1)/B(E2)$ ratios for each pair of doublet bands are similar and show odd-even staggering with the same phase as a function of spin. These behaviors are consistent with the fingerprints of chiral doublet bands, thus, these two pairs of bands were suggested as two pairs of chiral doublet bands. It indicates that chiral doublet bands can not only be formed in odd-odd nuclei but also in odd-A nuclei in the $A \approx 80$ mass region [15].

As well known, the transition probabilities carry more stringent information on the chiral geometry than excitation energies. The lifetime measurements are essential to extract the absolute electromagnetic transition probabilities. In the 80 mass region, the lifetime measurements for the candidate chiral doublet bands were performed in ^{80}Br [25] and ^{76}Br [12] isotopes. The behaviors of reduced transition probabilities observed in the partner bands of the ^{76,80}Br are consistent with the fingerprints of chiral doublet bands. Moreover, the lifetime measurements in ^{76}Br also revealed the interplay between nuclear chiral and reflection symmetry breakings [12].

In 2016, two pairs of positive- and negative-parity doublet bands together with eight strong electric dipole transitions have been identified in ^{78}Br, which were interpreted as multiple chiral doublet bands with octupole correlations [9]. The first example of simultaneous breaking of chiral and reflection symmetries reported in this nuclei, which indicates that chirality and octupole correlations can coexist in a single nucleus. After that, how do chirality and octupole correlations interact becomes a question of concerned. Subsequently, the experimental observations based on the precise lifetime measurements in ^{76}Br have been interpreted as the coexistence of chirality and octupole correlations, and play a key role to explore the simultaneous breaking of chiral and reflection symmetry and their interplay [12]. From the analyses of these two phenomena in Br isotopes, it could lead to a conclusion that the reflection symmetry breaking catalyzes rather than destroys chiral symmetry breaking in nuclear systems.

Very recently, the first chiral wobbler have been suggested in ^{74}Br [11], which indicates that nuclear chirality can be robust against wobbling excitation. This finding provides a unique candidate to study chirality and the wobbling mode in a single nucleus, which manifests the diversity and complexity of the angular momentum coupling modes of nuclei and opens a new arena to investigate the fundamental symmetry breaking in the $A \approx 80$ mass region.

15.4 SUMMARY

We review the studies of nuclear chirality in A≈80 mass region in Shandong university. The triaxial deformations with corresponding configurations had been investigated using the triaxial relativistic-mean-field calculations, the possible existence of chiral doublet bands and the MχD bands is demonstrated in the chain of bromine and rubidium isotopes. The experimental exploration of the chiral nuclei had been performed, and 8 candidate chiral nuclei have been reported in this mass region so far.

The $A \approx 80$ mass region is the lightest known chiral region in nuclide chart, which shows that the chiral symmetry properties are of a general nature and not related only to a specific nuclear mass region. It is of highly scientific interest to further explore the boundaries of chiral islands and new chiral mass regions. In addition, it is also important to search for the coexistence and interplay of chirality and other symmetries, like chirality-parity quartet bands in nucleus with both stable triaxial and octupole deformations and chiral-wobbling nucleus.

Bibliography

1. S. Frauendorf and J. Meng, Nucl. Phys. A **617**, 131 (1997).
2. S. Frauendorf, Rev. Mod. Phys. **73**, 463 (2001).
3. K. Starosta, T. Koike, C. J. Chiara, D. B. Fossan, *et al.*, Phys. Rev. Lett. **86**, 971 (2001).
4. J. Meng, J. Peng, S. Q. Zhang and S. G. Zhou, Phys. Rev. C **73**, 037303 (2006).
5. A. D. Ayangeakaa, U. Garg, M. D. Anthony, S. Frauendorf, *et al.*, Phys. Rev. Lett. **110**, 172504 (2013).
6. S. Y. Wang, C. Liu, B. Qi, W. Z. Xu and H. Zhang, Frontiers of Physics **18**, 64601, (2023).
7. B. Qi, H. Jia, C. Liu and S. Y. Wang, Sci. China, Phys. Mech. Astron. **62**, 012012 (2019).
8. B. Qi, H. Jia, C. Liu and S. Y. Wang, Phys. Rev. C **98**, 014305 (2018).
9. C. Liu, S. Y. Wang, R. A. Bark, S. Q. Zhang, *et al.*, Phys. Rev. Lett. **116**, 112501 (2016).
10. S. Y. Wang, B. Qi, L. Liu, S. Q. Zhang, *et al.*, Phys. Lett. B **703**, 40 (2011).
11. R.J. Guo, S. Y. Wang, C. Liu, R. A. Bark, *et al.*, Phys. Rev. Lett. **132**, 092501 (2024).
12. W. Z. Xu, S. Y. Wang, C. Liu, X. G. Wu, *et al.*, Phys. Lett. B **833**, 137287 (2022).
13. C. Liu, S. Y. Wang, B. Qi, S. Wang, *et al.*, Phys. Rev. C **100**, 054309 (2019).
14. X. Xiao, S. Y. Wang, C. Liu, R. A. Bark, *et al.*, Phys. Rev. C **106**, 064302 (2022).
15. L. Mu, S. Y. Wang, C. Liu, B. Qi, *et al.*, Phys. Lett. B **827**, 137006 (2022).
16. X.C. Han, S. Y. Wang, B. Qi, C. Liu, *et al.*, Phys. Rev. C **104**, 014327 (2021).
17. J. Meng, H. Toki, S. G. Zhou, S. Q. Zhang, W. H. Long, L. S. Geng, Prog. Part. Nucl. Phys. **57**, 470 (2006).
18. J. Meng, *Relativistic Density Functional for Nuclear Structure* (World Scientific, Singapore, 2015).
19. J. M. Yao, B. Qi, S. Q. Zhang, J. Peng, S. Y. Wang and J. Meng, Phys. Rev. C **79**, 067302 (2009).

20. W. H. Long, J. Meng, N. Van Giai and S. G. Zhou, Phys. Rev. C **69**, 034319 (2004).
21. J. Peng, H. Sagawa, S. Q. Zhang, J. M. Yao, *et al.*, Phys. Rev. C **77**, 024309 (2008).
22. P. Möller, J. R. Nix, W. D. Myers and W. J Swiatecki, At. Data Nucl. Data Tables, **109**, 1 (2016).
23. J. Döring, J. W. Holcomb, T. D. Johnson, M. A. Riley, *et al.*, Phys. Rev. C **47**, 2560 (1993).
24. I. Kuti, Q. B. Chen, J. Timár, D. Sohler, *et al.*, Phys. Rev. Lett. **113**, 032501 (2014).
25. R.J. Guo, S. Y. Wang, R. Schwengner, W. Z. Xu, *et al.*, Phys. Lett. B **833**, 137344 (2022).

16 Recent progress in nuclear chirality of cesium isotopes

Jian Li and Duo Chen
Jilin University, Changchun, China

16.1 INTRODUCTION

Chirality is a fundamental symmetry in nature, such as in chemistry, biology, and physics. The occurrence of chirality in atomic nuclear structure was suggested for triaxially deformed nuclei in 1997 [1]. The patterns of energy spectra exhibiting chirality i.e., chiral doublet bands, were also predicted with one pair of nearly degenerate $\Delta I = 1$ bands with the same parity. Since 1997, lots of efforts including both theoretical and experimental investigations have been made to understand these novel phenomena and explore the manifestations thereof in the nuclear chart; for a brief review, see Refs. [2–7]. Thus, nuclear chirality becomes a hot topic, and related issues such as MχD (multiple chiral doublet bands in one nucleus) [8–13], the nuclear Chirality-Parity (ChP) violation [14, 15], chiral wobblers (chirality and the wobbling mode can coexist in a single nucleus) [16–19] and so on have been widely discussed. Up to now, more than 60 candidate chiral nuclei have been reported in the $A \sim 80, 100, 130,$ and 190 mass regions of the nuclear chart, see, e.g., data tables [20].

In $A \sim 130$ mass region, the reported candidates for chiral nuclei form a large chiral island as shown in Fig. 16.1, where the cesium isotopes have the most chiral candidates. Chiral doublet bands were first identified in the $N = 75$ isotones including ^{130}Cs [21]. Subsequently, the chiral doublet bands have been claimed in ^{126}Cs [22] and ^{124}Cs [23]. In addition, the chiral doublet bands found in ^{128}Cs were proposed as the best example to reveal the chiral symmetry breaking [24], and the succeeding g-factor measurements can give important information on the relative orientation of the three angular momentum vectors [25, 26]. Meanwhile, the lifetimes of excited states belonging to the chiral partner bands have been reported in ^{126}Cs, and for the first time the large set of the experimental transition probabilities is in qualitative agreement with all selection rules predicted for the strong chiral symmetry-breaking limit [27].

At present, the candidate for chiral doublet bands has been observed in ^{122,124,126,128,130,132}Cs [21, 23, 24, 27–30] with configuration $\pi h_{11/2}^1 \otimes \nu h_{11/2}^{-1}$, and corresponding chirality is also well explained by many theoretical models, such as tilted axis cranking model [21, 28, 30, 31], triaxial particle-rotor model [24, 32–36], projected shell model [29, 37–39] and so on. In neighboring odd-odd isotopes ^{120,134}Cs, only one rotational band with the configuration $\pi h_{11/2}^1 \otimes \nu h_{11/2}^{-1}$ was reported [35, 40]. For these cesium isotopes, the valence protons and neutrons primarily originate from the $h_{11/2}$ orbit, both being involved in the same configuration $\pi h_{11/2}^1 \otimes \nu h_{11/2}^{-1}$. Therefore, it is worthwhile to investigate the rotational structures in ^{120,134}Cs to explore the evolution of chirality in cesium isotopes. Recently, based on the fully self-consistent and microscopic three-dimensional tilted axis cranking covariant density functional theory (3DTAC-CDFT), the possible chiral candidates and the evolution of chirality in $^{120-134}$Cs have been studied [41].

For the odd-A cesium isotopes, two types of three quasi-particle configurations can be expected to be chiral [42–47]. The first one occurs with a low-j neutron ($g_{7/2}/d_{5/2}$ or $s_{1/2}/d_{3/2}$ orbits) coupled to the neighboring odd-odd chiral configuration $\pi h_{11/2}^1 \otimes \nu h_{11/2}^{-1}$; the second one is formed by a

DOI: 10.1201/9781032691633-16

"

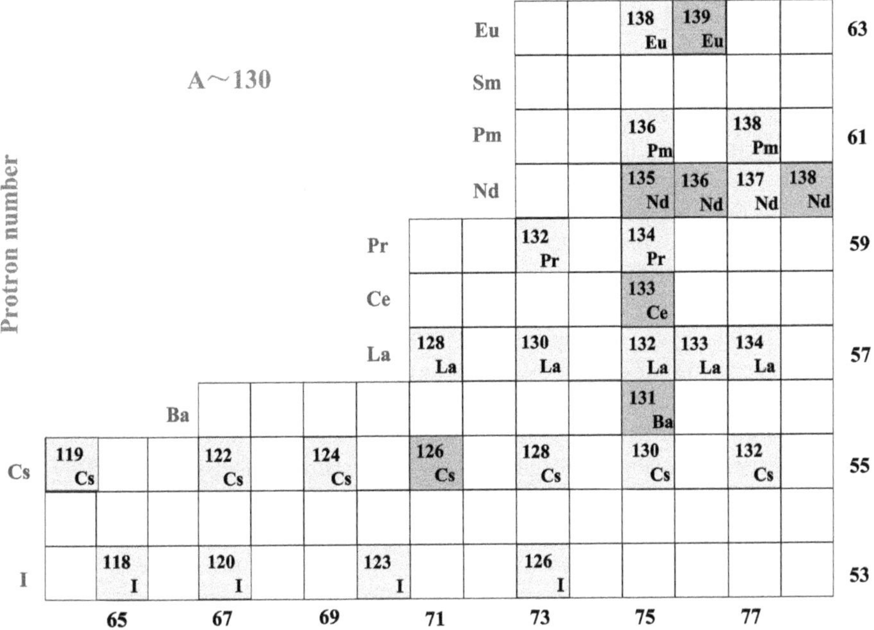

Figure 16.1 The nuclides with chiral doublet bands (red squares) and MχD (blue squares) observed in $A\sim130$ mass region. The data are taken from the Refs. [7, 20].

high-j $h_{11/2}$ proton particle and an aligned pair of high-j neutron $h_{11/2}$ holes, i.e., $\pi h_{11/2} \otimes v(h_{11/2})^{-2}$. The available results in ^{125,127,129,131}Cs, show the existence of a pair of doublet band structures involving the $\pi h_{11/2}^1 \otimes v h_{11/2}^{-1} g_{7/2}/d_{5/2}$ or $\pi h_{11/2}^1 \otimes v h_{11/2}^{-1} s_{1/2}/d_{3/2}$ configurations, and of only one single chiral band with the $\pi h_{11/2} \otimes v(h_{11/2})^{-2}$ configuration [42–45]. Therefore, the MχD is also worth studying in ^{125,127,129,131}Cs. In addition, only one single chiral band with the configuration $\pi h_{11/2}^1 \otimes v h_{11/2}^{-1} g_{7/2}/d_{5/2}$ or $\pi h_{11/2}^1 \otimes v h_{11/2}^{-1} s_{1/2}/d_{3/2}$ has been observed in ^{121,123,133}Cs [48]. It is also interesting to investigate the chirality and MχD in other odd-A cesium isotopes. Recently, through the adiabatic and configuration-fixed constrained covariant density functional theory (CDFT) calculations, the possible existence of MχD in ^{125,127,129,131}Cs has been investigated in Refs. [31, 49].

In this chapter, we will introduce two recent progresses in detail. First, we will investigate the chirality and explore the evolution of chirality in $^{120-134}$Cs. Second, the possible existence of MχD in ^{125,127,129,131}Cs will be discussed.

16.2 THE EVOLUTION OF THE CHIRAL SYMMETRY IN CESIUM ISOTOPES

The two quasi-particle configuration $\pi h_{11/2}^1 \otimes v h_{11/2}^{-1}$ and the three quasi-particle configurations $\pi h_{11/2}^1 \otimes v h_{11/2}^{-1}(gd)^{-1}$ and $\pi h_{11/2}^1 \otimes v h_{11/2}^{-1}(sd)^1$ have been observed in $^{120-134}$Cs [21, 23, 24, 27–31, 35, 40, 42–46, 48–50]. To discuss the possible chiral candidates and the evolution of chirality in $^{120-134}$Cs, the calculated results [41] from 3DTAC-CDFT with point-coupling functional PC-PK1 [51] are presented here. The 3DTAC-CDFT could describe the chirality fully self-consistently and microscopically, and it has been successfully applied to investigate the chirality in ^{106}Rh [52],

^{106}Ag [53], and ^{135}Nd [54] and $^{102-107}$Rh [55]. The positive-parity rotational bands have been observed in $^{120-134}$Cs [21, 23, 24, 27–31, 42–45, 49], and the calculated excitation energies are given in Fig. 16.2 in comparison with the available experimental data. In the present mean-field level, either the chiral vibrations or the tunneling between the left- and right-handed sectors are not taken into account. Therefore, only the band with lower excitation energies can be reproduced. However, it can be seen in Fig. 16.2 that the experimental excitation energies of the lower band can be reproduced well. To describe the partner band, one needs to go beyond the mean-field calculations by combining, for example, the methods of the random phase approximation [56] or the collective Hamiltonian with the CDFT [57, 58].

In Fig. 16.3, the available experimental data of $B(M1)/B(E2)$ in $^{120-134}$Cs [21, 23, 24, 27–31, 42–45, 49] are displayed in comparison with the 3DTAC-CDFT results, which are derived in the semiclassical approximation from the magnetic and electric quadrupole moments. Note that the deformation parameters (β, γ) change only slightly along the band, so the corresponding $B(E2)$ values are almost constant. However, the $B(M1)$ values decrease smoothly due to the continuous variation of rotational frequency and the closing of the neutron and proton angular momentum vectors, which mainly align along the short and long axes, respectively. This leads to the smooth-decreasing tendency for the $B(M1)/B(E2)$ ratios. Therefore, the $B(M1)/B(E2)$ ratios calculated by 3DTAC-CDFT are different from the experimental data, and only the smooth-decreasing tendency for the $B(M1)/B(E2)$ ratios can be given. In particular, the magnetic moments are derived from the relativistic expression of the effective current operator as in Ref. [59]. As shown in Fig. 16.3, the 3DTAC-CDFT results for $^{121,123,125-134}$Cs show coincident agreements with the data, while the calculation overestimates the data for ^{120,122,124}Cs. For ^{120,122,124}Cs, the effective pairing correlation and beyond mean-field effects which are not considered here may play important roles, which has been confirmed in Refs. [60, 61]. The falling tendency shows steeper and steeper with increasing neutron number. Further efforts to include the neutron and proton pairing correlations will be helpful to justify the present results.

In the 3DTAC-CDFT calculations, the orientation of the angular velocity ω with respect to the principal axis can be determined in a self-consistent manner either by minimizing the total Routhian or by requiring that ω is parallel to the total angular momentum J at a fixed ω value. Here, the polar angle θ and the azimuth angle φ are used to denote the direction of the angular velocity $\omega = \omega(\sin\theta\cos\varphi, \sin\theta\sin\varphi, \cos\varphi)$. The θ is the angle between the angular velocity ω and the long (l) axis, while φ is the angle between the projection of ω onto the short-medium (sm) plane and the short (s) axis. It has been noted that φ can be used to characterize the chirality of a rotational system [57].

To examine the possible presence of chiral geometry, the self-consistently obtained orientation angles θ and φ of the total angular momentum J in the intrinsic frame is shown as a function of the rotational frequency $\hbar\omega$ in Fig. 16.4. The polar angle θ in $^{120-134}$Cs has similar behavior, i.e., they increase with the rotational frequency. Nevertheless, the polar angle θ is always larger than $50°$ in all cesium isotopes. It is attributed that the angular momentum alignment along the s axis coming from the proton particles in the $h_{11/2}$ orbit is much larger than that along the l axis from the neutron holes in the $h_{11/2}$ orbit. For comparison, the azimuth angle φ for $^{121-133}$Cs is zero at low rotational frequencies, corresponding to the planar rotation in the s-l plane. Above the limited rotational frequency corresponding to the so-called critical frequency ω_{crit} of chiral rotation [62, 63], the values of φ become nonzero and it results in the transition from planar to aplanar rotation. Note that the kinks appear at the curves of θ for $^{121-133}$Cs with the critical frequency.

The azimuth angle φ is always zero for ^{120,134}Cs corresponding to the planar rotation in the s-l plane, even the rotational frequency goes up to $\hbar\omega = 0.55$ MeV for ^{120}Cs and $\hbar\omega = 0.45$ MeV for ^{134}Cs. By increasing the rotational frequency, the results do not converge for the configuration of $\pi h_{11/2}^1 g_{7/2}^4 \otimes \nu h_{11/2}^7 (gd)^8$ in ^{120}Cs and $\pi h_{11/2}^1 g_{7/2}^4 \otimes \nu h_{11/2}^9 (sd)^6$ in ^{134}Cs. Namely, it might mean that $\hbar\omega_{\mathrm{crit}}$ in the present 3DTAC-CDFT calculations is larger than $\hbar\omega = 0.55$ MeV for ^{120}Cs and

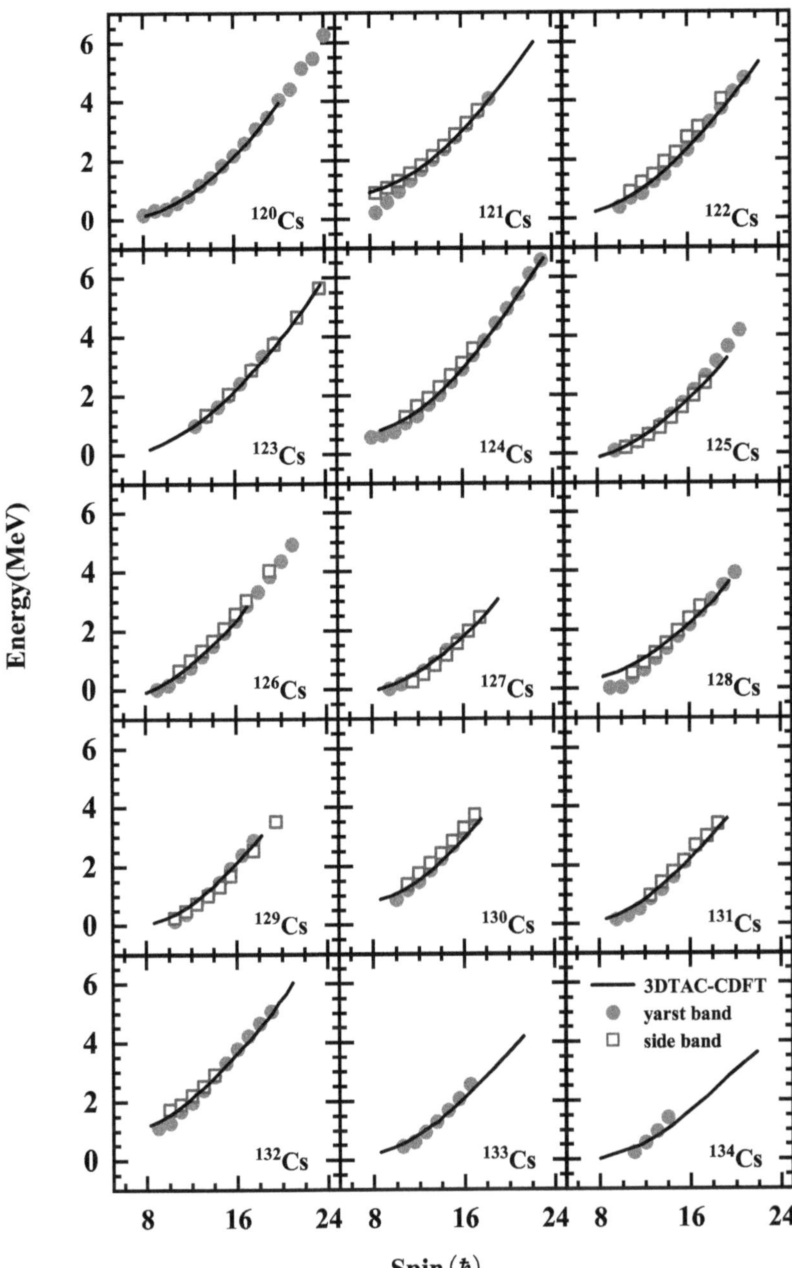

Figure 16.2 The calculated rotational excitation energies in $^{120-134}$Cs by 3DTAC-CDFT, as a function of the spin in comparison with the data for the chiral bands observed in Refs. [21, 23, 24, 27–31, 35, 40, 42–46, 48–50]. Taken from Ref. [41].

$\hbar\omega = 0.45$ MeV for ^{134}Cs even if it exists, which corresponds to $I \approx 24\,\hbar$, out of the spin range observed currently [35, 46, 64]. Thus, the present theoretical analysis does not support the existence of chirality in ^{120,134}Cs.

The chirality in nuclei with stable triaxial deformation is due to the aplanar rotation formed by the valence particle(s), valence hole(s), and collective core angular momentum vectors [1]. Therefore,

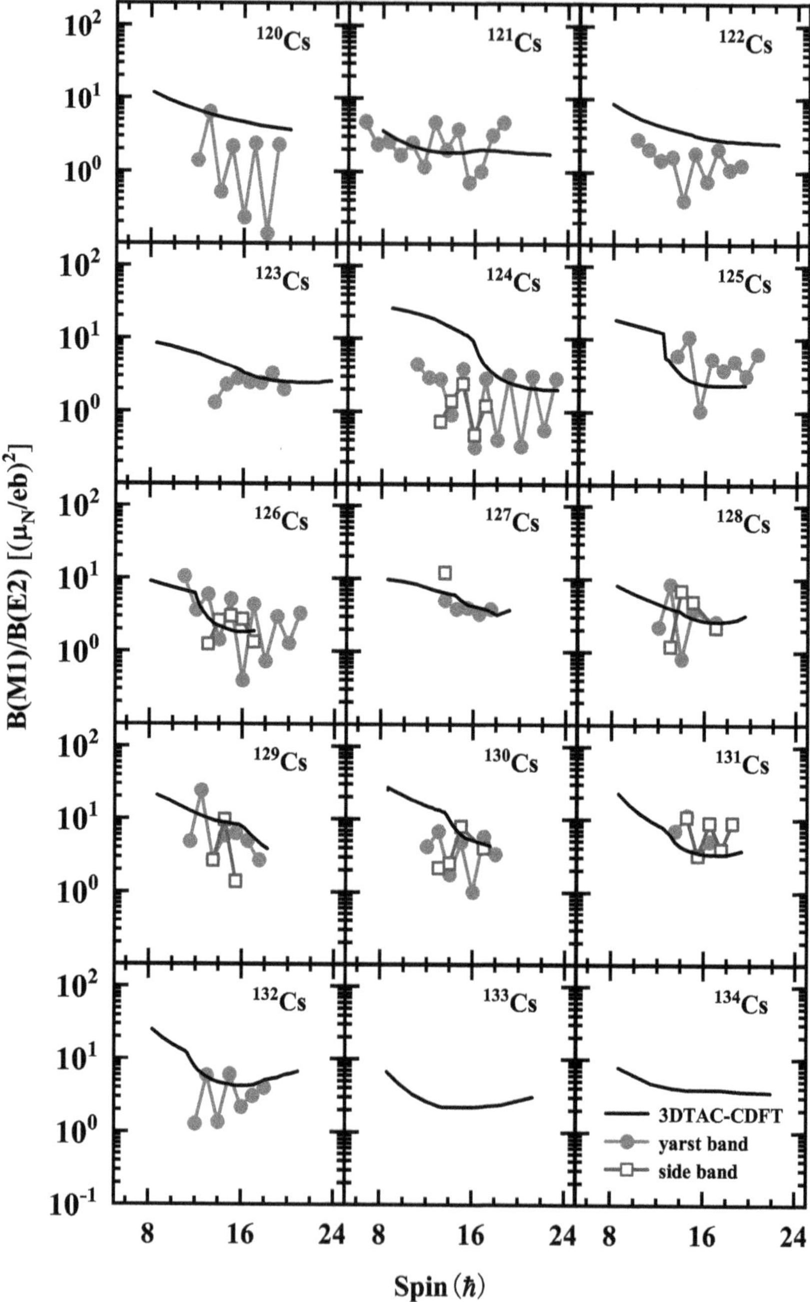

Figure 16.3 The calculated $B(M1)/B(E2)$ ratios in $^{120-134}$Cs by 3DTAC-CDFT, as a function of the spin in comparison with the data for the chiral bands observed in Refs. [21, 23, 24, 27–31, 35, 40, 42–46, 48–50]. Taken from Ref. [41].

it is important to calculate the frequency at which the transition from planar to chiral rotation occurs, i.e., the critical frequency ω_{crit}. As shown in Fig. 16.4, the calculated critical frequencies ω_{crit} all decrease with increasing mass number for three different configurations in $^{120-126}$Cs, ^{127,128}Cs and $^{129-134}$Cs. To understand this interesting behavior, one has to analyze the angular momentum

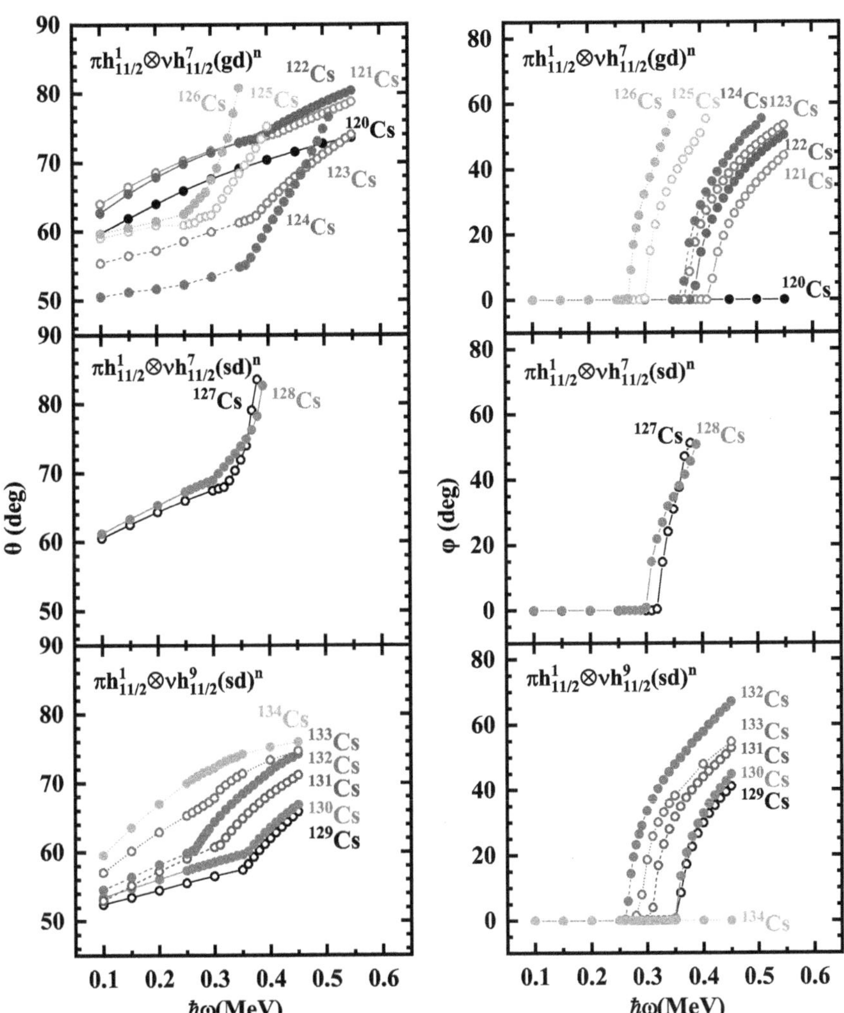

Figure 16.4 The evolution of the polar angle θ and azimuth angle φ for the total angular momentum J as driven by the increasing rotational frequency $\hbar\omega$ for configurations $\pi h_{11/2}^1 g_{7/2}^4 \otimes \nu h_{11/2}^7 (gd)^n (n = 8-14)$ in $^{120-126}$Cs, $\pi h_{11/2}^1 g_{7/2}^4 \otimes \nu h_{11/2}^7 (sd)^n (n = 1-2)$ in $^{127-128}$Cs and $\pi h_{11/2}^1 g_{7/2}^4 \otimes \nu h_{11/2}^9 (sd)^n (n = 1-6)$ in $^{129-134}$Cs. Taken from Ref. [41].

geometry. In 3DTAC-CDFT calculations, the total angular momentum comes from the individual nucleons in a coherent superposition manner.

In $^{120-134}$Cs, only one proton particle occupying the bottom of the $h_{11/2}$ orbit contributes to the angular momentum along the s axis by roughly 5.5 $\hbar$. It should be emphasized that the angular momentum contribution from the proton part is approximately along the s axis at low rotational frequency. Therefore, the angular momentum components along the l and m axes are mainly from the neutron part. In contrast, the neutron holes located at the top of the $h_{11/2}$ orbit contribute about 4 $\hbar$ to the angular momentum along the l axis since four of them are anti-aligned. The only difference is that the neutrons are distributed over the (gd) or (sd) orbits and results in different configurations for $^{120-134}$Cs. The available results show that the angular momentum increment of the (gd) or (sd) orbits along the s and l axes becomes smaller with increasing neutron number and the angular momentum increment along the m-axis becomes larger. Therefore, the corresponding critical frequency

ω_{crit} decreases, and it is easier to form the chiral rotation, which is similar to the earlier studies in Ref. [55]. It should be noted that in ^{133}Cs, the character of the (sd) neutron orbits are transformed from particle to hole.

16.3   POSSIBLE MULTIPLE CHIRAL DOUBLET BANDS IN CESIUM ISOTOPES

Two types of rotational bands have been observed in ^{125,127,129,131}Cs [42–45], as mentioned above. Based on the corresponding two configurations, the potential energy surfaces as a function of β in adiabatic constrained CDFT calculation with PK1 [65] for ^{125,127,129,131}Cs are given in Fig. 16.5, where the minima observed in the potential energy surfaces are labeled with A, B, C, D, and so on. The β^2-constrained calculations have been performed, in which the triaxial deformation γ is automatically obtained by minimizing the energy. At first, the adiabatic constrained calculation is used to obtain the states with different configurations and then the configuration-fixed constrained calculation is performed. In the configuration-fixed constrained calculation, the same configuration is guaranteed during the procedure of constraint calculation with the help of "parallel-transport" [66]. In addition to the β^2-constrained calculation [8], the constraints on the axial and triaxial mass quadrupole moments are also performed to obtain the potential energy surfaces (PES) in the two-dimensional β-γ plane [67, 68]. On the energy surfaces obtained from adiabatic constrained calculations, there are some irregularities, and some local minima are too obscure to be recognized. In comparison, the configuration-fixed constrained calculation avoids these irregularities, thus yielding a continuous, smooth, energy surface for each configuration. Meanwhile, the local minima become more obvious, and they are represented by stars labeled by letters of the alphabet. In addition, the particle-hole excitations based on the ground-state configuration are also performed to obtain two types of chiral three-quasiparticle configurations, i.e., a low-j neutron particle coupled to $\pi h_{11/2} \otimes \nu h_{11/2}^{-1}$ and $\pi h_{11/2} \otimes \nu (h_{11/2})^{-2}$, which are labeled as A1 and A2, respectively. Here, the minima of state A in each panel represent the ground state of the corresponding nucleus. In addition to the ground state, there are several excited minima with triaxial deformation. From Fig. 16.5, one can find that except for ^{127}Cs, ^{125,129,131}Cs have several states with minima for triaxial deformation suitable for chirality and predict triaxial shape coexistence. The only triaxial local minima for ^{127}Cs is just for state A1 $(0.24, 14.4°)$ [1, 69]. As discussed in Ref. [8], the existence of MχD can be expected in ^{125,129,131}Cs, but not in ^{127}Cs, i.e., chiral doublet bands can be built on these triaxial deformed states.

Apart from the triaxial deformation, the proper high-j particle and hole configurations are also necessary for the appearance of the chiral doublet bands. However, for all these nuclei, only the states A1 and A2 have proper high-j particle and hole configurations suitable for chirality. In table 16.1, the total energies E_{tot}, triaxial deformation parameters (β, γ), the corresponding valence nucleons and unpaired nucleons configurations of minima, as well as excitation energies E_x in ^{125,127,129,131}Cs within the configuration-fixed constrained CDFT calculations are given, in comparison with the experimental band-head energies of the rotational bands based on the corresponding configurations. From Table 16.1, one finds that the ground state for all these nuclei has one quasi-proton configuration and the last unpaired proton occupies the $d_{5/2}$ orbit in ^{125}Cs and $g_{7/2}$ orbit in ^{127,129,131}Cs.

Taking ^{125}Cs as an example, the minima of state A represents the ground state, with prolate deformation $\beta = 0.24$, $\gamma = 0.04°$ and the valence nucleon configuration $\pi g_{7/2}^4 d_{5/2}^1 \otimes \nu (sd)^6 h_{11/2}^6$. The corresponding unpaired nucleon configuration is $\pi d_{5/2}^1$. In comparison, the corresponding valence nucleon configuration for state B is $\pi g_{7/2}^4 h_{11/2}^1 \otimes \nu (sd)^6 h_{11/2}^6$ and unpaired nucleon configuration is $\pi h_{11/2}$. By exciting one neutron occupying the (sd) orbit to the $h_{11/2}$ orbit based on configuration B, the configuration of state A1 is obtained; or exciting two neutrons occupying (sd) orbit to $h_{11/2}$ orbit unpaired, the configuration of state A2 is obtained. For the minima of state A1, the triaxial deformation parameter is $(\beta = 0.25, \gamma = 26.3°)$, with corresponding valence nucleon

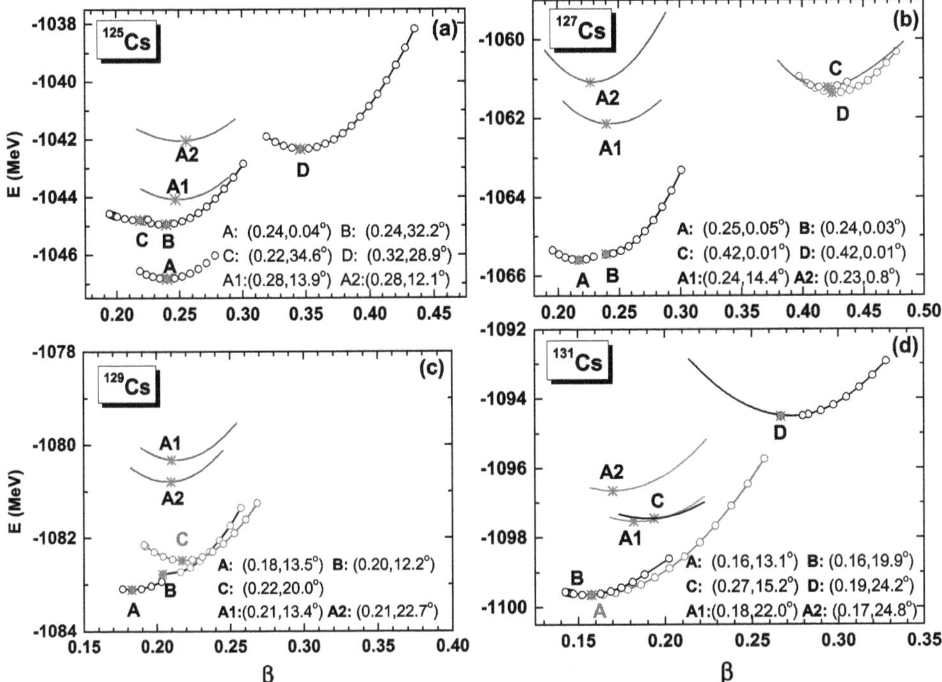

Figure 16.5 The energy surfaces in adiabatic (open circles) and configuration-fixed (solid lines) constrained CDFT calculation using effective interaction PK1 for ^{125}Cs (a), ^{127}Cs (b), ^{129}Cs (c) and ^{131}Cs (d). The minima in the energy surfaces for fixed configuration are represented as stars and labeled respectively as A, B, C, D, E, and so on. Their corresponding triaxial deformation parameters β and γ are also given. Taken from Ref. [49].

configuration $\pi g_{7/2}^4 h_{11/2}^1 \otimes \nu(sd)^5 h_{11/2}^7$ and unpaired nucleon configuration $\pi h_{11/2}^1 \otimes \nu h_{11/2}^{-1}(sd)^1$. For the minima of state A2, the triaxial deformation parameter is ($\beta = 0.26$, $\gamma = 24.3°$), with corresponding valence nucleon configuration $\pi g_{7/2}^4 h_{11/2}^1 \otimes \nu(sd)^2 h_{11/2}^8$ (six neutrons in the $h_{11/2}$ orbit are pairwise and the other two neutrons are aligned or unpaired) and unpaired nucleon configuration $\pi h_{11/2}^1 \otimes \nu(h_{11/2})^{-2}$. Although states C and D have different valence nucleon configurations, they have the same unpaired nucleon configuration $\pi g_{7/2}^1$. Therefore, the triaxial deformation suitable for chirality is found for A1 and A2 in ^{125}Cs, namely $26.3°$ and $24.3°$ respectively, which together with the corresponding high-j proton hole and high-j neutron particle configurations will lead to the MχD phenomenon [8] in ^{125}Cs.

The excitation energy in CDFT calculations is 2.75 MeV for minima A1, rather in agreement with the excitation energy of band head ($\frac{17}{2}^+$) for the candidate chiral rotational bands, i.e., 2.322 MeV, while the excitation energy is 4.78 MeV for minima A2. In Ref. [42], a strongly coupled band has been reassigned a high-K $\pi h_{11/2} \otimes \nu g_{7/2} \otimes \nu h_{11/2}$ three-quasiparticle configuration and a new side band likely to be its chiral partner has been identified. The hybrid version of the TAC model [70] also have been performed to demonstrate that the energy minimization fixes the deformation parameters at $\beta = 0.202, \gamma = 4°$ (nearly prolate) with an average tilt angle $\theta = 60°$ for the $\pi h_{11/2} \otimes \nu g_{7/2} \otimes \nu h_{11/2}$ configuration. The possible reason for such small triaxial deformation is that one unpaired neutron occupies low-j orbit $\nu g_{7/2}$, while the unpaired neutron occupies the (sd) orbit in CDFT calculations.

In ^{127}Cs, the ground state A, with prolate deformation $\beta = 0.22, \gamma = 0.05°$ and the corresponding valence nucleon configuration $\pi g_{7/2}^5 \otimes \nu(sd)^6 h_{11/2}^8$. By performing the similar particle-hole

Table 16.1

The total energies E_{tot}, triaxial deformation parameters β, γ, corresponding valence nucleons and unpaired nucleons configurations of minima as well as excitation energies E_x in ^{125,127,129,131}Cs within the configuration-fixed constrained CDFT calculations. The experimental excitation energy of the band head for corresponding rotational bands is taken from Refs. [42–45]. See text for more details.

	State	Configuration		E_{tot}	(β, γ)	E_x(cal.)	E_x(exp.)
		Valence nucleons	Unpaired nucleons	(MeV)		(MeV)	(MeV)
^{125}Cs	A	$\pi g_{7/2}^4 d_{5/2}^1 \otimes \nu(sd)^6 h_{11/2}^6$	$\pi d_{5/2}^1$	-1046.82	(0.24,0.04°)	0	
	B	$\pi g_{7/2}^4 h_{11/2}^1 \otimes \nu(sd)^6 h_{11/2}^6$	$\pi h_{11/2}^1$	-1044.94	(0.24,32.2°)		
	C	$\pi g_{7/2}^5 \otimes \nu(sd)^6 h_{11/2}^6$	$\pi g_{7/2}^1$	-1044.79	(0.22,34.6°)		
	D	$\pi g_{7/2}^5 h_{11/2}^2 p_{3/2}^{-2} \otimes \nu(sd)^4 h_{11/2}^8$	$\pi g_{7/2}^1$	-1042.33	(0.35,34.5°)		
	A1	$\pi g_{7/2}^4 h_{11/2}^1 \otimes \nu(sd)^5 h_{11/2}^7$	$\pi h_{11/2}^1 \otimes \nu h_{11/2}^{-1}(sd)^1$	-1044.07	(0.25,26.3°)	2.75	2.322
	A2	$\pi g_{7/2}^4 h_{11/2}^1 \otimes \nu(sd)^4 h_{11/2}^8$	$\pi h_{11/2}^1 \otimes \nu(h_{11/2})^{-2}$	-1042.04	(0.26,24.3°)	4.78	
^{127}Cs	A	$\pi g_{7/2}^5 \otimes \nu(sd)^6 h_{11/2}^8$	$\pi g_{7/2}^1$	-1065.59	(0.22,0.05°)	0	
	B	$\pi g_{7/2}^4 d_{5/2}^1 \otimes \nu(sd)^6 h_{11/2}^8$	$\pi d_{5/2}^1$	-1065.45	(0.24,0.03°)		
	C	$\pi g_{9/2}^{-2} g_{7/2}^4 h_{11/2}^1 s_{1/2}^2 \otimes \nu g_{7/2}^{-2}(sd)^6 h_{11/2}^8 f_{5/2}^2$	$\pi h_{11/2}^1$	-1061.23	(0.42,0.01°)		
	D	$\pi g_{9/2}^{-2} g_{7/2}^4 h_{11/2}^2 s_{1/2}^1 \otimes \nu g_{7/2}^{-2}(sd)^6 h_{11/2}^8 f_{5/2}^2$	$\pi s_{1/2}^1$	-1061.37	(0.42,0.01°)		
	A1	$\pi g_{7/2}^4 h_{11/2}^1 \otimes \nu(sd)^5 h_{11/2}^9$	$\pi h_{11/2}^1 \otimes \nu h_{11/2}^{-1}(sd)^1$	-1062.14	(0.24,14.4°)	3.45	3.002
	A2	$\pi g_{7/2}^4 h_{11/2}^1 \otimes \nu(sd)^6 h_{11/2}^8$	$\pi h_{11/2}^1 \otimes \nu(h_{11/2})^{-2}$	-1061.08	(0.23,0.8°)	4.51	3.816
^{129}Cs	A	$\pi g_{7/2}^5 \otimes \nu(sd)^8 h_{11/2}^8$	$\pi g_{7/2}^1$	-1083.10	(0.18,13.5°)	0	
	B	$\pi g_{7/2}^4 d_{5/2}^1 \otimes \nu(sd)^8 h_{11/2}^8$	$\pi d_{5/2}^1$	-1082.78	(0.20,12.2°)		
	C	$\pi g_{7/2}^4 h_{11/2}^1 \otimes \nu(sd)^6 h_{11/2}^8$	$\pi h_{11/2}^1$	-1082.48	(0.22,20.0°)		
	A1	$\pi g_{7/2}^4 h_{11/2}^1 \otimes \nu(sd)^7 h_{11/2}^9$	$\pi h_{11/2}^1 \otimes \nu h_{11/2}^{-1}(sd)^1$	-1080.32	(0.21,13.4°)	2.78	2.676
	A2	$\pi g_{7/2}^4 h_{11/2}^1 \otimes \nu(sd)^8 h_{11/2}^8$	$\pi h_{11/2}^1 \otimes \nu(h_{11/2})^{-2}$	-1080.79	(0.21,22.7°)	2.31	3.518
^{131}Cs	A	$\pi g_{7/2}^5 \otimes \nu(sd)^8 h_{11/2}^{10}$	$\pi g_{7/2}^1$	-1099.65	(0.16,13.1°)	0	
	B	$\pi g_{7/2}^5 \otimes \nu(sd)^{10} h_{11/2}^8$	$\pi g_{7/2}^1$	-1099.64	(0.16,19.9°)		
	C	$\pi g_{7/2}^4 h_{11/2}^1 \otimes \nu(sd)^8 h_{11/2}^8 h_{9/2}^2$	$\pi h_{11/2}^1$	-1094.51	(0.27,15.2°)		
	D	$\pi g_{7/2}^4 h_{11/2}^1 \otimes \nu(sd)^8 h_{11/2}^{10}$	$\pi h_{11/2}^1$	-1097.46	(0.19,24.2°)		
	A1	$\pi g_{7/2}^4 h_{11/2}^1 \otimes \nu(sd)^9 h_{11/2}^9$	$\pi h_{11/2}^1 \otimes \nu h_{11/2}^{-1}(sd)^1$	-1097.54	(0.18,22.0°)	2.11	2.55
	A2	$\pi g_{7/2}^4 h_{11/2}^1 \otimes \nu(sd)^{10} h_{11/2}^8$	$\pi h_{11/2}^1 \otimes \nu(1h_{11/2})^{-2}$	-1096.65	(0.17,24.8°)	3.00	3.465

excitations as in ^{125}Cs based on the ground state configuration A, the high-j proton hole and high-j neutron particle configurations of states A1 and A2 are obtained. However, only the minimum of A1 has considerable triaxial deformation ($\beta = 0.24$, $\gamma = 14.4°$), while the triaxial deformation for the minimum of A2 is very tiny ($\beta = 0.23$, $\gamma = 0.8°$). Therefore following the discussion in Ref. [8], the existence of MχD with the above two different configurations can not be expected in ^{127}Cs. In Ref. [45], two bands with possible configurations $\pi h_{11/2} \otimes \nu g_{7/2} \otimes \nu h_{11/2}$ and $\pi h_{11/2}^1 \otimes \nu(h_{11/2})^{-2}$ have been proposed, respectively. Furthermore, as discussed in Ref. [45], both the TRS calculations and the signature splitting support a triaxial shape for the band with assigned $\pi h_{11/2}^1 \otimes \nu(h_{11/2})^{-2}$ configuration. Therefore, the triaxial deformation for this configuration should be further investigated within the rotating mean field. In addition, the excitation energy in CDFT calculations is 3.45 MeV for the minimum of A1 and 4.51 MeV for the minimum of A2, in good agreement with the band-head excitation energy of the corresponding three-quasiparticles bands 3.002 ($\frac{23^+}{2}$) and 3.816 MeV($\frac{19^-}{2}$), respectively.

In ^{129}Cs, the ground state A, with prolate deformation $\beta = 0.18$, $\gamma = 13.5°$, the corresponding valence nucleon configuration is $\pi g_{7/2}^5 \otimes \nu(sd)^8 h_{11/2}^8$. The corresponding deformation parameters for the high-j proton hole and high-j neutron particle configurations of states A1 and A2 are $\beta = 0.21$, $\gamma = 13.4°$ and $\beta = 0.21$, $\gamma = 22.7°$, respectively. They together support the existence of MχD. In Ref. [44], a pair of strongly coupled band structures exhibiting similar features have been observed with the $\pi h_{11/2} \otimes \nu h_{11/2} \otimes \nu(g_{7/2}/d_{3/2}/s_{1/2})^1$ configuration. Self-consistent TAC calculations [44] with the $\pi h_{11/2} \otimes \nu h_{11/2} \otimes \nu g_{7/2}^1$ configuration have resulted in the deformation parameters $\beta = 0.15$ and $\gamma = 33°$, and an average tilt angle of the angular momentum with the principal axis of $\theta = 53°$, which provides further supports for chirality. In addition, the excitation energy of the $(\frac{19}{2}^+)$ band head is 2.676 MeV, also in rather good agreement with the CDFT calculations of 2.78 MeV. Another single band with assigned configuration $\pi h_{11/2}^1 \otimes \nu(h_{11/2})^{-2}$ was also observed in Ref. [44], with the corresponding excitation energy of the $(\frac{25}{2}^-)$ band head of 3.518 MeV, to be compared with the calculated CDFT value of 2.31 MeV for the same configuration.

In ^{131}Cs, the ground state A has prolate deformation $\beta = 0.16$, $\gamma = 13.1°$ and the corresponding valence nucleon configuration is $\pi g_{7/2}^5 \otimes \nu(sd)^8 h_{11/2}^{10}$. The corresponding deformation parameters for the high-j proton hole and high-j neutron particle configurations of minima A1 and A2 are $\beta = 0.18$, $\gamma = 22.0°$ and $\beta = 0.17$, $\gamma = 24.8°$, respectively. They also together support the existence of the MχD phenomenon in ^{131}Cs. Experimentally, the observation of the nearly degenerate dipole bands with assigned high-K three-quasiparticle configurations $\pi h_{11/2} \otimes \nu h_{11/2} \otimes \nu(g_{7/2}/d_{3/2}/s_{1/2})^1$ has been confirmed in Ref. [43]. A self-consistent TAC calculation with this choice of configuration has resulted in the deformation parameters $\beta = 0.122$, $\gamma = 30°$, and an average tilt angle $\theta = 57°$ for the angular momentum with the principal axis. In addition, the excitation energy of the $(\frac{17}{2}^+)$ band head is 2.55 MeV, in good agreement with the calculated CDFT value of 2.11 MeV. Another single band with configuration $\pi h_{11/2}^1 \otimes \nu(h_{11/2})^{-2}$ was also observed in Ref. [43] with the corresponding excitation energy of the $(\frac{25}{2}^-)$ band head of 3.465 MeV, in agreement with the calculated CDFT value of 3.00 MeV for the same configuration.

16.4 SUMMARY

In summary, a fully self-consistent and microscopic method 3DTAC-CDFT has been used to investigate the evolution of chirality in $^{120-134}$Cs. The obtained energy spectra and $B(M1)/B(E2)$ values show reasonable agreement with the available experimental data. By investigating the evolution of the polar angle θ and the azimuth angle φ as a function of the rotational frequency $\hbar\omega$, the transition from the planar rotation to the chiral rotation has been found in $^{121-133}$Cs. Moreover, the decrease of the critical frequency ω_{crit} in $^{121-133}$Cs with increasing neutron number can be attributed to the neutrons in (gd) and (sd) orbits having smaller angular momentum components along both the short and long axes, and larger components along medium axis, respectively. In comparison, only planar rotation has been obtained in ^{120,134}Cs, indicating the possible boundaries of chirality in the cesium isotopes with $\pi h_{11/2}^1 \otimes \nu h_{11/2}^{-1}$ configuration, similar to the result in ^{120}I [71].

Meanwhile, the adiabatic and configuration-fixed constrained CDFT approaches have been applied to investigate the possible MχD in ^{125,127,129,131}Cs. According to the suggested high-j proton hole and high-j neutron particle configurations, their triaxial deformations which can favor the existence of the chiral doublet bands were tested from the configuration-fixed constrained CDFT calculations. The present calculations suggest that the existence of MχD phenomenon is expected in ^{125,129,131}Cs, but not in ^{127}Cs.

Recently, in Ref. [72], the chiral bands of ^{119}Cs based on configurations with only protons are reported for the first time in the $A \sim 130$ mass region, i.e., a configuration with three protons, one in the strongly coupled $g_{9/2}^{-1}$ orbital which does not change orientation with increasing rotational frequency, and two in the $h_{11/2}$ orbital which reorients to the rotation axis, and it is different from the common chiral geometry picture from both protons and neutrons. It is interesting to perform more investigations, including 3DTAC-CDFT, particle-rotor model and so on. Investigations in these directions are in progress. In Ref. [73], two pairs of chiral doublet bands based on the $\pi h_{11/2}^{1} \otimes \nu h_{11/2}^{-1}$ configuration have been identified in ^{126}Cs. For the first time, the MχD phenomenon was reported in odd-odd cesium isotopes. Thus, more efforts on the investigation of nuclear chirality in the $A \sim 130$ mass region are still needed, such as the boundaries of the chirality region, the evolution of chirality in isotones, the possible chirality in Xe or Sm nuclei and so on, which will be helpful to understand the underlying rotational structure in these chiral nuclei.

16.5 ACKNOWLEDGMENTS

The authors would like to thank Prof. J. Meng, S. Q. Zhang, P. W. Zhao, and Q. B. Chen for their helpful discussions. This work was supported by the National Natural Science Foundation of China (No. 11675063) and the Natural Science Foundation of Jilin Province (No. 20220101017JC) as well as the Key Laboratory of Nuclear Data Foundation (JCKY2020201C157).

Bibliography

1. S. Frauendorf and J. Meng, Nuclear Physics A **617**(2), 131 (1997).
2. J. Meng, S.Q. Zhang and J. Phys. G: Nucl. Part. Phys. **37**, 064025 (2010).
3. J. Meng, Q.B. Chen and S.Q. Zhang, Int. J. Mod. Phys. E **23**, 1430016 (2014).
4. J. Meng and P. Zhao, Phys. Scripta **91**, 053008 (2016).
5. A. A. Raduta, Progress in Particle and Nuclear Physics **90**, 241 (2016).
6. K. Starosta and T. Koike, Physica Scripta **92**(9) (2017).
7. S. Y. Wang, C. Liu, B. Qi, W. Xu and H. Zhang, Front. Phys. **18**, 64601 (2023).
8. J. Meng, J. Peng, S. Q. Zhang and S. G. Zhou, Phys. Rev. C **73**, 037303 (2006).
9. A.D. Ayangeakaa, *et al.*, Phys. Rev. Lett. **110**, 172504 (2013).
10. S. Y. Wang, Chinese Physics C **44**(11), 112001 (2020).
11. J. Li, S.Q. Zhang and J. Meng, Phys. Rev. C **83**, 037301 (2011).
12. C.M. Petrache, *et al.*, Phys. Rev. C **97**, 041304 (2018).
13. I. Kuti, *et al.*, Phys. Rev. Lett. **113**, 032501 (2014).
14. Y. Wang, X. Wu, S. Zhang, P. Zhao and J. Meng, Science Bulletin **65**(23), 2001 (2020).
15. E.O. Lieder, *et al.*, Phys. Rev. Lett. **112**, 202502 (2014).
16. T. Koike, K. Starosta, I. Hamamoto, D.B. Fossan and C. Vaman, AIP Conference Proceedings **764**(1), 87 (2005).
17. Q.B. Chen and S. Frauendorf, Eur. Phys. Jour. A **58**, 75 (2022).
18. H. Jia, S. Wang, B. Qi, C. Liu and L. Zhu, Physics Letters B **833**, 137303 (2022).
19. R.J. Guo, *et al.*, Phys. Rev. Lett. **132**, 092501 (2024).
20. B. Xiong and Y. Wang, Atom Data Nucl Data **125**, 193 (2019).
21. K. Starosta, *et al.*, Phys. Rev. Lett. **86**, 971 (2001).
22. S. Wang, Y. Liu, T. Komatsubara, Y. Ma and Y. Zhang, Phys. Rev. C **74**, 017302 (2006).
23. D. Yang,*et al.*, Chin. Phys. Lett. **26**(8), 082101 (2009).
24. E. Grodner, *et al.*, Phys. Rev. Lett. **97**, 172501 (2006).
25. E. Grodner, *et al.*, Phys. Rev. Lett. **120**, 022502 (2018).
26. E. Grodner, *et al.*, Phys. Rev. C **106**, 014318 (2022).
27. E. Grodner, *et al.*, Phys. Lett. B **703**, 46 (2011).
28. U. Yong-Nam, *et al.*, J. Phys. G: Nucl. Part. Phys. **31**, B1 (2005).
29. F.Q. Chen, Q.B. Chen, Y.A. Luo, J. Meng and S.Q. Zhang, Phys. Rev. C **96**, 051303 (2017).
30. G. Rainovski, *et al.*, Phys. Rev. C **68**, 024318 (2003).
31. R. Guo, W.J. Sun, J. Li, D. Yang, Y. Liu, C. Ru and J. Chi, Phys. Rev. C **100**, 034328 (2019).
32. S.Y. Wang, B. Qi and D.P. Sun, Phys. Rev. C **82**, 027303 (2010).
33. S.Y. Wang, S.Q. Zhang, B. Qi and J. Meng, Phys. Rev. C **75**, 024309 (2007).

34. B. Qi, J. Meng, S.Q. Zhang, S.Y. Wang and J. Peng, Chin. Phys. C **33**(S1), 43 (2009).
35. T. Koike, K. Starosta, C.J. Chiara, D.B. Fossan and D.R. LaFosse, Phys. Rev. C **67**, 044319 (2003).
36. O. Shirinda and E.A. Lawrie, The European Physical Journal A **48**, 118 (2012).
37. G. Bhat, R. Ali, J. Sheikh and R. Palit, Nuclear Physics A **922**, 150 (2014).
38. F.Q. Chen, J. Meng and S.Q. Zhang, Physics Letters B **785**, 211 (2018).
39. P. Siwach, P. Arumugam, L.S. Ferreira and E. Maglione, Phys. Rev. C **103**, 024327 (2021).
40. Y. Liu, J. Lu, Y. Ma, G. Zhao, H. Zheng and S. Zhou, Phys. Rev. C **58**, 1849 (1998).
41. D. Chen, J. Li and R. Guo, The European Physical Journal A **59**, 142 (2023).
42. K. Singh, *et al.*, Eur. Phys. J. A **27**, 321 (2006).
43. S. Sihotra, *et al.*, Phys. Rev. C **78**, 034313 (2008).
44. S. Sihotra, *et al.*, Phys. Rev. C **79**, 044317 (2009).
45. Y. Liang, R. Ma, E.S. Paul, N. Xu, D.B. Fossan and R.A. Wyss, Phys. Rev. C **42**, 890 (1990).
46. C.B. Moon, T. Komatsubara, K. Furuno and J. Korean Phys. Soc. **38**, 83 (2001).
47. K. Singh, *et al.*, The European Physical Journal A **25**, 345 (2005).
48. S. Biswas, *et al.*, Phys. Rev. C **95**, 064320 (2017).
49. J. Li, Phys. Rev. C **97**, 034306 (2018).
50. C.B. Moon, T. Komatsubara and K. Furuno, Nucl. Phys. A **674**, 343 (2000).
51. P.W. Zhao, Z.P. Li, J.M. Yao and J. Meng, Phys. Rev. C **82**, 054319 (2010).
52. P. Zhao, Phys. Lett. B **773**, 1 (2017).
53. P.W. Zhao, Y.K. Wang and Q.B. Chen, Phys. Rev. C **99**, 054319 (2019).
54. J. Peng and Q. Chen, Phys. Lett. B **810**, 135795 (2020).
55. J. Peng and Q.B. Chen, Phys. Rev. C **105**, 044318 (2022).
56. D. Almehed, F. Donau and S. Frauendorf, Phys. Rev. C **83**, 054308 (2011).
57. Q.B. Chen, S.Q. Zhang, P.W. Zhao, R.V. Jolos and J. Meng, Phys. Rev. C **87**, 024314 (2013).
58. Q.B. Chen, S.Q. Zhang, P.W. Zhao, R.V. Jolos and J. Meng, Phys. Rev. C **94**, 044301 (2016).
59. P. Ring, Progress in Particle and Nuclear Physics **37**, 193 (1996).
60. Z.H. Zhang, M. Huang and A.V. Afanasjev, Phys. Rev. C **101**, 054303 (2020).
61. Y. Wang and J. Meng, Phys. Lett. B **841**, 137923 (2023).
62. P. Olbratowski, J. Dobaczewski and J. Dudek, Phys. Rev. C **73**, 054308 (2006)
63. P. Olbratowski, J. Dobaczewski, J. Dudek and W. Płóciennik, Phys. Rev. Lett. **93**, 052501 (2004)
64. H. Pai, *et al.*, Phys. Rev. C **84**, 041301 (2011).
65. W. Long, J. Meng, N.V. Giai and S.G. Zhou, Phys. Rev. C **69**, 034319 (2004).
66. T. Bengtsson, Nuclear Physics A **496**(1), 56 (1989).
67. Z.P. Li, J. Meng, Y. Zhang, S.G. Zhou and L.N. Savushkin, Phys. Rev. C **81**, 034311 (2010).
68. Y. Wang, J. Li, J. Bin Lu and J. Ming Yao, Progress of Theoretical and Experimental Physics **2014**(11), 113D03 (2014).
69. J. Meng, H. Toki, S. Zhou, S. Zhang, W. Long and L. Geng, Progress in Particle and Nuclear Physics **57**(2), 470 (2006).
70. V.I. Dimitrov, S. Frauendorf and F. Dönau, Phys. Rev. Lett. **84**, 5732 (2000).
71. R. Guo, Y.H. Liu, J. Li, W.J. Sun, L. Li and Y.J. Ma, Chin. Phys. C **44**(7), 074102 (2020).
72. K.K. Zheng, *et al.*, Eur. Phys. J. A **58**, 50 (2022).
73. T.J. Gao, *et al.*, Phys. Rev. C **109**, 024307 (2024).

17 Collective model for multiple chiral doublets bands and wobbling bands

Yu Zhang and Hong-Di Jiang
Liaoning Normal University, Dalian, China

Bin Qi
Shandong University, Weihai, China

17.1 INTRODUCTION

For odd-odd nuclei, a very important phenomenon is just the chiral doublet bands originating from chiral symmetry breaking in triaxially deformed systems, which was predicted by Frauendorf and Meng in 1997 [1]. Theoretically, chiral symmetry breaking is suggested to occur in a stable triaxially-deformed nucleus with high-j valence particles (holes) configurations. Its experimental signatures, chiral doublet bands, were first observed in the $N = 75$ isotopes [2] and then in around 50 nuclei in the $A \sim 80$, 100, 130, 190 regions with the data summarized in [3]. The multiple chiral doublet bands (MχD) based on different single-particle configuration [4] were further predicted to exist in a single nucleus. Since then, searching for MχD in experiments has been attracted extensive interests and the evidences were reported for nuclei with both odd and even mass [5–15]. For odd-A nuclei, triaxiality can be alternatively related to the appearance of wobbling bands. Wobbling motion in nuclei is described as small amplitude oscillation of the total angular momentum vector with respect to the principal axis with the largest moment of inertia [16]. The relevant evidences were mostly observed from the odd mass systems with $A \sim 100$, 130, 160, 190 [17–26]. In short, chiral rotation and wobbling motion of triaxial nuclei have become two significantly important subjects in nuclear structure researches.

The particle-rotor model (PRM) may be the most convenient framework to discuss chiral rotations in odd-odd nuclei or wobbling motion in odd-A nuclei. Different aspects of chiral bands and wobbling bands have been well studied within this model [1, 17, 26–29]. Nonetheless, both the β and γ degrees of freedom in the rotor core have been frozen so that only rotational excitations are allowed in the PRM [30]. A more general consideration of collective motions should start from the Bohr Hamiltonian [31] instead of a pure rotor [32]. A theoretical attempt has been presented in [33], where both the β and γ degrees of freedom are treated as the dynamical variables in the collective Hamiltonian. Although solving the Bohr Hamiltonian often needs to employ certain approximations [31], new insights into collectivity of nuclei can often be achieved from the model solutions (exact or approximate). A good example for this point is just the recent-years development of the concept, critical point symmetry (CPS), for the description of nuclear shape phase

DOI: 10.1201/9781032691633-17

transition [34–44], where different CPSs are associated with solutions of the Bohr Hamiltonian with different potential forms [34, 35].

In this work, we will briefly recall several typical CPSs to show how to model a transitional even-even system in a solvable way since these solvable models can be potentially taken to describe the adjacent odd systems through the core-particle coupling scheme. In particular, we will propose a γ-stable collective potential in the Bohr Hamiltonian to describe the core dynamics of an odd-odd or odd-A system. The resulting model is then expected to generate a collective vibrational picture for the MχD or multiple wobbling bands.

17.2　COLLECTIVE MODES FOR TRANSITIONAL SYSTEMS

The CPS modes have attracted considerable attentions since they provide benchmarks for studying nuclear structures in different shape phase transitions (SPTs) [45–47]. Typical examples for even-even nuclei include CPS of the spherical to γ-unstable SPT E(5) [34], CPS of the spherical to prolate SPT X(5) [35], CPS of the prolate to oblate SPT Z(5) [37] and CPS of the spherical to triaxially deformed SPT T(5) [42]. These CPSs are all established in the framework of Bohr model with their solutions being associated with the five-dimensional Euclidean dynamical symmetry [34, 48, 49]. Building upon successful applications in even-even systems, the CPS method was further extended to odd-A and odd-odd systems and became a standard way to model transitional structures. The original Bohr Hamiltonian is written as

$$H = -\frac{\hbar^2}{2B}\left\{\frac{1}{\beta^4}\frac{\partial}{\partial\beta}\beta^4\frac{\partial}{\partial\beta} + \frac{1}{\beta^2}\left(\frac{1}{\sin3\gamma}\frac{\partial}{\partial\gamma}\sin3\gamma\frac{\partial}{\partial\gamma}\right.\right.$$
$$\left.\left. -\frac{1}{4}\sum_k\frac{L_k^2}{\sin^2(\gamma - \frac{2}{3}k\pi)}\right)\right\} + V(\beta,\gamma),\tag{17.1}$$

where β and γ are the deformation variables, B is the collective mass parameter and L_k ($k = 1,\ 2,\ 3$) are the projections of the angular momentum on the body-fixed k-axis. As described below, different potential form will be employed for different CPS mode [50].

17.2.1　THE E(5) SOLUTION

The E(5) CPS is designed to describe the spherical to γ-unstable SPT (algebraically called the U(5)-SU(6) transition). The corresponding potential is assumed to depend only on β in a form of infinite square well [34], i.e.

$$V(\beta,\gamma) = V(\beta) = \begin{cases} 0, & \beta \leq \beta_W, \\ \infty, & \beta > \beta_W. \end{cases}\tag{17.2}$$

Since the potential is only the function of β, the Bohr Hamiltonian is separable with the wave function being expressed as

$$\Psi(\beta,\gamma,\theta_i) = f(\beta)\Phi(\gamma,\theta_i),\tag{17.3}$$

where θ_i represent the Euler angles. The eigenvalue equation $H\Psi = E\psi$ can be then divided into two parts, one for the β variable

$$\left[-\frac{\hbar^2}{2B}\left(\frac{1}{\beta^4}\frac{\partial}{\partial\beta}\beta^4\frac{\partial}{\partial\beta} - \frac{\Lambda}{\beta^2}\right) + V(\beta) - E\right]f(\beta) = 0\tag{17.4}$$

and the other for the γ variable

$$\left[-\frac{1}{\sin3\gamma}\frac{\partial}{\partial\gamma}\sin3\gamma\frac{\partial}{\partial\gamma} + \frac{1}{4}\sum_k\frac{L_k^2}{\sin^2(\gamma - \frac{2}{3}k\pi)} - \Lambda\right]\Phi(\gamma,\theta_i) = 0\tag{17.5}$$

with the eigenvalues $\Lambda = \tau(\tau+3)$ and $\tau = 0,\ 1,\ 2\cdots$. Since $V(\beta)$ is a square well as given in (17.2) the eigenvalue equations can be analytically solved with the eigenfunctions expressed by

$$f(\beta) = c_{\xi,\tau}\beta^{-3/2}J_{\tau+3/2}(k_{\xi,\tau}\beta),\tag{17.6}$$

$$\Phi(\gamma,\theta_i) = \sum_K g_K^{L,\tau,\nu}(\gamma)D_{M,K}^L(\theta_i).\tag{17.7}$$

In Eq. (17.6), $c_{\xi,\tau}$ denotes the normalization constant determined by

$$\int_0^\infty \beta^4 d\beta f^2(\beta) = 1,\tag{17.8}$$

$J_{\tau+3/2}(z)$ is the Bessel function with the order $\tau+3/2$, and $k_{\xi,\tau} = \frac{x_{\xi,\tau}}{\beta_W}$ with $x_{\xi,\tau}$ representing the ξ-th zero of $J(z)$. In Eq. (17.7), g_K is a symmetric function that can be expanded in terms of trigonomeric functions of γ [34, 39] and $D_{M,K}^L(\theta_i)$ is the Wigner function. The eigenvalues are accordingly obtained as $E = \frac{\hbar^2}{2B}k_{\xi,\tau}^2$, by which one can easily work out the entire level pattern. In short, the E(5) CPS can be solved in an analytical way, and its solution can be applied to benchmark critical structures in the second-order SPT [45] in the mass region $A \sim 130$, where chiral symmetry breaking is also expected to occur [27]. Therefore, how to extend the E(5) CPS to describe the relevant phenomena would be interesting.

17.2.2 THE X(5) SOLUTION

The X(5) CPS [35] was proposed to describe the spherical to prolate SPT (algebraically called the U(5)-SU(3) transition). The corresponding potential function is assumed to be separated into two parts,

$$V(\beta,\gamma) = V(\beta) + V(\gamma).\tag{17.9}$$

To simulate a transitional situation, $V(\beta)$ is also taken as an infinite square well as in (17.2), but $V(\gamma)$ is assumed to be a harmonic oscillator around $\gamma = 0°$ and given by

$$V(\gamma) = \frac{1}{2}a^2\gamma^2,\tag{17.10}$$

where a is a strength parameter. Then the eigenvalue equation can be approximately separated into the one for the β variable

$$\left[-\frac{\hbar^2}{2B}\left(\frac{1}{\beta^4}\frac{\partial}{\partial\beta}\beta^4\frac{\partial}{\partial\beta} - \frac{L(L+1)}{3\beta^2}\right) + V(\beta)\right]f'(\beta)$$
$$= E_\beta' f'(\beta)\tag{17.11}$$

and the other for the γ variable

$$-\frac{\hbar^2}{2B\langle\beta^2\rangle}\left[\frac{1}{\gamma}\frac{\partial}{\partial\gamma}\gamma\frac{\partial}{\partial\gamma} - \frac{K^2}{4\gamma^2} + \frac{K^2}{3} - C\frac{\gamma^2}{2}\right]\Phi'(\gamma,\theta_i)$$
$$= E_\gamma'\Phi'(\gamma,\theta_i),\tag{17.12}$$

where $\langle\beta^2\rangle$ is the average of β^2 over $f'(\beta)$ and $C = 2Ba^2\langle\beta^2\rangle/\hbar^2$. In the derivations, the following approximation in solving the Hamiltonian bas been adopted. Since the potential has a minimum at $\gamma = 0°$, the last term in the Bohr Hamiltonian (17.1) can be written by

$$R \simeq \sum_{k=1}^3 \frac{\hat{L}_k^2}{4\sin^2(\gamma - \frac{2}{3}k\pi)}\bigg|_{\gamma=0}$$
$$= \frac{1}{3}(\hat{L}_1^2 + \hat{L}_2^2 + \hat{L}_3^2) + \hat{L}_3^2\left(\frac{1}{4\gamma^2} - \frac{1}{3}\right).\tag{17.13}$$

Then, the total energy $E = E'_\beta + E'_\gamma$ can be approximately obtained as

$$E = \frac{\hbar^2}{2B}\left(\frac{x_{s,L}}{\beta_W}\right)^2 + \frac{\hbar^2}{2B}\left(\frac{a}{\sqrt{\langle\beta^2\rangle}}(n_\gamma+1) - \frac{K^2}{3\langle\beta^2\rangle}\right), \tag{17.14}$$

where $x_{s,L}$ is the sth zero of the Bessel function $J_\nu(x)$ with $\nu = \sqrt{\frac{L(L+1)}{3} + \frac{9}{4}}$, and n_γ and K are two quantum numbers with

$$n_\gamma = 0, \quad K = 0; \quad n_\gamma = 1, \quad K = \pm 2; \tag{17.15}$$

$$n_\gamma = 2, \quad K = 0, \pm 4; \cdots . \tag{17.16}$$

Accordingly, the total wave function is given by

$$\Psi(\beta, \gamma, \theta_i) = f'(\beta)\Phi'(\gamma, \theta_i) \tag{17.17}$$

with

$$f'(\beta) = c_{s,L}\beta^{-3/2}J_\nu(k_{s,L}\beta); \quad k_{s,L} = \frac{x_{s,L}}{\beta_W} \tag{17.18}$$

and

$$\Phi'(\gamma, \theta_i) = c_{n,K}\gamma^{|K/2|}e^{a\gamma^2/2}L_n^{|K|}(a\gamma^2) \tag{17.19}$$

$$\times\sqrt{\frac{2L+1}{16\pi^2(1+\delta_{K,0})}}\left[D_{u,K}^L(\theta_i) + (-1)^L D_{u,-K}^L(\theta_i)\right],$$

$$n = \left(\frac{n_\gamma - |K|}{2}\right), \tag{17.20}$$

where $c_{s,L}$ and $c_{n,K}$ represent the normalization constants of the corresponding eigenfunctions and $L_n^{|K|}$ is the Laguerre polynomial. In short, the X(5) CPS can be approximately solved, and its solution can be applied to benchmark the critical structures in the first-order SPT [45] in the mass region $A \sim 150$. Although the X(5) solution is supposed to describe an axially symmetric situation corresponding to $\gamma \sim 0°$, the model techniques can be applied to develop the axially asymmetric models with $\gamma \neq 0°$ as described bellow.

17.2.3   THE Z(5) AND T(5) SOLUTIONS

The Z(5) CPS is designed to describe the prolate to oblate SPT (algebraically called the SU(3) $-$ $\overline{SU(3)}$ transition) [37]. Its Hamiltonian is similar to the X(5) one [35] except that $V(\gamma)$ in Z(5) is assumed to be a harmonic oscillator around $\gamma = \frac{\pi}{6}$,

$$V(\gamma) = \frac{1}{2}c(\gamma - \frac{\pi}{6})^2 = \frac{1}{2}c\tilde{\gamma}^2, \quad \tilde{\gamma} = \gamma - \frac{\pi}{6}, \tag{17.21}$$

where c denotes the strength parameter. Since the potential has a minimum at $\gamma = \pi/6$, the rotational term in the Bohr Hamiltonian can be approximately expanded as

$$R \simeq \sum_{k=1}^{3}\frac{\hat{L}_k^2}{4\sin^2(\gamma - \frac{2}{3}k\pi)}\bigg|_{\gamma=\frac{\pi}{6}}$$

$$= \frac{1}{4}(\hat{L}_1^2 + \hat{L}_2^2 + \hat{L}_3^2) - \frac{3}{4}\hat{L}_1^2. \tag{17.22}$$

The similar expansion is used in solving the X(5) Hamiltonian as seen in (17.13). Then, the eigenvalue equation can be also separated into two parts, one for the β variable

$$\left[-\frac{\hbar^2}{2B}\left(\frac{1}{\beta^4}\frac{\partial}{\partial\beta}\beta^4\frac{\partial}{\partial\beta}-\frac{1}{\beta^2}\left(L(L+1)-\frac{3}{4}\alpha\right)\right)\right.$$
$$\left.+V(\beta)\right]f''(\beta)=E_\beta''f''(\beta) \tag{17.23}$$

with α denoting the projection of the angular momentum on the body-fixed x'-axis and the other for the γ variable

$$\frac{\hbar^2}{2B\langle\beta^2\rangle}\left[-\frac{\partial^2}{\partial\tilde{\gamma}^2}+\frac{1}{2}c'\tilde{\gamma}^2\right]\Phi''(\gamma,\theta_i)=E_\gamma''\Phi''(\gamma,\theta_i) \tag{17.24}$$

with $\langle\beta^2\rangle$ being the average of β^2 over $f''(\beta)$ and $c'=2Bc\langle\beta^2\rangle/\hbar^2$. Accordingly, the eigenvalue can be analytically expressed as

$$E=\frac{\hbar^2}{2B}\left(\frac{x_{s,\nu}}{\beta_W}\right)^2+\frac{\hbar^2}{2B}\left(\sqrt{2c\langle\beta^2\rangle}\left(n_{\tilde{\gamma}}+\frac{1}{2}\right)\right), \tag{17.25}$$

where $x_{s,\nu}$ is the s-th zero of the Bessel function $J_\nu(x)$ with

$$\nu=\sqrt{\frac{L(L+1)}{3}-\frac{3}{4}\alpha^2+\frac{9}{4}} \tag{17.26}$$

and $n_{\tilde{\gamma}}=0,\ 1,\ 2,\cdots$ is the quantum number of one-dimensional harmonic oscillator. The total wave function is given by

$$\Psi(\beta,\gamma,\theta_i)=f''(\beta)\Phi''(\gamma,\theta_i) \tag{17.27}$$

with

$$f''(\beta)=c_{s,\nu}\beta^{-3/2}J_\nu(k_{s,\nu}\beta);\ \ k_{s,\nu}=\frac{x_{s,\nu}}{\beta_W} \tag{17.28}$$

and

$$\Phi''(\gamma,\theta_i)=c_{n_{\tilde{\gamma}}}H_{n_{\tilde{\gamma}}}(b\tilde{\gamma})e^{-b^2\tilde{\gamma}^2/2} \tag{17.29}$$

$$\times\sqrt{\frac{2L+1}{16\pi^2(1+\delta_{\alpha,0})}}\left[D^L_{u,\alpha}(\theta_i)+(-1)^L D^L_{u,-\alpha}(\theta_i)\right],$$

$$b=\left(\frac{c\langle\beta^2\rangle}{2}\right)^{1/4}, \tag{17.30}$$

where $c_{s,\nu}$ and $c_{n_{\tilde{\gamma}}}$ represent the normalization constants of the corresponding eigenfunctions and $H_{n_{\tilde{\gamma}}}$ denotes the hermite polynomial. It is clear that the Z(5) CPS is suitable to describe a triaxial systems with $\gamma\sim30°$, which thus satisfies the core deformation condition required for the chiral or wobbling bands.

The T(5) CPS [42] further extends the Z(5) mode from $\gamma=30°$ to any γ-deformations by taking the γ potential form with

$$V(\gamma)=\frac{1}{2}c(\gamma-\gamma_e)^2=\frac{1}{2}c\tilde{\gamma}^2,\ \ \tilde{\gamma}=\gamma-\gamma_e \tag{17.31}$$

and $0\le\gamma_e\le\pi/3$. As the potential has a minimum at $\gamma=\gamma_e$, the rotational term can be approximately expanded as

$$R\ \simeq\ \sum_{k=1}^{3}\frac{\hat{L}_k^2}{4\sin^2\left(\gamma-\frac{2}{3}k\pi\right)}\bigg|_{\gamma=\gamma_e}$$

$$=\ \frac{1}{4}\left(\frac{\hat{L}_1^2}{\sin^2\left(\gamma_e-\frac{2\pi}{3}\right)}+\frac{\hat{L}_2^2}{\sin^2\left(\gamma_e-\frac{4\pi}{3}\right)}+\frac{\hat{L}_3^2}{\sin^2(\gamma_e)}\right). \tag{17.32}$$

The process of solving the T(5) Hamiltonian is very similar to that in the Z(5) case except that rotational wave function will be obtained by numerically solving the eigenvalue equation

$$R_L \varphi_{M,s}^L(\theta_i) = r_{L_s} \varphi_{M,s}^L(\theta_i) \,. \tag{17.33}$$

with

$$\varphi_{M,s}^L(\theta_i) = \sum_K C_{s,K}^L \chi_{M,K}^L(\theta_i), \tag{17.34}$$

$$\chi_{M,K}^L(\theta_i) = \sqrt{\frac{2L+1}{16\pi^2(1+\delta_{K,0})}}$$
$$\times [D_{M,K}^L(\theta_i) + (-1)^L D_{M,-K}^L(\theta_i)] \,, \tag{17.35}$$

where the expanding coefficients $C_{s,K}^L$ are determined from the eigenvalues equation (17.33). The total energy is then given by

$$E = \frac{\hbar^2}{2B} \left(\frac{x_{\tilde{s},v}}{\beta_W}\right)^2 + \frac{\hbar^2}{2B}\left(\sqrt{2c\langle\beta^2\rangle}\left(n_{\tilde{\gamma}}+\frac{1}{2}\right)\right), \tag{17.36}$$

where $x_{\tilde{s},v}$ is the $\tilde{s}$-th zero of the Bessel function $J_v(x)$ with

$$v = \sqrt{r_{L_s} + \frac{9}{4}} \tag{17.37}$$

and $n_{\tilde{\gamma}} = 0, 1, 2, \cdots$ is the quantum number of one-dimensional harmonic oscillator. Similarly, the total wave function is given by

$$\Psi(\beta,\gamma,\theta_i) = \eta(\beta)K(\gamma,\theta_i) \tag{17.38}$$

with

$$\eta(\beta) = c_{\tilde{s},v}\beta^{-3/2}J_v(k_{\tilde{s},v}\beta); \quad k_{\tilde{s},v} = \frac{x_{\tilde{s},v}}{\beta_W} \tag{17.39}$$

and

$$K(\gamma,\theta_i) = c_{n_{\tilde{\gamma}}}H_{n_{\tilde{\gamma}}}(b\tilde{\gamma})e^{-b^2\tilde{\gamma}^2/2}\varphi_{M,s}^L(\theta_i), \tag{17.40}$$

$$b = \left(\frac{c\langle\beta^2\rangle}{2}\right)^{1/4}, \tag{17.41}$$

where $c_{\tilde{s},v}$ and $c_{n_{\tilde{\gamma}}}$ represent the normalization constants of the corresponding eigenfunctions.

Undoubtedly, the T(5) CPS provides a more general description of axially-asymmetric systems than the Z(5) CPS and it is therefore more suitable to take the T(5) mode as the even-even core to describe the triaxiality-related phenomena in odd nuclei. In fact, a CPS model for odd-odd nuclei has already been developed [33] within the core-particle coupling scheme by adopting the T(5) Hamiltonian to describe the even-even core dynamics. The resulting model may generate collective vibrational excitations of the chiral-doublet bands in odd-odd systems. In the following, we will propose a collective Hamiltonian with γ-stable potential to describe more details on how to build a collective model for odd nuclei (both odd-odd and odd-A cases) through the core-particle scheme [30].

17.3   A γ-VIBRATIONAL MODEL FOR ODD SYSTEMS

In the core-particle coupling scheme [30], the Hamiltonian for odd-odd systems can be written as

$$\hat{H} = \hat{H}_c + \hat{H}_p + \hat{H}_n \,, \tag{17.42}$$

where $\hat{H}_c$ presents the Bohr Hamiltonian for the even-even core and $\hat{H}_\sigma$ with $\sigma =$ p, n represents the single-particle Hamiltonian for the odd nucleons (unpaired proton or neutron). The total angular momentum operator is given by $\hat{I} = \hat{L} + \hat{j}_p + \hat{j}_n$, with $\hat{L}$, $\hat{j}_p$, $\hat{j}_n$ and $\hat{I}$ denoting the ones for the core, odd proton, odd neutron and whole odd-odd system. Specifically, the core Hamiltonian can be generally written as [31]

$$\hat{H}_c = -\frac{\hbar^2}{2B}\left[\frac{1}{\beta^4}\frac{\partial}{\partial\beta}\beta^4\frac{\partial}{\partial\beta} + \frac{1}{\beta^2}\left(\frac{1}{\sin3\gamma}\frac{\partial}{\partial\gamma}\sin3\gamma\frac{\partial}{\partial\gamma}\right.\right.$$
$$\left.\left. -\frac{1}{4}\sum_k\frac{\hat{L}_k^2}{\sin^2(\gamma-\frac{2}{3}k\pi)}\right)\right] + V(\beta,\ \gamma). \tag{17.43}$$

For different core deformation, one should adopt different potential form to describe the corresponding odd systems. But it is often taken $V(\beta,\ \gamma) = V(\beta) + V(\gamma)$ like in the T(5) CPS in order to achieve the variables separation. In [33], a β-soft but γ-stable potential form was employed to describe the transitional odd-odd nuclei. It is shown that the model parameters can be well determined from fitting the low-lying properties of the adjacent even-even nuclei [33]. In contrast, we will discuss below a case associated with β-rigid to illustrate how to build a solvable collective model for odd systems through the core-particle coupling scheme.

If "freezing" the β deformation (β rigid) but with the γ variable stabilized around γ_e, the core Hamiltonian in (17.43) will be reduced to [51]

$$\hat{H}_c = -\frac{\hbar^2}{2B\beta^2}\left[\frac{1}{\sin3\gamma}\frac{\partial}{\partial\gamma}\sin3\gamma\frac{\partial}{\partial\gamma} - \frac{1}{4}\sum_k\frac{\hat{L}_k^2}{\sin^2(\gamma-\frac{2}{3}k\pi)}\right]$$
$$+ V(\gamma), \tag{17.44}$$

in which the γ potential is further assumed to be a harmonic oscillator as in T(5),

$$V(\gamma) = \frac{\bar{c}}{4B\beta^2}(\gamma-\gamma_e)^2. \tag{17.45}$$

In Eq. (17.45), $\bar{c}$ represents the strength parameter and γ_e is used to measure the core deformation. For the unpaired valence nucleons, we consider the triaxial single particle Hamiltonian for a high j shell [27, 28], which is written as

$$\hat{H}_{sp} = \hat{H}_p + \hat{H}_n \tag{17.46}$$
$$= \sum_{\sigma=p,n} C_\sigma\left\{\left[\hat{j}_{\sigma3}^2 - \frac{j_\sigma(j_\sigma+1)}{3}\right]\cos(\gamma)\right.$$
$$\left. +\frac{1}{2\sqrt{3}}\left[\hat{j}_{\sigma+}^2 + \hat{j}_{\sigma-}^2\right]\sin(\gamma)\right\}$$

with $\hat{j}_{\sigma\pm} = \hat{j}_{\sigma1} \pm i\hat{j}_{\sigma2}$. If applying this single-particle Hamiltonian to odd-A systems, the summation in (17.46) will only involve either the unpaired proton or the unpaired neutron. In the concrete calculations, the parameter C_σ (in MeV) is often taken the form [27, 28]

$$C_\sigma = \pm\frac{195}{2j_\sigma(j_\sigma+1)}A^{-1/3}\beta, \tag{17.47}$$

in which the plus sign refers to a particle and the minus to a hole.

As seen above, both the collective and single particle degrees of freedom are involved in the model, which makes it hard to solve the Hamiltonian without any approximations. Since the γ-stable potential in (17.45) has a minimum at $\gamma = \gamma_e$, the rotational term in (17.44) can be approximated as [35, 37, 42, 51]

$$
\begin{aligned}
R &\equiv \sum_{k=1}^{3} \left. \frac{\hat{L}_k^2}{4\sin^2(\gamma - \frac{2}{3}k\pi)} \right|_{\gamma \simeq \gamma_e} \\
&= \frac{(\hat{I}_1 - \hat{j}_{p1} - \hat{j}_{n1})^2}{4\sin^2(\gamma_e - \frac{2}{3}\pi)} + \frac{(\hat{I}_2 - \hat{j}_{p2} - \hat{j}_{n2})^2}{4\sin^2(\gamma_e - \frac{4}{3}\pi)} + \frac{(\hat{I}_3 - \hat{j}_{p3} - \hat{j}_{n3})^2}{4\sin^2(\gamma_e - 2\pi)} .
\end{aligned}
\tag{17.48}
$$

The same approximation $\gamma \simeq \gamma_e$ is also applied for the single-particle Hamiltonian in (17.46). By further introducing the reduced potential $u = 2B\beta^2 V/\hbar^2$ [35, 37] and single-particle energy $\hat{h}_{sp} = 2B\beta^2 \hat{H}_{sp}/\hbar^2$, the Schrödinger equation $\hat{H}\Psi = E\Psi$ can be separated into two parts, i.e. the rotational equation

$$
\left[R + \hat{h}_{sp} \right] \varphi(\theta_i, a_p, a_n) = r\varphi(\theta_i, a_p, a_n) ,
\tag{17.49}
$$

and the γ-vibrational equation

$$
\left[-\frac{1}{\sin3\gamma}\frac{\partial}{\partial\gamma}\sin3\gamma\frac{\partial}{\partial\gamma} + u(\gamma) \right] f(\gamma) = \lambda f(\gamma) .
\tag{17.50}
$$

The total wave function is then written as

$$
\Psi = f(\gamma)\varphi(\theta_i, a_p, a_n) ,
\tag{17.51}
$$

where θ_i $(i = 1, 2, 3)$ are the Euler angles and $a_{p(n)}$ represents generally the coordinates of the odd proton (neutron).

As seen from Eq. (17.49), the reduced single particle Hamiltonian is directly involved in collective rotation of the system with the eigenvalue equation form being actually as same as the one constructed from the PRM [30]. To solve this equation, one can expand the rotational wave function as

$$
\varphi_{M,s}^{I}(\theta_i, a_p, a_n) = \sum_{K,\Omega_p,\Omega_n} C_{s,\Omega_p,\Omega_n}^{I,K} \chi_{M,\Omega_p,\Omega_n}^{I,K}(\theta_i, a_p, a_n)
\tag{17.52}
$$

with

$$
\begin{aligned}
&\chi_{M,\Omega_p,\Omega_n}^{I,K}(\theta_i, a_p, a_n) \\
&= \sqrt{\frac{2I+1}{16\pi^2}} \left[D_{M,K}^{I}(\theta_i)\phi_{\Omega_p}^{j_p}(a_p)\phi_{\Omega_n}^{j_n}(a_n) \right. \\
&\quad + \left. (-1)^{I-j_p-j_n} D_{M,-K}^{I}(\theta_i)\phi_{-\Omega_p}^{j_p}(a_p)\phi_{-\Omega_n}^{j_n}(a_n) \right] ,
\end{aligned}
\tag{17.53}
$$

where $D_{M,K}^{I}(\theta_i)$ is the Wigner D-function and $\phi_{\Omega}^{j}(a)$ is the single-particle wave function with Ω denoting the component of the spin j on the intrinsic z axis. The additional quantum number s with $s = 1, 2, 3,...$ is introduced to distinguish among the states with the same I, M. The expansion coefficients $C_{s,\Omega_p,\Omega_n}^{I,K}$ are then determined by the eigenvalue Equation (17.49) with the eigenvalues denoted by r_s^I.

To solve the γ-vibrational equation, $\sin(3\gamma)$ in Eq. (17.50) will be approximately replaced by $\sin(3\gamma_e)$ [42, 51]. By making a variable replacement with $\gamma' = \gamma - \gamma_e$, the equation can be transformed into

$$
\left[-\frac{\partial^2}{\partial\gamma'^2} + \frac{1}{2}\bar{c}\gamma'^2 \right] f(\gamma') = \lambda f(\gamma') ,
\tag{17.54}
$$

which is just the harmonic oscillator equation with the eigenvalues [37]

$$\lambda = \sqrt{2\bar{c}}(n_\gamma + \frac{1}{2}), \quad n_\gamma = 0,1,2,\cdots. \tag{17.55}$$

It should be mentioned that Eq. (17.54) can be alternatively obtained from the direct quantizing the classical Hamiltonian corresponding to γ vibration of nuclear shape [31, 51]. The corresponding eigenfunctions are then obtained as

$$f(\gamma) = N_{n_\gamma} H_{n_\gamma}(b\gamma)e^{-b^2\gamma^2/2}, \quad b = (\bar{c}/2)^{1/4}, \tag{17.56}$$

where H_{n_γ} is the hermite polynomial and N_{n_γ} is the normalization constant given by

$$N_{n_\gamma} = \sqrt{\frac{b}{\sqrt{\pi}2^{n_\gamma}n_\gamma!}}. \tag{17.57}$$

The total energy is finally expressed as

$$E(s,I,n_\gamma) = E_0 + A'n_\gamma + B'r_s^I \tag{17.58}$$

with A' and B' denoting the corresponding parameters and E_0 representing the constant. It is clear that the contributions from the γ and rotational degrees of freedom have been completely separated under the adiabatic approximation [51].

To calculate $B(E2)$ transitional rates, the $E2$ operator is chosen by [27, 28]

$$T_u^{E2} = \sqrt{\frac{5}{16\pi}}Q_0 \tag{17.59}$$
$$\times \left[D_{u,0}^{(2)}(\theta_i)\cos(\gamma) + \frac{1}{\sqrt{2}}\left(D_{u,2}^{(2)}(\theta_i) + D_{u,-2}^{(2)}(\theta_i)\right)\sin(\gamma)\right],$$

where Q_0 denotes the intrinsic quadrupole moment. Then the $B(E2)$ transitional rates will be calculated via

$$B(E2;\alpha_i,I_i \to \alpha_f,I_f) = \frac{|\langle\alpha_f,I_f \| T^{E2} \| \alpha_i,I_i\rangle|^2}{2I_i+1}. \tag{17.60}$$

In the concrete calculations, the same approximation $\gamma = \gamma_e$ is adopted for the $E2$ operator, which means that the integral over γ will give $\delta_{n_{\gamma_i},n_{\gamma_f}}$. Similarly, one can choose the $M1$ operator [27, 28]

$$T_u^{M1} = \sqrt{\frac{3}{4\pi}}\frac{e\hbar}{2Mc}\left[(g_p - g_R)\hat{j}_{pu} + (g_n - g_R)\hat{j}_{nu}\right] \tag{17.61}$$

to calculate the $B(M1)$ transitional rates. In the $M1$ operator defined in (17.61), $\hat{j}_u$ represents the spherical tensor in the laboratory frame while g_p, g_n, and g_R denote the effective g factors for proton, neutron and even-even core, respectively. For the odd-A systems, only the first or second term in (17.61) is needed, depending on whether it is an odd-proton or odd-neutron nucleus [26]. The $B(M1)$ transitional rates can be evaluated via

$$B(M1;\alpha_i,I_i \to \alpha_f,I_f) = \frac{|\langle\alpha_f,I_f \| T^{M1} \| \alpha_i,I_i\rangle|^2}{2I_i+1}. \tag{17.62}$$

Clearly, different rotational bands in the model will be distinguished by different values of n_γ and s. For example, the lowest doublet bands are characterized by $n_\gamma = 0$ but with $s = 1$ and $s = 2$,

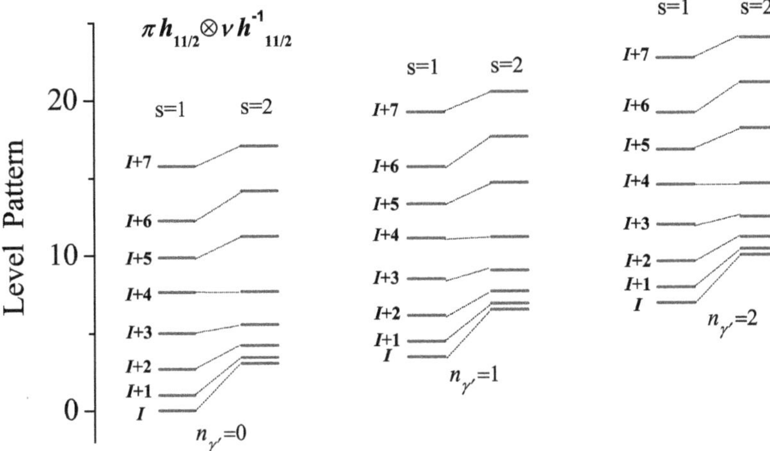

Figure 17.1 Level pattern for the MχD produced from the model calculations for the particle-hole configuration with $\pi h_{11/2} \otimes v h_{11/2}^{-1}$ and triaxial deformation with $\gamma_e = 30°$. All the results have been normalized to the yrast level $E(I+1) = 1.0$ with $I = 9$.

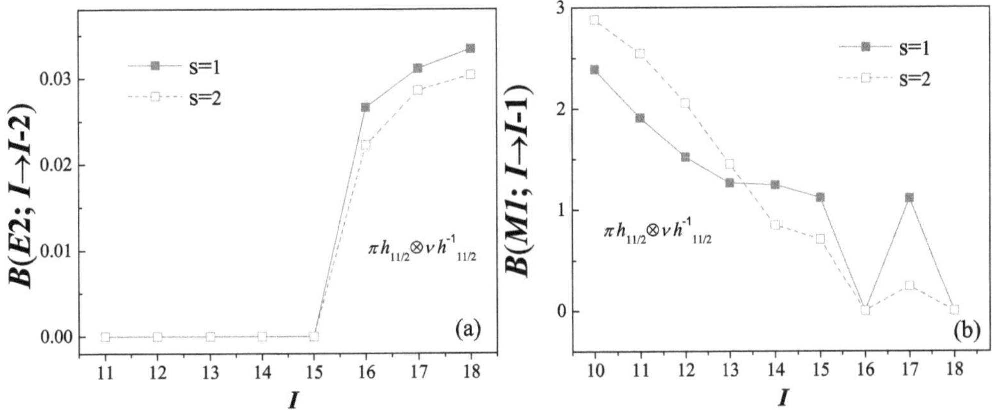

Figure 17.2 The intraband $B(E2)$ (units in Q_0^2) and $B(M1)$ (units in μ_N^2) results with $s = 1$, $s = 2$ denoting two bands in each doublets. In the calculations, the g factor are taken as $g_p - g_R = 0.7$ and $g_n - g_R = -0.6$, respectively. In addition, the quantum number n_γ has been ignored since its values will not affect the $B(E2)$ and $B(M1)$ results (see the text).

respectively, while the doublets based on γ vibration will be characterized by $n_\gamma = 1, 2, 3, \cdots$. In addition, the intraband $B(E2)$ or $B(M1)$ transitions may have the same results for the bands with the same s because the integral over γ may contribute only the orthonormalization constant under the given approximation. For odd-A systems, the core-particle coupling Hamiltonian given in (17.42) may only contain H_p or H_n with the corresponding total angular momentum given by $\hat{I} = \hat{L} + \hat{j}_{p(\text{or } n)}$. The other steps for solving the Hamiltonian would be very similar to those adopted in the odd-odd case as discussed above, which means that the eigenvalues and eigenfunctions for odd-A systems have the same forms as those given in (17.58) and (17.51) except that single-proton or single-neutron degree of freedom is only considered. So, we will omit the derivations for odd-A cases and only discuss the concrete results in the following.

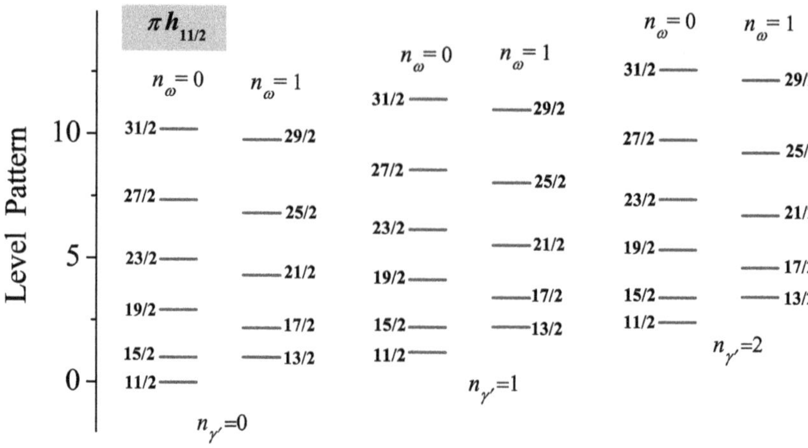

Figure 17.3   Level pattern for the multiple wobbling bands produced from the model calculations with the $\pi h_{11/2}$ configuration and $\gamma_e = 30°$. All the results have been normalized to the yrast level $E(15/2_1) = 1.0$.

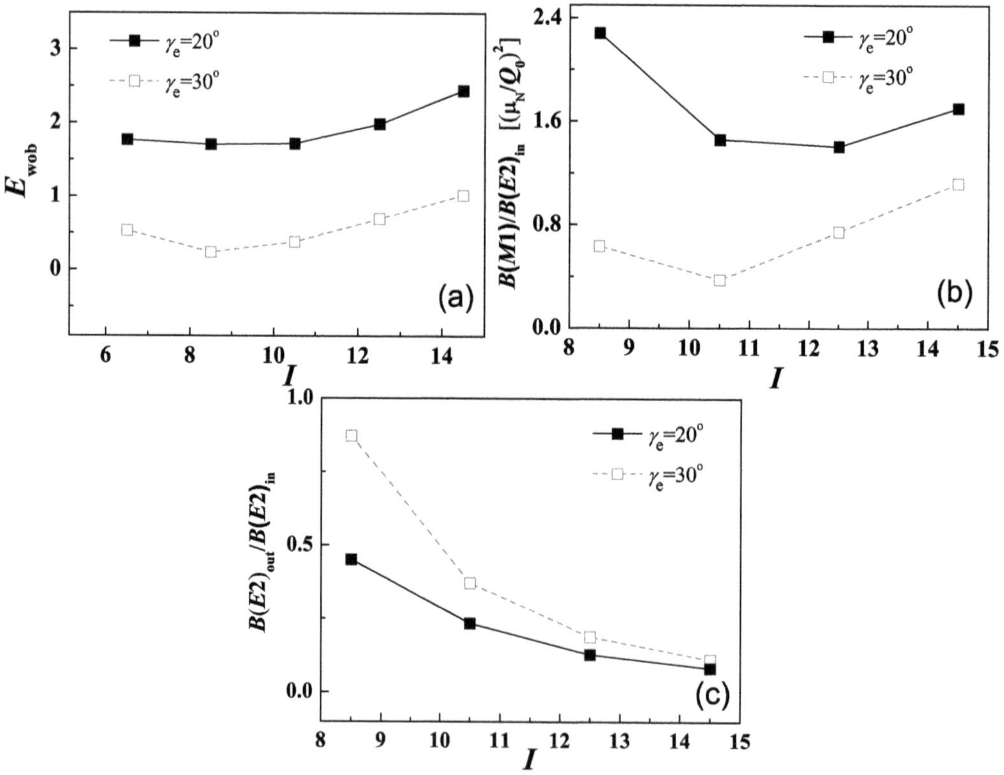

Figure 17.4   (a) The wobbling energies extracted from the doublet corresponding to $n_\gamma = 0$ given in FIG. 17.3. (b) The calculated transitional ratios $B(M1)/B(E2)$. (c) The calculated transitional ratios $B(E2)_{\text{out}}/B(E2)_{\text{in}}$. All the values are given as functions of I with the quantum number n_γ being ignored since its values affect nothing to the results (see the text).

17.3.1 DISCUSSIONS

Although the model dynamics can cover cases from $\gamma_e = 0°$ (axial symmetry) to $\gamma_e = 30°$ (maximally triaxial), the present model is mainly designed to describe chiral bands and wobbling bands that are supposed to appear in triaxially-deformed systems. In the following, we will focus on a theoretical discussion of the model dyanmics by taking $\gamma_e = 30°$, which is further divided into two parts according to the odd-odd and the odd-A cases. For the odd-odd system, the single-particle configuration is adopted as $\pi h_{11/2} \otimes v h_{11/2}^{-1}$. The other parameters are simply fixed as $C_\pi = -C_v = 0.307 \frac{1}{B\beta^2}$ and $\bar{c} = 12.5$. With these parameter, the level pattern together with the intraband $B(E2)$ and $B(M1)$ transitions are calculated and the results are shown in FIG. 17.1 and FIG. 17.2, respectively. For the odd-A system, only the single proton configuration $\pi h_{11/2}$ is included in the calculations and the related parameters are taken same as those for odd-odd systems. The calculated level pattern and other quantities are presented in FIG. 17.3 and FIG. 17.4, respectively.

17.3.2 CHIRAL ROTATIONS IN ODD-ODD SYSTEMS

As seen in FIG. 17.1, the model exhibits a MχD pattern based on the γ vibrations with $n_\gamma = 0, 1, 2, \cdots$. The near degeneracies in each doublets ($s = 1, s = 2$) can be clearly observed above spin $I = 11$. This feature can be considered as a fingerprint of chiral rotations in triaxial systems. Moreover, the doublets with different n_γ show nearly the same rotational pattern, which means that γ vibrations here will influence more the band head energies than the rotational excitation energies. As further seen in Fig. 17.2(a), the $B(E2)$ results for the doublet bands ($s = 1, s = 2$) are very close to each other. Specifically, the $B(E2)$ values as a function of I remain zero for both bands until $I = 15$ followed by a rapid and monotonic increment. These features agree well with the previous PRM analysis of the chiral bands based on the same single-particle configuration [28]. It is thus justified that the doublet bands at $\gamma_e = 30°$ indeed belong to a pair of chiral bands with the static chiral rotations starting near $I = 15$. The $B(M1)$ results in Fig. 17.2(b) further confirm the similarities in between the doublet bands. Different from $B(E2)$, the $B(M1)$ values as a function of spin may monotonically decrease until $I \sim 15$ and then turn to fluctuate with I. Note that the calculated $B(E2)$ and $B(M1)$ transitional values are independent of the n_γ quantum number under the adiabatic approximation adopted in the calculations. It means that the results given in Fig. 17.2 actually represent those for all the doublets ($s = 1,\ s = 2$) with different n_γ. In addition, other types of doublet bands such as those for $s = 3,\ s = 4$ are also allowed in the model. Nevertheless, the numerical calculations indicate that the $B(E2)$ and $B(M1)$ transitional features in such kind of doublets are different from those for $s = 1,\ s = 2$. This point has been already confirmed by the PRM and CPS model calculations [33]. In other words, the best candidates of chiral doublets in theory just come from the ones for $s = 1,\ s = 2$.

As mentioned above, a CPS model for odd-odd nuclei was recently proposed by us to describe the possible MχD in transitional systems [33]. Different from the β-rigid assumption adopted here, a β-soft potential is employed in the CPS to simulate nuclear structure near critical point and the model solutions can be approximately expressed in terms of Bessel functions, by which both the γ-vibrational and β-vibrational excitations of the chiral doublets are generated. In comparison to the CPS, the present method suggests a simpler picture of the MχD based on collective vibration by extending the PRM-like structure from γ rigid to γ stable. In [52], the analysis based on the Core-Particle-Hole Coupling model [53] indicates that even a Wilets-Jean type [54] of γ-soft core can generate a pair of chiral doublet bands similar to that generated by the PRM calculations with $\gamma \sim 30°$. This point actually supports the present conclusion that the γ-stable core adopted here can indeed sustain chiral rotations in odd-odd systems if only it has a relatively large equilibrium γ-deformation with $\gamma_e \sim 30°$.

17.3.3   WOBBLING BANDS IN ODD-A SYSTEMS

Wobbling motion provides another way to see triaxiality in nuclei [16]. Specifically, the wobbling mode in an odd-A system is characterized by the $\Delta I = 2$ rotational bands connected with $\Delta I = 1$ $E2$ transitions, which requires in theory a triaxially-deformed core and a high-j single-particle configuration as indicated in the present model. As shown in Fig. 17.3, the results indicate that the $\Delta I = 2$ rotational band ($n_\omega = 0$) and its wobbling partner ($n_\omega = 1$) group into a rotation-wobbling doublet (or to say a wobbling-bands pair). Collective excitations further yield a picture of multiple rotation-wobbling doublets with $n_\gamma = 1,\ 2,\ 3,\cdots$, which is very similar to the MχD pattern shown in FIG. 17.1 but associated with different dynamical mechanism. Note that chiral double bands can also appear in triaxial odd-A systems but the spin difference in adjacent levels in each chiral band may vary with $\Delta I = 1$ rather than $\Delta I = 2$ as exhibited by the rotation-wobbling doublets shown in FIG. 17.3. In addition, one can adopt some quantities to further identify the wobbling mode in odd-A nuclei. The typical ones include the wobbling energy defined by

$$E_{\text{wob}} = E(1,I) - \frac{1}{2}[E(0,I-1) + E(0,I+1)] \tag{17.63}$$

and the $B(\sigma,\lambda)$ transitional ratios defined by

$$B(M1)/B(E2)_{\text{in}} = \frac{B(M1;(1,I) \rightarrow (0,I-1))}{B(E2;(1,I) \rightarrow (1,I-2))}, \tag{17.64}$$

$$B(E2)_{\text{out}}/B(E2)_{\text{in}} = \frac{B(E2;(1,I) \rightarrow (0,I-1))}{B(E2;(1,I) \rightarrow (1,I-2))}, \tag{17.65}$$

where the rotational states with $(0,I)$ and $(1,I)$ denote those corresponding to $n_\omega = 0$ and $n_\omega = 1$, respectively. It should be mentioned that the quantum number n_ω is taken from the harmonic approximation that is traditionally used to work out the wobbling frequency from the triaxial rotor model [16]. In other words, n_ω may play a role similar to that of the quantum number K. Likewise, the intraband $B(E2)$ and $B(M1)$ transitional values in odd-A cases may be independent of the n_γ quantum number under the adiabatic approximation adopted in the calculations. So, we only present the calculated results for the first doublet corresponding to $n_\gamma = 0$. To make a comparison, both the results for $\gamma = 30°$ and those for $\gamma = 20°$ are provided.

As seen from the FIG. 17.4(a), the wobbling energies may increase as a function of the angular momentum when $I > 19/2$, which suggests a longitudinal wobbling mode in the present model. The similar spectral feature has been also predicted by the PRM calculations based on the hydrodynamical formulas of momentums of inertia given in [26]. The results for $\gamma_e = 20°$ further indicate that decreasing axial asymmetry may enhance the normalized wobbling energies but unchange the evolutional trend with increasing I. Notably, the wobbling energies shown in the FIG. 17.4(a) may simultaneously represent those for different n_γ if the band head energy for each rotation-wobbling doublets is set by $E(11/2) = 0$. As further seen in FIG. 17.4, one can deduce from the results for the $B(\sigma,\lambda)$ ratios that the wobbling band ($n_\omega = 1$), which connects the $n_\omega = 0$ band through $\Delta I = 1$ transitions, are indeed dominated by the $E2$ transitions. In fact, these results also agree well with the PRM calculations given in [26]. In addition, it is shown that less γ deformation will enhance the ratio $B(M1)/B(E2)_{\text{in}}$ but weaken the ratio $B(E2)_{\text{out}}/B(E2)_{\text{in}}$. Anyway, the results confirm that a picture of multiple rotation-wobbling bands can be similarly generated by the γ vibration in the present model.

17.4   SUMMARY

In summary, chiral doublet bands and wobbling bands provide two unambiguous fingerprints for triaxiality in low-energy nuclear structures [1]. Emergence of MχD in a single nucleus even suggests that chiral rotation may be more robust against single-particle or collective excitations than n expectation. Theoretically, the PRM may be the most convenient framework to describe different aspects of chirality and wobbling, by which many related concepts and model techniques have been developed and applied. However, further developments of the PRM description of chirality and wobbling should go beyond the rigid assumptions on β and γ. One way of directly extending the PRM dynamics and meanwhile keeping the model solvability is to generalize the solvable Bohr Hamiltonian based on the core-particle coupling scheme. To exemplify that, a collective model with a γ-stable potential has been constructed in this work. The model solutions for high-j single-particle configurations can be easily obtained under the adiabatic approximation. The case with the core deformation $\gamma_e = 30°$ and high-j single-particle configurations is worked out to illustrate the model descriptions of the possible MχD in odd-odd systems and multiple rotation-wobbling doublets in odd-A systems. The results indicate that these multiple doublets can be induced by the γ-vibrational excitations with $n_\gamma = 1, 2, 3, \cdots$. Since the early prediction of MχD based on multiple single-particle configurations [4], searching for the possible candidates has become an active issue in experiments [5–15]. Although the present results are achieved based on a very simple assumption of γ potential, the predicted MχD caused by γ vibration [33] is clearly different from those previously predicted based on several single-particle configurations [4, 55] or based on an identical configuration [6, 56, 57], therefore adding a new mechanism for MχD in the odd-odd systems. Similarly, the present analysis also predicts the multiple rotation-wobbling doublets based on γ-vibrational excitations. It thus demonstrates a new possibility of producing multiple-bands pattern in odd systems. However, whether or not the present predictions can be revealed in experiments needs more careful analyses and more reliable theoretical treatments. The related work is in progress.

17.5 ACKNOWLEDGMENTS

Fruitful discussions with Q. B. Chen, Z. P. Li, J. Meng, Y. Y. Wang, S. Q. Zhang and P. W. Zhao are acknowledged. This work is supported by the National Natural Science Foundation of China (12375113,11875158,11675094).

Bibliography

1. S. Frauendorf and J. Meng, Nucl. Phys. A **82**, 617 (1997).
2. K. Starosta, *et al.*, Phys. Rev. Lett. **86**, 971 (2001).
3. B. W. Xiong and Y. Y. Wang, Atomic Data and Nuclear Data Tables, **125**, 193 (2019).
4. J. Meng, J. Peng, S. Q. Zhang and S. G. Zhou, Phys. Rev. C **73**, 037303 (2006).
5. A. D. Ayangeakaa, *et al.*, Phys. Rev. Lett. **110**, 172504 (2013).
6. I. Kuti, *et al.*, Phys. Rev. Lett. **113**, 032501 (2014).
7. C. Liu, *et al.*, Phys. Rev. Lett. **116**, 112501 (2016).
8. D. Tonev, *et al.*, Phys. Rev. Lett. **112**, 052501 (2014).
9. E. O. Lieder, *et al.*, Phys. Rev. Lett. **112**, 202502 (2014).
10. N. Rather, *et al.*, Phys. Rev. Lett. **112**, 202503 (2014).
11. C. M. Petrache, *et al.*, Phys. Rev. C **97**, 041304(R) (2018).
12. T. Roy, *et al.*, Phys. Lett. B **782**, 768 (2018).
13. B. Qi, H. Jia, C. Liu and S. Y. Wang, Sci. China-Phys. Mech. Astron. **62**, 012012 (2019).
14. S. Guo, *et al.*, Phys. Lett. B **807**, 135572 (2020).
15. B. F. Lv, *et al.*, Phys. Rev. C **100**, 024314 (2021).
16. A. Bohr and B. R. Mottelson, *Nucear Structure* (Benjamin, New York, 1975), Vol. II.
17. S. Frauendorf and F. Dönau, Phys. Rev. C **89**, 014322 (2014).
18. S. W. Ødegård, *et al.*, Phys. Rev. Lett. **86**, 5866 (2001).
19. G. Schönwaßer, *et al.*, Phys. Lett. B **552**, 9 (2003).
20. H. Amro, *et al.*, Phys. Lett. B **553**, 197 (2003).

21. P. Bringel, *et al.*, Eur. Phys. J. A **24**, 167 (2005).

22. D. J. Hartley, *et al.*, Phys. Rev. C **80**, 014304(R) (2009).

23. J. T. Matta, *et al.*, Phys. Rev. Lett. **114**, 082501 (2015).

24. J. Timár, *et al.*, Phys. Rev. Lett. **122**, 062501 (2019).

25. S. Biswas, *et al.*, Eur. Phys. J. A **55**, 159 (2019).

26. H. Zhang, B. Qi, X. D. Wang, H. Jia and S. Y. Wang, *et al.*, Phys. Rev. C **105**, 034339 (2022).

27. J. Peng, J. Meng and S. Q. Zhang, Phys. Rev. C **68**, 044324 (2003).

28. S. Q. Zhang, B. Qi, S. Y. Wang and J. Meng, Phys. Rev. C **75**, 044307 (2007).

29. B. Qi, S. Q. Zhang, S. Y. Wang, J. M. Yao and J. Meng, Phys. Rev. C **79**, 041302(R) (2009).

30. P. Ring and P. Schuck, *The Nuclear Many-Body Problem*, Springer-Verlag, Berlin 1980.

31. A. Bohr and Mat. Fys. Medd. Danske Vid. Selsk. **26**, No. 14 (1952)

32. A. S. Davydov, Nucl. Phys. **8**, 237 (1958).

33. Y. Zhang, B. Qi and S. Q. Zhang, Sci. China-Phys, Mech, Astron. **64**, 122011 (2021).

34. F. Iachello, Phys. Rev. Lett. **85**, 3580 (2000).

35. F. Iachello, Phys. Rev. Lett. **87**, 052502 (2001).

36. F. Iachello, Phys. Rev. Lett. **91**, 132502 (2003).

37. D. Bonatsos, D. Lenis, D. Petrellis and P. A. Terziev, Phys. Lett. B **588**, 172 (2004).

38. N. Pietralla and O. M. Gorbachenko, Phys. Rev. C **70**, 011304(R) (2004).

39. L. Fortunato, Phys. Rev. C **70**, 011302(R) (2004).

40. D. Bonatsos, D. Lenis, D. Petrellis, P. A. Terziev and I. Yigitoglu, Phys. Lett. B **621**, 102 (2005)

41. D. Bonatsos, D. Lenis, D. Petrellis, P. A. Terziev and I. Yigitoglu, Phys. Lett. B **632**, 238 (2006)

42. Y. Zhang, F. Pan, Y. A. Luo and J. P. Draayer, Phys. Lett. B **751**, 423 (2015).

43. R. Budaca and A. I. Budaca, Phys. Lett. B **759**, 349 (2016).

44. R. Budaca and A. I. Budaca, Phys. Rev. C **94**, 054306 (2016).

45. P. Cejnar, J. Jolie and R. F. Casten, Rev. Mod. Phys. **82**, 2155 (2010).

46. R. F. Casten, Nat. Phys. **2**, 811 (2006).

47. R. F. Casten and E. A. McCutchan, J. Phys. G **34**, R285 (2007).

48. Y. Zhang, Y. X. Liu, F. Pan, Y. Sun and J. P. Draayer, Phys. Lett. B **732**, 55 (2014).

49. Y. Zhang, F. Pan, Y. X. Liu, Y. A. Luo and J. P. Draayer, Phys. Rev. C **90**, 064318 (2014).

50. Eur. Phys. J. A **26**, 1 (2005).

51. A. S. Davydov, Nucl. Phys. **24**, 682 (1961).

52. Ch. Droste, S. G. Rohoziński, K. Starosta, L. Próchniak and E. Grodner, Eur. Phys. J. A **42** 79-89 (2009).

53. F. Dönau and S. Frauendorf, Phys. Lett. B **71**, 263 (1977).

54. L. Wilets and M. Jean, Phys. Rev. **102**, 788 (1956).

55. S. Y. Wang, Chin. Phys. C **44**, 112001 (2020).

56. Q. B. Chen, J. M. Yao, S. Q. Zhang and B. Qi, Phys. Rev. C **82**, 067302 (2010).

57. I. Hamamoto, Phys. Rev. C **88**, 024327 (2013).

Index

For Product Safety Concerns and Information please contact our EU
representative GPSR@taylorandfrancis.com
Taylor & Francis Verlag GmbH, Kaufingerstraße 24, 80331 München, Germany